*Neural Mechanisms of
Goal-Directed Behavior
and Learning*

Neural Mechanisms of Goal-Directed Behavior and Learning

Edited by

RICHARD F. THOMPSON

Department of Psychobiology
University of California, Irvine
Irvine, California

LESLIE H. HICKS

Department of Psychology
Howard University
Washington, D.C.

V. B. SHVYRKOV

Institute of Psychology
Academy of Sciences of the USSR
Moscow, USSR

ACADEMIC PRESS   1980
A Subsidiary of Harcourt Brace Jovanovich, Publishers

New York   London   Toronto   Sydney   San Francisco

COPYRIGHT © 1980, BY ACADEMIC PRESS, INC.
ALL RIGHTS RESERVED.

NO PART OF THIS PUBLICATION MAY BE REPRODUCED OR
TRANSMITTED IN ANY FORM OR BY ANY MEANS,
ELECTRONIC OR MECHANICAL, INCLUDING PHOTOCOPY,
RECORDING, OR ANY INFORMATION STORAGE AND
RETRIEVAL SYSTEM, WITHOUT PERMISSION IN WRITING
FROM THE PUBLISHER. THIS WORK MAY BE REPRODUCED
IN WHOLE OR IN PART FOR THE OFFICIAL USE OF THE U.S.
GOVERNMENT.

ACADEMIC PRESS, INC.
111 Fifth Avenue, New York, New York 10003

United Kingdom Edition published by
ACADEMIC PRESS, INC. (LONDON) LTD.
24/28 Oval Road, London NW1 7DX

Library of Congress Cataloging in Publication Data
Main entry under title:

Neural mechanisms of goal-directed behavior and
    learning.

    Based on the first joint US-USSR symposium in
psychology and the neurosciences, held Apr. 11-18,
1978 at the University of California, Irvine.
    Includes bibliographies and index.
    1. Learning--Physiological aspects--Congresses.
2. Motivation (psychology)--Congresses. 3. Goal
(Psychology)--Congresses. 4. Neuropsychology--
Congresses. 5. Human behavior--Congresses.
I. Thompson, Richard Frederick, Date.
II. Hicks, Leslie H. III. Shvyrkov, Viacheslav
Borisovich. [DNLM: 1. Learning--Congresses.
2. Motivation--Congresses. 3. Neurophysiology--
Congresses. 4. Psychophysiology--Congresses.
WL103 N493 1978]
QP408.N48   153.1'5   79-6775
ISBN 0-12-688980-5

PRINTED IN THE UNITED STATES OF AMERICA

80 81 82 83   9 8 7 6 5 4 3 2 1

Contents

List of Contributors   xi

Preface   xv

PART I
Theory and Experimental Issues in the Psychobiology of Motivation and Learning

CHAPTER 1
Introductory Remarks to the Soviet–American Symposium on Neurophysiological Mechanisms of Goal-Directed Behavior   3
   B. F. LOMOV

CHAPTER 2
Reflex Mechanisms of Motivational Behavior   11
   E. A. ASRATYAN

CHAPTER 3
Pavlovian Mechanisms of Goal-Directed Behavior   39
   I. GORMEZANO

CHAPTER 4
Different Ways in Which Learning Is Involved in Homeostasis   57
   NEAL E. MILLER and BARRY R. DWORKIN

CHAPTER 5
Central and Peripheral Catecholamine Function in Learning and Memory Processes   75
JAMES L. McGAUGH, JOE L. MARTINEZ, JR., ROBERT A. JENSEN, RITA B. MESSING, and BEATRIZ J. VASQUEZ

CHAPTER 6
A Neurophysiological Model of Purposive Behavior   93
E. R. JOHN

PART II
Motivation and Recovery of Function

CHAPTER 7
A Comparison of Instinct and Motivation with Emphasis on Their Differences   119
ALAN N. EPSTEIN

CHAPTER 8
Motor Subsystems in Motivated Behavior   127
PHILIP TEITELBAUM, TIMOTHY SCHALLERT, MARC DE RYCK, IAN Q. WHISHAW, and ILAN COLANI

CHAPTER 9
Synaptic Growth as a Plasticity Mechanism in the Brain   145
CARL W. COTMAN

CHAPTER 10
The Basal Ganglia and Psychomotor Behavior   153
LESLIE H. HICKS

CHAPTER 11
Basal Ganglia Dopaminergic Control of Sensorimotor Functions Related to Motivated Behavior   167
JOHN F. MARSHALL

CHAPTER 12
Hierarchical Organization of Physiological Subsystems in Elementary Food-Acquisition Behavior   177
YU. I. ALEKSANDROV and YU. V. GRINCHENKO

Contents

CHAPTER 13
Independence of Behavioral and Autonomic
Thermoregulatory Responses   189
EVELYN SATINOFF

PART III
Neuronal Processes of Learning

CHAPTER 14
Goal as a System-Forming Factor in Behavior and Learning   199
V. B. SHVYRKOV

CHAPTER 15
Brain Mechanisms of Learning   221
RICHARD F. THOMPSON, THEODORE W. BERGER,
and STEPHEN D. BERRY

CHAPTER 16
Neurophysiological Studies of Learning in Association
with the Pupillary Dilation Conditioned Reflex   241
NORMAN M. WEINBERGER

CHAPTER 17
Mechanisms of Classical Conditioning of Spinal Reflexes   263
MICHAEL M. PATTERSON

CHAPTER 18
Analysis of Neuron Activity in the Rabbit's Olfactory Bulb
during Food-Acquisition Behavior   273
A. P. KARPOV

CHAPTER 19
The Functional Neuroanatomy of a Conditioned Response   283
DAVID H. COHEN

CHAPTER 20
Unit Activity in Cingulate Cortex and Anteroventral Thalamus
during Acquisition and Overtraining of Discriminative
Avoidance Behavior in Rabbits   303
MICHAEL GABRIEL, KENT FOSTER, and EDWARD ORONA

PART IV
Perception and Information Processing

CHAPTER 21
Image, Information, and Episodic Modes of Central Processing   319
KARL H. PRIBRAM

CHAPTER 22
A Study of Neuron Systems Activity in Learning   341
U. G. GASANOV, A. G. GALASHINA, and A. V. BOGDANOV

CHAPTER 23
The Activity of Neuronal Networks in Cognitive Function   353
M. VERZEANO

CHAPTER 24
Activity of Visual Cortex Neurons in Systems Processes of Behavioral Act Interchange   375
D. G. SHEVCHENKO

CHAPTER 25
Brain Mechanisms of Attention and Perception   387
DAVID F. LINDSLEY, KENT M. PERRYMAN, and DONALD B. LINDSLEY

CHAPTER 26
Behavioral Modulation of Visual Responses of Neurons in Monkey Superior Colliculus and Cerebral Cortex   397
MICHAEL E. GOLDBERG and DAVID LEE ROBINSON

PART V
Hormonal, Pharmacological, and Developmental Factors

CHAPTER 27
Limbic System Contributions to Goal-Directed Behavior   409
ROBERT L. ISAACSON

CHAPTER 28
A Psychopharmacological Approach to Memory Processing   425
LINDA PATIA SPEAR

CHAPTER 29
A Case Study in the Neuroendocrine Control of Goal-Directed Behavior: The Interaction between Angiotensin II and Prostaglandin E_1 in the Control of Water Intake   437
NANCY J. KENNEY

CHAPTER 30
Goal-Directed Behavior in Ontogenesis   447
K. V. SHULEIKINA-TURPAEVA

CHAPTER 31
The Ontogenesis of Suckling, A Goal-Directed Behavior   461
ELLIOTT M. BLASS

CHAPTER 32
Infantile Forgetting of Acquired Information   471
RALPH R. MILLER

CHAPTER 33
An Animal Model of "Cooperation" Learning   481
BARRY D. BERGER, DANIEL MESCH, and RICHARD SCHUSTER

PART VI
Human Psychophysiology, Information Processing, and Language

CHAPTER 34
The Specific Role of Heart Rate in Sensorimotor Integration   495
JOHN I. LACEY and BEATRICE C. LACEY

CHAPTER 35
Control of Reflex Blink Excitability   511
FRANCES K. GRAHAM

CHAPTER 36
Modification of Goal-Directed Behavior in Discourse   521
VALENTINA ZAVARIN

CHAPTER 37
Neurophysiological Mechanisms of Processing Speech Information in Man   541
T. N. USHAKOVA

CHAPTER 38
Spatial Synchronization of Brain Electrical Activity
Related to Cognitive Information Processing   555
    ROBERT W. THATCHER

CHAPTER 39
A Mathematical Model for Human Visual Information,
Perception, and Storage   569
    A. N. LEBEDEV

APPENDIX A
Individual and General Discussions   585

APPENDIX B
General Discussion   625
    JOHN LACEY

Index   637

List of Contributors

Numbers in parentheses indicate the pages on which the authors' contributions begin.

Yu I. ALEKSANDROV (177), Institute of Psychology, Academy of Sciences of the USSR, Moscow, USSR

E. A. ASRATYAN (11), Institute of Higher Nervous Activity and Neurophysiology, Academy of Sciences of the USSR, Moscow, USSR

BARRY D. BERGER (481), Department of Psychology, University of Haifa, Mount Carmel, Haifa 31 999, Israel

THEODORE W. BERGER, (221), Department of Psychology, University of Pittsburgh, Pittsburgh, Pennsylvania 15260

STEPHEN D. BERRY (221), Department of Psychology, Miami University, Oxford, Ohio 45056

ELLIOTT M. BLASS (461), Psychology Department, Johns Hopkins University, Baltimore, Maryland 21218

A. V. BOGDANOV (341), Institute of Higher Nervous Activity and Neurophysiology, Academy of Sciences of the USSR, Moscow, USSR

DAVID H. COHEN (283), Department of Neurobiology and Behavior, State University of New York at Stony Brook, Long Island, New York 11794

ILAN COLANI (127), Department of Psychology, University of Illinois at Urbana-Champaign, Champaign, Illinois 61820

CARL W. COTMAN (145), Department of Psychobiology, University of California, Irvine, Irvine, California 92717

MARC DERYCK (127), Department of Psychology, University of Illinois at Urbana-Champaign, Champaign, Illinois 61820

BARRY R. DWORKIN (57), Department of Psychology, Rockefeller University, New York, New York 10021

ALAN EPSTEIN (119), Department of Biology, University of Pennsylvania, Leidy Laboratory, Philadelphia, Pennsylvania 19104

KENT FOSTER (303), Department of Psychology, University of Texas, Austin, Texas 78712

MICHAEL GABRIEL (303), Department of Psychology, University of Texas, Austin, Texas 78712

A. G. GALASHINA (341), Institute of Higher Nervous Activity and Neurophysiology, Academy of Sciences of the USSR, Moscow USSR

U. G. GASANOV (341), Institute of Higher Nervous Activity and Neurophysiology, Academy of Sciences of the USSR, Moscow, USSR

MICHAEL E. GOLDBERG (397), Sensorimotor Research, National Institute of Health, Bethesda, Maryland 20205

I. GORMEZANO (39), University of Iowa, Department of Psychology, Iowa City, Iowa 52240

FRANCES K. GRAHAM (511), Departments of Pediatrics and Psychology, University of Wisconsin, Madison, Wisconsin 53706

YU. V. GRINCHENKO (177), Institute of Psychology, Academy of Sciences of the USSR, Moscow, USSR

LESLIE HICKS (153), Department of Psychology, Howard University, Washington, D.C. 20059

ROBERT L. ISAACSON (409), Department of Psychology, State University of New York at Binghamton, Binghamton, New York 13901

R. A. JENSEN (75), Department of Psychobiology, University of California, Irvine, Irvine, California 92717

E. R. JOHN (93), Department of Psychiatry, New York University Medical Center, New York, New York 10016

A. P. KARPOV (273), Institute of Psychology, Academy of Sciences of the USSR, Moscow, USSR

NANCY J. KENNEY (437), Department of Psychology, University of Washington, Seattle, Washington 98195

BEATRICE LACEY (495), Section of Behavioral Physiology, The Fels Research Institute, Wright State University School of Medicine, Yellow Springs, Ohio 45387

JOHN LACEY (495), Section of Behavioral Physiology, The Fels Research Institute, Wright State University School of Medicine, Yellow Springs, Ohio 45387

A. N. LEBEDEV (569), Institute of Psychology, Academy of Sciences of the USSR, Moscow, USSR

DAVID LINDSLEY (387), Physiology Department, University of Southern California Medical School, Los Angeles, California 90033

DONALD LINDSLEY (387), Psychology Department, University of California, Los Angeles, California 90024

B. F. LOMOV (3), Institute of Psychology, Academy of Sciences of the USSR, Moscow, USSR

JAMES L. MCGAUGH (75), Department of Psychobiology, University of California, Irvine, Irvine, California 92717

JOHN F. MARSHALL (167), Department of Psychobiology, University of California, Irvine, Irvine, California 92717

J. L. MARTINEZ, JR. (75), Department of Psychobiology, University of California, Irvine, Irvine, California 92717

DANIEL MESCH (481), Department of Psychology, University of Haifa, Mount Carmel, Haifa 31 999, Israel

R. B. MESSING (75), Department of Psychobiology, University of California, Irvine, Irvine, California 92717

NEAL E. MILLER (57), Department of Psychobiology, University of California, Irvine, Irvine, California 92717

RALPH R. MILLER (471), Department of Psychology, Brooklyn College, Brooklyn, New York 11210

EDWARD ORONA (303), Department of Psychology, University of Texas, Austin, Texas 78712

MICHAEL M. PATTERSON (263), College of Osteopathic Medicine, Ohio University, Athens, Ohio 45701

KENT PERRYMAN (387), Physiology Department, University of Southern California Medical School, Los Angeles, California 90033

KARL H. PRIBRAM (319), Department of Psychology, Stanford University, Stanford, California 93405

DAVID L. ROBINSON (397), Laboratory of Sensorimotor Research, National Institute of Health, Bethesda, Maryland 20205

EVELYN SATINOFF (189), Departments of Psychology and Physiology and Biophysics, University of Illinois at Urbana-Champaign, Champaign, Illinois 61820

TIMOTHY SCHALLERT (127), Department of Psychology, University of Texas, Austin, Texas 78712

RICHARD SCHUSTER (481), Department of Psychology, University of Haifa, Mount Carmel, Haifa 31 999, Israel

D. G. SHEVCHENKO (375), Institute of Psychology, Academy of Sciences of the USSR, Moscow, USSR

K. V. SHULEIKINA-TURPAEVA (447), Institute of Higher Nervous Activity and Neurophysiology, Academy of Sciences of the USSR, Moscow, USSR

V. B. SHVYRKOV, (199), Institute of Psychology, Academy of Sciences of the USSR, Moscow, USSR

LINDA P. SPEAR (425), Department of Psychology, State University of New York at Binghamton, Binghamton, New York 13901

PHILIP TEITELBAUM (127), Department of Psychology, University of Illinois at Urbana-Champaign, Champaign, Illinois 61820

ROBERT THATCHER (555), Psychiatry Department, New York Medical College, New York, New York 10029

RICHARD F. THOMPSON (221), Department of Psychobiology, University of California, Irvine, Irvine, California 92717

B. J. VASQUEZ (75), Department of Psychobiology, University of California, Irvine, Irvine, California 92717

MARCEL VERZEANO (353), Department of Psychobiology, University of California, Irvine, Irvine, California 92717

T. N. VSHAKOVA (541), Institute of Higher Nervous Activity and Neurophysiology, Academy of Sciences of the USSR, Moscow, USSR

NORMAN M. WEINBERGER (241), Department of Psychobiology, University of California, Irvine, Irvine, California 92717

IAN Q. WHISHAW (127), Department of Psychology, University of Illinois at Urbana-Champaign, Champaign, Illinois 61820

VALENTINA ZAVARIN (521), Langley Porter Neuropsychiatric Institute, San Francisco, California 94127

Preface

This volume is an historic first—a product of the first Joint Seminar in Psychology and the Neurosciences between the National Academy of Sciences of the USA and the Academy of Sciences of the USSR, held April 11–18, 1978, at the University of California, Irvine. It is not, however, simply a reproduction of the talks given at the seminar; the final chapter of each section has been written expressly for this volume. We hope that the book will serve as a survey of the current status of knowledge and as a guide to future research directions in key topics in the broad and critically important area of neural mechanisms of motivation and learning. As such, it will be of considerable interest to scientists and graduate students in behavioral neuroscience.

Participating in the seminar were 10 Soviet scientists and 31 American scientists, including leading senior scientists and promising younger scientists from both countries. The volume thus provides a good cross section of the work being done here and in the USSR. The volume is of particular value because it is perhaps the first publication in English of the contemporary approaches of Soviet scientists to the broad field of psychobiology. Readers will note a distinct difference between the general approaches of the Soviet and American scientists: The Soviets emphasize broad theoretical frameworks (but not just one framework) within which they analyze specific problems; Americans focus more on analysis of mechanisms. Both approaches are clearly of value.

The book is divided into six major topic areas corresponding to the organization of the seminar: Theory and Experimental Issues in the Psychobiology of Motivation and Learning; Motivation and Recovery of Function; Neuronal Processes of Learning; Perception and Information Processing; Hormonal, Pharmacological, and Developmental Factors; and Human Psychophysiology, Information Processing, and Language.

Although exchange programs in various scientific disciplines have existed between the Soviet Union and the United States for some time, exchange programs in psychology and the social sciences are more recent. As a result of discussions between representatives of the National Academy of Sciences of the USA and the Academy of Sciences of the USSR, a delegation of American psychologists traveled to Moscow in July 1976 to discuss implementing a series of joint seminars in experimental psychology. The American delegation, sponsored by the Assembly of Behavioral and Social Sciences of the National Academy of Sciences/National Research Council, was headed by R. Duncan Luce of Harvard University. An agreement for the conduct of the seminar series was signed by Luce, representing the National Academy of Sciences of the USA and Boris L. Lomov, director of the Institute of Psychology, representing the Academy of Sciences of the USSR

After the agreement was concluded, it was decided to hold the first seminar in the broad topic area of neuronal mechanisms of motivation and learning, a field in which the Soviets have a very distinguished tradition dating back to Sechenov and Pavlov. It was also decided to hold the seminar at the University of California, Irvine, in part because of that university's strong interdisciplinary Department of Psychobiology. The seminar was co-chaired by Richard F. Thompson, Department of Psychobiology, University of California, Irvine and V. B. Shvyrkov, Institute of Psychology, Academy of Sciences of the USSR. (An informative background article on the seminar was published in *Science,* 1978, *200,* May 12, pp. 631–633).

Acknowledgments

A number of individuals and institutions provided help and support for the seminar and for this book. At the National Academy of Sciences, we are indebted to David A. Goslin, executive director, and Sarah M. Streuli, of the Assembly of Behavioral and Social Sciences, for continuing, strong, and often ingenious support; to Alan Campbell and Denise Surber, of the Commission on International Relations, for truly remarkable efforts; and to the president of the Academy, Philip Handler, for his constant support. Financial support for the seminar was provided by a grant from the National Science Foundation to the Assembly of Behavioral and Social Sciences (C 310 Task Order 368). At the University of California, Irvine we are indebted to Chancellor Daniel Aldrich and Executive Vice-Chancellor James

McGaugh for providing encouragement and financial support; to the Department of Psychobiology; and to Sharon Phillips, for outstanding efforts as the administrative assistant for the seminar; to Stephen Berry, who spent endless hours to ensure that the seminar was on tape; and to Judith Thompson, for her extraordinary efforts as hostess to the seminar.

This book was brought to completion by the two American editors, Richard F. Thompson and Leslie H. Hicks, during their tenure as fellows at the Center for Advanced Study in the Behavioral Sciences at Stanford. We are indebted to Gardner Lindzey, director of the Center, for helping us to obtain financial support from the Center, from the National Institute of Mental Health (5 T 32 MH1481-03), and from the National Science Foundation (BNS 76-22943 AO2). Special thanks are also due to Katherine Jenks, assistant director of the Center. Robert Goldberg, a student at Stanford University, California, provided us with invaluable help on the tape transcripts. We are also indebted to Paula DeLuigi for initial typing of the tape transcripts at the University of California, Irvine, and to Sharon Phillips, Carol Hibbert, and Margaret Butler for manuscript work. Last, but not least, we are deeply indebted to our co-editor from the Soviet Union, V. B. Shvyrkov, for his exceptional efforts in co-chairing the seminar, ensuring the delivery of all Soviet chapters, and providing strong moral support.

Neural Mechanisms of Goal-Directed Behavior and Learning

Theory and Experimental Issues in the Psychobiology of Motivation and Learning

PART I

B. F. LOMOV

Introductory Remarks to the Soviet–American Symposium On Neurophysiological Mechanisms of Goal-Directed Behavior

1

In 1976, the National Academy of Sciences proposed that its exchange agreement with the USSR Academy of Sciences include a paragraph that would provide for joint research projects in the field of psychology. The background story to this action is as follows: Talks about several questions pertaining to furthering cooperation were held during a visit to the Soviet Union by a group of American psychologists that included Dr. David Goslin. David Goslin very appropriately suggested that we begin our joint work by discussing the most important problems in the various fields of psychology. Understandably, I, as Director of the newly organized Institute of Psychology, supported that idea, inasmuch as our Institute has been exceedingly interested in developing international scientific ties. We believe that such ties constitute a most important condition for the advancement of science, and psychology particularly, in which there is still so much that remains unclear. There are many more questions than answers.

Following that meeting, Professor Philip Handler, President of the National Academy of Sciences, wrote to the President of the USSR Academy of Sciences, Professor A. P. Aleksandrov. In the summer of 1976, a group of American scientists headed by Professor D. Luce came to Moscow, where we agreed to hold a series of joint seminars on problems of mutual interest. This, the first of those seminars, is concerned with the neurophysiological mechanisms of goal-directed behavior.

In our view, the problem of goal-directed behavior occupies perhaps a central position in psychology. There is hardly anyone who doubts that human behavior is of a goal-directed character. This is a fact that each person encounters in everyday life. Both lay people and practicing psychologists are able to take good account of this fact in their own behavior. In everyday life they generate their own goals in a superb fashion and learn, in an equally superb fashion, to identify the goals of others. Common sense has given us remarkable results here. Science, however, is still far behind common sense when it comes to understanding this problem. From where are goals obtained? Why do they appear? In what way do they govern our behavior? What is the connection between goal and cause? We still do not have strictly scientific answers to these questions.

The determination of goal-directed behavior is quite a complex problem that has many aspects, including the psychobiological one, which is the principal topic for discussion at this symposium.

If we look at history, we can see that it is only relatively recently that the category of goal has become regarded as a scientific one. For a long time, determinism, with which science has been primarily concerned, seemed incompatible with the concept of goal-direction.

Let us take a short trip into history.

As we know, the understanding of goal in the form of "telos" or "entelichy" in the teaching of ancient philosophers, especially Plato and Aristotle, was very central to their philosophical concepts. Interestingly, closely associated with this understanding were the concepts of "direction," "integrity," "organization," and "orderliness." Also connected with this idea was the concept of soul and psyche. Unfortunately, both Aristotle and especially Plato elevated the concept of goal to the very highest rank in their attempt to explain all phenomena (including the cosmos) from the viewpoint of goal-direction, that is, teleologically.

Although even in ancient times contradictions were noted between teleology and determinism, "goal" and "cause" were still placed in opposition to one another just as strongly as in later times. This contrast became rather distinct somewhat later, when mechanics began to evolve as a science. The successes of mechanics, particularly in the solution of practical tasks, were possible only on the basis of strict determinism, with the complete exclusion of the "goal" concept from both descriptive and explanatory models. Those successes seemed to make mechanical determinism omnipotent. Bacon, Hobbes, and Déscartes were the classic exponents of mechanistic determinism, according to which the only scientific explanation of phenomena is one made in terms of linearly connected causes and effects. (I would call this method of explanation linear determinism.) On the other hand, when they had to deal with biological and psychological phenomena, they were not completely consistent and were forced to deviate from the principle of determinism.

Philosophical teleology was banished from science (which was, of course, a progressive development), and along with it so was the very concept "goal." The question of goal became generally ignored. There were, at times, attempts to deny goal-direction even in human acts. One may encounter such "telephobia" even in our time. The advancement of biology, however, could not bypass the problem of goal-conformity in living things. We are obliged to Darwin for having goal-conformity in living nature recognized as a real fact requiring explanation.

The attempts to interpret the advances of biology from idealistic viewpoints brought about new teleological philosophical systems. Kant and Hegel once again raised the question of goal and organization. Here, the problem of goal was bound to the question of regulating and organizing parts into a whole. Hegel was perhaps the first to formulate concisely the distinction between the concepts "mechanism" (technical device) and "organism." The concept of goal in this case became an important criterion of distinction. However, causal and goal explanations turned out to be mutually exclusive in these systems.

But the scientific task—that of offering a causal explanation for the emergence and evolution of goal—lagged behind. The questions "why" and "for what purpose" are not mutually exclusive in the study of the behavior of living things, and especially in the study of human behavior. A causal explanation must be found for goal-directed behavior.

This is, of course, one of the most complicated tasks in science. It can hardly be solved by using concepts of linear determinism in which causes and effects form a single-dimensional chain. Other approaches are required here. We need an understanding of the complexity, multidimensionality and multileveled nature of phenomena, their mutual transfers, contradictions, and so on. It seems to us that the approach that has come to be called the systems approach is a promising one for tackling this problem.

Goal-direction from this approach is viewed as a specific cause of causality. In fact, when we study behavior, we encounter phenomena of variable levels. For example, we can study behavior to learn how it encompasses the laws of mechanics (the transposition of a living body in space, of course, obeys the laws of mechanics). This seems to give us at least one slice in the analysis of cause–effect relationships. Incidentally, Newton, who formulated a very rigid system of mechanics that superbly explains mechanical movement without resorting to the concept of goal, asked the question: "In what way do movements of bodies comply with the will and what is the source of animal instinct?" He formulated this question as one that could not be solved within his system of mechanics. Strictly speaking, this question concerns the mechanics of a controlled body that, as a class of general mechanics, was developed much later than Newton's time. I would point out, by the way, that a book recently published in Russia by B. V. Korneev, *Introduction to Human Mechanics,* suggests an interesting approach to solving the problem raised by Newton.

The study of an organism's biochemical processes might also give us a "slice" of the causal–effect relationships. If we study the physiological processes occurring in an organism (including the nervous system), that, too, will give us some knowledge about the laws governing these processes. This also would give us one more slice of the cause–effect relationships.

The same can be said for the psychological processes. When we deal with human behavior, we must inevitably study social processes that, in the same determined manner as any other processes, take place in objective reality. This has been confirmed by modern science. Thus, each scientific discipline reveals some slice of the cause–effect relationships.

Incidentally, the unique nature of human beings as an object of scientific study lies in the fact that within them are somehow concentrated all forms of motion. They are so closely intertwined that specific changes, for example, in chemical processes, bring about changes in physiological, psychological, and other processes, and influence human behavior. This has been demonstrated by studies of the biochemical basis of behavior and psychopharmacology.

When we speak about the various slices observed in the process of analyzing a living being, this does not mean that we are simply looking at phenomena from different viewpoints. The phenomena themselves have different levels of organization. The teaching of dialectical materialism has formulated various forms of material movement that are being studied by different scientific disciplines. Those forms really exist, however, and have quite complex interrelationships. Of course, one of the most difficult problems in this regard is that of the interrelationships between various levels and of transfers from one level to the next, or from one form of material movement to another.

We find direct relationships between causes and effects on different levels. However, one can probably assume that there are also cause–effect relationships between different levels. That is to say, that one level that evolves by its own laws gives rise to a specific organization of another level. For example, we can identify the specific laws (i.e., cause–effects) of neurophysiological processes. But in the "pure form," so to speak, those laws are observed only under special experimental conditions (e.g., isolated neuron). We get quite another picture when we try to follow these processes under real conditions of animal behavior. Here, the physiological processes are organized in relation to the nature of behavior. The incorporation of one level (let us say, a lower one) into another (a higher one) inevitably places certain limitations on that level and transforms its dynamics. It would seem that the concept of goal has no place in the preceding enumerated levels of material movement that are governed by objective laws. The identification of the causes underlying the emergence of goal is hardly possible through an analysis of mechanical, neurochemical, and neurophysiological processes per se. Those causes must be found by examining behavior, that is, by looking at an organism within the system of its interrelationships with the environment. However, goal-directed behavior organizes the dynamics of neurophysiological processes in a specific fashion. Therefore, the study of neurophysiological processes must include an analysis of processes at different levels, both behavioral and neurophysiological. In our view, a scientific analysis of goal formation and goal direction first of all necessitates an explanation of the interrelationships between processes at different levels. This is, of course, a very different task, both theoretically and methodologically. An explanation of interrelationships at different levels requires the use of such concepts as organization, whole and partial, regulation, and others, which, as has been noted, have become part of teleological concepts, albeit in a mystifying form. It is now important to interpret that form in a strictly scientific manner.

As is known, it was only after a considerably long period of time and much difficulty that the concept "goal" was finally accepted in brain physiology research, which was then applied primarily to animal studies. One obstacle was the fear of falling into the trap of teleology and anthropomorphism. The other obstacle was that physiology in general and brain physiology in particular were dominated by the mechanistic determinism of Déscartes, who likened a living organism to a machine. The Déscartes concept of reflex reduced animal behavior to direct mechanical effects brought about by direct external causes. These obstacles are being gradually overcome in the wake of scientific progress.

One should note, of course, that in the strictest sense of the word, the concept of goal relates to human activity. However, that activity is evolved under conditions of human social life. There are such features in animal behavior, too, which allow us to say that behavior is directed by certain images of a future result (e.g., images which might be viewed as having a certain similarity to goal). Of course, when we look at the controlling effects exerted by images of future results, we must account for qualitative differences in animal adaptive behavior that is determined by biological laws and human, essentially social, transformative activity. Therefore, the direct transfer of data obtained from studies of human activity to animal behavior and vice versa is impermissible. However, the study of animal behavior mechanisms can be useful to the study of human behavior. This, of course, is particularly true in the case of the neurophysiological mechanisms of behavior.

In Russia, the evolution of psychology and physiology is associated with the name of I. M. Sechenov. He was the first to connect both physiological and psychological approaches to behavioral analysis, which, to a certain extent, helped to overcome mechanistic determinism.

In developing the theory of higher nervous activity, his successor, I. P. Pavlov, underscored the need to view that theory as a system. Although he attempted to eliminate psychological terminology from objective physiological studies in the early period of his research, Pavlov later considered the comparison of physiological and psychological phenomena to be a principal research problem. In particular, Pavlov tried to discover a process of goal formation from the reflex theory viewpoint. I have in mind his article "Goal Reflex." We find in Pavlov's works a number of very valuable ideas concerned with the systems approach. Investigators in that area included A. A. Ukhotomskii, who worked out the dominant principle; I. S. Beritashvili, who viewed behavior as image-directed; and N. A. Bernshtein, who proposed the reflex ring principle.

We particularly would like to note the research of P. K. Anokhin, who developed the functional system theory. In contrast to many versions of the systems approach that are based on formal models, the functional systems approach is based entirely on biological facts. According to P. K. Anokhin, "the term system can be applied only to a complex of selectively involved components whose interactions and interrelationships acquire the character of component interaction for the purpose of achieving a focused useful result." Anokhin always thought about spanning the "conceptual bridge," as he put it, between physiology and psychology. That problem was solved, to a certain extent, by his functional systems theory which allows one to view both physiological and psychological processes as a unified whole.

Anokhin devoted particular attention to analyzing the "organization of physiological processes" or "systems processes" that organize elementary physiological phenomena into an integral system. He introduced to physiology such concepts as action results acceptor, sensory synthesis, reverse afferentation, and decision making. All these concepts have now become a means of analyzing neurophysiological mechanisms of goal-directed behavior.

We hardly believe that goal can be understood by merely analyzing material processes (mechanical, chemical, etc.) taken by themselves. It is essential here to understand how an organism's environment is reflected in the brain (or nervous system). Goal is, of course, not a substance nor is it some specific property of matter. It is an image of something, such as an image of an organism's environment. An organism can hardly survive and behave adequately in an environment if it is somehow not reflected in the organism's brain. Thus, reflection (information being received) plays the role of a behavior regulator.

The concept "reflection" has been tied closely to the concept "information." Therefore, it is no accident that the problem of goal was formulated in connection with the study of information processes.

To represent reflection as a simple mirror image of the environment would be incorrect. In the process of an organism's interaction with its environment, not only is there reflection of the environment's status at any given moment, but also of tendencies toward change in the environment. This creates the possibility of advance reflection. The organism (its brain) reflects that which has not yet occurred, but is supposed to occur (expected to occur).

We view goal as a phenomenon of advance reflection.

The recognition of goal-directed behavior in the functional system theory does not contradict the principle of causality, but, on the contrary, develops it further. In fact, here the cause of a behavioral act is considered to be a goal which constitutes an image, a reflection, or a model of a future event (an event that is about to occur). But the goal exists prior to the action (or specific nervous activity, expressed neurophysiologically) that naturally and inevitably causes an action. At the same time, a goal is the natural consequence of selecting those elements from an organism's total memory that are essential to organization at a given moment. A model is constructed of what the organism needs by retrieving information from its memory, and this model then directs behavior. The formation of the model is influenced by both internal factors (motivation and needs that act as objective necessity) and external factors (environmental circumstances).

Behavioral mechanisms in the functional system theory are viewed as a hierarchy of systems that includes molecular, physiological, and psychological levels. Of course, there is still much in the mechanisms of goal-directed behavior that remains unclear and requires further study.

The chapters to be presented by associates of our Institute will report some specific results obtained in the area of functional system theory. Of course, we shall all be very grateful for critical analyses of those results.

Understandably, the functional system theory is not the only approach in Russia to investigating the problem of goal-directed behavior. There are others. In particular, there is the original and quite interesting concept developed by Professor E. A. Asratyan, who is participating in our seminar. Also interesting is the concept of a nerve stimulus model proposed by E. N. Sokolov. There are other approaches as well.

We are greatly interested in the work of American psychologists and physiologists engaged in this problem. We are very familiar with the work of G. A. Miller, G. Galanter, and K. Pribram, "Plans and Structures of Behavior." The T–O–T–E (Test–Operate–Test–Exit) concept is close to several concepts being developed in Russia. The hypothesis concerning the geographic principle involved in the processing of input information in the brain essentially confirms the systems principle of brain function. I believe that the research of Richard F. Thompson on the neurophysiology of learning is also quite promising with respect to understanding the problem we are investigating. I have in mind especially his data that indicate that many brain systems are included in the formation of an engram, and his data on the role of the hippocampus.

As has been demonstrated by N. E. Miller, motivation and reinforcement in learning are related directly to goal-directed learning. We are particularly impressed by N. E. Miller's ideas on the complex study of behavior, although we feel that psychology rather than physiology occupies a more central position among the disciplines involved in the study of behavior.

The theory of statistical configuration proposed by E. Roy John makes, in our view, a very significant contribution to understanding the problem under discussion inasmuch as it allows us to describe not local processes but systems of processes.

The same should be said about the research results obtained by D. B. Lindsley on the central mechanisms of motivation–activation effects (general principles and neural mechanisms).

Of great significance in the neurophysiology of goal-directed behavior are the data of Olds (the phenomenon of self-stimulation, the discovery of short-latent neuron discharges associated with information retrieval from memory); MacLean (neurophysiology of motivation and emotions); Evarts (neuron mechanisms of natural goal-directed movements); Cohen, Magoun, Gormazano, Isaacson, Kandel, Teitelbaum, and many other scientists. We are gratified that the American side of the symposium is represented by such prominent scholars.

I have taken the liberty to speak not only about Soviet but also about American symposium participants in order to take note of the fact that, in my estimation, we are following similar paths in elaborating a number of problems. This raises hopes for making the work of our seminar more useful.

If we are successful in concisely and accurately formulating and defining the problem at hand in the course of the seminar discussions, this will in itself constitute a significant result. We have a saying that one-half of a problem's solution lies in its correct formulation. If, in addition, we are successful in finding the most promising approaches toward the problem's solution and toward a program of joint research, then the results of the symposium will have been all the more valuable.

In my introductory remarks, I have had to talk mainly about very general philosophical questions. I did that because of determinism and also (goal-direction) to present the initial premises upon which the study of neurophysiological mechanisms of goal-directed behavior have been undertaken at the USSR Academy of Sciences Institute of Psychology, and to take note of a certain similarity in the approaches to this problem in both Russia and the United States.

E. A. ASRATYAN

Reflex Mechanisms of Motivational Behavior

2

In the course of his classic studies of the physiology of digestion, the great Russian scientist Ivan Petrovich Pavlov said that science moves ahead in spurts, depending on the progress made by method. This has become a generally recognized truth. The correctness of that thesis has been clearly proven by what has been happening in recent decades in experimental studies of the so-called problems of motivational behavior. As is known, this problem, under different names and in different aspects, has long been successfully investigated experimentally by a number of prominent psychologists and brain physiologists such as Thorndike, Pavlov, Köhler, and their followers. Experimental studies of motivational behavior received a powerful impetus for further development as a result of the elaboration, the improvement, and broad application of new methods that have made it possible to undertake direct, thorough, and detailed studies of the functions not only of superficial but also deep brain formations in chronic animal experiments (see survey by Doty, 1969), and even to conduct clinical studies of human patients.

By using methods that involve the stimulation of these formations, the recording of their electrical activity, their local electrolysis, and so on, the investigators of many countries, and particularly those of the United States, have obtained a mass of diversified, original, and scientifically highly significant experimental material that can by rights be included among the most significant modern scientific achievements of brain activity studies. However, these advances have also had some negative consequences. In particular, the problem under discussion has been isolated from the total complex of knowledge about brain functions, and many of the previously known facts and theoretical tenets of fundamental significance with respect to the behavior of highly developed organisms have been consigned to oblivion, or, in any case, not taken into consideration. This particularly holds true for the precise, rich, original, and exceptionally valuable factual data of I. P. Pavlov on cerebral activity and his teaching on higher nervous activity that has truly constituted an epoch in the developmental history of our knowledge about the functions of this higher organ of the central nervous system. This kind of inattentive, disdainful, and often negative attitude toward a valuable scientific legacy in the study of brain function, as well as the considerable isolation and unique self-isolation of the problem of motivational behavior from the total flow of modern research on the activity of this organ in other aspects, could not help but have their own negative effects.

In a background of contradictory concepts on many aspects of the problem, there has been a clearly dominating tendency to negate the reflex genesis of motivational behavior that is simply seen as a version of nonreflex forms of brain activity that are primarily activated and regulated by endogenic factors. In a milder form, that tendency attributes the principal role in motivational behavior to the nonreflexive and suprareflexive forms of brain activity and assigns a secondary, auxiliary role to the reflexes. One should note that it has become a bad tradition for most advocates of these kinds of concepts to accompany the defense of their viewpoint (which is fully their right) by distortions of the essence of the reflex theory and by ludicrous representations of the reflex, including the conditioned reflex. This deprives scientific discussion of its progressive significance and useful role, and transforms such a discussion into a source and means of misinformation. In spite of the truth and formulations of Pavlov, the conditioned reflex theory is qualified as an analytical theory that does not account for the importance of the brain's initial functional state, an organism's endogenic factors, its past individual experience, and the correlating role of action results accomplished by feedback mechanisms. Consequently, the reflex theory is viewed as unsuitable for understanding the synthetic acts and autoregulation phenomena in brain activity. Ignored are established facts concerning the existence of a large family of diversified conditional reflexes that differ in degree of complexity, level of execution, functional architecture, means of organization, their exclusive dependence on endogenic shifts and environmental factors, their clearly defined adaptive variability, and their capability to integrate behavioral acts into complex forms. Opponents of the reflex theory have persistently characterized the conditional reflex as a monotonous, banal, automatic, rigid, invariable phenomenon in brain activity.

I had the honor and good fortune of being a pupil of I. P. Pavlov and I am a confirmed supporter of his teaching on higher nervous activity (or behavior)—that teaching embodying the reflex principle in his profound, evolutionary, and dynamic understanding of the subject. My colleagues and I have been engaged for a long time in the experimental and theoretical elaboration of several aspects of his teaching, including those, which in a narrow sense, might be included in the so-called problem of motivational behavior. The results of our research and a critical analysis of the current status of experimental and theoretical work in motivational behavior have led us to conclude that the most solid and progressive scientific–theoretical basis of this vital problem is Pavlov's teaching whose organic component is so essential, that through his teaching one cannot only explain many intricate aspects of the problem most satisfactorily, but one can also discover the best prospects for the successful experimental and theoretical elaboration of that problem as a whole.

Due credit must be given to a group of prominent progressive investigators in this area (Bindra & Campbell, 1967; Cofer & Applay, 1964; Cohen, G. W. Brown & M. L. Brown, 1957; Delgado, 1954; Fonberg, 1967; Hull, 1943; Lissak et al., 1966; Miller, 1941; Milner, 1970; Mogenson & Stevenson, 1966; Roberts, 1958b; Valenstein, 1972; Wyrwicka, 1972, etc.), who, in their experiments on various species of animals, demonstrated the very important and oftentimes even the decisive role played by environmental factors in the emergence of various types of motivational behavioral acts. They also demonstrated the participation of the conditional reflex mechanism in the formation and realization of these acts. This was quite important from the viewpoint of using Pavlov's ideas to understand specific, partial aspects of the problem at the modern stage of its development through the use of the most recent experimental methods. But no matter how significant those studies were, they could only be viewed as the first steps in this area of research. Even though the role of the conditional reflex mechanism in motivational behavioral acts was studied within a limited framework, on a low methodological level or in a somewhat primitive form in a number of investigations, one should emphasize that Pavlov's teaching contains a number of other important theoretical tenets and ideas that can be effectively utilized for the fruitful experimental and theoretical elaboration of many knotty aspects of the motivation problem which may not have received the investigators' attention.

In recent years we have attempted to demonstrate the correctness of what has been already said in a number of publications, both in Russian and other foreign languages (Asratyan, 1974, 1976a, 1976b, 1977, 1978). They are based on data that have been obtained by us and other investigators and are supported by specific theoretical arguments. I am very happy to have the opportunity of summarizing these thoughts, ideas, and several new data to such an eminent audience, in the country moreover whose scientists have rendered such a special service in the experimental and theoretical elaboration of the problem under discussion.

I shall take the liberty of reminding you of one notion of fundamental significance, ascribed to Pavlov by a certain group of investigators, to the effect that, according to Pavlov, higher nervous activity or behavior is not at all limited by conditional reflexes and other forms of acquired reflexes. In his definition, this activity, which provides for the most perfect regulation of an organism's complex interrelationships with the environment, is the product of the interconnected activity of the cerebral cortex and the subcortical formations and consists of several stages that differ from each other in the genesis and character of their component reflexes and their anatomical substrata. What he meant by the first stage were the specialized, complex, unconditional reflexes, such as the alimentary, defensive, sexual, otherwise called instincts, and inclinations, and believed that they are primarily realized by subcortical formations. He understood the second stage to be the conditional reflexes that are developed on the basis of the aforementioned unconditional or innate reflexes. Pavlov believed the cerebral cortex in humans and in higher animals to be the substratum for this basic type of acquired reflex. There can be no doubt that he would have included in this stage new types of acquired reflexes or associations that he identified in anthropoids in the last period of his life and that he placed above the usual conditional reflexes. Pavlov believed that higher nervous activity in animals was limited to these two stages. He assumed that a third stage, together with these two, existed in humans; this was a stage of higher nervous or psychic activity that was specific to humans in the form of a speech or a second signal system of activity whose substratum he also believed was the neocortex.

According to Pavlov, the participation of the subcortical formations in behavior is not relatively limited by the independent role of the complex, specific, vital alimentary, defensive, sexual, and other unconditional reflexes that are realized by those formations, but also by the fact that the latter serve as a basis for developing corresponding conditional reflexes. He believed that these formations have a constant toning effect on the cortex and elevate its activity. Furthermore, with regard to the subcortical formations, Pavlov emphasized especially the exceptional importance of the initial functional state of brain structure for their activity and the important role of endogenic and humoral factors during changes of their functional state.

From what has been said, it is clear that Pavlov's teaching on higher nervous activity, based exclusively on the principles of reflex activity, can be a firm scientific basis for a correct understanding and interpretation of behavioral reactions (also called motivational reactions) and contains within itself all that is required to be considered characteristic of and specific to motivational behavior. This particularly applies to the relationship of the subcortical formations and endogenic humoral factors to such behavior; although Pavlov, understandably, could only discuss this in a general form without the degree of detail and specificity that characterizes today's research on this problem.

It would seem that the great potential possibilities of Pavlov's teaching in regard to motivational behavior could be demonstrated by a more satisfactory explanation of the rich, reliable, and valuable factual material that is available on a number of key aspects of the problem than could be done from the viewpoint of other theoretical concepts.

We shall discuss three such questions here: (*a*) the possibility of producing the same behavioral act by means of electrostimulation of different deep brain formations and the development of conditional reflexes when that stimulation is used either as a reinforcement or as a signal; (*b*) the possibility of producing different kinds of behavioral acts by electrostimulation of the same formation; (*c*) the goal-directed character of motivational behavioral acts.

Since the existing contradictions in each of these problems are well known, there is no need to repeat them here.

From our point of view, the essence of the first of these problems at the present stage might be best understood if one considered, together with the aforementioned fundamental thesis of Pavlov on the stages of higher nervous activity, his concept of the nerve center as an aggregrate of nerve structures broadly situated in various sections of the cerebrum and under the cerebrum that are closely interconnected and compose a uniform complex structural–functional whole and that accomplish a specific function. These concepts, in accord with the views of Sherrington (1948) and Magnus (1924) on various levels of integration of innate reflexes, and the results of our experiments (in which we studied the effects of surgical section or functional exclusion of the cerebral cortex in higher animals on the state of various unconditional reflexes and functions of organisms) were the basis of our hypothesis (1955) on the multistage arc of an innate reflex or nerve center and its schematic representation (Figure 2.1). Later, we began to use these concepts and schemes in our work on motivational behavior, and particularly in our attempts to understand and interpret the aforementioned phenomena. We have the impression that here they have turned out to be even more than adequate.

We assume that the branches of the central section of the arc of the unconditional reflex (I–V) that pass along different levels of the central nervous system are not the same in structure and function and therefore play different roles. Each branch is characterized by specific structural–functional features that accord a particular functional tone to the appropriate reflex, whereas the main arc branch of each type of reflex determines its basic features, and passes along any one of the hierarchical series of the system's levels. Although the main branch in the arc of many elementary motor and sympathetic reflexes is within the limits of the spinal cord (1st level), and is within the medulla oblongata (2nd level) in the arc of complex cardiovascular, motor, tonic cervical and labyrinthine reflexes, there are sufficient grounds to believe that the main branch of complex alimentary, drinking, sexual, aggressive, and other unconditional reflexes of in-

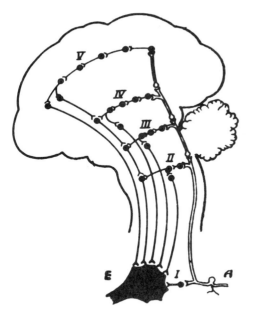

Figure 2.1. Sketch of an unconditional reflex arc. Designations: (I–V) branches of the central portion of the reflex's arc at various levels of the central nervous system (A) Afferent neuron; (E) efferent neuron. (From Asratyan.)

terest to us are within the hypothalamus (3rd level). In such a case, the structure of various limbic formations participating in the indicated complex, specialized, unconditional reflexes might be considered as satellite or supplementary branches to the hypothalamus. They are situated in the zone of the 3rd and 4th levels. The cortical branch of the arc of each of these reflexes (5th level), which corresponds to what Pavlov called the cortical representative of the unconditional reflex, provides for a high level execution of that reflex and serves as the basis for the formation of numerous and diversified conditional reflexes in response to all kinds of stimuli from the external world and to various stimuli of an endogenic nature.

In the light of this concept, the facts established by many investigators (Anand, 1971; Hess, 1949; Morgane, 1961; Robinson and Mishkin, 1962; Valenstein, 1972, and others) that electrostimulation of the hypothalamus and many formations of the limbic system (hippocampus, amygdala, septum) may cause a particular innate or motivational act, such as alimentary, drinking, aggressive, and sexual, which we understand as a result of the activation of elements in the complex arc of some particular complex, specialized, unconditional reflex that are contained in each of those formations. The possibility of producing conditional reflexes by stimulating those elements (as a reinforcing factor) in response to a situation or individual stimuli (Bindra, 1974; Delgado, 1964; Fonberg, 1967; Gengerelli, 1975; Larson, 1954; Mendelson, 1966; Miller, 1961; Mogenson, 1972; Roberts, 1958) is also easy to understand from this viewpoint. When those elements are stimulated, inevitably stimulated also is the cortical branch of the arc of the corresponding unconditional reflex. When the cortical points of situational factors or special stimulants are simultaneously excited, a corresponding conditional reflex is produced.

The development of conditional reflexes in response to the electrostimulation of hypothalamic structures and a number of limbic system formations in the form of a signal or conditional stimulation (Asratyan, 1977; Ellen & Powell, 1966; Mogenson, 1962; Nielsen, Knight & Porter, 1962) might be more satisfactorily understood and interpreted from the viewpoint we are developing here.

It would seem advisable to discuss these factors in somewhat greater detail in several aspects.

One should note that until recently the data on this question were significantly less convincing than the data underlying the basis of the aforementioned phenomena. Of particular significance here are several shortcomings in the experimental studies of conditional reflexes that were unfortunately characteristic of the research undertaken by many of our foreign colleagues in this area. This research concerns studies of these reflexes by using methods that by no means corresponded to the standards of Pavlov's laboratory. In the first place, the experiments that were conducted primarily on rats were performed without an objective graphic recording of the produced conditional reflexes, including the recording of such basic characteristics as latent period, magnitude, character, and duration. Furthermore, the experimentation undertaken in this area followed a rather intricate design. Most of these studies provided for a special series of experiments in which the phenomenon of autostimulation was formulated preliminarily, then followed by another series of experiments in which the animals acquired the skill of pressing a lever to obtain food, and, finally, followed by a third series of experiments in which autostimulation was combined ("reward stimulation," in the terminology of the investigators) with either the receipt of food or with the electrostimulation of the paws by the shuttle box method for the purpose of producing alimentary or defense conditional reflexes in the animals. Perhaps some kinds of conclusions about the interrelationships between autostimulation and the conditioned reflex might be drawn from the results of these kinds of experiments. However, it would be extremely difficult to obtain from them a clear and unequivocal answer to whether or not one can produce conditioned reflexes in response to electrostimulation of deep brain formations in the form of a conditional or a signal stimulation.

Among the studies on this problem undertaken by other investigators, we note that the work done by Nielsen, Knight, and Porter (1962) on cats are the only experiments that were more or less satisfactory with respect to method and results. These investigators succeeded in producing an instrumental, defense-conditioned reflex of the avoidance reaction type by combining electrostimulation of particular deep brain formations with local electro-pain stimulation of one of the animal's anterior paws in chronic experiments on cats in which more than 20 electrodes were placed in various deep brain formations, including several limbic formations.

We too have been engaged, together with a large group of colleagues in our laboratory, in experimental studies of this nature in the past few decades. Classic and instrumental alimentary and electrodefense-conditioned reflexes have been produced in dogs and cats by using the direct electrostimulation of various subcortical formations as a signal or conditional stimulus. All of the necessary factors were objectively and graphically recorded in those experiments and we carefully observed all of Pavlov's strict rules for producing conditioned reflexes (Asratyan, 1970). In recent years, we have been giving greater attention to the hypothalamus nuclei and certain formations of the limbic system, and the results we have obtained seem to be indisputable proof of the possibility of producing alimentary and defense motor-conditioned reflexes in response to electrostimulation of the indicated deep formations in the way of a conditional stimulus. Those results are briefly summarized as follows.

The experiments were conducted on dogs and cats with chronically implanted electrodes in the nuclei of the hypothalamus (lateral and ventromedial), the amygdala (lateral, cortical, and basal), and septum (lateral and medial), and the hippocampus (dorsal and ventral sections). Local instrumental-conditioned reflexes—alimentary or electrodefense—that were objectively and graphically recorded simultaneously with several other important functions of the organism were produced in dogs in traditional, shielded, soundproof chambers. The experiments on the cats were conducted in large shielded cages that allowed the animals to move around freely. Conditioned reflexes in the form of directed movement toward food and local instrumental food-acquisition movement were produced in the animals and were objectively and graphically recorded. The evoked potentials in the cortical points of the "instrumental" paws were also recorded in some of the experiments. A two-phase rectangular current of moderate intensity (0.2–0.8 ma) and a frequency of 10–100 Hz and stimulation time of 2–6 secs was used to stimulate the deep brain formations. Conditional responses to transitional external stimuli were produced along with these conditioned reflexes in a series of experiments on the same animals.

The data obtained in these experiments, partially represented in the following figures, indicate that both alimentary and electrodefense general motor and specialized local motor instrumental-conditioned reflexes can be produced by electrostimulation from ony of the aforementioned deep nerve formations. Furthermore, homofunctional conditional reflexes (e.g., alimentary conditional responses to electrostimulation of the lateral nucleus of the hypothalamus) are produced more quickly and easily than heterofunctional conditional reflexes (e.g., alimentary conditional responses to electrostimulation of the ventromedial nucleus of the hypothalamus). It is clear from our data that these reflexes do not differ in principle from conditioned responses to conventional extraneous external stimuli with respect to formation, specialization, and consolidation. Those reflexes appear after 20–30 combinations and are consolidated after 50–70 combinations, and their latent period is gradually shortened (Figures 2.2, 2.3, 2.4, 2.5). It is noteworthy that as the stimulated formations acquire a greater signal importance for alimentary and defense activity, the innate reactions that are initially caused by those stimuli are gradually attenuated and extinguished. This is particularly clear when the initial response to the electrostimulation of the formations is of a defense–aggressive nature (e.g., during the stimulation of the ventromedial nucleus of the hypothalamus or the lateral nucleus of the septum) and when alimentary conditional responses are produced by such stimulation. After a period, the defense-aggressive responses to such stimulation weaken, are extinguished, and are replaced by a clearly defined general motor alimentary reaction and local alimentary instrumental motor reflex. One cannot help but note the similarity between these data and the results of Erofeeva's famous experiments in Pavlov's laboratory in which alimentary conditioned responses to electropain stimulation of the paw were produced in dogs. The produced conditioned reflexes are extinguished and are restored (Figures 2.2, 2.3, 2.4, 2.5), and are generalized both within the framework of the described deep formation and within other formations, and, in both cases, differentiations can be produced in response to them. As I do not have the opportunity of relating all of this in detail, I shall limit this discussion to several graphic recordings (Figures 2.6, 2.7, 2.8, 2.9, 2.10).

The physiological mechanism of producing a conditioned response to the electrostimulation of the previously mentioned deep brain formations in the capacity of a signal or conditional stimulation we believe consists of the following. In essence, any primary conditioned reflex can be viewed as the result of combining the activity of two unconditioned reflexes, that is, it is the product of their synthesis. This is true both for the combination of a so-called indifferent stimulus (an orientational reflex caused by such a stimulus is also unconditioned) with some biologically significant stimulus, and the combination of two typical unconditioned stimuli, such as illustrated by the previously indicated experiment of Erofeeva (Pavlov, 1973). Of course, an important role here is played by the strength ratios of the stimuli being combined, the order of their action, and so on. In our previously described experiments, the cortical branch of the arc of one of the reflexes is activated by the stimulation of the subcortical elements of the same arc by a current of moderate intensity. This is followed by an even stronger excitation of the cortical branch of another unconditioned reflex —the alimentary or defense reflex—by a natural stimulation of its recep-

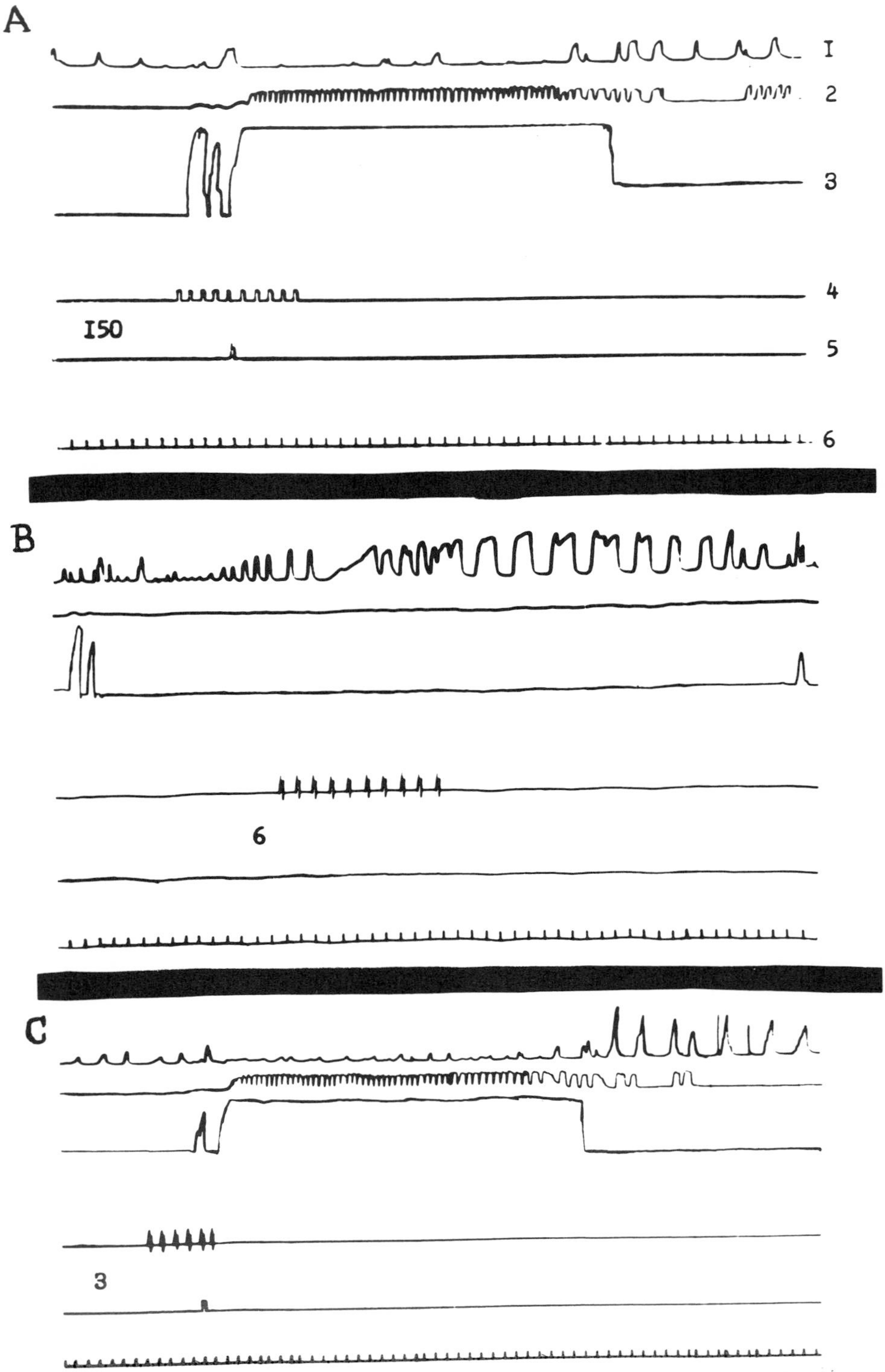
A
I
2
3
4
I50
5
6
B
6
C
3

tive field. We then have the basic conditions for producing the appropriate conditioned reflex.

We shall present our ideas about the effectiveness of fruitfully utilizing other concepts and theoretical tenets of Pavlov for a satisfactory and strictly scientific understanding and explanation of the remaining two key problems, and still somewhat knotty problem of motivational behavior, in a more concise form since they have already been frequently discussed in a number of our aforementioned publications (Asratyan, 1973–1978).

We are referring here to the fact, established by many investigators, that the most varied of motivational behavioral reactions, such as alimentary, drinking, aggressive, and sexual, can be induced by the electrostimulation of the same point in the same nucleus of the hypothalamus or of any limbic formation, depending on changes in the intensity, frequency, and other parameters of the stimulating current, and the functional state of the organism or situational factors of the experiments (Caggiuli, 1973; Gallistel, 1969; Grastyan, Gzopf, Angyan, & Izabo, 1966; Mogenson, 1972; Mogenson & Stevenson, 1966; Roberts, 1958a, b; Valenstein, 1972). This evidence is in clear contradiction to the traditional concept of static specialization and localization of functions in the central nervous system of the higher organisms. However, they could be easily understood and interpreted from the viewpoint of dynamic specialization and localization of functions as applied to the cerebral cortex by Luciani (1915) in an outline form and as thoroughly worked out by Pavlov (1973) in the form of a concise theoretical thesis which we (Asratyan, 1949) applied to the entire central nervous system as a general principle. This concept states essentially that the morphological substrate of specialized functions is represented in the form of limited nuclear zones in which the highly specialized nervous elements for a given function are concentrated in the form of broad peripheral regions in which the less specialized and diffuse nervous elements are represented in diminishing density. Furthermore, these regions of diffuse elements overlap each other, thereby creating a unique polyfunctional mosaic of structures.

If the hypothalamus and limbic system formations are not exceptions to this rule, then it is easy to understand how an electrostimulation of their individual points can simultaneously encompass, to a greater or less degree, the structure of several specific functions, and either activates them

---

Figure 2.2. *Alimentary instrumental-conditioned reflex in response to electrostimulation of the medical nucleus of the septum in dogs. Designations: (A) produced conditioned reflex; (B) extinguished conditioned reflex; (C) restored conditioned reflex. The numbers under the stimulus marker line are the numbers of conditioned reflex combinations with reinforcement. In the case of extinction, these numbers indicate the number of conditioned stimulus presentations without reinforcement. Symbols: (1) recording of respiration; (2) recording of mastication; (3) recording of paw movement; (4) stimulation of septum; (5) food delivery; (6) time in seconds. (From experiments of N. P. Balezina.)*

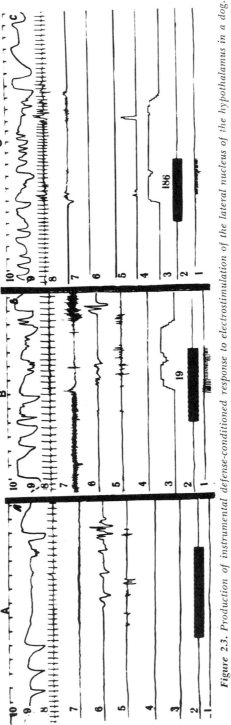

Figure 2.3. Production of instrumental defense-conditioned response to electrostimulation of the lateral nucleus of the hypothalamus in a dog. Designations: (A) prior to reflex production; (B) first stage of reflex production; (C) stage of consolidated reflex. Symbols: (1) unconditional stimulus (short lines—switch-off of electrocutaneous stimulation by raising of the paw); (2) conditional stimulus (the underlying numbers are the numbers of combinations); (3) and (4) mechanograms of the left and right anterior paw; (5) recording of head movement; (6) mechanogram of jaw movement; (7) EMG of left anterior paw; (8) ECG; (9) mechanogram of respiratory movements; (10) time in seconds. (From experiments of L. I. Chilingradyan.)

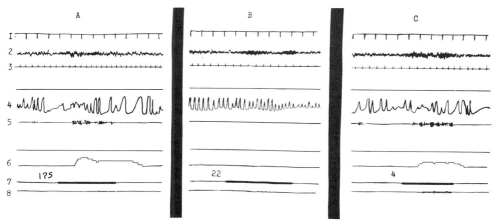

Figure 2.4. Instrumental defense conditioned response to electrostimulation of the ventral hyppocampus in the dog. Designations: *(A)* conditioned reflex; *(B)* extinguished conditioned reflex; *(C)* restored conditioned reflex. *(1)* Time marker; *(2)* electrical activity of the hyppocampus; *(3)* ECG; *(4)* recording of respiration; *(5)* head movement recordings; *(6)* mechanogram of the left anterior paw; *(7)* conditional stimulus marker; *(8)* unconditional stimulus marker. (From experiments of A. G. Crirog'yan.)

separately at different times, depending on the level of their excitability and parameters of the stimulating current, or elevates their excitability simultaneously and thereby creates the prerequisites and a unique operative readiness for activating any one of those functions, depending on the presence of an adequate external stimulus in the environment. One should note that several other investigators (Milner, 1970; Mogenson, 1972; Valenstein, 1972) have currently expressed a similar point of view from somewhat different positions.

The goal-directed nature of motivational behavior is the third key problem which we are attempting to approach from new angles. Goal-direction has been considered generally to be the most typical feature of motivational behavior, but the interpretation of that behavior has been regarded as the most complicated. The existing opinions on this problem have ranged from unrestrained narrations in a teleological spirit to simply describing the phenomenon itself or to an enumeration of the supposed circumstances and factors that condition goal-directed behavior. For example, Tolman (1948) ascribes the goal-directive nature of motivational behavior to a "knowledge map" that is formed in the individual life of a species; Beritashvili (1960) ascribes it to the directing influence of the so-called psychoneural image; Milner (1970) ascribes it to the presence of "predictor cellular aggregates," whose activity is adjusted by sensory signals. Anokhin (1968) ascribes it to the initial existence of goal and the programming role of the so-called action acceptor, Bindra (1974) ascribes it to

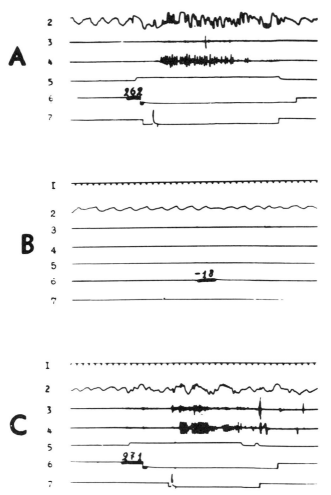

Figure 2.5. Alimentary-conditioned response to electrostimulation of the lateral nucleus of the amygdala in the cat. Designations: (A) positive conditioned reflex; (B) extinguished reflex; (C) restored conditioned reflex. Symbols: (1) time in seconds; (2) pneumogram; (3) left eyelid movement; (4) head movement and masticatory movements; (5) head turns toward food box and visual fixation on the food box (upward deflection); (6) electrostimulation marker (sketched line segment); start position vacated (downward deflection) and return to the start position after eating (upward deflection to original level); (7) approach to food box (downward line deflection) and delivery of food (sharp upward projection). The numbers over the electrostimulation marker indicate the order number of combination and the application of a nonreinforced stimulus during extinction. (From experiments of F. K. Dadurova and R. F. Kolotygina).

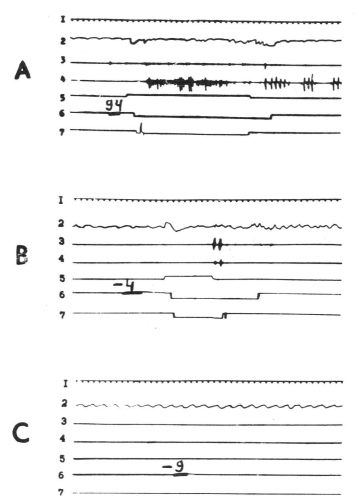

Figure 2.6. Generalization of a conditioned alimentary response to electrostimulation of the lateral nucleus of the amygdala in the cat. (A) Generalization of response to electrostimulation of the cortical nucleus of the amygdala. (B) Differentiation to stimulation of this nucleus (C) Designations are the same as Figure 2.5. (From experiments of F. K. Dadurova and R. F. Kolotygina.)

the action of environmental factors that have acquired a specific orienting significance as a result of learning. Deutch (1960) ascribes it to the formation of complex "subgoal" systems in the process of accumulating individual experience and the sequence of their realization that account for the results of each transpired stage. As I do not have the opportunity to comment here in detail on all of these views, we shall merely note that from our viewpoint most of those views lack a clear and concise answer to questions about the physiological mechanisms of the phenomenon itself.

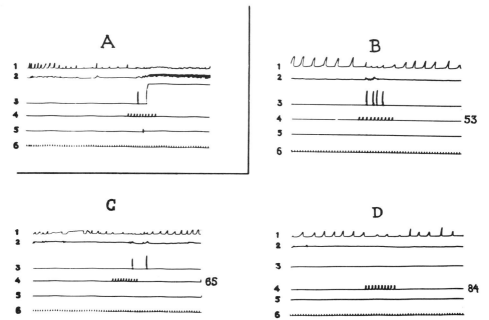

Figure 2.7. (A) Alimentary instrumental response to stimulation of the ventromedial hypothalamus in dogs. (B,C) Generalization of response to stimulation of the lateral hypothalamus. (D) Differentiation to the stimulation of this nucleus. Designations: (1) respiration; (2) mastication; (3) movement of experimental paw; (4) stimulation marker; (5) introduction of reinforcement; (6) time in seconds. Numbers on the right designate the numbers of the differentiated signal presentation. (From experiments of O. G. Pavlova.)

Pavlov's idea about the bilateral nature of the conditional link, which he hypothesized in reference to the instrumental variety of the conditioned reflex, is the basis of our understanding of this problem's essence. He believed that the development and realization of instrumental motion were conditioned by the formation and activation of conditional feedback. Whereas many investigators, by right, consider that instrumental-conditioned reflexes constitute a very successful mode for studying motivational behavior, Pavlov's new interpretation of this reflex's physiological mechanisms created especially favorable opportunities for effectively using this model to learn about the physiological mechanisms of motivational behavior as a whole. We have enthusiastically and boldly taken advantage of this opportunity because, as a result of many years of research undertaken by Pavlov's followers and particularly our research completed with colleagues (1970, 1977) and the work of the Czech scientist Dostalek (1964), our teacher's idea has now become a theoretical thesis that has been rather well substantiated experimentally.

A more specific illustration follows.

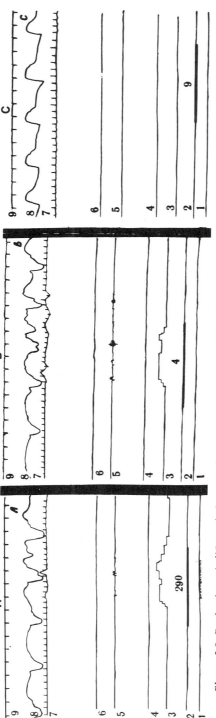

Figure 2.8. Production of differentiation. (A) instrumental defense-conditioned response to electrostimulation of the lateral nucleus of the hypothalamus; (B) generalization of response to electrostimulation of the lateral nucleus of the opposite side of the hypothalamus; (C) differentiation in response to stimulation of the same nucleus; numbers over the conditioned stimulus designate the number of differentiated stimulus applications. Remaining designations are the same as in Figure 2.3. (From experiments of L. I. Chilingradyan.)

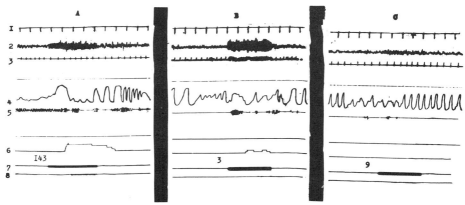

Figure 2.9. *Differentiation dynamics of electrostimulation effects of the right and left sections of the ventral hippocampus. (A) Instrumental defense conditioned response to electrostimulation of the right ventral section of the hippocampus. (B) Generalized conditioned response to electrostimulation of the symmetric point of the contralateral ventral hippocampus. (C) Differentiation of response to the same stimulation. Designations are the same as in Figure 2.4. (From experiments of G. A. Grigoryan.)*

In the process of producing an instrumental-conditioned reflex, a certain specific movement always precedes reinforcement and becomes the signal stimulus that is capable of inducing the reinforcing reflex through a direct conditional link. In accordance with Pavlov, it would be correct to consider that the realization of this movement, resulting from alimentary excitation induced by endogenic or exogenic factors, is conditioned by the activation of feedback that is produced in this process, that is, the link from the alimentary center to the cortical structures of instrumental movement.

Many years ago (Asratyan, 1967), we proposed a schematic representation of the arc of that reflex (Figure 2.11), which was based on an abundance of diversified data obtained in our laboratory and was guided by this idea. It seems to us that this scheme could be the key to understanding satisfactorily the physiological mechanisms not only of the generally known models of instrumental reactions but also of autostimulation (Olds & Milner, 1954). This scheme and its underlying principles can be used to understand and interpret other factors. If preliminary alimentary, drinking, or defense instrumental-conditioned reflexes are produced in animals, a

Figure 2.10. *(A) Instrumental food-acquisition response to electrostimulation of the ventromedial hypothalamus. (B) Generalization of response to electrostimulation of the lateral hypothalamus. (C) Differentiation of response to electrostimulation of the same nucleus. Symbols: (1) conditioned reflex; (2) position of cat at start position; (3) paw pressing level; (4) introduction of food box with meat; (5) cat taking meat from the food box; (6) time in seconds. (From experiments of G. L. Vanetsian.)*

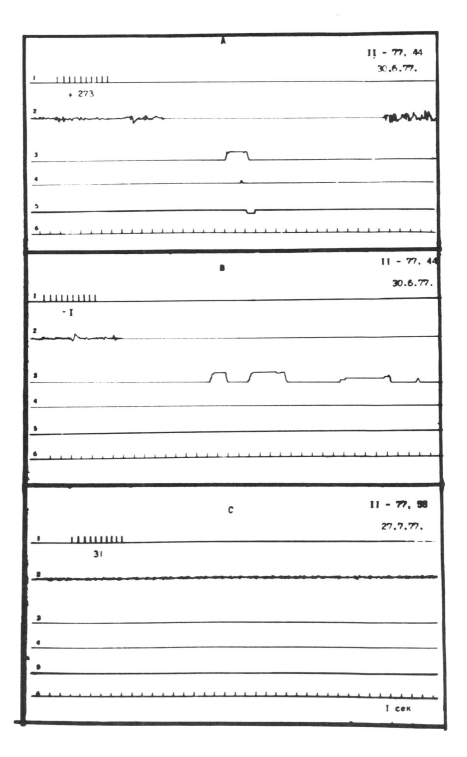

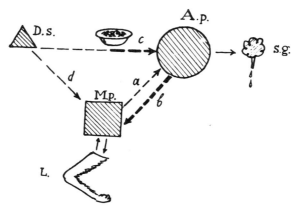

Figure 2.11. Arc diagram of instrumental alimentary conditioned reflex. Symbols: D.s.—distant conditioned stimulus; A.p.—cortical point of alimentary unconditioned reflex; M.p.—cortical point of experimental paw; L—paw; S.g.—salivary gland; (a), (b), (c), (d)—conditional links. (From E. A. Astryan.)

subsequent electrostimulation of the hypothalamic structures will cause these produced instrumental movements that can also be stimulated by corresponding natural, unconditional stimuli (Andersen & Grastyan et al., 1956; Delgado, 1964; Miller, 1961; Wyrwicka, 1957; Wyrwicka, 1972; and others).

We are also inclined to explain the positional preparation of experimental animals before the realization of a necessary local conditioned reflex movement in experiments conducted in special rooms as the result of conditional feedback activity (Kasyanov, 1950; Ioffe, 1975). Moreover, a thorough analysis of data that we obtained in studying alimentary-conditioned reflexes in dogs and cats moving about freely in a spacious experimental chamber or large room leads us to the conclusion that the animals' learned ability to remain for an extensive period in a specific start area in expectation of a subsequent trigger signal may be more satisfactorily explained as a manifestation of tonic activity of the very same conditional feedback. We are drawn to this conclusion not only by the data from special experiments performed by our colleagues M. I. Struchkov, L. P. Rudenko, E. K. Davydova, F. K. Daurova, R. F. Kolotygina, G. L. Vanetsian, and S. I. Shumikhina (Asratyan, 1977), but also by the data of Kupalov (1964) and his associates on the so-called situation-conditioned reflex. In both cases, in the process of producing conditioned reflexes, the animals receive food only when they go to the food box from the start position that subsequently becomes the constantly acting, specific spatial, conditional stimulus that is associated bilaterally with the reinforcing stimulus.

Thus, in our understanding, the formation of bilateral conditional links and the activation of feedback play an important role in the realization of the aforementioned reactions that are essentially elementary goal-directed acts with a clearly adaptive meaning.

From our viewpoint, these characteristics of conditioned reflex links also have an important place in the mechanism of complex behavioral acts called motivational acts. Our assumption should not be understood to be a mechanical extrapolation from the elementary to the complex. In accordance with Pavlov, we see complex behavior not as a simple sum or piling up of elementary behavioral reactions, but rather as a product of a higher integration or synthesis of the latter into something that is a uniform whole with its own specific characteristics. This is a synthesis in which the properties of the components can be preserved only to a known degree. This idea of Pavlov was most clearly and convincingly proven in experiments performed by him and his associates in a study of anthropoid behavior (Pavlov, 1973). These experiments showed in particular that the chimpanzee, in order to acquire objects that suitably satisfy its own urgent needs, actively interacts with the environment by "trial and error" and, in accord with laws of conditioned reflex development, gradually develops and perfects skills that are at first easy and simple, and, subsequently, through their integration, become increasingly complex and difficult, until such time as the animal forms a quite complex and long chain of goal-directed behavioral acts. We might cite the following as an example: A chimpanzee opens the door to a room with a key, walks into the room, extinguishes a hot alcohol flame in the window, goes through the window into a room where fruit is hanging beyond its reach, makes a stable pyramid out of the boxes lying on the floor, and then climbs up and takes the bait.

In contradiction to the views of Köhler (1929) and other gestalt psychologists on this subject, this theoretical thesis of Pavlov received subsequent support from the results of new experiments conducted by his followers, particularly those of Vazuro (1948) and Schastni (1906). We believe that this theoretical tenet of Pavlov is also very much in accord with the interesting data of Wolfe (1936) and others with respect to the dynamics of skill formation in anthropoids on the use of coins, as well as the exceptionally interesting data of Ulanova (1950), Gardner (1969), and others, concerning the dynamics of speech gesture skills in anthropoids, even though these very scientists adhere to different views with respect to their own data.

In analyzing the experimental data of Pavlov's own laboratories on the mechanisms of complex behavioral acts in higher animals, including the anthropoids, Pavlov did not utilize his idea about the bilateral nature of the conditional link as applied to a simple instrumental-conditioned reflex. This profound idea has now already become a theoretical thesis that has been rather well substantiated experimentally. And it seemed to us that the time had come to proceed cautiously in the direction of explaining the possible role of the bilateral conditional link in certain other phenomena of brain activity as well, and particularly in the physiological mechanisms of the goal-directed nature of motivational behavior.

Whether the case in point is the formation of a complex skill in an animal that requires the selection of a correct key to open a door, the building of a pyramid made of boxes, the dousing of a flame, the building of a bridge across a water barrier, and the use of coins or language gestures, the systems of specific movements in all such cases, accomplished in a specific sequence, are executed by the acquisition of needed biologically important objects and by their reinforcement.

It would be correct to say that the same thing occurs here as occurs in the development of a simple instrumental-conditioned reflex in response to a specific elementary movement, that is to say, that a complex chain of conditioned reflexes is produced in response to corresponding systems of movements, because in all cases they precede reinforcement. Furthermore, just as in the case of an elementary instrumental-conditioned reflex, a system of bilateral conditional links is formed in each of these systems. In the conditional direct link, this occurs from the cortical structures of the system of movements (as a chain conditional stimulus) to the cortical point of the reinforcing stimulus, and in conditional feedback—from the latter to the former. If, in accordance with Pavlov, and on the basis of considerable experimental material, we consider that a local instrumental-conditioned reflex (as a model of goal-directed arbitrary movement or elementary motivational act) is a consequence of activating conditional feedback, it would then seem correct to view the goal-directedness of motivational behavior as primarily a result of activating the aforementioned system of conditional feedback.

A schematic representation and brief account of the essence of our viewpoint with respect to the formation and realization of motivational behavior under natural habitat conditions follow. In order to satisfy its current urgent needs under the influence of endogenic and exogenic factors in an unfamiliar situation, the naive animal makes a number of search motions in all possible directions from the starting point (in Figure 2.12, the white triangle IA with lead-off continuous curved arrows). If the animal moves forward once or several times in a specific sequence along intermediate connecting points of a path (white geometric figures IA–IB–IC with one-way curved arrows) to the end point—the location of the desired object (white circle)—and thereby satisfies its current urgent need, a chain of spatial-time conditioned reflexes will be formed with bilateral conditional links between the cortical projections of the basic path points (in Figure 2.12, the bilateral fine dashed arrows between the crosshatched geometric figures IIA–IIB–IIC–IID, which represent not only the basic points of the effective path, but also their "images" in the cerebral cortex). Even in the beginning stage of their formation, these conditioned reflexes can orient the animal correctly. The animal is able to proceed along the

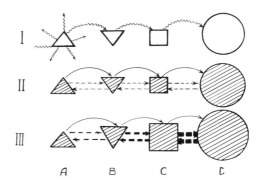

Figure 2.12. Schematic representation of the physiological mechanisms of goal-directed movement. (From E. A. Asratyan.) Explanations are found in the text.

effective path toward the location of the desired object without any special hunts or vacillation (one-way continuous arclike arrows between the cross-hatched figures). In the stage of consolidation and specialization of this developed chain of space-time conditioned reflexes, a stepwise growth in the strength and duration of these bilateral conditional links takes place, in accordance with long known laws, as the animal approaches the object's location and time of reinforcement (IIIA–IIIB–IIIC–IIID).

From our point of view, the basic drive of this complex integrated system of conditional reflexes in the form of a whole behavioral act is the heightened excitability or excitation of the central nervous structure of the reinforcing stimulus or the complex nerve center of complex, specialized, vital, conditioned reflexes in the Pavlovian understanding. Its "heightened excitability" or excitation ("central motivational state" according to the authors) may be produced by both endogenic neurohumoral factors and various remote or proximal signals, that is, by conditional stimuli ("acquired drive" or "conditioned incentive motivation" according to other investigators). Once this complex nerve center is activated by one method or the other, it sets into motion an entire complex integrated system of conditioned reflexes, primarily via the chain of conditional feedback links established between its cortical representative and various other motor and sensory cortical points, along with the direct conditional links. This activity may be in the form of a heightened excitability of the nervous structures of the entire complex system, that is, where those structures are made operatively ready for action, or that activity may be in the form of excitation, that is, in preparation for the realization of the entire system's complex reflex activity as a whole.

Also conceivable is the formation of direct linear and conditional feedback between the intermediate and final points of the path; but in order to keep the diagram from becoming too complicated, these links have not been included in the sketch. For these same reasons, the diagram does not show the feedback from the cortical point of the vital, complex, conditioned reflex to the cortical points of the signal stimulus through which, according to our data, the excitability of the latter is significantly heightened. That is of exceptional importance for behavior as a whole.

The rather modest experimental data we have thus far presented might be viewed as a direct reinforcement of our viewpoint.

The integration of separate conditioned reflexes into rather complex integral behavioral acts is one of the recent studies made by our associates with the use of animals under conditions of free movement. A characteristic methodological feature of these experiments has been an objective, graphic recording of the reflexes under study, as was also done in experiments on bound animals. In these kinds of experiments, our associates have studied the production of separate conditioned reflexes with bilateral conditioned links and their integration into complex behavioral reactions, that is, the production of the instrumental alimentary reflex of pressing a lever and the subsequent production of a second-order conditioned reflex in response to an extraneous stimulus and the further complex involvement of this chain by means of producing a conditioned response to the spatial factor. This entails a prolonged stay in a designated start position prior to the action of some trigger conditional stimulus and the synthesis of all present conditioned reflexes into a specific dynamic behavioral stereotype, and the switching of the entire system of conditional stimuli from the signalization of one type of activity to another, and so on (Davydova & Struchkov, 1978; Rudenko & Struchkov, 1978; and others).

Our following data will be of particular interest as regards our viewpoint on the physiological mechanisms of the goal-directed nature of motivational behavior. Rudenko and Struchkov, under the conditions of the previously described experiments on dogs, have studied the dependence of alimentary and drinking behavior of animals on the excitability level of the central nerve formations that govern these functions. This is a phenomenon that has been generally known since the experiments of Kendler (1946). Two separate strands are mounted in a spacious room at a specific distance from each other and with identical components: a start position, and 4 m away from that position is a box with an internal device for dispensing food or water; at a distance of 80 cm from the latter is a lever which, when pressed, produces food or water. A system of drinking-conditioned reflexes was produced in identical dogs and in identical experiments with one apparatus, and a system of alimentary-conditioned reflexes was produced with the other apparatus. A musical tone was used as the trigger signal for the drinking reflex system, and rhythmic beats of a metronome were used as the trigger signal for the alimentary reflex system.

Warranting attention are the following facts we obtained from these experiments in the stage of reflex specializations and consolidation. When preliminary thirst was amplified in the dogs (by depriving them of water), as a rule they strained to get to the start position of the drinking-conditioned reflex system and remained there until the trigger signal was sounded. But when their hunger was preliminarily amplified (by depriving them of food) after the dogs entered the room, they proceeded to the start position of the alimentary reflex system and were held there until an adequate trigger stimulus was activated. After the realization of a chain of movements and reinforcement by an appropriate product, the dogs returned to the appropriate start position. When they were satiated with one of the products, they would frequently go to the inappropriate start position. Some of the dogs, when very thirsty or hungry, often preferred to stand near the appropriate goal-box instead of the start position. In proceeding from the required start position, they often ran directly to the goal-box without first pressing the lever, and then pressed the lever only after they approached the box. The simultaneous satisfaction of thirst and hunger led to a breakdown in the reflex systems and confusion both with respect to the selection of the start positions and the selection of levers and boxes. Often the dogs would generally wander around the room, sometimes approaching one box or another, and pressing one lever or another.

These data clearly indicate the importance that heightened excitability of the central nerve structures of the drinking and alimentary-unconditioned reflexes has for the existence of corresponding behavioral acts. Furthermore, we believe that these data also indicate that bilateral, specialized conditional links are formed between the cortical structural elements of each of these unconditioned reflexes and cortical points of movement systems to corresponding start positions, and between each delay at those positions and at each pressing of appropriate levers. We mean here the formation of both direct conditional links from the cortical points of the aforementioned movements to the cortical points of each of the reinforcing stimuli, and the specialized conditional feedback links through whose activation the central nervous structures of the reinforcing stimulus cause primarily the previously mentioned goal-directed movements.

We view these data as preliminary, more or less adequate experimental support of our viewpoint about the physiological mechanisms of the goal-directed nature of motivational behavior.

In conclusion, I should like to note that we are perfectly aware of the fact that this chapter on the reflex mechanisms of motivational behavior contains much that is hypothetical and debatable, and consequently requires further systematic, thorough experimental, and theoretical elaboration. Nevertheless, it seems to us that even its present form, the concept that we have proposed differs beneficially from most of the existing indeterminate, eclectic, and at times teleological, concepts on the problem under examination to the extent that it strives to put the problem of the mechanisms of motivational behavior onto the tested tracks of the reflex theory and thereby include that problem within the orbit of the mighty monistic teaching of Pavlov on higher nervous activity.

REFERENCES

Anand, B. K. Regulation of nervous drives in feeding behavior. *Totus Homo,* 1971, *3,* 59–66.
Anderson, B., & Wyrwicka, W. The elicitation of drinking motor conditioned reaction by electrical stimulation of the hypothalamic "Drinking Area" in the goat. *Acta Physiologica Scandinvica,* 1957, *41,* 194–198.
Anokhin, P. K. *Biology and neurophysiology of the conditioned reflex.* Moscow, 1968. (In Russian)
Asratyan, E. A. Certain general features of nervous activity disturbance and restoration. *Bulletin of the USSR Academy of Sciences,* Biology Series, 1949, *6,* 726–733.
Asratyan, E. A. *Compensatory adaptation, reflex theory and the brain.* Oxford: Pergamon Press, 1965.
Asratyan, E. A. Some peculiarities of formation, functioning and inhibition of conditioned reflexes with two-way connections. In E. A. Asratyan (Ed.), *Brain reflexes.* Amsterdam: Elsevier, 1967, p. 8.
Asratyan, E. A. *Studies on the physiology of conditioned reflexes.* Moscow, 1970. (In Russian)
Asratyan, E. A. Certain features of the contemporary developmental stage of the conditioned reflex theory. *Journal of Higher Nervous Activity,* 1973, *23,* 262–278. (In Russian)
Asratyan, E. A. *Reflex theory and the problem of motivation. Basic problems in the electrophysiology of the brain.* Moscow, 1974, pp. 5–20. (In Russian)
Asratyan, E. A. Conditioned reflex theory and motivational behavior. *Acta Neurobiology Experiments,* 1974, *34,* 15–30.
Asratyan, E. A. "Physiological Mechanisms of Goal-Directed Movements." In T. Desiraju (Ed.), *Mechanisms in transmission of signals for conscious behavior.* Amsterdam: Elsevier, 1976a, pp. 81–95.
Asratyan, E. A. Some aspects of the problem of motivation in the light of pavlovian teaching. *Acta Physiologica Academy Science Hungaricae,* 1976b, *48*(4), 323–334.
Asratyan, E. A. *Studies on the higher nervous system.* Erevan, 1977.
Asratyan, E. A. Reflexes conditions et comportements. *La Recherche,* 1978, *87,* 237–244.
Beritashvili, I. S. *Nervous mechanisms of higher animal behavior.* Moscow, 1960.
Bindra, D. A motivational view of learning, performance and behavior modification. *Psychology Review,* 1974, *81,* 199–213.
Bindra, B., & Campbell, I. P. Motivational effects of rewarding intracranial stimulation. *Nature,* 1967, *215,* 375–376.
Caggiuli, A. R. Analysis of the copulation-reward properties of posterior hypothalamic stimulation in male rats. *Journal of Comparative and Physiological Psychology,* 1973, 70:399–412.
Cofer, C. P., & Applay, M. H. *Theory and research.* New York: Wiley, 1964.
Cohen, M. D., Brown, G. W., & Brown, M. L. Avoidance learning motivated by electric stimulation of the brain. *American Journal of Physiology,* 1954, *179,* 587–593.
Davydova, E. K., & Struchkov, M. I. Formation of conditional feedback in dogs during free movement. *Journal of Higher Nervous Activity,* 1978, *28,* 57–61.
Delgado, M. J. R., Rosvold, H. E., & Looney, E. Evoking conditioned fear by electrical stimulation of subcortical structures in the monkey brain. *Journal of Comparative and Physiological Psychology,* 1956, *49,* 373–380.
Delgado, M. J. R. Free behavior and brain stimulation. *International Review of Neurobiology,* 1964, *6,* 349–449.
Deutch, G. A. *The structural basis of behavior.* Chicago: Chicago Univ. Press, 1960.
Dostalek, E. Rücklaufige Bedingte Verbindungen. *Verl. Tschechoslow. Acad. Wiss.,* Praha, 1964.

Doty, R. Electrical stimulation of the brain in behavioral context. *Annual Review of Psychology,* 1969, *20,* 289–320.

Ellen, P., & Powell, E. W. Differential conditioning of septum and hippocampus. *Experimental Neurology,* 1966, *16,* 161–171.

Fonberg, E. The motivational role of the hypothalamus in animal behavior. *Acta Biol. Exp.,* 1967, *27,* 303–318.

Gallistel, C. R. Self-stimulation: Failure of protrial stimulation to affect rats electrode preference. *Journal of Comparative and Physiological Psychology,* 1969, *69,* 722–729.

Gardner, R. A., & Gardner, B. T. Teaching sign language to a chimpanzee. *Science,* 1969, *165,* 664–672.

Gengerelli, J. A. Studies in the neurophysiology of learning IX. Conditioning of evoked potentials due to cerebral stimulation. *Journal of Psychology,* 1975, *9,* 287–301.

Grastyan, E., Gzopf, J., Angyan, L., & Szabo. I. The significance of subcortical motivational mechanisms in the organization of conditioned connections. *Acta Physiologica Academy of Science Hungaricae,* 1965, *26,* 9–46.

Grastyan, E., Karmos, G., Vereczkey, L., & Gzopf, J. The possible mechanisms of subcortical motivational processes in conditioning. In D. G. Martin & F. Guma (Eds.), *Cortico-subcortical relationships in sensory regulation.* Havana: Acad. Sci., 1966, pp. 455–472.

Grastyan, E., Lissak, K., & Kekesi, K. Facilitation and inhibition of conditioned alimentary and defensive reflexes by stimulation of the hypothalamic and reticular formation. *Acta Physiologica Academy of Science Hungaricae,* 1956, *9,* 133–151.

Hess, W. R. *Das Zwischenhirn.* Schwade, Basel, 1949.

Hull, C. L. *Principles of behavior.* New York: Appleton-Century-Crofts, 1943.

Ioffe, M. E. *Cortico-spinal mechanisms of instrumental motor reactions.* Moscow, 1975. (In Russian)

Kasyanov, V. M. Mechanism of positional excitation in the conditioned defense reaction. *Bulletin of Experimental Biology and Medicine,* 1950, *2,* 405–410. (In Russian)

Kendler, H. H. The influence of simultaneous hunger and thirst of two opposite spatial responses of the two white rats. *Journal of Experimental Psychology,* 1946, *36,* 212–220.

Konorski, Y. M. *Integrative activity of the brain.* Chicago: Univ. of Chicago Press, 1967.

Köhler, W. *Gestalt psychology,* New York: Liveright, 1929.

Korzenev, V. V., & Slezin, B. S. Regulations of complex forms of animal behavior on the basis of electrostimulation of emotiogenic . . . [rest of sentence is missing, transl.]

Kupalov, P. S. *Situational conditioned reflexes in dogs under normal and pathological conditions.* Leningrad, 1964. (In Russian)

Larson, S. Hyperphagia from stimulation of the hypothalamus and medulla in sheep and goats. *Acta Physiologica Scandinvica,* 1954, *32* (Suppl. 8–40.)

Lissak, K., Grastyan, E., Karmos, G., Vereczkey, L., & Losonczy, H. The nature of orientation reaction viewed from the aspect of motivation. In D. G. Martin & E. Guma (Eds.), *Cortico-subcortical relationship in sensory regulation.* Acad. Sci. Havana, 1966, pp. 439–453.

Luciani, L. *Human physiology* (Vol. 3). London: Macmillan, 1915.

Magnus, (?). *Korperstellung.* Berlin: Springer, 1924.

Mendelson, J. Role of hunger in T-maze learning for food by rats. *Journal of Comparative and Physiological Psychology,* 1966, *62,* 341–350.

Miller, N. E. Learning and performance motivated by direct stimulation of the brain. In D. E. Sheer (Ed.), *Electrical stimulation of the brain,* 1961, 387–390.

Miller, N. E. An experimental investigation of acquired drives, *Psychology Bulletin,* 1941, *38,* 534–535.

Milner, P. *Physiological psychology,* New York: Holt, 1970.

Mogenson, G. I. Brain and behavior. *Abstract XXth International Congress in Psychology,* (Tokyo), 1972, p. 113.

Mogenson, C. J., & Morrison, M. I. Avoidance response to "reward" stimulation of the brain. *Journal of Comparative and Physiological Psychology*, 1962, *55*, 691–694.

Mogenson, C. J., & Stevenson, I. A. P. Drinking and self-stimulation with electrical stimulation of lateral hypothalamus. *Physiological Behavior*, 1966, *1*, 251–254.

Morgane, P. J. Distinct "feeding and hunger" motivating systems in the lateral hypothalamus of the rat. *Science*, 1961, *133*, 887–888.

Nielsen, H. C., Knight, J. M., & Porter, P. B. Subcortical conditioning generalization and transfer. *Journal of Comparative and Physiological Psychology*, 1962, *55*, 168–173.

Olds, I., & Milner, O. Positive reinforcement produced by electrical stimulation technics. *Journal of Comparative and Physiological Psychology*, 1954, *47*, 419–427.

Pavlov, I. P. *Twenty years of experience in the study of animal higher nervous activity (Behavior)*. Moscow, 1973. (In Russian)

Pavlov, I. P. *Unpublished and little known materials*. Leningrad, 1975. (In Russian)

Popova, E. I., & Pavlova, O. G. Effect of the direct stimulation of the lateral hypothalamus in dogs during the conversion of a passive-defense situational reflex into an alimentary reflex. *Journal of Higher Nervous Activity*, 1975, *25*, 477–485.

Roberts, W. W. Rapid escape learning without avoidance learning motivated by hypothalamic stimulation in cats. *Journal of Comparative and Physiological Psychology*, 1958a, *51*, 391–399.

Roberts, W. W. Both rewarding and punishing effects from stimulation of posterior hypothalamus of cat with same electrode at same intensity. *Journal of Comparative and Physiological Psychology*, 1958b, *51*, 400–407.

Robinson, B. W., Mishkin, M. Alimentary responses evoked from forebrain structures in Macaca mulata. *Science*, 1962, *136*, 260–262.

Rudenko, I. P., & Struchkov, M. I. Conditioned reflex switching in free dog behavior. *Journal of Higher Nervous Activity*, 1978, *28*, 457–464. (In Russian)

Sadowski, B., & Dembinska, M. Some characteristics of self-stimulation behavior of dogs. *Acta Neurobiology Experiments*, 1973, *33*, 769.

Schastni, A. I., & Sherrington, Ch. S. *The integrative action of the nervous system*. London: Cambridge Univ. Press, 1906.

Snowden, Ch. T. Motivation, regulation and the control of meal parameters with oral and intragastric feeding. *Journal of Comparative and Physiological Psychology*, 1969, *69*, 91–100.

Stutz, R. H., & Asdourian, D. Positively reinforcing brain shock as a CS in the acquisition and extinction of a shuttle box avoidance response. *Psychoneural Science*, 1965, *3*, 191–192.

Tolman, E. C. Cognitive maps in rats and man. *Psychological Review*, 1948, *55*, 189–208.

Ulanova, L. I. Formation of conditional symbols in monkeys. In *Studies of higher nervous activity in natural experiments*. Kiev, 1950, pp. 132–153.

Valenstein, E. S. On the activity of behavior elicited by brain stimulation. *Abstract XXth International Congress in Psychology* (Tokyo), 1972, p. 1155.

Valenstein, E. S., Cox, V. C., & Kakolewski, J. W. Modification of motivated behavior elicited stimulation of the hypothalamus. *Science*, 1968, *159*, 1118–1121.

Vazuro, E. G. *Study of higher nervous activity in anthropoids*. Moscow, 1948. (In Russian)

Wolfe, J. B. Effectiveness of token rewards for chimpanzees. *Comparative Psychology Monographs*, 1936, *12*, 1–72.

Wyrwicka, W. *The mechanisms of conditioned behavior*. Springfield, Ill.: Thomas, 1972.

I. GORMEZANO

Pavlovian Mechanisms of Goal-Directed Behavior[1]

3

Introduction

Many of the behavioral investigations from my laboratory have been directed at delineating the role of contiguity and determining the possible mechanisms of reinforcement in classical defense and reward conditioning. Moreover, on appropriate occasions we have attempted to relate our findings to those theoretical accounts of instrumental conditioning that have been based upon the axiomatic application of the presumed laws of classical conditioning. In all of these efforts we have been guided by the assumption that manipulation of the most basic parameters of the classical conditioning paradigm would best serve our scientific objectives. Accordingly, although over the years our experimental efforts have provided us with theoretical insights, an important residual consequence is that we have been able to detail a large number of robust empirical relationships governing simple conditioning in our classical defense (rabbit nictitating membrane) and classical reward (rabbit jaw movement) conditioning preparations. With these empirical laws for simple classical conditioning in hand, we have recently become interested in studying extensions of the classical conditioning paradigm to determine its ability to fulfill its long-term promise as a source of insight into the behavior of organisms in more complex stimulus environments. In particular, we have become interested in examining the

[1] The preparation of this chapter and the research reported was supported by the National Science Foundation, Grant BNS 76-84561.

implications of CR-mediational accounts of goal-directed instrumental behavior.

CR-Mediational Accounts of Goal-Directed Instrumental Performance

Theoretically, any instrumental learning situation has within it the stimulus conditions for the development of classically conditioned responses. The reinforcing event (e.g., food), although dependent upon the occurrence of the instrumental response, nevertheless occurs in a specific stimulus situation; and since these situational stimuli are regularly associated with reinforcement in the "manner" of classical conditioning, it has been speculated that situational CRs may concurrently develop. American learning theorists have given formal recognition to such possible CR accompaniments of goal-directed instrumental performance by concepts such as secondary reinforcements, secondary drive, and incentive motivation. In the application of these concepts to goal-directed instrumental performance, it is assumed that two different learning processes are involved. One process consists of the acquisition of the specific instrumental response, whereas the second consists of the acquisition of collateral CRs conditioned to situational stimuli and/or proprioceptive stimuli arising from the instrumental responses of the organism. It is then assumed that these CRs "goal-direct" the instrumental response tendency by their action as motivators (incentive motivators and secondary drives) or reinforcers (secondary reinforcers) of the instrumental response.

Although a number of American learning theorists have envisioned a simple instrumental-conditioning situation as consisting of compounds or sequences of stimulus elements that could serve as effective situational CSs for the elicitation of collateral CRs, it was Hull (1930, 1931, 1934) in a provocative series of articles who first developed the thesis that such collateral CRs serve to goal-direct instrumental behaviors. Specifically, Hull attempted to provide a mechanistic account of seemingly purposeful and insightful goal-directed behavior through the application of Pavlov's (1927) classical conditioning principles. Hull's most important theoretical construct for the elucidation of goal-directed behavior was the fractional anticipatory goal response, which became known as the r_g–s_g mechanism, and was hypothesized to be a classically conditioned response learned in the instrumental situation. The r_g–s_g mechanism was not treated extensively by Hull in his book published in 1943; nevertheless, it had a major impact on learning theorists, who postulated a dynamic role for situational CRs on goal-directed instrumental performance (e.g., Amsel, 1958; Konorski, 1948; Logan & Wagner, 1965; Miller, 1948, 1963; Mowrer, 1947, 1960; Spence,

1956). Then, over a period of time, two research strategies evolved to investigate the purported influences of situational CRs on goal-directed instrumental performance (see Gormezano & Moore, 1969). Some investigators concurrently recorded presumed situational CRs and instrumental responses to ascertain whether the empirical relationships between situational CRs and instrumental responses corresponded to theoretical expectations. Other investigators resorted to the use of "classical-instrumental" transfer paradigms. However, in using transfer designs, usually no attempt is made to isolate mediating situational CRs. Instead, changes in instrumental performance resulting from the presentation of the previously established CS is attributed to the effects of covert CRs presumed to be elicited by the explicit CS.

A PROVOCATIVE INITIAL EXPERIMENT

Under the concurrent measurement procedure, observations are made of the instrumental response and a situational CR selected from one or more of the response systems elicited as UCRs to the reinforcing event (e.g., food or shock). Accordingly, to assess initially the face validity of CR-mediational accounts, we selected our classical reward rabbit jaw movement conditioning preparation as a possible situational CR to study in a specially devised runway. Figure 3.1 portrays the preparation. In the experiment, a group of rabbits ($N = 12$) was trained over the course of 14 days of acquisition to traverse a runway (8 m long, 15 cm wide, and 33 cm high). At the goal-box (45 cm long) the activation of a photocell delivered a 1-ml squirt of water, via a cheek fistula, to the rabbit's oral cavity, and unconditioned jaw movement responses were detected by the transducer mounted off the animal's headmount. During the course of the experiment, all animals were maintained on a water deprivation regime of 90 cc per day, and on each day the animals received four reinforced acquisition trials. (The limitation of space precludes specifying further details of the apparatus.) Figures 3.2, 3.3, and 3.4 present the mean rate of jaw movement responding and running speeds in each of five segments of the alley for Days 1-4, Days 6-9, and Days 11-14, respectively. As can be clearly seen across the panels, the frequency of jaw movement responses and speed of running increased over the days of training in the theoretically expected manner. Most interestingly, the shape of the gradient for the frequency of jaw movement responses in segments of the alley closely corresponds to Hull's speculation that "r_g" would display a runway gradient that would be an increasing concave upward function with increasing proximity to the goal-box.

As interesting as the pattern of rabbit runway results are for the CR-mediational accounts of goal-directed instrumental behaviors, the study has the logical limitation of being unable to reject the hypothesis that

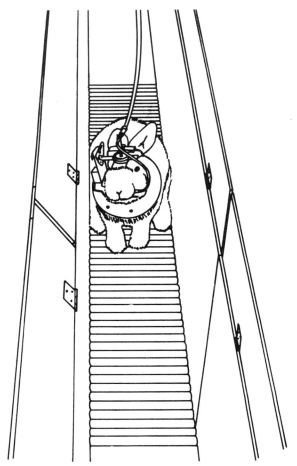

Figure 3.1. *The rabbit in the runway is equipped with a headmount containing a transducer for recording jaw movements of a squirt of water delivered, via a cheek fistula, into the oral cavity. A Plexiglass collar prevents the rabbit from interfering with the water delivery and recording system.*

situational CRs are inconsequential concomitants of the instrumental-conditioning situation (see Gormezano & Kehoe, 1975). That is, although the study is provocative, the correlational nature of all such concurrent measurement studies suffer the unavoidable limitation that they cannot by themselves indicate causal relationships between CRs and instrumental responses or vice versa. Nevertheless, the study was of considerable heuristic value in directing our research efforts. Specifically, reflecting upon the need to specify exactly the stimulus determinants for the systematic change in

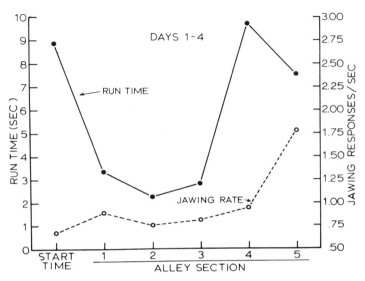

Figure 3.2. Run time and rate of jaw movements in successive sections of the runway.

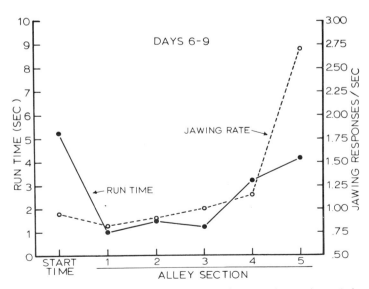

Figure 3.3. Run time and rate of jaw movements in successive sections of the runway.

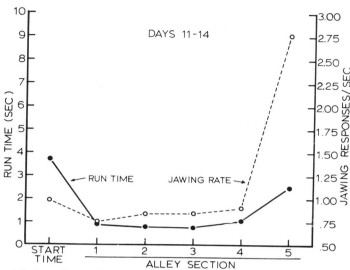

Figure 3.4. Run time and rate of jaw movements in successive sections of the runway.

jaw movement responses observed over the length of the runway, we were led to reexamine most carefully the stimulus assumptions made by CR-mediational theories. Clearly, these theories assume that the effective CS in goal-directed instrumental behavior consists of an extended sequence of stimulus elements antedating reinforcement, and that the sequence of elements will become conditioned to hypothetical CRs whose stimulus consequences serve reinforcing, motivational, or response-evoking functions for the overt instrumental responses. Commonly, CR mediational theories also assume that the stimulus elements most contiguous with reinforcement will acquire maximal CR-evoking capability, but that elements occurring well in advance of reinforcement will also acquire substantial CR-evoking properties through higher order conditioning or stimulus generalization from the more contiguous element. It is apparent, therefore, that CR-mediational theories have employed *axiomatically* the laws (or presumed laws) of classical conditioning to account for the goal-directed nature of instrumental behavior. However, empirical demonstrations within sequences of CSs of the operation of higher order conditioning and stimulus generalization in classical conditioning have been neither systematic nor compelling to date. To the extent that no solid empirical base exists for the principal axioms of CR-mediational theories of instrumental performance, such axioms logically constitute *hypotheses* of CR acquisition and transfer under serial CS training. Hence, it became compellingly obvious to us that the logical basis for assessing the fundamental assumptions of CR-mediational theories involves determining the empirical laws of classical serial compound conditioning.

Pavlovian Conditioning to Serial CS Compounds

There is a vast Soviet serial CS literature, but their major emphasis has been to detail the phylogeny of, and the integrative structures involved in, the perceptual synthesis of complex stimulus sequences (cf. Razran, 1971). On the other hand, in the American literature the effects of serial stimuli on response acquisition in Pavlovian conditioning are exceedingly few in number and with the exception of the experimental program of Wickens and his associates (e.g., Wickens, Nield, Tuber, & Wickens, 1973), the research has been fragmentary at best. In any event, the search for the laws of transfer of associative strength of CRs from CSs just anticipating UCS onset to more remote CSs within the Pavlovian paradigm has not been actively investigated. Accordingly, our studies, conducted with our rabbit nictitating membrane response (NMR) preparation (portrayed in Figure 3.5), were guided by the assumption that examination of serial compound CSs may lead to identifying mechanisms through which CRs may be evoked by component CSs lying outside of the empirically determined single-CS contiguity gradient for a given response system as demanded by CR-mediational theories of goal-directed instrumental behaviors.

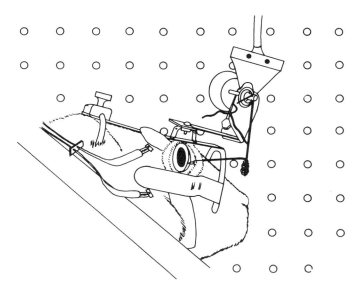

Figure 3.5. Close-up view of the rabbit nictitating membrane preparation. (For further details see Gormezano, 1966.)

Since our introduction of the rabbit's classically conditioned NMR preparation (Gormezano, Schneiderman, Deaux, & Fuentes, 1962), it has been widely employed in the investigation of learning and, more recently, it has been promoted as an ideal vehicle for the investigation of neurophysiological substrates of learning (Thompson, 1976). The advantages of the preparation are (*a*) the contributions to CR measurement of baseline, sensitized, and pseudoconditioned responses are minimal, generally of the order of 2% or less; and (*b*) the effects of a number of parametric manipulations have been extensively documented. In particular, for present purposes, our NMR preparation has consistently revealed a tightly bound range of effective CS–UCS interstimulus intervals (ISIs) over a broad range of parameters in which conditioning is most rapid when the ISI is between 200 and 400 msec but declines to negligible levels as the ISI approaches 3–4 sec (cf. Gormezano, 1972). Accordingly, as a "model" preparation with a clearly delineated contiguity gradient, employment of the NMR preparation in the initial phases of our serial compound conditioning research has markedly facilitated our ability to search for Pavlovian conditioning mechanisms (e.g., higher order conditioning, stimulus generalization) that could operate to yield responding over an extended series of stimuli and thus mediate goal-directed behavior. Hence, we became interested in identifying associative transfer mechanisms among components of serial compounds by means of which Pavlovian conditioning could be stretched beyond the bounds of a *known* contiguity gradient to a single-CS to yield responding to those members of a serial stimulus sequence long antedating the UCS and, in so doing, to gain some insight as to how such mechanisms might operate to permit CR-mediation of an extended sequence of goal-directed behavior.

Operationally, our concerns in the initial series of Pavlovian serial CS conditioning investigations were directed at determining whether in the conditioning of the rabbit's NMR to a two-component serial compound (CS1–CS2–UCS), the conditioning to its components would be differentially affected by manipulations of the CS1–CS2 interval. Specifically, these studies were directed at (*a*) determining the capacity of serial compounds to yield responding to CS1 at CS1–UCS intervals well beyond the bounds of a single-CS contiguity gradient, as expected by CR-mediational accounts of goal-directed instrumental performance; and (*b*) identifying the possible mechanisms underlying the capacity of serial compounds to bridge long CS1–UCS intervals.

EXPERIMENT 1

In the experiment, the NMR was conditioned to a two-component serial compound (CS1–CS2–UCS) in which CS1 was a 400-msec, 1000-Hz tone, and CS2 was a 400-msec, 10-Hz interruption of the house lights. On reinforced compound trials, CS2 overlapped and terminated with a 50-

msec, 3-mA, 60-Hz paraorbital shock UCS; hence, the CS2–UCS interval was 350 msec for all groups, an efficacious interval for conditioning of this response system. The trace interval between CS1 offset and CS2 onset was varied across the four groups ($N = 12$) to include the values of 0, 500, 1000, and 2000 msec, respectively. Accordingly, the intercomponent interval between the onsets of CS1 and CS2 was 400, 900, 1400, and 2400 msec, respectively; and the CS1–UCS interval was 750, 1250, 1750, and 2750 msec, respectively. For expository purposes, the four groups were labeled in terms of the modality of the first component (Tone), their respective trace interval expressed in seconds (0, .5, 1, or 2), and the modality of the second component (Light). Thus, the four groups were designated as T–0–L, T–.5–L, T–1–L, and T–2–L, respectively. All subjects received 16 days of acquisition training, with each day consisting of 60 reinforced trials to the compound CS interspersed with two nonreinforced test trials to each component and to the compound. (On test trials, a constant 2800-msec observation interval for recording CRs was employed.)

The CS1 test trial results are presented in Figure 3.6 as the mean percentage of CRs in two-day blocks for each group and indicates that, despite the rather large differences in CS1–UCS intervals across groups, acquisition to CS1 was at first rapid and similar across groups. Moreover, all groups attained at least an 80% level of responding before the groups that were trained under the longer CS1–UCS intervals (Groups T–1–L and T–2–L) of 1750 and 2750 msec showed pronounced declines in performance. Hence, the fact that CS1, when conditioned in a two-component serial compound

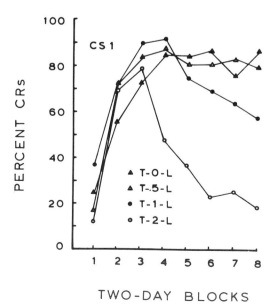

Figure 3.6. The mean percentage of CRs in two-day blocks for each group on CS1.

at ISIs ranging up to 2750 msec, evidence CR-evoking capabilities far exceeding those observed at comparable CS–UCS intervals in single-CS studies, suggested the operation of a mechanism of associative transfer from CS2 to CS1 such as higher order conditioning, sensory conditioning, or stimulus generalization. However, irrespective of the mechanism, the high levels of responding to CS1 under the longer CS1–UCS intervals would appear to indicate that serial compounds can bridge long temporal gaps between CSs and UCSs.

EXPERIMENT 2

The unequivocal determination of whether the high levels of responding to CS1 observed in the previous experiment is attributable to the presence of CS2 requires demonstrating serial compound conditioning effects on CS1 responding above and beyond the effects of direct conditioning to CS1 and/or generalized responding from CS2 to CS1. Accordingly, for purposes of identifying enhanced responding to CS1, a second experiment was conducted containing serial compound Groups T–0–L and T–2–L, each representing replications of the identically labeled groups in the previous experiments. Matched to each serial compound group were two single-CS control groups that received training with only a single component (CS1 or CS2), but were tested to both components and the compound in the same fashion as the corresponding groups. Therefore, Group T–0–L was matched in terms of component CS parameters and testing procedure with Group T–0–O, which was trained with only CS1 at a CS–UCS interval of 750 msec, and Group O–0–L, which was trained with only CS2 at a CS–UCS interval of 350 msec. Similarly, Group T–2–L was matched with Group T–2–O, which was trained with CS1 at a CS–UCS interval of 2750 msec, and Group O–2–L, which was trained with CS2 at a CS–UCS interval of 350 msec. Groups O–0–L and O–2–L received the same single CS2–UCS training and test trials to both components; but to serve as complete contrast controls, each received serial compound test trials corresponding to that received by Groups T–0–L and T–2–L, respectively.

The results of the experiment are shown in Figure 3.7, where the mean percentage of CRs on CS1 test trials over two-day blocks for all six groups is presented. Although the CS1 test trial performance function of Group T–0–L rose to a high level, Group T–0–O revealed an even somewhat higher level of responding throughout most of the training. Thus, under the CS1–UCS interval of 750 msec, serial compound training appeared to interfere slightly with acquisition to CS1. On the other hand, examination of the right-hand panel reveals that Group T–2–L, as in the previous experiment, demonstrated rapid acquisition to CS1 to a level of 88% before showing a subsequent decline, whereas Group T–2–O showed a level of responding averaging only about 15%. Hence, at this

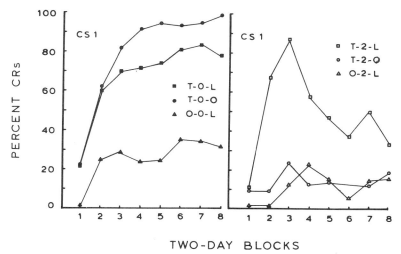

Figure 3.7. *The mean percentage of CRs on CS1; test trials are plotted as a function of the trace and CS1–UCS interval.*

2750-msec CS1–UCS interval, serial compound training appeared to have been essential to the extraordinarily high level of responding to CS1 exhibited by Group T–2–L. Moreover, examination of the CS1 test trial performance functions of groups trained with only CS2 (Groups O–0–L and O–2–L) indicates that maximal and overall levels of responding to CS1 for Group T–2–L far exceeded that attributable to the cross-modal generalization effects seen in Groups O–0–L and O–2–L. Accordingly, the most striking feature of the results is the finding that the low level of CR acquisition to CS1 at the CS1–UCS interval of 2750 msec (Group T–2–O) was extraordinarily facilitated under serial compound training (Group T–2–L). Furthermore, even if the estimated contribution of cross-modal generalization from CS2 to CS1 were added to the direct conditioning effects of the 2750 msec CS1–UCS interval, the substantial CR acquisition to CS1 in Group T–2–L would only be partially explained. Hence, the mechanism(s) by which enhanced CS1 responding occurs under serial compound training requires experimental delineation.

As an initial step in the delineation of mechanisms of associative transfer within serial CSs, higher order, sensory conditioning, and stimulus generalization hypotheses may be contrasted on the basis of various aspects of the serial CS conditioning situation presumed crucial to such transfer. Specifically, both higher order and sensory conditioning hypotheses contend that the temporal contiguity, or pairing of CS1 and CS2 in compound, is a necessary condition for transfer from CS2 to CS1. The sensory conditioning hypotheses (Wickens, 1965) assumes that CS1–CS2 pairings lead to the formation of a forward association such that presentation of CS1 alone "reinstates" CS2. Hence, to the extent that CS2 elicits a robust CR, the reinstatement of CS2 by CS1 should produce substantial responding. On the other hand, a higher order conditioning account of serial CS associative transfer would contend that CS2–UCS pairings enable CS2 to serve as the "UCS" for CS1 on CS1–CS2 paired trials. Whereas a generalization hypothesis contends that associative transfer from CS2 to CS1 in compound is based solely on the physical or mediated similarity between CS2 and CS1 irrespective of their temporal contiguity. That is, the generalization account, in contrast to the sensory and higher order conditioning hypotheses, would assume that CS1–CS2 pairings are neither necessary nor sufficient for associative transfer.

EXPERIMENT 3

This experiment was conducted to determine whether there are serial compound conditioning effects on CS1 responding that are above and beyond the combined direct conditioning effects of the CS1–UCS and CS2–UCS pairings inherent in the compound. The demonstration of augmented CS1 responding above and beyond these combined direct conditioning effects would indicate that CS1–CS2 pairings determine facilitated CS1 responding under serial compound training and hence provide support for the operation of a higher order or sensory conditioning mechanism. In the experiment, rabbits were assigned to five groups ($N = 8$) and were trained with either a serial compound or a single CS. There was one compound group, Group T–1–L, that differed only slightly from the identically labeled group of the first experiment in that the CS2–UCS interval was 400 msec and training consisted of eight days of 60 reinforced compound trials per day. Group T–1–L was matched in terms of component CS parameters and testing procedure with Group T–1–O, which was trained with only CS1 at a CS–UCS interval of 1800 msec, and Group O–1–L, which was trained with only CS2 at a CS–UCS interval of 400 msec. In addition, to assess the contribution of CS1–CS2 pairings to responding to CS1 in compound, two contrast conditions were employed that maintained the CS1–UCS and CS2–UCS intervals in effect for Group T–1–L. Group T/L (8) received 8 sessions of 60 CS1–UCS and 60 CS2–UCS trials to equate the number of CS1 and CS2 presentations per session with Group T–1–L. On the other hand, Group T/L (16) received 16 sessions of 30 CS1–UCS and 30 CS2–UCS trials to equate the number of UCS presentations per session with Group T–1–L. For Groups T/L (8) and T/L (16) the order of presentation of CS1–UCS and CS2–UCS trials was restrictively randomized for each block of 14 trials. In addition, for all groups, except Group T/L (16), each day of acquisition consisted of two nonreinforced test trials to CS1, whereas Group T/L (16) received one daily test trial.

Figure 3.8 presents the mean percentage of CRs in two-day blocks on

3. Pavlovian Mechanisms of Goal-Directed Behavior

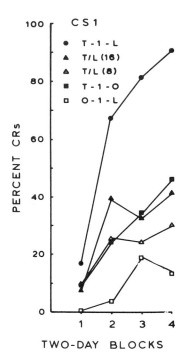

Figure 3.8. The mean percentage of CRs in two-day blocks on CS1 test trials for all five groups.

CS1 test trials for all five groups. The CS1 test trial performance function for Group T–1–L rose rapidly to a level above 90% CRs, with an overall mean level of responding of 63%. In contrast, the overall mean level of CS1 responding in Group T–1–O (26%) revealed only a moderate, direct conditioning effect of the 1800-msec CS1–UCS interval, and the low mean performance level for Group O–1–L (9%) revealed a very modest degree of cross-modal generalization from CS2 to CS1. Moreover, the mean CS1 performance levels of 22% and 30% of CRs observed for Groups T/L (8) and T/L (16), respectively, in reflecting the combined direct conditioning effects of the 1800-msec CS1–UCS and 400-msec CS2–UCS intervals of the serial compound, was low. Accordingly, the substantially higher level of CS1 responding for Group T–1–L, relative to the contrast control groups, indicates rather unequivocally that CS1–CS2 pairings in compound are necessary for the enhanced responding to CS1 under serial compound training, as required by higher order and sensory conditioning accounts of associative transfer from CS2 to CS1.

EXPERIMENT 4

To provide a more direct demonstration that the rather substantial level of responding to CS1 observed under serial compound training in the previous experiment was a function of CS1–CS2 pairings, another investigation was conducted in which subjects received nonreinforced CS1–CS2 pairings interspersed with CS2–UCS trials. Four of the six conditions comprised the cells of a 2×2 factorial design in which pairing or explicit unpairing of CS1 and CS2 was orthogonal to the pairing or unpairing of CS2 and UCS. The ISIs on CS1–CS2 and CS2–UCS trials were 1400 and 400 msec, respectively, and during the 16 daily acquisition sessions each of the four groups ($N = 12$) received 30 paired or explicitly unpaired CS1–CS2 and CS2–UCS trials. Group P–P received both CS1–CS2 and CS2–UCS paired trials, Group U–P explicitly unpaired CS1/CS2 presentations and CS2–UCS paired trials, Group P–U paired CS1–CS2 trials and explicitly unpaired CS2/UCS trials, and, finally, Group U–U received explicitly unpaired CS1/CS2 and CS2/UCS trials. In addition, Group SC (serial compound) received 30 reinforced serial compound trials per session, thereby providing a baseline from which to assess the potential deleterious effects of nonreinforced CS1–CS2 pairings on CS1 responding in Group P–P; whereas Group PR (partial reinforcement) received 30 reinforced and 30 nonreinforced serial compound trials per session in order to assess the extent to which performance differences between Groups SC and P–P would be attributable to the 50% PR schedule in effect for CS2 in Group P–P.

The mean percentage of CRs for all six groups *during the 400-msec application of CS1* on CS1–CS2 trials within two-day blocks of training is presented in Figure 3.9. Inspection of the figure reveals that, in contrast to the low levels of responding to CS1 exhibited by Groups U–P, P–U, and U–U, Groups P–P, SC, and PR revealed systematic increases in performance across acquisition training to substantial levels. Moreover, although the response level for Group SC was reliably higher than that of Group P–P, the terminal performance level for Group P–P was higher than the levels exhibited by Groups P–U and U–U. Furthermore, a marginally significant difference was obtained between Groups P–P and U–P. Hence, the study reveals clearly the necessity of temporal pairing of CS1 and CS2 for the transfer of associative strength from CS2 to CS1. In fact, the particular CS1–CS2 relationship employed, either in compound (Group SC) or on trials intermixed with reinforced CS2 presentations, produced systematic increases in CS1 responding to levels substantially higher than the low levels observed for appropriate unpaired, single-CS and cross-modal generalization controls.

EXPERIMENT 5

The findings of the previous two studies we have considered clearly indicate that CS1–CS2 pairings in compounds can promote CR acquisition to a CS1 located near the outer boundary of effective CS–UCS intervals for conditioning the rabbit NMR to a single-CS. Accordingly, another ex-

3. Pavlovian Mechanisms of Goal-Directed Behavior

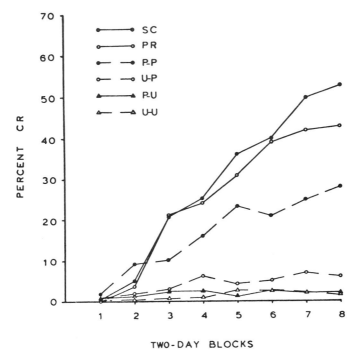

Figure 3.9. *The mean percentage of CRs to CS1 over two-day blocks for all six groups.*

periment was conducted to determine whether serial compound conditioning would yield facilitated CS1 responding at CS1–UCS (and CS1–CS2) intervals located well beyond the single-CS contiguity gradient for the rabbit NMR preparation. In the experiment, rabbits were assigned to three groups ($N = 12$) and were trained with the serial compound in which the trace interval between the offset of CS1 and the onset of CS2 was either 4000, 8000, or 18,000 msec. The three groups were labeled Groups T–4–L, T–8–L, and T–18–L, and had CS1–UCS intervals of 4750, 8750, and 18,750 msec, respectively.

Figure 3.10 presents the mean percentages of CRs in two-day blocks to CS1 during the 400 msec of its occurrence on reinforced compound trials for all three groups. As can be seen, all groups showed orderly CR acquisition functions to CS1, reaching maximum levels of 34%, 27%, and 21% for Groups T–4–L, T–8–L, and T–18–L, respectively, before revealing declines in performance. Clearly, these groups acquired levels of responding to CS1 far in excess of the 2% baseline frequency of responding for the NMR preparation. Moreover, a determination of the frequency of responding during only the 400-msec duration of CS1 for single-CS and crossmodal generalization control groups considered in earlier experiments re-

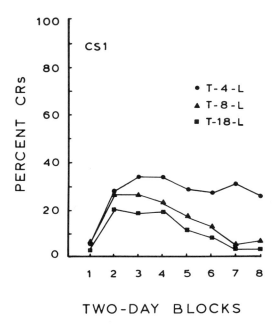

Figure 3.10. *The mean percentage of CRs in two-day blocks during CS1 on compound test trials for all three groups.*

vealed overall performance levels not exceeding 2.4% and 7.0%, respectively, Accordingly, the levels of acquisition to CS1 for Groups T–4–L, T–8–L, and T–18–L all exceeded the expected effects of single-CS training and cross-modal generalization.

Conclusion

The series of investigations presented demonstrate that serial compound training can produce extraordinary CR acquisition to CS1 under CS1–UCS intervals approaching and far exceeding the CS–UCS contiguity gradient for conditioning the rabbit NMR to a single CS. If the single-CS contiguity gradient can be considered to have an outer boundary of no more than 3–4 sec under the conditions presently employed (cf. Gormezano, 1972), then the observation of CR acquisition to CS1 at CS1–UCS intervals in excess of 18 sec indicates that serial compound conditioning functionally expanded the range of effective CS–UCS intervals more than fourfold. In addition, these investigations clearly revealed that the augmentation of CS1 responding was attributable to the CS1–CS2 pairings within serial compounds. Therefore, these findings strongly support the general proposition of CR-mediational theories that a stimulus relatively contiguous to a reinforcing event will serve as a source of associative transfer back to temporally more remote stimuli and mediate the acquisition and maintenance of an extended chain of goal-directed instrumental acts. Moreover, consistent with CR-mediational accounts, the role of CS1–CS2 pairings in augmented CS1 responding suggests that higher order or sensory conditioning might underlie the capacity of serial compounds to bridge long CS1–UCS intervals. In summary, these results lend substance to CR-mediational accounts that postulate a crucial role for the temporal contiguity, and not stimulus generalization, between sequential stimuli antedating reinforcing events.

REFERENCES

Amsel, A. The role of frustrative nonreward in noncontinuous reward situations. *Psychological Bulletin*, 1958, *55*, 102–119.

Gormezano, I. Classical conditioning. In J. B. Sidowski (Ed.), *Experimental methods and instrumentation in psychology*. New York: McGraw-Hill, 1966.

Gormezano, I. Investigations of defense and reward conditioning in the rabbit. In A. H. Black & W. F. Prokasy (Eds.), *Classical conditioning II: Current research and theory*. New York: Appleton-Century-Crofts, 1972.

Gormezano, I., & Kehoe, E. J. Classical conditioning: Some methodological-conceptual issues. In W. K. Estes (Ed.), *Handbook of learning and cognitive processes* (Vol. 2). Hillside, N.J.: Lawrence Erlbaum Associates, 1975.

Gormezano, I., & Moore, J. W. Classical conditioning. In M. M. Marx (Ed.), *Learning: Processes*. Toronto: Macmillan, 1969.

Gormezano, I., Schneiderman, N., Deaux, E. B., & Fuentes, I. Nictitating membrane: Classical conditioning and extinction in the albino rabbit. *Science*, 1962, *138*, 33–34.

Hull, C. L. Knowledge and purpose as habit mechanisms. *Psychological Review*, 1930, *37*, 511–525.

Hull, C. L. Goal attraction and directing ideas conceived as habit phenomena. *Psychological Review*, 1931, *38*, 487–506.

Hull, C. L. The rat's speed of locomotion gradient in the approach to food. *Journal of Comparative Psychology*, 1934, *17*, 393–422.

Hull, C. L. *Principles of behavior*. New York: Appleton, 1943.

Konorski, J. *Conditioned reflexes and neuron organization*. Cambridge, England: Cambridge Univ. Press, 1948.

Logan, F. A., & Wagner, A. R. *Reward and punishment*. Boston: Allyn & Bacon, 1965.

Miller, N. E. Studies of fear as an acquirable drive: I. Fear as motivation and fear reduction as reinforcement in the learning of new response. *Journal of Experimental Psychology*, 1948, *38*, 89–101.

Miller, N. E. Some reflections on the law of effect produce a new alternative to drive reduction. In M. A. Jones (Ed.), *Nebraska symposium on motivation, 1963*. Lincoln: Univ. of Nebraska Press, 1963.

Mowrer, O. H. On the dual nature of learning—a re-interpretation of "conditioning" and "problem solving." *Harvard Educational Review*, 1947, *17*, 102–148.

Mowrer, O. H. *Learning theory and behavior*. New York: Wiley, 1960.

Pavlov, I. P. *Conditioned reflexes*. (Trans. by G. V. Anrep.) London: Oxford Univ. Press, 1927.

Razran, G. *Mind in evolution: An east-west synthesis of learned behavior and cognition*. Boston: Houghton Mifflin, 1971.

Spence, K. W. *Behavior theory and conditioning*. New Haven, Conn.: Yale Univ. Press, 1956.

Thompson, R. F. The search for the engram. *American Psychologist*, 1976, *31*, 209–227.
Wickens, D. D. Compound conditioning in humans and cats. In W. F. Prokasy (Ed.), *Classical conditioning: A symposium*. New York: Appleton-Century-Crofts, 1965.
Wickens, D. D., Nield, A. F., Tuber, D. S., & Wickens, C. D. Stimulus selection as a function of CS1-CS2 interval in compound classical conditioning of cats. *Journal of Comparative and Physiological Psychology*, 1973, *85*, 295–303.

NEAL E. MILLER
BARRY R. DWORKIN

Different Ways in Which Learning Is Involved in Homeostasis

4

The homeostatic regulation of many different functions, such as temperature, fluid and electrolyte balance, nutrition, respiration, and blood pressure, is essential to the life of mammals. Thus, any increase in our understanding of homeostatic mechanisms is likely to have deep implications for health. In this chapter, we shall discuss traditional ways in which learned goal-directed skeletal responses are involved in homeostasis, some less conventional ways in which learned skeletal responses can be involved in homeostasis, and a novel way in which learned visceral responses may be involved. In discussing this last possibility, we shall be frankly speculative. Incidentally, we shall mention an improved technique for maintaining rats completely immobilized by paralysis in an alert and healthy condition for several days. In addition to the study of visceral responses, we believe that this technique may be useful in experiments using techniques such as single-cell recording to investigate the neurophysiological mechanisms of learning.

Traditional Role of Learned Skeletal Responses in Homeostasis

Figure 4.1 illustrates the traditional role of instrumentally learned skeletal responses in homeostasis. For example, an internal cue from osmoreceptors and/or volume receptors signals a deficiency of water. The animal learns a skeletal response such as running to the proper location or pressing a lever that has the effect on the external environment of making water

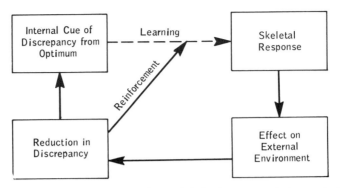

Figure 4.1. Traditional role of learned skeletal responses in homeostasis.

available. Drinking reduces the discrepancy from optimal fluid balance and the cues elicited by this discrepancy. This process reinforces the learning. Probably the arrow indicating reinforcement should be drawn from the cues rather than from the reduction in discrepancy, but that is not important for our immediate purpose.

Homeostatic drives of this type—for example, food deprivation—have been used in many experiments on learning. But the roles that learned skeletal responses play in homeostasis are more extensive than most people realize. In fact, for certain types of regulation, such as that of temperature, the behavioral type of regulation is phylogenetically older than the physiological one. Thus, fish, which are often thought of as devoid of thermal regulation, will tend to swim to the location in which the water has the optimal temperature for their growth. That this is not merely a tropism is shown by the fact that they can learn an arbitrary response of pressing a lever in order to regulate the temperature of the water within a few degrees (Rozin & Mayer, 1961). They also will learn an arbitrary response to regulate the oxygen content of the water (van Sommers, 1962). In another chapter in this volume, Dr. E. Satinoff will point out that mammals have a lower center in the brain that is involved only in behavioral regulation of temperature and a higher center that is involved in both behavioral and reflexive regulation. However, the older type of behavioral regulation is not necessarily less sophisticated. In fact, it is the learned behavioral regulation of body temperature—the use of fire, clothing, houses, space heating, and air conditioning—that has allowed *Homo sapiens* to extend its geographical range from the frozen Arctic to the hot desert.

Homeostatic Function of Classically Conditioned Visceral Responses

It is clear that learned skeletal responses play an important role in maintaining homeostasis. Can learned visceral responses play a similar role? Evidence from Pavlov (1927) and his many students in the Soviet

Union (e.g., Bykov, 1957) has shown the large variety of visceral responses that are subject to classical conditioning. For example, to take only the digestive system, there is conditioning of secretion of saliva in the mouth, digestive juices by the stomach, pancreatic juice by the pancreas, bile by the liver, and contractions of the gallbladder, stomach, and intestines. It seems unreasonable that the ability to learn these and many other conditioned visceral responses would have evolved if they had not proved to be adaptive in the normal environment. And it seems reasonable that being prepared for food by anticipatory conditioned responses lessens the burden of maintaining homeostasis by enabling digestion and elimination to proceed more efficiently. Perhaps some of the clinically observed ill effects on the gastrointestinal system of changing from regular to irregular routines are due to interference with classically conditioned responses (Miller, 1977). It should be possible experimentally to investigate this problem.

Experiments by Booth, Lee, and McAleavey (1976) are especially pertinent and convincing. It has been known that animals suddenly exposed to a diet of a new caloric density will fail to eat meals of the correct size, but over a period of days will appropriately adjust their intake. By associating an arbitrary cue with a given caloric density, Booth has shown that this regulation involves learning, presumably conditioned satiation. In experiments on both rats and people, one flavor was associated with a concentrated starch solution and another with a dilute starch solution. Over a short period, the subjects adjusted their intake appropriately so that, as they shifted back and forth on different days between the two solutions, they immediately drank the appropriate amount without having to wait for the effects of digestion and absorption that, because of the delay that these processes involve, would have caused them to consume a meal of inappropriate size. That the anticipatory adjustment was indeed a learned one to the arbitrary cues of the flavors (the roles of which were reversed for different groups) was shown by the fact that when the rats were fed a meal of intermediate density, they consumed less if it were flavored with the cue associated with high caloric density and more if it were flavored with the cue associated with low caloric density. One can see that such learning to circumvent the delay of digestion by immediately adjusting the meal size to the caloric density of the diet increases the efficiency of the homeostatic control of energy balance.

Classical Conditioning versus Instrumental Learning

In the traditional concept of classical conditioning, the association by contiguity of a neutral conditioned stimulus with reinforcement by an unconditioned stimulus causes the unconditioned response elicited by it to be transferred to the conditioned stimulus. Thus, the learning is limited to the response elicited by the reinforcing unconditioned stimulus.

In instrumental learning (also called trial-and-error learning or operant conditioning or type II conditioning), any (or almost any) response that is regularly followed by reinforcement is learned. Thus, one reinforcer, for example, food or a reduction in pain, can produce the learning of any one of a variety of different responses, or one response can be strengthened by a variety of reinforcers. The advantage of classical conditioning is that the unconditioned stimulus immediately elicits the correct response; the advantage of instrumental learning is that it can occur in a situation in which there is no unconditioned stimulus available to elicit the correct response.

Are visceral responses subject to the more flexible kind of instrumental learning? The traditional answer has been "No!" If this answer should be wrong, it would extend greatly the possible role of learning in homeostasis; learned visceral responses could function in homeostasis in the same way as do goal-directed skeletal ones, except that in this case the action would not be via the external environment but directly on the internal environment.

Evidence for Effects of Instrumental Learning on Visceral Responses

The results of an early experiment by Miller and Carmona (1967) are presented in Figure 4.2. We can see that thirsty dogs will increase their rate of salivation if increases are rewarded by water or will decrease their rate of salivation if decreases are rewarded by water; there was a fourteenfold difference between the groups rewarded for changes in opposite directions. This study has been confirmed by Shapiro and Herendeen (1975), who used food as a reward to train hungry dogs to reduce their rate of salivation, which was a change opposite to that that would be expected from the classical conditioning of salivation to the experimental situation.

Similarly, an early experiment in which rats were trained to increase or decrease their heart rate (DiCara & Miller, 1969) in order to avoid shock

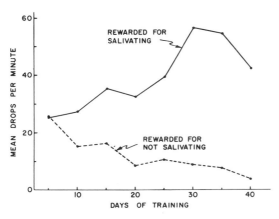

Figure 4.2. Instrumental learning of salivation by thirsty dogs rewarded by sips of water. (From Miller and Carmona, 1967.)

has been confirmed by van Kalmthout (1979) as well as in experiments on monkeys (Engel & Gottlieb, 1970) and on baboons (Harris, Gilliam, Findley, & Brady, 1976). Bidirectional changes in blood pressure have been produced in rats (Pappas *et al.,* 1970) and monkeys (Benson, Herd, Morse, & Kelleher, 1969), and baboons have been trained to maintain a 30 mm Hg elevation in blood pressure for a 12-hr period (Harris, Gilliam, & Brady, 1973). Experiments on human subjects summarized by Kimmel (1974) have shown that instrumental learning can produce increases or decreases in vasomotor responses, galvanic skin response, heart rate and rhythm, blood pressure, and salivation.

In the foregoing experiments, clearly the instrumental training changed the visceral response, but it is not clear whether the effect was a direct one or was mediated by a skeletal response that stimulated the receptive field of a visceral reflex (Dworkin & Miller, 1977; Miller, 1978).[1]

We shall next discuss the role in homeostasis of mediated effects of instrumental learning on visceral responses. Then we shall discuss the possible homeostatic role of direct effects of instrumental learning on visceral responses.

Mediated Effect of Instrumental Learning

Figure 4.3 illustrates how instrumental learning can produce a mediated effect on homeostasis. This may be illustrated by a case of paroxysmal tachycardia, a condition in which the heart suddenly starts beating much too rapidly. Some patients with this condition can learn to arrest an attack by taking a sudden deep breath. In this case, the internal cue is the sudden beating of the heart, which elicits the learned skeletal response of taking a sudden deep breath. Instead of acting on the external environment as in Figure 4.1, this skeletal response acts on the internal environment to stimulate the receptive field of a vagal inhibitory reflex that breaks up the tachycardia and restores the normal heart rate. Relief from the discomfort and anxiety of the tachycardia reinforces the performance of the learned response.

Another example is a case of paroxysmal bigeminy studied by Pickering and Miller (1977). This patient's heart would suddenly start alternating between normal beats and premature ventricular contractions. This made him

[1] In order to rule out the effects of the overt performance of skeletal responses, a series of experiments was performed on rats paralyzed by curare (Miller, 1969). At first these were replicated in three other laboratories (Banuazizi, 1972; Brener & Hothersall, 1966; Slaughter, Hahn, & Rinaldi, 1970), but later it became impossible to repeat them (Miller & Dworkin, 1974). Since then we have made many improvements in the technique of respirating rats paralyzed by curare, producing a preparation that remains in excellent condition for several days as indicated by discriminated classically conditioned responses (Dworkin & Miller, 1977). We are testing the ability of this preparation to show instrumental learning of visceral responses.

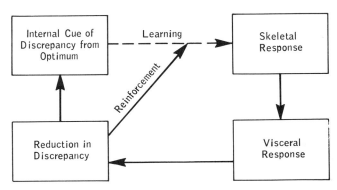

Figure 4.3. How the instrumental learning of a skeletal response can act on the internal environment to elicit a visceral response that restores homeostasis.

feel faint. But he learned that he could correct this condition by exercise that stimulated in various ways the receptive fields of reflexes that speeded up the heart, which in turn interrupted the bigeminy and returned his heart to a normal sinus rhythm.

The foregoing two conditions were brought to our attention because they involved conspicuous abnormalities. We wonder how many other examples there are in which the learned adjustment occurs so smoothly and automatically that it is not noticed until a special effort is made to look for it.

The patient with paroxysmal bigeminy felt faint so that the therapeutic exercise was difficult for him. Furthermore, it was not always acceptable in social situations. Therefore, he was connected to equipment that displayed small fluctuations in heart rate as perspicuous movements of a needle on a meter. He was instructed to try to cause the needle on the meter to move to the right. Such movements, indicating that he was succeeding, rewarded his learning. Eventually, he was able to make large enough increases to stop an attack of bigeminy. He also became able to perceive changes in heart rate so that he could practice by himself, with the reward of feeling his heart speed up and then return to normal sinus rhythm. In this case, he was not obviously changing his breathing or tensing his muscles to stimulate the receptive field of a visceral reflex, but, since we did not meticulously study his skeletal responses, he may have been using them. On the other hand, he may have been achieving the type of direct visceral learning that is discussed next.

Instrumentally Learned, Direct Visceral Response

Figure 4.4 illustrates how a visceral response that corrects a discrepancy may be instrumentally learned as a direct response to the cues indicating that discrepancy. If direct responses can indeed be learned, they are likely

4. Different Ways in Which Learning Is Involved in Homeostasis

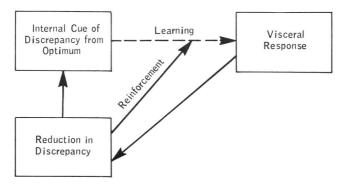

Figure 4.4. *How the direct learning of a visceral response may play a role in homeostasis.*

to be more important in homeostasis than the mediated ones, because they will not be limited to those situations in which there is a skeletal response that is able to stimulate the receptive field of a reflex that produces the needed change.

We can illustrate this last type of homeostatic correction by examples that apparently involve the direct learning of a visceral response, although we cannot yet be absolutely certain that all possibilities for skeletal mediation have been ruled out completely. A patient paralyzed by a gunshot that had severed his spinal cord at T4 had a strong ambition to learn to walk with crutches and braces. But whenever he was helped into an upright position, the necessary cardiovascular homeostatic adjustments were not made; his blood pressure fell so much that he fainted; he suffered from orthostatic hypotension. After working with him for almost three years, physical therapy had given up.

This patient heard that one of our collaborators, Bernard Brucker, was training people to control blood pressure and came to him for help. To everyone's surprise, when this patient was given information on how his blood pressure was changing, he quickly learned to produce large increases (Brucker and Ince, 1977). He also learned to identify changes in blood pressure so that he could practice by himself. The results are presented in Figure 4.5. We can see that when he was told to exert no learned control over his blood pressure and then was helped into an upright posture, his systolic pressure fell within 2 min to a level of 50 mm Hg, at which point he was about to faint so that he had to be returned to the sitting position. But when he was asked to make a learned increase in blood pressure, he did so and was able to maintain his pressure high enough to avoid fainting in a standing position for at least 5 min. Such learned corrections of low blood pressure enabled him to walk with crutches and braces. At first he occasionally would feel his blood pressure falling and have to stop walking to concentrate on raising it. Eventually, this learned correction became so automatic—as do many skills, such as riding a bicycle—that he did not

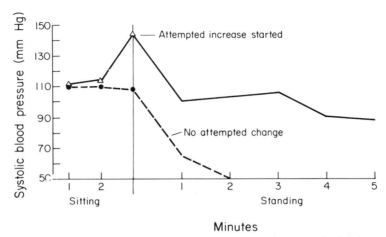

Figure 4.5. *The use of a learned response to correct a homeostatic deficiency. When a patient with a spinal lesion at T4 is helped to stand up, his systolic blood pressure falls to a dangerously low level, but if he produces a voluntary increase first, he can maintain an adequate pressure. (From Brucker and Ince, 1977.)*

have to think about it. In short, the instrumentally learned control corrected for the defect in homeostatic regulation caused by the spinal lesion.[2]

Figure 4.6 shows the learning curve of another patient whose neck had been broken in an automobile accident, severing the cord at C4/5 two years before the beginning of training. This patient, and a similar one to be described next, suffered from such severe orthostatic hypotension that she had to go around in a wheelchair in a semireclining position with her feet elevated to approximately the level of her head in order to avoid fainting. This seriously interfered with many activities such as riding in a car, attending a concert, or going to a motion picture. She was given 25 training trials, after which she was able to perceive changes and hence to practice by herself. The interesting thing about this additional period of practice is that during it the response gradually became much more specific; the unnecessary response of increasing heart rate dropped out so that she performed only the increase in blood pressure that was needed to correct the

---

[2] In this example, the reinforcement for correcting low blood pressure was complex; it involved success in achieving a number of goals: acquiring voluntary control over blood pressure, avoiding fainting, and walking with crutches and braces. But Miller, DiCara, and Wolf (1968) have shown that a simpler response of a visceral organ that restores homeostasis can serve as a reward. Hypophysectomized rats injected with antidiuretic hormone (ADH) on one side of a T-maze and with isotonic saline on the other side will, if loaded with excess salt, choose the side where the ADH enables the kidney to get rid of the excess salt by forming more concentrated urine; but, if loaded with excess water, the rats will choose the opposite side where the absence of ADH enables the kidney to get rid of the excess water. Presumably, such a simple reward produced directly by the action of a visceral organ could function also to reward any learnable visceral response.

4. *Different Ways in Which Learning Is Involved in Homeostasis* 65

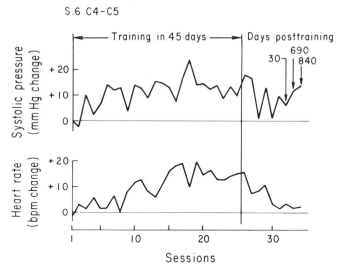

Figure 4.6. A patient with a spinal transection at C4/5 learns to produce voluntary increases in blood pressure during 25 sessions of training. Initially, the response involves increases in both heart rate and blood pressure; during additional practice, it becomes specific to the increase in blood pressure that is needed to correct orthostatic hypotension. (From Brucker, 1977, with additional data added.)

homeostatic discrepancy (Brucker, 1977; Miller & Brucker, 1979). In the learning of motor skills, unnecessary responses usually do drop out.

Another patient had a complex lesion producing complete paralysis of the skeletal muscles from C5 down, a rim of destruction interrupting segmental reflexes between C4 and C7, partial disconnection of tracts for the T1–T4 levels, and apparent total disconnection of all long tracts below T4. This lesion had occurred 2.5 years before the start of training. He showed a similar increase in specificity during the additional practice that he gave himself after the 25 formal training trials. Before training, this patient always fainted when his feet were lowered. After training, Figure 4.7 shows him with his feet down, maintaining his blood pressure constant and sufficiently high for over an hour during which he was carrying on a normal conversation without apparently having to pay any attention to maintaining homeostasis. That this was not a purely physiological adjustment that had developed as a result of being exposed to a normal sitting position during practice is shown by the results in Figure 4.8. When he was given the novel, distracting task of counting rapidly backward from 100, the systolic blood pressure fell over 35 mm Hg, and was restored by an increase of over 50 mm Hg as soon as the strong distraction had been removed so that he was able to sense that his pressure was dangerously low.

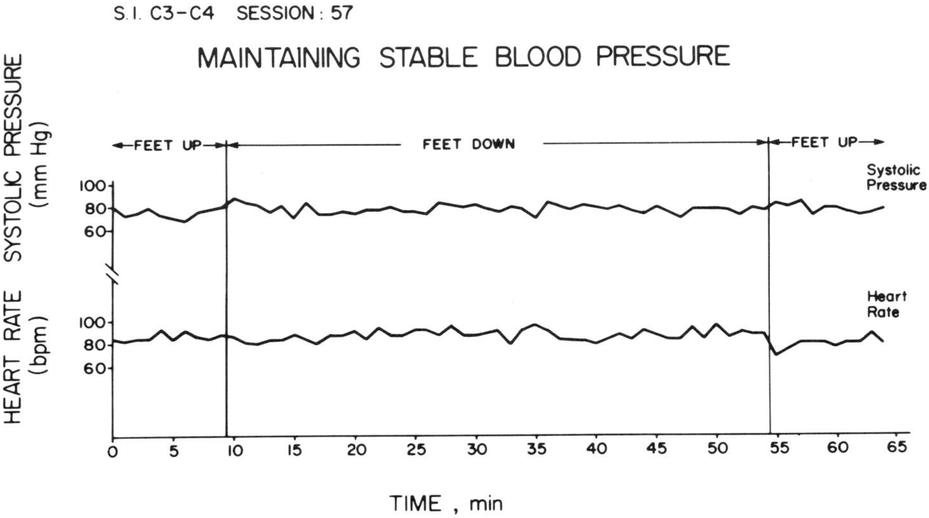

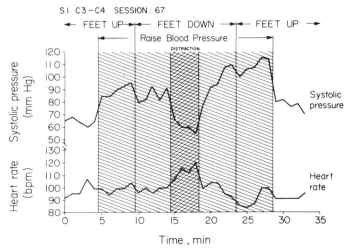

Figure 4.7. The patient with complete paralysis of skeletal muscles from C5 down, whose blood pressure fell to dangerously low levels whenever his feet were lowered, has learned to overcome orthostatic hypotension. He is able to keep his blood pressure constant during changes in position of the feet and while carrying on a conversation. (From Miller and Brucker, 1979.)

Figure 4.8. Effect of a strong distraction on learned maintenance of blood pressure by the patient whose performance is shown in Figure 4.7. (From Miller and Brucker, 1979.)

Tests on this and other similar patients show but little evidence of changes in muscular tension, in breathing rate, or in the pCO_2 of expired air during large voluntary increases in blood pressure. Much larger increases in muscle tension plus commands to contract even paralyzed muscles as strongly as possible and far greater changes in breathing and pCO_2 produce smaller changes in blood pressure (Brucker, 1977; Miller & Brucker, 1979). Therefore, these particular types of skeletal responses do not seem to be involved in the voluntary change. We are continuing the task of trying to rule out other conceivable skeletal responses, such as those of the diaphragm and glottis, in the limited repertoire available to these patients.

One might ask why these patients do not learn by themselves without special training. The answer is that probably some do learn the homeostatic adjustment by themselves. One patient with a lesion at C5 described how he got the nurse to lower his legs slowly, stopping whenever he began to feel faint while he fought off feeling faint and dizzy (Brucker, personal communication). From his description, it appeared that the success in fighting off faintness served as reinforcement for learning the compensatory increase in blood pressure. Since these paralyzed patients cannot move their own legs, they need help in order to have any opportunity to learn, but there is a shortage of nursing time. Furthermore, fainting can be extremely dangerous for them. They will not slump down like a normal person to a position that brings more blood to the brain. It seems reasonable that the foregoing peculiar circumstances prevented the easier learning of a homeostatic adjustment and made the process more conspicuous. If learning had easily and automatically corrected their problems, neither they nor we would have been likely to notice them. If we now look more carefully, how many other instances will we find in which visceral learning plays a significant role in homeostasis?

Paradoxical Conditioned Responses

In looking for additional examples of the role of visceral learning in maintaining homeostasis, certain paradoxical responses produced by the training procedure of classical conditioning merit further investigation. According to the traditional theory of classical conditioning, a response originally elicited by the unconditioned stimulus is transferred to the conditioned one. But there are certain paradoxical cases in which the response to the conditioned stimulus is opposite to the unconditioned response and hence could not have been produced by mere simple transfer. For example, in many situations the unconditioned response (UCR) to an electric shock is an increase in heart rate, whereas the conditioned response (CR) is the opposite one of a decrease.[3]

With many drugs the CR is opposite to the UCR. For example, the UCR to insulin is hypoglycemia, but the CR is hyperglycemia (Siegel, 1972, 1975). The UCR to atropine is inhibition of salivation, but the CR is salivation (Lang, Brown, Gershon, & Korol, 1966). The UCR to epinephrine is decreased gastric secretion, but the CR is an increase (Guha, Dutta, & Pradhan, 1974). In these and other studies, some of the authors (especially Siegel and Lang) have been aware of the compensatory, homeostatic function of these paradoxical CRs. Furthermore, Siegel (1977, 1978) has shown that such compensatory learned responses, rather than traditional, purely pharmacological effects, are involved in habituation to the hyperthermic and to the analgesic effects of mild doses of morphine. But none of the foregoing studies has analyzed the mechanism for the learning of compensatory responses. Among the main possibilities are the following:

1. The subject may learn a central response (image or expectancy) anticipatory to the UCS and the response innately wired to that central response may be opposite to the one innately wired to the UCS (e.g., see the analysis of fear, Miller, 1951, p. 443; 1971, p. 134). In this case, preventing the overt occurrence of the UCR would not prevent the acquisition of the compensatory response.

2. The occurrence of the UCR may produce a homeostatic disturbance that stimulates the receptive field of an innate compensatory reflex that functions as the UCR for a traditional CR similar to the compensatory UCR but opposite to the primary one.[4] In this case, as contrasted with the previous one, preventing the occurrence of the primary UCR should prevent stimulation of the receptive field for the compensatory UCR and hence prevent the acquisition of any compensatory CR.[5]

3. The compensatory response may be relatively high in the innate hierarchy of responses to the homeostatic disturbance produced by the pri-

---

[3] We have found that the slowing to the CS shows up especially clearly when the complicating effects on heart rate of overt muscular responses are controlled by paralysis by curare, using a vastly improved procedure for respirating paralyzed rats (Dworkin & Miller, 1977).

[4] Why does not a CR to the primary UCR overpower the compensatory CR? In some cases, the treatment described as the primary UCS may produce its disturbing effect without involving the central nervous system; it may not elicit a true UCR. This could be the case with insulin. Another possibility is that the homeostatic value of the compensatory response may, through natural selection, have caused it to become more readily conditionable innately than the primary UCR.

[5] Preventing the overt occurrence of the UCR should not necessarily interfere with the learning of a traditional classical conditioned response, however, because the activation of the central connection involved in that response is direct, that is, does not depend on its overt occurrence.

mary UCR, but it may be an instrumental response that is reinforced by the reduction that it produces in this disturbance. In this case, the learning of the paradoxical CR could be prevented by (a) preventing the overt occurrence of the primary UCR, or (b) allowing that response to occur but preventing the overt occurrence of any compensatory response, or (c) merely completely neutralizing any equilibrium-restoring effects of the compensatory response. For example, if the mechanism of conditioned hyperglycemia to insulin was of the third type, it could be prevented by servo-controlling the blood sugar level and putting the animal in a "glucose clamp," or perhaps by merely using such a large dose of insulin that any effects of a compensatory response would be completely overwhelmed. But, by contrast, if the response is elicited by mechanism 2, the latter two treatments should intensify the stimulation of the receptive field for the compensatory UCR and hence, if anything, facilitate its conditioning.

Adjusting Priorities

Under the radically changed circumstances of a high spinal transection, we have seen the advantages of a mechanism for producing learned changes in blood pressure as a corrective mechanism for orthostatic hypotension. Consider the difficulties of designing a hard-wired system to make the appropriate adjustments in priorities among the competing requirements of a large number of homeostatic mechanisms during radical changes in the demands on and effectiveness of the systems that are regulated. Great changes in surface to volume ratio and in many other requirements occur during growth. Great changes in the effectiveness of different systems occur during aging. There are demands of extreme environments. To help deal with these and many other changing demands, one can see the desirability of having available some mechanisms for producing learned adjustments.

One way of achieving the adjustments required by changes and of reallocating priorities in an effective manner would be to have the various types of homeostatic imbalances produce a distress such as pain and to have the behavior that minimized the total distress reinforced by a reduction in that distress. Such a mechanism would fit in with the fact that although many parts of the viscera are relatively ill-equipped with highly specific types of sense organs to monitor exact levels of each specific variable, most of them seem to be well endowed with relatively nonspecific nerve endings responsive to the particular types of deviation from optimal, for example, ischemia in the cardiac muscle, that are vital to their function.

One simple example of pain motivation producing a learned shift in the relative reliance on two alternative homeostatic mechanisms comes from the experiment described in Figure 4.9. Here it can be seen that if drinking from a waterspout was made to deliver strong shocks, homeostasis

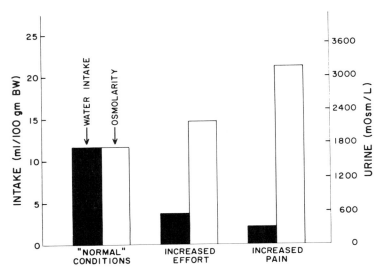

Figure 4.9. When obtaining water is made more effortful or painful, the rat relies less on drinking to maintain fluid balance and more on reabsorption of water by the kidney to form urine that has higher osmolarity. (From Miller, 1967.)

would be maintained by drinking less water and making more use of reabsorption of water by the kidney to form a more concentrated urine.

There is evidence that people can learn to affect the temperature of highly specific parts of the skin by changing the flow of blood to them (Taub, 1977). There is evidence that learning to warm the fingers can reduce the atypically strong vasoconstriction that patients with Raynaud's disease show in response to laboratory exposure to cold stress (Surwit, Pilon, & Fenton, 1977). There is also evidence that people working in cold environments on tasks that cannot be performed while wearing heavy gloves gradually become able to work better and to suffer less pain from their hands, and they actually stay warmer. Experiments have shown that such adjustments (changing the original balance of priorities in a direction that conserves less heat but preserves more comfort and dexterity in the hand) can be made so specific that they involve a finger of one hand but not of the other one and that they apparently involve the central nervous system (Adams & Smith, 1962). But critical tests have not yet been made to determine whether or not such adjustments involve learning.

Another place to look for learned homeostatic effects is among the radical changes that are required in many functions during long exposures to 0 g in space and to the normal g at the end of a long space flight. Biomedical scientists were surprised that astronauts returning from longer space flights seemed to make many of the adjustments to normal g with less difficulty than those who had returned from shorter ones (Cooper, 1978).[6]

[6] Unfortunately, the differences in duration were confounded with differences in exercise and other factors.

One would not expect this to be true of purely physiological readjustments. But the person who has thoroughly learned two conflicting types of responses can shift from one to the other with much less interference than someone who is just beginning to learn one of them. To the extent that some of the adjustments to changes in g can be learned, one would expect the skill of adjusting to be improved by practice.

REFERENCES

Adams, T., & Smith, R. E. Effect of chronic local cold and exposure on finger temperature responses. *Journal of Applied Physiology*, 1962, *17*, 317–322.

Banuazizi, A. Discriminative shock-avoidance learning of an automatic response under curare. *Journal of Comparative and Physiological Psychology*, 1972, *81*, 336–346.

Benson, H., Herd, A. J., Morse, W. H., & Kelleher, R. T. Behavioral induction of arterial hypertension and its reversal. *American Journal of Physiology*, 1969, *217*, 30–34.

Booth, D. A., Lee, M., & McAleavey, C. Acquired sensory control of satiation in man. *British Journal of Psychology*, 1976, *67*, 137–147.

Brener, J., & Hothersall, D. Heart rate control under conditions of augmented sensory feedback. *Psychophysiology*, 1966, *3*, 23–28.

Brucker, B. S. Learned voluntary control of systolic blood pressure by spinal cord injury patients. Ph.D. thesis, New York Univ., 1977.

Brucker, B. S., & Ince, L. P. Biofeedback as an experimental treatment for postural hypotension in a patient with a spinal cord lesion. *Archives of Physical Medicine and Rehabilitation*, 1977, *58*, 49–53.

Bykov, K. M. *The cerebral cortex and the internal organs.* (Ed. and trans. by W. H. Gantt.) New York: Chemical Publ., 1957.

Cooper, S. F., Jr. *A house in space.* New York: Bantam, 1978.

DiCara, L. V., & Miller, N. E. Heart-rate learning in the noncurarized state, transfer to the curarized state, and subsequent retraining in the noncurarized state. *Physiological Behavior*, 1969, *4*, 621–624.

Dworkin, B. R., & Miller, N. E. Visceral learning in the curarized rat. In G. E. Schwartz & J. Beatty (Eds.), *Biofeedback: Theory and research.* New York: Academic Press, 1977, pp. 221–242.

Engel, B. T., & Gottlieb, S. H. Differential operant conditioning of heart rate in the restrained monkey. *Journal of Comparative and Physiological Psychology*, 1970, *73*, 217–225.

Guha, D., Dutta, S. N., & Pradhan, S. N. Conditioning of gastric secretion by epinephrine in rats. *Proceedings of Society of Experimental Biology and Medicine*, 1974, *147*, 817–819.

Harris, A. H., Gilliam, W. J., Findley, J. D., & Brady, J. V. Instrumental conditioning of large-magnitude, daily, 12-hour blood pressure elevations in the baboon. *Science*, 1973, *182*, 175–177.

Harris, A. H., Gilliam, W. J., & Brady, J. V. Operant conditioning of large-magnitude, 12-hour duration, heart rate elevations in the baboon. *Pavlovian Journal of Biological Science*, 1976, *11*, 86–92.

Kimmel, H. D. Instrumental conditioning of autonomically mediated responses in human beings. *American Psychology*, 1974, *29*, 325–335.

Lang, W. J., Brown, M. L., Gershon, S., & Korol, B. Classical and physiologic adaptive conditioned responses to anticholinergic drugs in conscious dogs. *International Journal of Neuropharmacology*, 1966, *5*, 311–315.

Miller, N. E. Learnable drives and rewards. In S. S. Stevens (Ed.), *Handbook of experimental psychology.* New York: Wiley, 1951, pp. 435–472.

Miller, N. E. Behavioral and physiological techniques: Rationale and experimental designs for combining their use. In C. F. Code and W. Heidel (Eds.), *Handbook of physiology* (Section 6, Vol. 1). Washington, D.C.: American Physiological Society, 1967, pp. 51–61.

Miller, N. E. Learning of visceral and glandular responses. *Science,* 1969, *163,* 434–445.

Miller, N. E. *Neal E. Miller: Selected papers.* Chicago: Aldine, 1971.

Miller, N. E. Effect of learning on gastrointestinal functions. *Clinics in Gastroenterology,* 1977, *6,* 533–546.

Miller, N. E. Biofeedback and visceral learning. *Annual Review of Psychology,* 1978, *29,* 373–404.

Miller, N. E., & Brucker, B. S. Learned large increases in blood pressure apparently independent of skeletal responses in patients paralyzed by spinal lesions. In N. Birbaumer and H. D. Kimmel (Eds.), *Biofeedback and self-regulation.* Hillside, N.J.: Lawrence Erlbaum Associates, 1979, pp. 287–304.

Miller, N. E., & Carmona, A. Modification of a visceral response, salivation in thirsty dogs, by instrumental training with water reward. *Journal of Comparative and Physiological Psychology,* 1967, *63,* 1–6.

Miller, N. E., & Dworkin, B. R. Visceral learning: Recent difficulties with curarized rats and significant problems for human research. In P. A. Obrist, A. H. Black, J. Brener, and L. V. DiCara (Eds.), *Cardiovascular psychophysiology.* Chicago: Aldine, 1974, pp. 312–331.

Miller, N. E., DiCara, L. V., & Wolf, G. Homeostasis and reward: T-maze learning induced by manipulating anti-diuretic hormone. *American Journal of Physiology,* 1968, *215,* 684–686.

Pappas, B. A., DiCara, L. V., & Miller, N. E. Learning of blood pressure responses in the noncurarized rat: Transfer to the curarized state. *Physiology and Behavior,* 1970, *5,* 1029–1032.

Pavlov, I. P. *Conditioned reflexes.* (Trans. by G. V. Anrep.) London: Oxford Univ. Press, 1927. (Reprinted New York: Dover, 1960.)

Pickering, T. G., & Miller, N. E. Learned voluntary control of heart rate and rhythm in two subjects with premature ventricular contractions. *British Heart Journal,* 1977, *39,* 152–159.

Rozin, P. N., & Mayer, J. Thermoreinforcement and thermoregulatory behavior in the goldfish, *Carassius auratus. Science,* 1961, *134,* 942–943.

Shapiro, M. M., & Herendeen, D. L. Food-reinforced inhibition of conditioned salivation in dogs. *Journal of Comparative and Physiological Psychology,* 1975, *88,* 628–632.

Siegel, S. Conditioning of insulin-induced glycemia. *Journal of Comparative and Physiological Psychology,* 1972, *78,* 233–241.

Siegel, S. Conditioning insulin effects. *Journal of Comparative and Physiological Psychology,* 1975, *89,* 189–199.

Siegel, S. Morphine tolerance acquisition as an associative process. *Journal of Experimental Psychology: Animal Behavior Processes,* 1977, *3,* 1–13.

Siegel, S. Tolerance to the hyperthermic effect of morphine in the rat is a learned response. *Journal of Comparative and Physiological Psychology* 1978, *92,* 1137–1149.

Slaughter, J., Hahn, W., and Rinaldi, P. Instrumental conditioning of heart rate in the curarized rat with varied amounts of pretraining. *Journal of Comparative and Physiological Psychology,* 1970, *72,* 356–359.

Surwit, R. S., Pilon, R. N., & Fenton, C. H. Behavioral treatment of Raynaud's disease. Paper presented at 11th Annual Association for Advancement of Behavior Therapy Convention, December 9–11, 1977, Atlanta, Ga.

Taub, E. Self-regulation of human tissue temperature. In G. E. Schwartz & J. Beatty (Eds.), *Biofeedback: Theory and research.* New York: Academic Press, 1977, pp. 265–300.

van Kalmthout, M. Operant heart-rate conditioning in freely-moving rats. In N. Birbaumer & H. D. Kimmel (Eds.), *Biofeedback and self-regulation*. Hillside, N.J.: Lawrence Erlbaum Associates, 1979, pp. 305–320.
van Sommers, P. Oxygen-motivated behavior in the goldfish, *Carassius auratus*. *Science*, 1962, *137*, 678–679.

JAMES L. McGAUGH
JOE L. MARTINEZ, JR.
ROBERT A. JENSEN
RITA B. MESSING
BEATRIZ J. VASQUEZ

Central and Peripheral Catecholamine Function in Learning and Memory Processes[1]

5

Introduction

Our research focuses on treatments that modulate the storage or consolidation of newly acquired responses. The guiding hypothesis is that gaining an understanding of the effects of experimental treatments on memory, together with a knowledge of the neurobiological bases of the treatments, will provide important leads to understanding the neural bases of memory.

Findings from our laboratory, as well as those of other investigators, have provided evidence that retention of newly learned responses is influenced by peptide hormones, catecholamines, and drugs that affect hormone and catecholamine metabolism (de Wied & Bohus, 1979; McGaugh, et al., 1979; Telegdy & Kovács, 1979). These findings provide support for the view that processes associated with affective states modulate memory storage (Kety, 1970; Gold & McGaugh, 1977). Our recent findings emphasize, in particular, the role of peripheral hormonal systems in the modulation of memory storage.

In our studies, rats or mice are trained on a simple learning task and then, shortly after training, are administered drugs or hormones that affect central or peripheral neuronal and endocrine activity. A retention test is

[1] This research was supported by research grants MH 12526 from the U.S. Public Health Service; BNS 76-17370 from the National Science Foundation, Grant AG 00538 from the National Institute on Aging, and a grant from the McKnight Foundation (all awarded to James L. McGaugh).

administered later to determine whether the retention is altered by the posttraining treatment. We use tasks such as one-trial inhibitory avoidance, one-way active avoidance, or Y-maze brightness discrimination because training in these tasks requires only a few seconds or minutes and we can vary rather precisely the time between acquisition and posttraining treatment (McGaugh & Herz, 1972). Ordinarily the retention test is given within a week after training.

Brain Catecholamines and Memory

Many studies, including our own, have shown that retention of newly learned responses is impaired by either pre- or posttraining parenteral administration of drugs, such as diethyldithiocarbamate (DDC) or fusaric acid, that interfere with the synthesis of norepinephrine (Haycock, van Buskirk, & McGaugh, 1977a; Randt, Quartermain, Goldstein, & Anagnoste, 1971). Studies in our laboratory showed that DDC causes anterograde and retrograde amnesia in several learning tasks (Haycock et al., 1977a; Spanis, Haycock, Handwerker, Rose, & McGaugh, 1977). In general, the degree of memory impairment varies directly with the dose and inversely with the time, pretraining or posttraining, of drug administration. These findings have been interpreted as indicating that these memory effects are due to drug-induced alterations of brain catecholamines. In support of this view we found, in mice, that retention is enhanced by posttraining intracerebroventricular administration of catecholamines (Haycock, van Buskirk, Ryan, & McGaugh, 1977b). We also found in rats that retention impairment is produced by direct intracerebroventricular administration of DDC (Jensen et al., 1977). In these experiments, DDC (.4–4.0 mg in volumes of 1–10 μl) was administered into the left lateral ventricle of rats either before or after they were trained in a one-trial inhibitory avoidance task. Figure 5.1 shows the retention latencies of animals given intracerebroventricular DDC 15 min before they were trained. Retention tested three days after training was markedly impaired by the 2.0 and 4.0 mg doses. Administration of DDC 24 hr before training did not affect retention latencies. Control animals given 10 μl of Ringer's or NaOH displayed good retention. Figure 5.2 shows the retention latencies of animals given posttraining unilateral intracerebroventricular DDC. Significant retention impairment was produced only by the 4 mg dose. Impairment was produced by administration immediately, but not 6 or 24 hr after training. Thus, the effects of intracerebroventricularly administered DDC on retention are dose dependent and time dependent. The finding that the retention effects can be produced by such low doses, in comparison with the typical parenteral doses, supports the view that DDC effects on memory are due to central, rather than peripheral, effects of the drug. Whether the effects are du to influences

5. Catecholamine Function in Learning and Memory Processes

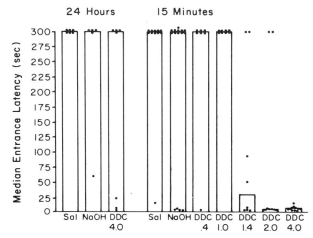

Figure 5.1. *Effect of pretrial intracerebroventricular administration of diethyldithiocarbamate (DDC) on retention in rats. The black dots indicate scores of individual animals. Doses are in mg/kg. The pH of the NaOH control group was 11.4. (From Jensen et al., 1977.)*

on brain catecholamines or other influences (Danscher & Fjerdingstad, 1975) remains to be determined.

Additional evidence, suggesting that central catecholamines are involved in modulating memory storage processes, was provided by Stein, Belluzzi, and Wise (1975), who showed that the acquisition impairment produced by DDC was attenuated by intracerebroventricular administration of norepinephrine shortly after training. These findings suggest that the learning deficit and its attenuation are due to reduction and replenishment of brain norepinephrine.

In recent studies we have replicated and extended these experiments.

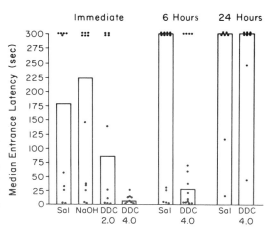

Figure 5.2. *Effect of posttraining intracerebroventricular administration of DDC on retention in rats. Black dots indicate scores of individual animals. Doses are in mg/kg. The pH of the NaOH control group was 11.4. (From Jensen et al., 1977.)*

However, our findings question the view that replenishment of central norepinephrine is required for attenuating DDC-induced learning deficits. Rats were given a Ringer's control injection or DDC (680 mg/kg) intraperitoneally approximately .5 hr before they were trained in a one-trial inhibitory avoidance task (Meligeni, Ledergerber, & McGaugh, 1978). Two footshock levels were used, a low footshock (.5 mA, 1 sec) and a high footshock (2 mA, 2 sec). Either immediately or 4 hr after training, Ringer's solution or norepinephrine was administered intracerebroventricularly through chronically implanted cannulae. A retention test was given one week later. With low footshock, DDC produced retention impairment in all groups, and posttraining administration of .1, 1, or 100 μg norepinephrine did not attenuate the effects of DDC. With high footshock, DDC produced a retention deficit in animals given Ringer's solution after training, as well as in animals given the high dose of norepinephrine (100 μg). However, DDC did not produce impairments in animals given the lower doses of norepinephrine immediately after training. In other words, intraventricularly administered norepinephrine attenuated the retention impairment produced by DDC only in the high footshock condition but not in the low footshock group and then only with the lower doses. Attenuation of the DDC effects was not produced by any dose if the norepinephrine was administered 4 hr after training. These findings based on experiments using central administration of norepinephrine are quite consistent with those reported by Stein, Belluzzi, and Wise (1975), and provide additional support for the view that the retention impairment is caused by alterations in central norepinephrine biosynthesis.

Modulating Influences of Peripheral Catecholamines

Other recent experiments have demonstrated that in rats retention can be enhanced by parenteral injections of norepinephrine and epinephrine (Gold & van Buskirk, 1975; 1976a). It is unlikely that these effects are due to a direct action of these amines in the brain because there is little evidence that they readily pass the blood brain barrier. In view of these findings, we examined the effects of posttraining parenteral administration of norepinephrine and epinephrine on retention in rats given DDC before training. The effects of norepinephrine are shown in Figures 5.3 and 5.4. When low footshock was used (Figure 5.3), the DDC injections produced retention impairment in rats given either subcutaneous injections of Ringer's solution or a low dose of norepinephrine (5 μg/kg) after training.

Impairment was not found in animals given higher posttraining doses of norepinephrine (50 and 500 μg/kg). In groups that received higher footshock (Figure 5.4), significant attenuation of DDC-induced retention deficits was found only with the lowest dose of norepinephrine (5 μg/kg). No effects

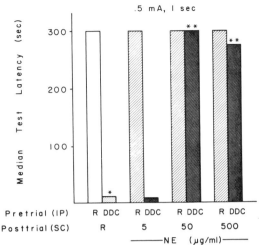

Figure 5.3. Effects of immediate posttrial subcutaneous injections of norepinephrine on rats given DDC (680 mg/kg) 30 min before training (with low footshock) on an inhibitory avoidance task. R, Ringer's control; *, significantly different from Group R-R; **, significantly different from Group DDC-R. (From Meligeni et al., 1978.)

were produced by norepinephrine administration 4 hr after training. Clearly, the effective dose of norepinephrine required to attenuate the impairing effects of DDC depends on the footshock level. With high footshock, only the lowest dose attenuated the DDC effect, whereas with a low footshock, higher doses of norepinephrine were effective. Given the nature of the dose-footshock interaction observed in this experiment, it is not surprising that intraventricularly administered norepinephrine was ineffective in attenuating the DDC-induced retention deficit reported in the previous experiment.

The effects of parenterally administered epinephrine are comparable to those produced by norepinephrine. Figure 5.5 shows the effects of sub-

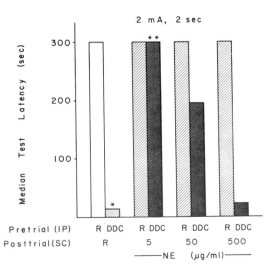

Figure 5.4. Effects of immediate posttrial subcutaneous injections of norepinephrine on rats given DDC (680 mg/kg) 30 min before training (with high footshock) on an inhibitory avoidance task. R, Ringer's control; *, significantly different from Group R-R; **, significantly different from Group DDC-R. (From Meligeni et al., 1978.)

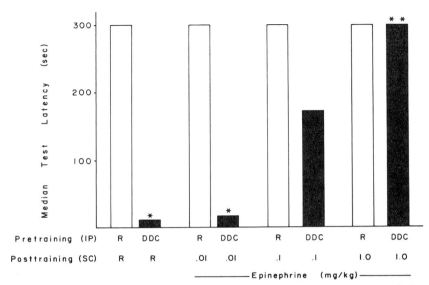

Figure 5.5. Effects of immediate posttrial subcutaneous injection of epinephrine on rats given DDC (680 mg/kg) 30 min before training (with high footshock) on an inhibitory avoidance task. R, Ringer's control; *, significantly different from Group R-R; **, significantly different from Group DDC-R. (From McGaugh et al., 1979.)

cutaneously administered epinephrine, using procedures identical to those used in the previous experiment except that only the high footshock was used (McGaugh et al., 1979). DDC administered one-half hour before training impaired retention in rats given subcutaneous injections of Ringer's solution or the lowest dose (.01 mg/kg) of epinephrine. Good retention was displayed by rats given Ringer's solution or DDC followed immediately by subcutaneous injections of epinephrine at a dose of 1 mg/kg. No effects were found with epinephrine injections given 4 hr after training. This finding is interesting in view of the findings of Gold and van Buskirk (1975) that, in normal rats, retention is enhanced by immediate posttraining subcutaneous injections of .01 and .10 mg/kg epinephrine. Therefore, when poor retention is produced by DDC, a higher dose of epinephrine is required to enhance retention.

Our findings are consistent with those of other studies showing that retention can be modulated by posttraining peripheral administration of substances that are normally released by emotional experiences. ACTH, norepinephrine, and epinephrine affect retention when administered subcutaneously (Gold & van Buskirk, 1975, 1976b). Norepinephrine and epinephrine attenuate DDC-induced retention impairments (McGaugh et al., 1979; Meligeni et al., 1978). The peptide fragment $ACTH_{4-10}$ attenuates the learning deficit produced by anisomycin, an inhibitor of protein synthesis

(Flood, Jarvik, Bennett, Orme, & Rosenzweig, 1977). Because none of these hormones is known to readily pass the blood brain barrier, it seems unlikely that they act directly on the brain when administered peripherally. However, it is worth noting that the doses of norepinephrine injected subcutaneously in our experiment were extremely high in comparison with those doses found to be effective when administered intracerebroventricularly (50 and 500 μg/kg versus .01 and 1 μg/rat). Thus, the possibility that the effects are due to small amounts of the hormone passing through the blood brain barrier cannot be completely ruled out. There is, of course, substantial evidence that these substances affect the brain when they are administered into the brain or ventricles (Dunn & Gispen, 1977). Thus, when these substances are present in the brain, they may act directly to modulate neuronal processes involved in memory storage (Gold & McGaugh, 1977; Kety, 1970).

Peripherally administered hormones might act indirectly through influences on the cardiovascular system. However, dramatic alterations of cardiovascular function not initiated directly by hormones or neurotransmitters do not seem to affect memory (Martinez, Jensen, Vasquez, Lacob, McGaugh, & Purdy, 1979). Alternatively, the modulating influences of peripherally administered hormones might be due to the activating influences of such changes on brain systems involved in arousal. Peripherally induced changes in arousal may act just as electrical stimulation of the mesencephalic reticular formation to produce brain changes that modulate memory storage processes (Bloch, 1970).

Amphetamine Influences on Memory Storage

Our view that memory storage is modulated by activation of peripheral systems is supported by the findings of a series of experiments performed in my laboratory designed to assess the effects of amphetamine on memory storage.

It is well known that learning and memory are enhanced by amphetamine in a number of learning tasks (for review see McGaugh [1973]). The common interpretation of these effects is that amphetamine acts on brain catecholamines in some way to enhance memory storage. This view is consistent with the hypothesis that catecholamines may act directly to modulate neuronal processes involved in memory storage. However, amphetamine has a variety of peripheral effects, including some known to affect brain functioning. For example, amphetamine alters dramatically central blood flow and oxygen utilization in the brain (Carlsson, Hägerdal, & Siesjö, 1975). Further, our findings that retention is influenced by parenterally administered catecholamines suggests that amphetamine may affect learning by acting on peripheral catecholamines.

In order to address this question, we investigated the effects on retention of immediate posttrial intracerebroventricular administration of d-amphetamine. Rats were infused intraventricularly with one of several doses (50, 100, 300, and 500 μg/rat) of d-amphetamine sulfate immediately following training in a one-trial inhibitory avoidance task (Martinez et al., 1980). A 750 μA, 1 sec footshock and a 72-hr training-retention test interval was employed. Intraventricular administration of amphetamine did not influence retention with any of the doses tested. Given the wide range of doses used, these findings suggest that it is unlikely that memory storage is affected by centrally administered amphetamine.

In order to determine whether amphetamine has any behavioral effects at all when administered centrally in the doses used, we examined the effects of intracerebroventricularly administered amphetamine on locomotor activity in an open field test. The subjects were given infusions of amphetamine and then placed in a 1 m² open field divided into squares. The number of line crossings that each animal made during 31 1-min observation periods was recorded. As Figure 5.6 shows, activity increased as a direct function of the dose of intracerebroventricularly administered amphetamine.

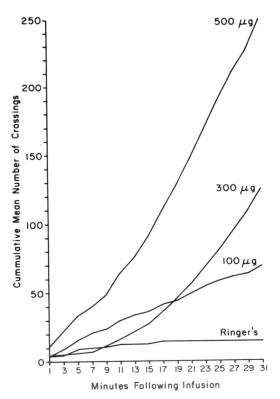

Figure 5.6. Effects of an intracerebroventricular infusion of d-amphetamine on open field activity of rats. Note the dose-dependent increase in open field activity that was evident from the first minute of observation. (From Martinez et al., *1980*.)

Clearly, the amphetamine affected locomotor activity. The finding that centrally administered amphetamine does not influence retention suggests that its effects on memory are due to influences on the peripheral sympathetic division of the autonomic nervous system as well as the adrenal glands. This interpretation suggests that the facilitatory effect of amphetamine on memory should be eliminated or altered by destruction of presynaptic sympathetic neurons with 6-hydroxydopamine (6–OHDA) and/or by adrenal demedullation. The first possibility was examined in a series of studies on the effects of amphetamine on learning in animals previously treated with 6–OHDA. We first investigated the effects of 6–OHDA on learning and retention. The 6–OHDA dissolved in .1% ascorbic acid and normal saline was injected intravenously in rats 24 hr prior to training in an inhibitory avoidance task. The rats were tested 72 hr later. The 24-hr treatment-training interval was used to allow acute toxic effects of the 6–OHDA to subside, and the 72-hr training-testing interval was chosen to minimize the possibility that the results would be influenced by compensatory developments following 6–OHDA treatment such as neuronal regeneration (Kostrzewa & Jacobowitz, 1974).

Within 3 hr after the retention test, all animals were sacrificed by decapitation and a portion of their hearts was assayed for norepinephrine concentrations. Additional untrained rats were administered either physiological saline or 100 mg/kg 6–OHDA. Their brains were removed four days following treatment and assayed for norepinephrine. Figure 5.7 shows the retention latencies of the animals receiving 6–OHDA. An impairment of retention was observed in the groups that received 5 mg/kg 6–OHDA. The results of the norepinephrine assay indicated that 6–OHDA reduced norepinephrine concentrations in the heart by 75% in the 5 mg/kg group, 87% for 25 mg/kg, and 93% in the 100 mg/kg group. Individual correlations between norepinephrine concentrations in heart tissue and retention performance indicated that these two variables were not significantly related in any of the groups. Finally, the norepinephrine assays of whole brain homogenates (excluding the cerebellum) showed that the highest dose (100 mg/kg) produced a slight, but nonsignificant decrease in brain norepinephrine.

We then examined the effect of amphetamine on learning in 6–OHDA-treated rats. In this experiment, 6–OHDA was given 24 hr prior to training, and amphetamine was given immediately after training. We reasoned that if the peripheral sympathetic system is involved in the mediation of amphetamine effects on learning, then, following 6–OHDA treatment, amphetamine should either not affect learning or affect learning when given in lower doses.

Twenty-four hours prior to training, rats were injected (i.v.) with either vehicle solution, 5 mg/kg or 100 mg/kg 6–OHDA. Immediately after training, rats were injected (i.p.) with either saline, .25, 1, or 4 mg/kg

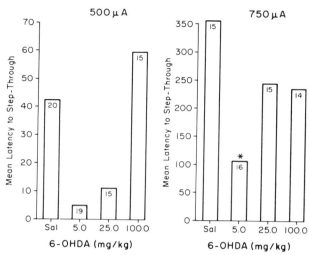

Figure 5.7. Effects of 6-hydroxydopamine (6-OHDA) given 24 hr before training in an inhibitory avoidance task on retention measured 72 hr after training. The left panel shows data for animals trained with a 500 µA, 1 sec footshock, and the right panel shows data for animals trained with a 750 µA, 1 sec footshock. Only the 5.0 mg/kg dose of 6-OHDA produced a significant impairment in retention at the higher footshock although the pattern of results was the same for the lower footshock. (From Martinez et al., 1980.)

d-amphetamine. An additional group of rats received no injection. A 500 µA, 1 sec footshock was used with a 72-hr training-testing interval.

As is shown in Figure 5.8, 1.0 mg/kg amphetamine facilitated retention in rats not treated with 6–OHDA (saline pretreatment). In rats pretreated with 5 mg/kg 6–OHDA, the facilitatory dose of amphetamine was lowered to .25 mg/kg. The pattern of effect (seen in Figure 5.8) of immediate posttrial treatments with amphetamine was the same for both the 5 and 100 mg/kg 6–OHDA pretreatment groups. In both cases the highest retention latencies were observed in the .25 mg/kg amphetamine group; that is, the dose-response curve was shifted to the left.

Several findings have emerged from this series of studies. First, intraventricular administration of a wide dose range of amphetamine (50–500 µg/rat) was ineffective in altering retention performance in an inhibitory avoidance task. In contrast, peripherally administered amphetamine did facilitate retention performance. It could be that the centrally administered amphetamine was ineffective in altering memory simply because the amphetamine did not reach the proper place in the brain, at the appropriate concentration, at the critical time. Although this interpretation remains a possibility, it seems unlikely in view of the clear-cut, dose-dependent increase in open field activity following intraventricular administration of amphetamine. These behavioral changes were evident from the first minute of observation and lasted over the half-hour observation period.

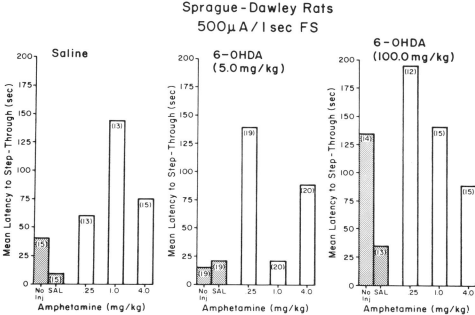

Figure 5.8. Inhibitory avoidance retention latencies of rats receiving either saline, 5 mg/kg, or 100 mg/kg 6-OHDA 24 hr before training and one of several doses of d-amphetamine immediately following training. 1 mg/kg of d-amphetamine in the saline condition (panel furthest to the left) significantly enhanced retention performance. In the case of 5.0 mg/kg 6-OHDA (middle panel) administered 24 hr before training, a .25 mg/kg dose of amphetamine significantly enhanced retention, thus indicating that pretreatment with 6-OHDA shifts the dose response curve to the left. There were no significant differences if 100 mg/kg 6-OHDA was administered 24 hr before training and d-amphetamine immediately after training (panel furthest to the right). (From Martinez et al., 1980.)

We also found that a small dose of 6–OHDA (5 mg/kg), which produces a 75% depletion of peripheral catecholamines, tends to impair acquisition of an inhibitory avoidance response. This effect must be secondary to the amount of depletion produced by 6–OHDA because there was no significant correlation between norepinephrine depletion in heart tissue and retention performance. This conclusion is supported by the fact that a 100 mg/kg dose of 6–OHDA, which produces an even greater depletion of norepinephrine, did not significantly affect acquisition of the response. It is possible, although unlikely, that the difference between the effectiveness of 5 mg/kg and 100 mg/kg 6–OHDA is due to the slight, but nonsignificant reduction in brain norepinephrine concentrations produced by the high dose of 6–OHDA.

The third major finding of these studies is that following pretreatment with 6–OHDA, retention is enhanced by lower posttrial doses of amphetamine. This finding further supports the view that amphetamine influences learning and memory processes by acting on peripheral catecholamines. The dose response shift might result from sensitivity to the agonist action of d-amphetamine (Inness & Nickerson, 1975) and catecholamines released from the adrenal medulla by the amphetamine (Kostrzewa & Jacobowitz, 1974). That is, the agonist action of these amines would be greater following denervation because reuptake and inactivation processes that are normally performed by the presynaptic membranes would have been destroyed by the 6–OHDA. Alternatively, the effect could result from imbalances produced between central and peripheral catecholamine systems, or between the parasympathetic and sympathetic divisions of the autonomic nervous system.

These results provide evidence that the memory modulatory actions of amphetamine are, at least in part, peripherally mediated. There are at least two obvious ways in which this may be accomplished. The first is through agonist action at peripheral end-organ postsynaptic receptors and resultant feedback directly to the brain. The second is through an action of amphetamine or adrenal catecholamines on the end organs themselves. For example, Carlsson et al. (1975) found that in rats a dose of 5.0 mg/kg d-amphetamine increases cerebral blood flow 400% above normal and increases cerebral metabolic rate for oxygen by 40%. Thus, the role of peripheral α- and β-receptors, as well as adrenal catecholamines, remains to be elucidated in the investigation of the facilitatory effect of amphetamine on memory storage processes. In an attempt to address this question more directly, we conducted a series of experiments to study the effects on learning of 4–OH amphetamine, a drug that affects primarily peripheral stores of catecholamines (Martinez et al., 1979).

The 4–OH amphetamine was injected into the rats in a dose range of .21–13.1 mg/kg immediately following training in an inhibitory avoidance task. A 72-hr training-retention testing interval was used with a 500 μA, 1 sec footshock. The retention latencies of the animals receiving 4–OH amphetamine are shown in Figure 5.9. A dose of .82 mg/kg enhanced retention. This finding provides additional evidence that the facilitatory effects of amphetamines are normally mediated through their actions on peripheral systems.

In a further attempt to characterize the action of the peripherally acting 4–OH amphetamine on memory, a study was performed to see whether pretreatment with 6–OHDA would shift the dose response function of 4–OH amphetamine as it does for d-amphetamine. Twenty-four hours prior to training, rats were injected (i.v.) with 100 mg/kg 6–OHDA. Immediately following training in an inhibitory avoidance task, the rats were injected (i.p.) with either saline, .21, .82, or 3.3 mg/kg 4–OH amphetamine. A 750 or 500 μA, 1 sec footshock and a 72-hour training-testing interval were used. With the 500 μA footshock, no dose of 4–OH amphetamine significantly facilitated retention. However, with the 750 μA footshock, a .21

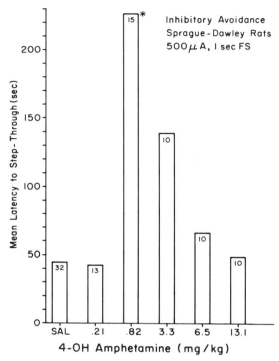

Figure 5.9. *Effects of immediate posttrial intraperitoneal injection of 4-OH amphetamine on retention performance of rats. A .82 mg/kg dose significantly enhanced retention. (From Martinez et al., 1979.)*

mg/kg dose of 4–OH amphetamine significantly enhanced retention. Thus, if the rats were pretreated with 6–OHDA, the dose response curve for 4–OH amphetamine-induced facilitation is shifted to the left just as it is for d-amphetamine. These results are shown in Figure 5.10.

Taken together, the results of studies using 6–OHDA suggest that presynaptic stores of norepinephrine found in the sympathetic system are not necessary for the enhancement of learning produced by either d-amphetamine or 4–OH amphetamine. These data suggest that the other major store of peripheral catecholamines, the adrenal medulla, may be important for the effect of amphetamine on memory. Thus, a series of studies was undertaken to examine the effect on memory of 6–OHDA and 4–OH amphetamine in adrenal medullectomized (ADXM) rats.

A preliminary study was conducted to compare the retention of ADXM rats to that of sham-operated controls and to ADXM rats given 100 mg/kg of 6–OHDA 24 hr prior to training. As before, 500 μA and 750 μA footshocks, and a 72-hr training-testing interval were used. With the 750 μA footshock, the combined treatment of ADXM and pretreatment with 6–OHDA enhanced retention performance. The retention performance of the ADXM animals, compared to the sham operates, was superior but not sig-

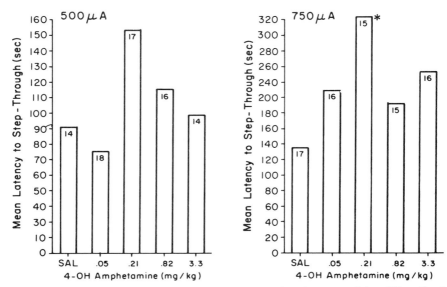

Figure 5.10. Inhibitory avoidance retention latencies of rats receiving 100mg/kg 6-OHDA 24 hr before training and one of several doses of 4-OH amphetamine immediately following training. The rats were trained with either a 500 or 750 μA, 1 sec footshock. In the case of the 750 μA footshock, a .21 mg/kg dose of 4-OH amphetamine was found to enhance significantly retention. Note that the pattern of results was the same for the 500 μA footshock condition. (From Martinez et al., 1979.)

nificantly so. Thus, peripheral catecholamines are not necessary for learning, and their elimination may under some conditions even produce enhanced retention.

In the final series of experiments, we investigated the effects of 4–OH amphetamine (.21–3.3 mg/kg) and d-amphetamine (.25–4 mg/kg) in ADXM animals using a 500 μA footshock. As before, the training-testing interval was 72 hr. The results of these experiments are shown in Table 5.1. None of the individual comparisons approached significance. In addition, ADXM rats pretreated with 6–OHDA (100 mg/kg) also showed no amphetamine-induced facilitation of memory.

These data provide strong evidence that the memory-enhancing effect of amphetamine is mediated through adrenal catecholamines since adrenal demedullation abolishes the memory-enhancing effect of both 4–OH amphetamine and d-amphetamine.

Conclusion

Most research on catecholamines and learning, including our own, has been concerned with understanding the role of brain catecholamines in learning. And, as we have noted, there is much evidence to support the

Table 5.1
Mean Retention Latencies (Mean ± SEM) of Adrenal Demedullated Rats (ADXM) Receiving 4-OH Amphetamine or d-Amphetamine Immediately Following Training in an Inhibitory Avoidance Task

Saline (.9%)	59 ± 30.82
d-amphetamine (mg/kg)	
.25	39 ± 25.67
1.0	92 ± 43.45
4.0	82 ± 33.55
4-OH amphetamine (mg/kg)	
.21	36 ± 11.50
.82	65 ± 28.36
3.3	102 ± 43.33

view that learning is altered by treatments that influence central catecholamines. However, there is also much evidence from recent experiments indicating that learning and memory are influenced by alterations in peripheral catecholamines as well as other peripheral hormones. This evidence is, of course, quite consistent with the general hypothesis that physiological processes associated with affective states modulate memory storage (e.g., Kety, 1970). However, the finding that memory is influenced by experimentally induced alterations in peripheral hormonal systems emphasizes the possibility that learning normally involves the release of peripheral hormones that then influence brain processes involved in memory storage. Such a view does not assume that learning requires such feedback. Rather, it seems that the feedback does occur and that involvement of peripheral hormonal systems may be important for regulating the degree or strength of retention of recent experiences (Gold & McGaugh, 1977). Viewed in this way, the importance of an experience may be defined by its consequences for peripheral hormonal systems.

REFERENCES

Bloch, V. Facts and hypotheses concerning memory consolidation. *Brain Research,* 1970, *24,* 561–575.

Carlsson, C., Hägerdal, M., & Siesjö, B. K. Influence of amphetamine sulphate on cerebral blood flow and metabolism. *Acta Physiologia Scandinavica,* 1975, *94,* 128–129.

Danscher, G., & Fjerdingstad, E. J. Diethyldithiocarbamate (antabuse): Decrease of brain heavy metal staining pattern and improved consolidation of shuttle box avoidance in goldfish. *Brain Research,* 1975, *83,* 143–155.

de Wied, D., & Bohus, B. Modulation of memory processes by neuropeptides of hypothalamic-neurohypophyseal origin. In M. A. B. Brazier (Ed.), *Brain mechanisms in memory and learning* (Vol. 4, IBRO Monograph Series). New York: Raven Press, 1979, pp. 139–149.

Dunn, A. J., & Gispen, W. H. How ACTH acts on the brain. *Biobehavioral Reviews,* 1977, *1,* 15-23.

Flood, J. F., Jarvik, M. E., Bennett, E. L., Orme, A. E., & Rosenzweig, M. R. Effects of ACTH peptide fragments on memory formation. *Pharmacology Biochemistry and Behavior,* 1977, *5,* 41-51.

Gold, P. E., & McGaugh, J. L. Hormones and memory. In L. H. Miller, C. A. Sandman, & A. J. Kastin (Eds.), *Neuropeptide influences on the brain and behavior.* New York: Raven Press, 1977, pp. 127-143.

Gold, P. E., & van Buskirk, R. B. Facilitation of time-dependent memory processes with posttrial epinephrine injections. *Behavioral Biology,* 1975, *13,* 145-153.

Gold, P. E., & van Buskirk, R. B. Effects of posttrial hormone injections on memory processes. *Hormones and Behavior,* 1976a, *7,* 509-517.

Gold, P. E., & van Buskirk, R. B. Enhancement and impairment of memory processes with posttrial injections of adrenocorticotrophic hormone. *Behavioral Biology,* 1976b, *16,* 387-400.

Haycock, J. W., van Buskirk, R., & McGaugh, J. L. Effects of catecholaminergic drugs upon memory storage processes in mice. *Behavioral Biology,* 1977a, *20,* 281-310.

Haycock, J. W., van Buskirk, R., Ryan, J. R., & McGaugh, J. L. Enhancement of retention with centrally administered catecholamines. *Experimental Neurology,* 1977b, *54,* 199-208.

Innes, I. R., & Nickerson, M. Norepinephrine, epinephrine, and the sympathomimetic amines. In L. S. Goodman & A. Gilman (Eds.), *The pharmacological basis of therapeutics.* New York: Macmillan, 1975, 477-513.

Jensen, R. A., Martinez, Jr., J. L., Vasquez, B., McGaugh, J. L., McGuiness, T., Marrujo, D., & Herness, S. Amnesia produced by intraventricular administration of diethyldithiocarbamate. *Neuroscience Abstracts,* 1977, *3,* 235.

Kety, S. S. The biogenic amines in the central nervous system: Their possible roles in arousal, emotion, and learning. In F. O. Schmitt (Ed.), *The neurosciences.* New York: Rockefeller Univ. Press, 1970.

Kostrzewa, R. M., & Jacobowitz, D. M. Pharmacological actions of 6-hydroxydopamine. *Pharmacological Reviews,* 1974, *26,* 200-287.

Martinez, Jr., J. L., Jensen, R. A., Vasquez, B. J., Lacob, J. S., McGaugh, J. L., & Purdy, R. E. Acquisition deficits induced by sodium nitrite in rats and mice. *Psychopharmacology,* 1979, *60,* 221-228.

Martinez, Jr., J. L., Jensen, R. A., Messing, R. B., Vasquez, B. J., Soumireu-Mourat, B., Geddes, D., Liang, K. C., & McGaugh, J. L. Central and peripheral actions of amphetamine on memory. *Brain Research,* 1980, *182,* 157-166.

Martinez, Jr., J. L., Vasquez, B. J., Jensen, R. A., Messing, R. B., Rigter, H., Liang, K. C., & McGaugh, J. L. Adrenal medullary catecholamines are necessary for amphetamine-induced enhancement of learning in rats. *Neuroscience Abstracts,* 1979, *5,* 320.

McGaugh, J. L. Drug facilitation of learning and memory. *Annual Review of Pharmacology,* 1973, *13,* 229-241.

McGaugh, J. L., & Herz, M. J. *Memory consolidation.* San Francisco: Albion, 1972.

McGaugh, J. L., Gold, P. E., Handwerker, M. J., Jensen, R. A., Martinez, J. L., Meligeni, J. A., & Vasquez, B. J. Altering memory by electrical and chemical stimulation of the brain. In M. A. B. Brazier (Ed.), *Brain mechanisms in memory and learning* (Vol. 4, IBRO Monograph Series). New York: Raven Press, 1979, pp. 151-164.

Meligeni, J. A., Ledergerber, S. A., & McGaugh, J. L. Norepinephrine attenuation of amnesia produced by diethyldithiocarbamate. *Brain Research,* 1978, *149,* 155-164.

Randt, C. T., Quartermain, D., Goldstein, M., Amagnoste, B. Norepinephrine biosynthesis inhibition: Effects on memory in mice. *Science,* 1971, *172,* 498-499.

Spanis, C. W., Haycock, J. W., Handwerker, M. J., Rose, R. P., & McGaugh, J. L. Impairment of retention of avoidance responses in rats by posttraining diethyldithiocarbamate. *Psychopharmacology*, 1977, *53*, 213–215.

Stein, L., Belluzzi, J. D., & Wise, C. D. Memory enhancement by central administration of norepinephrine. *Brain Research*, 1975, *84*, 329–335.

Telegdy, G., & Kovács, G. L. Role of monoamines in mediating the action of hormones on learning and memory. In M. A. B. Brazier (Ed.), *Brain mechanisms in memory and learning* (Vol. 4, IBRO Monograph Series). New York: Raven Press, 1979, pp. 249–268.

E. R. JOHN

A Neurophysiological Model of Purposive Behavior

6

Goal-directed behavior is teleologically purposive. It often seems to be a search for a goal previously defined by a model or *idea* in the brain (Granit, 1977). Exactly *how* a goal is achieved, however, can vary. There are two approaches to account for purposive behavior. One is the cybernetic approach, which views behavior as homeostatic and largely reflexive (Wiener, 1961). According to this model, an organism is endowed with innate patterns of behavior explained as reflexes triggered by the stimulus or as the reduction of drives. Numerous observations have established the great power of this approach to account for many complex as well as simple behaviors in humans and other mammals, as well as in insects, fish, and birds.

As we ascend the phylogenetic scale, the cybernetic approach becomes unsatisfactory. Behaviors emerge that cannot be explained plausibly as innate or conditioned reflexes. For example, responses learned to a specified stimulus can be elicited by generalization to a novel stimulus that activates very different afferent pathways; learned responses can be executed by using muscles that achieve the desired purpose but which were never before used for that behavior; animals and humans can learn new skills by watching the behavior of another individual.

As one tries to understand such behaviors, another approach must be considered that assumes that higher animals possess consciousness, have ideas, and can think about the significance of information from the environment. The newborn individual can survive only by the action of species characteristic reflexes and homeostatic processes. Initially, these invariant processes are probably relatively well localized in cortical and lower brain regions that have especially high signal-to-noise ratios for specific functions. As the individual develops, other mechanisms may serve to distribute information to additional regions of the brain. The signal-to-noise ratio for such information is lower in these newly responsive regions than in more committed regions. In this manner, multisensory and multivariate transactions begin to modulate genetically specified processes that were initially more simply determined and a cognitive model of the environment is built gradually, incorporating features of individual experience as well as species characteristic features.

One can view much behavior as resulting from a cognitive process, which involves an interaction between neural events representing the previous experience, the present state of the individual, and the occurrence of particular features in the environment. Such behavior consists of the attempt to match new experience against an idea reflecting past experience. It is cognitive rather than reflective, involving thinking rather than activation of specific neural pathways constituting stimulus-response circuits. This chapter is intended to provide a tentative neurophysiological model for such behavior and to make more explicit some of the issues that must be confronted.

The most fundamental problems in building such a model are what types of information are essential for goal-directed behavior and where and how are such information represented in the brain. At least three kinds of information can be identified: (*a*) information about the presence of a goal in the environment, (*b*) information about behaviors that might attain that goal, and (*c*) information constituting the idea to seek some previously defined goal.

Information about the presence of a goal in the environment must be made available by afferent sensory input. How is such information represented? Much evidence shows that the activity of single cells in "sensory-specific" structures, especially in the striate cortex, is maximum when stimuli are presented with particular attributes. Many workers have interpreted such findings to mean that single neurons are "feature detectors" that decompose events in the visual world into a limited variety of perceptual elements. A hierarchical system of neurons represents individual percepts by combining the reports from detectors of the separate unitary features of each percept.

There are serious problems with both the logical conclusions and the experimental basis of these ideas (John, 1972; John & Schwartz, 1978; Thatcher & John, 1977). The logical problem was recognized early by Sherrington (1906), who introduced the notion of "one ultimate pontifical neuron . . . the climax of the whole system of integration," and promptly abandoned this concept in favor of mind as a "million fold democracy whose each unit is a cell." Other serious problems arise when one tries to specify the sensory attributes that will trigger firing of a single cell. Study of a single "feature extractor" cell reveals that it sometimes fires "spontaneously" in the absence of the supposed trigger feature, that it displays an extremely variable response to presentations of this feature, and that many other stimulus features elicit increased rates of discharge (Fishman & Michael, 1973; Hoeppner, 1974; Horn & Hill, 1969; Hubel & Wiesel, 1962; Morrell, 1972; Rose, 1974).

How then is it logically possible from the firing rate of any single neuron to determine the presence in the visual field of a high contrast edge at one orientation, a lower contrast edge at a more favored orientation, a short edge moving fast, a long edge moving more slowly but in a more favored direction, or influences from the auditory, oculomotor, or vestibular systems? Trigger features for which a cell will increase its firing rate are readily found; the converse operation, inferring the presence of unique features in the environment from the fact that a given cell is firing at a particular rate, seems logically impossible.

Using movable microelectrodes chronically implanted into unrestrained behaving cats, we found that single cells in the lateral geniculate body and other regions display highly variable responses to single conditioned visual cues (John & Morgades, 1969a,b). However, long-term, average post-stimulus histograms (PSHs) of different single cells often converged to the same temporal pattern. Firing patterns of single cells isolated from multiple unit responses by spike height discrimination, and averaged over very large numbers of stimulus presentations, often converged to the same PSH waveshape as the firing patterns of small groups of cells averaged over a relatively small number of stimuli.

In differentially trained cats, PSHs from single or multiple units and average evoked responses (AERs) were simultaneously recorded from the same movable microelectrode during multiple trials requiring discrimination between visual cues. After an adequate sample of data was obtained over several days at that position, the microelectrode was advanced slightly and the same data were gathered from the neural ensembles at the new position.

As seen in Figure 6.1, which shows the PSH and AER recorded at successively deeper positions along an electrode traverse through the lateral geniculate body, these different neural groups displayed extremely similar temporal patterns of firing to a conditioned visual stimulus. At each position, the two discriminated visual cues caused different average firing patterns. From such data it was possible to construct gradients of each PSH or AER component along the microelectrode tracks. In trained animals, such gradients were flat across relatively large anatomical domains, showing that a characteristic average temporal firing pattern was diffusely

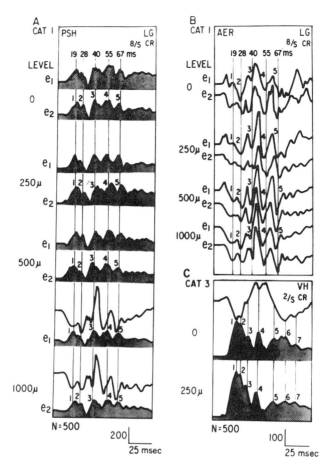

Figure 6.1. A. Poststimulus time histograms (PSH) from two electrodes (e_1 and e_2) separated by 125 μ, recorded at four different levels spanning 1000 μ. These are averages of responses to 500 stimuli and were each computed during five or six trials, resulting in correct performance (CR) to the 8 Hz CS. The five guide lines show the latency of five peaks common to all of these responses. The average evoked responses (AER) simultaneously obtained from e_1 and e_2 are shown as solid curves above the shaded PSH at the fourth level. B. AER obtained from e_1 and e_2 simultaneously with the PSH shown in A. Guide lines are at the same latency. Note both the general similarity and the small differences between records from adjacent regions. The details of these waveshapes were well reproducible over periods of weeks. C. PSH recorded from two levels separated by 250 μ in the ventral hippocampus of a different cat. These averages were computed during 16 trials, resulting in CR to the 2 Hz CS (N-500). Guide lines show the peaks considered similar at the two levels. Note the correspondence between the latency of components in the AER and peaks in the PSH at level 0. The evoked potential waveshape at 250 μ (not shown) closely resembled that at level 0. (Data from John and Morgades, 1969b.)

distributed across these large neural ensembles. In untrained animals similar procedures produced steep gradients, revealing far greater heterogeneity of firing patterns in different neural groups.

Computation of the grand average PSH and AER across all positions, the variance of responses to the two signals within each position, and the variance of responses to each signal between many positions generated the results shown in Figure 6.2. The variance of responses to the same signal between different cell groups across the ensemble was smaller than the variance in responses to two different signals within a single position. The information about a sensory signal seems to be encoded by a statistically invariant spatiotemporal pattern of departure of the ensemble from random firing, rather than by the variable firing pattern of any neuron in the ensemble. I will refer to this as the *ensemble information,* or EI.

The next problem is where information about the goal is represented.

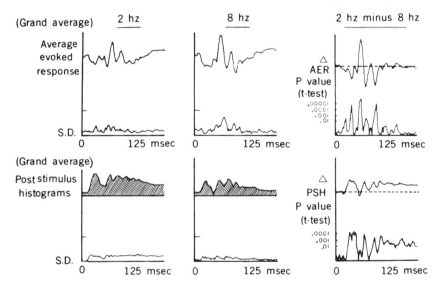

Figure 6.2. Top left: *The top curve shows the grand average of the AERs elicited by the 2 Hz CS across all electrode positions in the mapped region, whereas the lower curve shows the standard deviation (S.D.) of the groups of AERs.* Bottom left: *The top curve shows the grand average of the PSHs elicited by the 2 Hz CS across the same electrode positions and the lower curve shows the S.D.* Top center: *This shows the grand average of the AERs elicited by the 8 Hz CS and the corresponding S.D.* Bottom center: *This shows the grand average PSH elicited by the 8 Hz CS and its S.D.* Top right: *The top curve shows the difference waveshape resulting from the subtraction of the grand average AER elicited by the 8 Hz CS from the grand average AER elicited by the 2 Hz CS. The lower curve shows the p value as computed by the t-test for each point of difference wave.* Bottom right: *The top curve shows the difference waveshape resulting from the subtraction of the grand average PSH elicited by the 8 Hz CS from the grand average PSH elicited by the 2 Hz CS. The lower curve shows the p value for each point of the difference. (From John, 1972.)*

Much evidence shows that information (and function) must be distributed across extensive anatomical domains. Many studies of single and multiple stage lesions, which compare animals with the same residual brain tissue but different temporal sequences of lesion, reveal an impressive capability of the brain to reorganize information processing after localized damage (Stein, Rosen, & Butters, 1974). For example, rats that were allowed to explore actively a patterned visual environment between successive lesions of left and right visual cortices retained visually guided behavior after the second lesion, but rats that were passively moved through the same environment between the two lesions became functionally blind (Thatcher & John, 1977). One must wonder why such visually guided motor behavior between the two lesions was essential for this functional reorganization. Other recent evidence also indicates that in some mammals brightness and patterns of visual stimuli can be discriminated in the absence of the striate cortex (Doty, 1973; Pasik & Pasik, 1973a, 1973b; Sprague, Berlucchi, & Rizzolatti, 1973; Weiskrantz, 1972), thus showing that multiple brain regions possess the relevant information.

We have obtained electrophysiological evidence that the representation of information about a sensory cue becomes more extensive as a result of learning. This is illustrated in Figure 6.3, which shows bipolar AERs recorded simultaneously from a variety of brain structures in a performing cat learning successively more discrete meanings for a visual signal. AER waveshapes change as a cat trained to lever press for food perceives a meaningless flicker, begins to lever press only when the contingent flicker cue is present, learns to differentiate between one flicker frequency for food and a different frequency signaling shock avoidance, and finally achieves almost automatic response after overtraining in differentiation. The arrows in the figure point to AER components of endogenous origin, which will be discussed later.

During learning, initial relative localization of the most vigorous response to the classically defined visual system is replaced by much more anatomically extensive and similar responses and new, long latency components appear. I believe that this spread is the result of contiguity of neural activity in different brain regions, due to the action of what Ukhtomski called the "dominant focus" (Ukhtomski, 1945). A wide variety of studies, reviewed elsewhere (John, 1967), show that when brain regions are repeatedly active at the same time they become functionally associated into what I have called a *representational system* (RS). Subsequently, appropriate activation of one of the anatomical regions in the RS will result in activation of the other neural ensembles in the RS.

Similar temporal patterns of discharge in different anatomical regions are seen in Figure 6.4, which shows the identical latency of AER components and some similar PSH component latencies in the left lateral geniculate body and the right dorsal hippocampus of a trained cat (Livanov

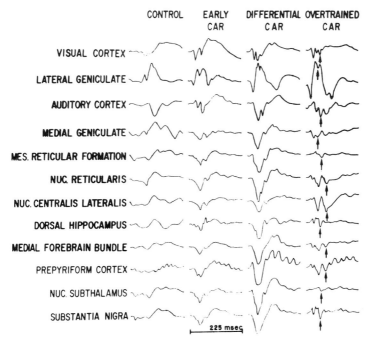

Figure 6.3. Evolution of visual evoked responses. Control: *These are average responses evoked in different brain regions of a naive cat by presentation of a novel flicker stimulus. Several regions show little or no response, and different regions display differing types of response.* Early CAR: *Responses to the same stimulus shortly after elaboration of a simple conditioned avoidance response (CAR). A definite response with similar features can now be discerned in most regions.* Differential CAR: *Changes in the response evoked by the flicker CS shortly after establishment of differential approach-avoidance responses to flicker at two different frequencies. As usual, discrimination training has greatly enhanced the response amplitude, and the similarity between responses in different structures has become more marked.* Overtrained CAR: *After many months of overtraining on the differentiation task, the waveshapes undergo further changes. The arrows point to a component usually absent or markedly smaller in behavioral trials on which this animal failed to perform.* (Nuc. Retic., nucleus reticularis; MFB, median forebrain bundle; Prep. Cx., prepyiform cortex; Nuc. Subthal., nucleus subthalamus; Subst. N., substantia nigra.) *(From John, Science, 1972.)*

& Poliakov, 1945). The response patterns of neural elements at different positions in these structures remained essentially constant as the microelectrode was advanced.

Figure 6.4 also shows that part of this observed pattern of neural activity was endogenous, and that the correspondence in the activity of the two regions was not inherent but depended upon their participation in the same RS. Both the AER and the PSH from these two regions are displayed. When the animal responded correctly, as shown on the left of the figure, the lateral geniculate and dorsal hippocampus showed maxima at the same

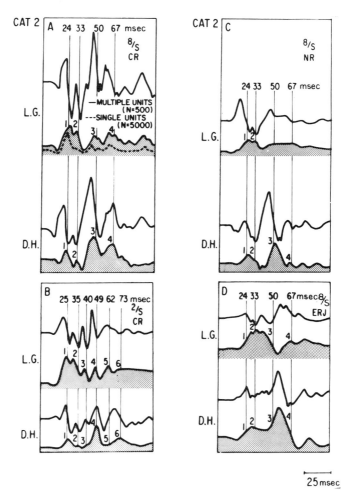

Figure 6.4. (A) AERs (solid curves) and PSHs (shaded areas) simultaneously recorded from microelectrodes in the lateral geniculate (L.G.) body on the left side and the dorsal hippocampus (D.H.) on the right side during correct performance (CR) to the 8 Hz CS by cat 2. Numbered vertical lines indicate components considered to correspond with respect to relative latency. These and all other responses illustrated in this figure are computed from 500 stimulus presentations, except for the PSH derived from a single unit in LG, shown as a dotted line (N-5000). (Note the correspondence between the curve describing the probability of firing of this single neuron observed over a long period of time and the PSH for the neural ensemble observed for one-tenth that time.) (B) AERs and PSHs simultaneously recorded from L.G. and D.H. during CR to the differential 2 Hz CS. (C) AERs and PSHs simultaneously recorded from L.G. and D.H. during presentations of the 8 Hz CS that resulted in no behavioral performance (NR). (D) AERs and PSHs simultaneously recorded from L.G. and D.H. during presentation of a novel stimulus illuminated by the 8 cps flicker. (From John and Morgades, 1969b.)

latencies. When the animal committed an error, as seen in the upper right, the firing patterns in the two structures became obviously different. When a novel stimulus was presented, as seen in the lower right, the two regions displayed radically disparate firing patterns (John & Morgades, 1969a,b).

These data illustrate the spread of information into more extensive neural regions during learning. Different regions adopt similar temporal patterns of discharge as a result of incorporation into an RS. I believe information about a sensory stimulus is represented initially in the brain regions of the corresponding sensory system. As stimuli in different sensory modalities coincide, as movements are organized in response to those stimuli, and as positive and negative consequences to those movements are experienced, the neural representation of the stimuli expands, and anatomically extensive RSs develop.

The initial localization of function is thus replaced by multivariate participation of many regions in the representation of information. Typically, an RS will include regions that represent information about such modality of sensory input, describing features of goal stimuli, central states reflecting arousal level, emotional valence and drive levels, behavioral responses, and their outcomes. In each region this information is represented by the spatiotemporal patterns of departure from random or baseline activity in large ensembles of neurons. This is the EI. Because each active region constitutes a dominant focus, the characteristic EI of each is propagated to every other one. These reciprocal transactions enhance and preserve those features of the local patterns that can resonate between regions, whereas other features are damped out. The result is the emergence of a *common mode* of EI, an anatomically extensive and characteristic activity pattern that can reverberate throughout the various anatomical regions belonging to that RS for the period of time required for consolidation of memory to occur. The activity of any individual neuron is important only insofar as it contributes to the local EI. The same neuronal ensembles can participate in numerous RSs, each with its characteristic EI pattern. Chemical changes that occur during the consolidation period store each common mode EI as an increased probability for neurons in distributed ensembles to fire with the EI pattern characteristic of each RS.

Firing of any substantial portion of an RS in the EI pattern characteristic of activity during the prior experience propagates to all other portions, causing resonance that results in activation of the whole RS, releasing the common mode of EI that was consolidated. Thus, the anatomically extensive neural ensembles representing a whole memory can be activated by sight, sounds, smells, moods, needs, or movements that occurred during the earlier experience. Although no individual neuron necessarily repeats its previous firing pattern, the released EI is a statistical facsimile of the EI during the actual experience. This EI is a model of the absent event, and can be called the *idea* of that event.

An idea may generate expectancies or may initiate a search of the environment for events that match the model produced by the RS. Among the electrophysiological correlates of such processes in humans are the contingent negative variation that appears when an event is expected, the late positive components such as P300 related to uncertainty about an event, the even later positive event at about 450 ms related to delayed matching from sample, the positive waves emitted when an expected event fails to occur, and the positive wave that appears when an anticipated target is detected (Cooper, McCallum, Newton, Papakostopoulos, Pocock, & Warren, 1977; Klinke, Fruhstorfer, & Finkenzeller, 1968; Sutton, Braren, Zubin, & John, 1965; Thatcher, 1976; Walter, Cooper, McCallum, & Cohen, 1965; Weinberg, Walter, & Crow, 1970). Characteristically, these phenomena are large in amplitude, long in duration and latency, and anatomically widespread. These features seem to implicate the nonsensory specific system as an important component of RSs. With respect to the long latency of these processes, it is of interest to note that conscious awareness of direct electrical stimulation of the brain takes almost half a second (Libet, 1966).

Recurrence of events in the environment similar to a prior experience produces two kinds of effects. Part of the resulting EI is *exogenous,* caused directly by afferent information about the environment. Part of the EI which ensues is *endogenous,* resulting from the appropriate or inappropriate activation of a representational system.

Electrophysiological evidence for released facsimiles of previous EI was first presented by Livanov and Poliakov (1945), who described "assimilation of the rhythm" (Cooper *et al.,* 1977; Klinke *et al.,* 1968; Sutton *et al.,* 1965; Walter *et al.,* 1965; Weinberg *et al.,* 1970). The EEG of animals being trained with a CS that had a characteristic repetition rate became dominated by waves at the frequency of the absent stimulus during the intertrial interval. Assimilation, which has been observed by numerous workers using a wide variety of experimental animals and procedures, appears when the animal enters the training situation but is absent in the home cage, and is often accompanied by behavioral rehearsal (John, 1961; Thatcher & John, 1977). Assimilated activity has also been observed in unit activity recorded from behaving animals (Ramos & Schwartz, 1976).

Released EI facsimiles have been obtained when animals generalize to a novel stimulus, performing behavior learned to other cues, or when differentially trained animals commit errors. Under these conditions, some brain elements often display activity as if the absent CS for the performed behavior were present. EI facsimile release during generalization or errors has been seen in studies of the EEG, the evoked potential or unit activity (John & Killam, 1960; John & Morgades, 1969a; Majkowski, 1958; Thatcher & John, 1977).

The AERs elicited in various brain regions when a novel test stimulus

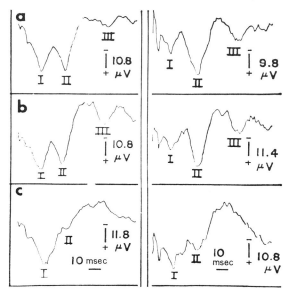

Figure 6.5. Computations of average responses obtained from the lateral geniculate nucleus and nucleus reticularis of the cat under various conditions during the same experimental session. First row of averages (a) is based upon 100 repetitions, and second (b) and third (c) rows are based upon 42 repetitions of the same stimulus applied during a number of behavioral trials. Analysis epoch was 90 msec. a. Average responses evoked in structures by the 10 Hz CS (flicker) actually used in training during repeated correct behavioral performances. b. Average responses evoked by a novel 7.7 Hz CS during repeated generalization behavior. Test trials with the 7.7 Hz stimulus were interspersed among trials with the actual 10 Hz CS, and were never reinforced. c. Average responses evoked by the 7.7 Hz flicker on presentations when no generalization behavior was elicited. The waveshape elicited by the actual CS is similar to the response evoked by the novel stimulus during generalization behavior. Notice the absence of the second positive component in the evoked potential when generalization behavior failed to occur. (From Ruchkin and John, 1966.)

fails to elicit a CR have been compared with the AERs recorded when the same neutral stimulus elicited behavior as if the CS were present and with the AERs to the actual CS. The top row of waves in Figure 6.5 shows AERs usually elicited in the lateral geniculate and nucleus reticularis by a 10/S flicker CS. The bottom row of waves shows AERs elicited by a 7/S test flicker when no CR occurred. The middle row shows the response to the 7/S test stimulus when it elicits the CR appropriate to the 10/S CS. The facsimile released by the test stimulus during generalization is quite accurate (Bartlett, John, Shimokochi, & Kleinman, 1975; John, 1972; John, Bartlett, Shimokochi, & Kleinman, 1973, 1975; Ruchkin & John, 1966). Subtraction of the AERs when generalization fails to occur from AERs when generalization takes place yields a difference wave that has a similar time course and latency in many brain regions (John, Ruchkin, Leiman, Sachs, & Ahn, 1965; John, 1967). An example of this released common mode EI is shown in Figure 6.6. Note that the released facsimile appears earliest in a cortical-reticular system and then appears to be sent centrifugally to the

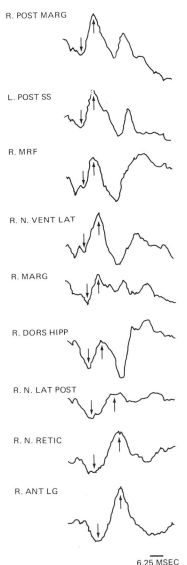

Figure 6.6. Difference waveshapes constructed by subtraction of averaged responses evoked by 7.7 Hz test stimulus during trials resulting in no behavioral performance from average responses evoked by the same stimulus when generalization occurred. Each of the original averages was based on 200 evoked potentials providing a sample from five behavioral trials. Analysis epoch was 62.5 msec. These difference waveshapes begin 10 msec after the stimulus. The onset and maximum of the difference wave have been marked by two arrows on each waveshape. The structures have been arranged from top to bottom in rank order with respect to latency of the difference wave. Note that the latency and shape of the initial component of the difference wave are extremely similar in the first four structures, and then appear progressively later in the remaining regions. (Post Marg., posterior marginal gyrus; Post SS, posterior suprasylvian gyrus; MRF, mesencephalic reticular formation; N. Vent. Lat., nucleus ventralis lateralis; Marg., marginal gyrus; Dors. Hipp, dorsal hippocampus; N. Lat. Post., nucleus lateralis posterior; N. Retic., nucleus reticularis; Ant. Lg., anterior lateral geniculate; R, right side; L, left side.) (From John, Mechanisms of Memory, Academic Press, New York, 1967. Reproduced by permission.)

lateral geniculate body. Such facsimiles are released only on the trained side of split brain cats (Majkowski, 1967).

The clearest evidence that RSs can produce precise EI facsimiles of different absent events comes during differential generalization (John, Shimokochi, & Bartlett, 1969). In such studies, animals are differentially trained to perform two different CRs to discriminated visual or auditory cues at two different repetition rates. After overtraining to automatic behavior, occasional trials with a neutral test stimulus midway between the frequencies of CS_1 and CS_2 are interspersed in a random sequence of those two stimuli.

The animal sometimes performs one and sometimes the other CR to the test stimulus. Since the test stimulus is always physically identical, differences in AERs when it is interpreted in two different ways must be due to endogenous processes. Figure 6.7 shows the results of differential generalization in 14 cats. In each set of 4 AERs, the top waveshape was elicited by a flicker at one frequency (V_1), which was the cue to press the left lever on a work panel to get food (CR_1). On the bottom is the AER elicited by a flicker at a second frequency (V_2), which was the cue to press the right lever to get food or to avoid shock (CR_2). Both the second and third waveshapes were elicited by a test flicker at a third frequency (V_3) midway between V_1 and V_2. The second AER was averaged from trials in which the left lever was pressed in response to the test stimulus (V_3CR_1), whereas the third AER was obtained when the same stimulus resulted in pressing of the lever on the right side (V_3CR_2). V_3CR_1 and V_3CR_2 are significantly different. More important, V_3CR_1 closely resembles V_1CR_1, whereas V_3CR_2 closely resembles V_2CR_2. The numbers to the right of each set of waves show the correlation coefficients between the indicated pairs of AERs. The broken lines between the V_3CR_1 and V_3CR_2 waveshapes are at the latencies at which the t-test between the 2 AERs reached the $P = .01$ level.

In differential generalization trials, the sequence of flashes elicited a variety of individual EP waveshapes within each behavioral trial. These waveshapes could be classified into *modes* highly predictive of subsequent behavioral outcomes, using computer pattern recognition methods (Bartlett et al., 1975; John, 1972; John et al., 1973; John et al., 1975). A striking feature of these modes was that many or all of the aspects of fine structure were faithfully repeated in the waveshapes elicited by individual EPs. Single EPs to test stimuli were classified into the same modes as single EPs from conditioned stimuli. This further evidence that a representational system capable of producing a detailed facsimile of previous experience in a single EP has been established. A wide variety of controls permits nonspecific factors to be excluded as the origin of these released facsimiles of responses to absent events (Bartlett et al., 1975; John, 1972, John et al., 1973; John et al., 1975). Abundant evidence from scalp recorded AER's shows

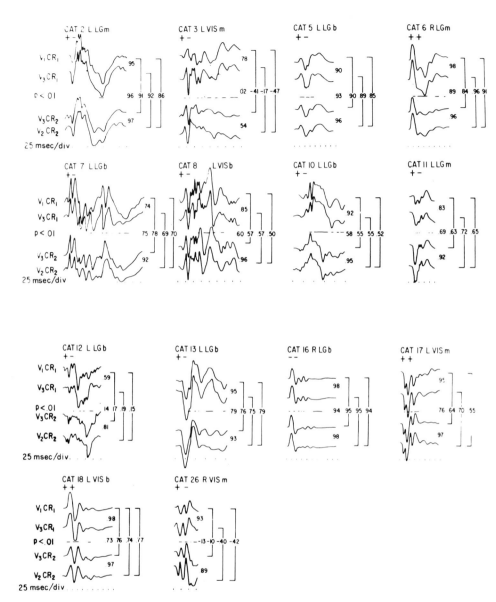

Figure 6.7. Examples of similar waveshapes evoked by different visual stimuli when they elicit the same behavioral response from 14 different trained cats. In each set of four waveshapes the top curve shows the waveshap elicited by flicker stimulus V_1 during trials resulting in correct performance of CR_1, and the bottom curve shows the response elicited by flicker stimulus V_2 during trials resulting in correct performance of CR_2. The second and third curves were elicited during differential generalization to flicker at a frequency, V_3, midway between V_1 and V_2. The second curve shows the evoked response waveshape during trials when V_3 elicited performance of CR_1, appropriate to V_1, and the third curve shows the waveshape evoked by V_3 during trials resulting in CR_2, appropriate to V_2. The intermittent line between the second and third curves in each set shows latencies at

analogous endogenous processes in human subjects (Begleiter, Porjesz, Yerre, & Kissin, 1973; Brown, Marsh, & Smith, 1973; Clynes, Kohn, & Gradijan, 1967; Grinberg & John, 1974; Herrington & Schneidau, 1968; John, Herrington, & Sutton, 1967; Johnston & Chesney, 1974; Picton, Hillyard & Galambos, 1973; Sutton, Tueting, Zubin, & John, 1967; Tepas, Guiteras & Klingaman, 1974; Weinberg, Grey-Walter, Cooper, & Aldridge, 1974; Weinberg, Grey-Walter, & Crow, 1970). Ideas seem to influence the occurrence and nature of these released potentials.

We postulated that each EP contained an *exogenous* component reflecting the sensory environment and an *endogenous* component reflecting activation of an RS related to interpretation of the afferent input. We constructed simultaneous equations that allowed isolation of terms representing the postulated exogenous and endogenous processes. Corresponding computer operations on AERs elicited by V_1, V_2, or V_3 in trials with CR_1 or CR_2 outcomes yielded residuals with similar waveshapes from independent combinations of AERs. Next, we computed the portion of the variance represented by the exogenous and endogenous components of AERs from each of many brain regions. Figure 6.8 shows the amount of exogenous activity in the AER plotted on the horizontal axis versus the logarithm of the amount of endogenous activity plotted on the horizontal axis versus the logarithm of the amount of endogenous activity plotted on the vertical axis. Each point represents data from two to eight structures in that anatomical system, averaged across 4 to 18 cats per structure, with 37 to 305 residuals per point. Data from visual cues and auditory cues are plotted as open or closed circles (Bartlett & John, 1973).

Figure 6.7. *(continued)*
which the difference between data averaged in V_3CR_1 and in V_3CR_2 reached significance at better than the .01 level, assessed by the t-test. Numbers to the right of each set of curves represent the Pearson product moment correlation coefficients between the bracketed curves. Inspection of data shows that in every case, V_3CR_1 closely resembles V_1CR_1, whereas V_3CR_2 closely resembles V_2CR_2. In most cases this visual evaluation is corroborated by the values of the correlation coefficients. In some cases (e.g., cats 2, 5, 7) the correlation between V_3CR_1 and V_3CR_2 is higher than the correlation between V_3CR_1 and V_1CR_1, thus contradicting the impression given by visual evaluation of similarity. These examples illustrate the inadequacy of the correlation coefficients as a pattern-recognition procedure. The EPs combined into these averaged waveshapes were selected using subjective evaluation of the experimenter in half the cases (cats 2, 3, 7, 8, 11, 12, 26) and using computer-sorting methods in the other half (cats 5, 6, 10, 13, 16, 17, 18). Sample sizes ranged from 6–30 EPs for the subjectively selected samples and from 25–200 for the computer-sorted data, and were composed from several behavioral trials in most cases. This variability was dictated by choices made by the animal in differential generalization, a factor beyond control of the experimenter. The cat number, recording derivation, and type of discrimination performed are indicated above each set of data. The signs + − signify approach-avoidance; + +, approach-approach; − −, avoidance-avoidance; subscript m donates monopolar and b, bipolar derivation. Time scale is 25 msec/division. *(From John et al., 1973).*

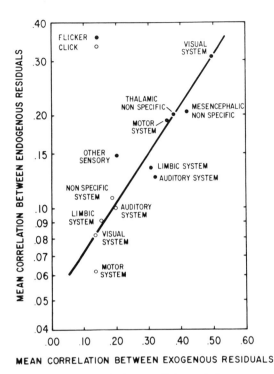

Figure 6.8. Plot of mean correlation coefficients between exogenous residuals for different neural systems and for different cue modalities. Closed circles: *Flicker frequencies as stimuli.* Auditory system: N = 305; Aud. Cx. (16 cats), Med. Genic. (16), Brach. Inf. Coll. (1). Limbic system: N = 303; Hippocampus (16), Dentate (5), Cingulate (5), Septum (5), Prepyriform (6), Med. Forebrain Bundle (6), Mamm. Bodies (5), Hypothalamus (7). Mesencephalic nonspecific: N = 158; Retic. Form. (18), Cent. Gray (1), Cent. Teg. Tract (1). Motor system: N − 146; Motor Cx. (4), Subs. Nigra (10), Nuc. Ruber (4), Nuc. Vent. Ant. (9), Subthal. (5). Other sensory: N = 54; Sensorimotor Cx. (4), Nuc. Post. Lat. (1), Nuc. Vent. Post. Lat. (5), Nuc. Vent. Post. Med. (1). Thalamic nonspecific: N = 139; Cent. Lat. (13), Nuc. Retic. (6), Nuc. Reuniens (1), Med. Dors. (5), Pulvinar (1). Visual system: N = 394; Visual Cx. (18), Lat. Genic. (18), Sup. Coll. (2). Open circles: *Click frequencies as stimuli.* Auditory system: N = 48; Aud. Cx. (5 cats), Med. Genic. (5). Limbic system: N = 69; hippocampus (5), Dentate (3), Cingulate (3), Septum (3), Prepyriform (2), Med. Forebrain Bundle (3), Mamm. Bodies (3), Hypothalamus (2). Motor system: N = 37; Motor Cx. (1), Subs. Nigra (4), Nuc. Ruber (1), Nuc. Vent. Ant. (5), Subthal. Nonspecific system: N = 50; Mesen. Retic. Form. (6), Cent. Gray (1), Cent. Teg. Tract (1), Cent. Lat. (3), Nuc. Retic. (3), Visual system: N = 55; Visual Cx. (6), Lat. Genic. (6), Sup. Coll. (1). Data from monopolar and bipolar derivations were combined. Replications varied across cats and structures. (From Bartlett et al., 1975).

6. A Neurophysiological Model of Purposive Behavior

Endogenous activity is logarithmically proportional to exogenous activity in any region. Although all regions studied responded to the sensory signals, the signal-to-noise ratio varied greatly between regions. Structures in the sensory system corresponding to the CS modality had the highest signal-to-noise ratio. Endogenous activity reflecting participation in an RS was also widely distributed throughout the brain. The lawful logarithmic relationship between the two kinds of processes in every region should come as no surprise. The greater the regional EI about an event, the more likely it seems that neurons in that region will undergo consolidation and be recruited into the RS for that event.

Using chronically implanted movable microelectrodes, we studied single units and small groups of multiple units during differential generalization. Two types of neurons were found close together in several brain regions, as shown in Figure 6.9. The activity of "stable" neurons was statistically exogenous, with PSHs showing no significant difference during CR_1 or CR_2, independent of how the signal was interpreted. The activity of "plastic"

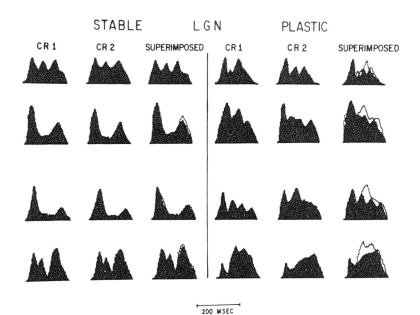

Figure 6.9. Simultaneously recorded stable and plastic cells from the lateral geniculate nucleus (LGN) of four different cats. All PSHs on the same row come from a single cat. The PSHs in the first column were obtained during trials resulting in V_3CR_1, and those in the second column came from V_3CR_2 trials. The PSHs in the first and second column are superimposed in the third column. Note that stable cells display essentially identical PSHs to the same stimulus no matter what behavior ensues. The PSHs in the fourth column were obtained from plastic cells recorded during V_3CR_2 trials at the same time as the stable cells in column 6. Note that plastic cells display significantly different PSHs in trials resulting in different behavioral outcomes, although the sensory stimuli are identical. (From Ramos, Schwartz, and John, 1976c.)

neurons was statistically endogenous, with PSHs showing significant differences to the test signal, which were predictive of the subsequent behavioral outcome, independent of which signal was present (Ramos et al., 1976a,b; Schwartz et al., 1976). Representational systems must depend heavily upon plastic cells to generate facsimiles. Figure 6.10 shows statistical facsimile generation by plastic cells, demonstrated by similar PSHs from trials with similar outcomes to different stimuli.

Our belief that information about an experience is stored statistically as an EI pattern in anatomically extensive ensembles has received more direct support. Different brain regions were electrically stimulated in cats overtrained to discriminate different repetition rates of auditory as well as visual signals (John & Kleinman, 1975; Kleinman & John, 1975). We used high frequency pulse trains modulated by low frequencies to simulate EI patterns. These cats immediately performed differential CRs when a variety of brain regions was electrically stimulated. Greatest accuracy of discrimination was found with direct stimulation of the mesencephalic reticular formation (MRF). MRF stimulation could also be "titrated" against the visual or auditory CSs and could usually preempt control over the animal's behavior.

If a temporal pattern of electrical stimulation was divided so that two different brain regions each received a part of the temporal pattern, the CR was almost always to the temporal pattern received by the whole brain rather than to the pattern delivered to any single region. When behavioral integration of electrical inputs to different regions occurred, electrophysiological summation could be seen in the intralaminar thalamic nuclei (John, 1976).

The conscious awareness of the neural models of past events is information about information. Consciousness must be mediated in the same statistical fashion as other types of information. There is no evidence that suggests that qualitatively *unique* neural processes represent ideas. The multimodal content of consciousness somehow reflects EIs in different parts of the brain. Although "backup systems" exist, extensive lesions of the MRF often produce profound disturbances of consciousness. This system may play a unique role in interpretation of the environment and mediation of consciousness.

Conscious awareness of the multimodal EI might arise in three ways. The first way is that the local EI in each ensemble produces an emergent cooperative property, with consciousness of the information corresponding to the primary local afferent input. This explanation fails to deal with the multisensory nature of consciousness. It does not explain how the EI in different regions is integrated.

The second explanation is that activation of an RS causes a *common mode of EI* to appear in many brain regions. This multimodal information converges upon the dorsal thalamus and MRF. Outflow from this system,

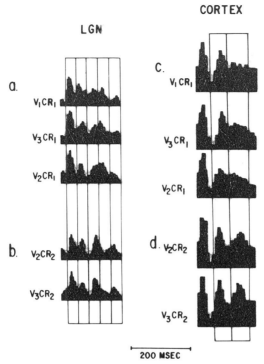

Figure 6.10. Examples of facsimile cell activity. On the left are shown five PSHs obtained from a cell in the LGN of one cat during trials resulting in (a) V_1CR_1, V_3CR_1, and V_2CR_1 and (b) V_2CR_2 and V_3CR_2. The three PSHs depicted in (a) were not significantly different from one another, and the two PSHs in (b) were similarly indiscriminable. All five PSHs show clear peaks in both the first and second intervals demarcated by the vertical lines. The PSHs from trials resulting in CR_2, however, show a peak during the third interval that is absent from the PSHs observed during trials resulting in CR_1. Conversely, the PSHs observed during CR_2 trials show a trough during the fourth interval instead of the peak observed in the PSHs from CR_1 trials. On the right are shown PSHs from a facsimile cell in the association cortex. All five PSHs show a comparable early peak. The second peak occurs signigficantly later in (c), the top three PSHs all obtained during trials resulting in CR_1, than in (d), the bottom two PSHs obtained during CR_2 trials. In the last interval demarcated by the vertical lines, a marked peak appears in the bottom two PSHs that is absent from the three PSHs above. In both sets of records, it is noteworthy that the shape of the PSH appears to depend upon the behavior subsequently performed by the cat and not upon the stimulus which was actually present. The different firing patterns thus cannot be attributed to the physical stimuli, but seem to reflect the meaning attributed to the afferent input in the context of previous experience. No data from V_1CR_2 errors were obtained in either of these instances. (From Ramos et al., 1976c.)

modulated by nonspecific thalamus, feeds back to regions that contributed high signal-to-noise ratios in the incoming barrage. This integrated, reverberating cortico-thalamo-reticular EI pattern produces unusually high levels of firing in the common mode in participating ensembles. Without this resonance, comparable ratios of organized signal-to-random firing or "noise" would be difficult to achieve. A multimodal, integrated common mode EI is established in a densely packed aggregate of cells in this reverberating system. In these various structures, there exist many local circuit neurons (LCNs), with axons that are shorter than the wavelength of an action potential (Rakic, 1976). Such an aggregate of cells firing in the common mode EI pattern might be called a "multineuron," a system of cells whose subgroups contained all of the features of the multimodal EI. Phase-locked changes in membrane potential in masses of such cells might be peculiarly favorable to mediate the emergence of consciousness, a new cooperative property of the multineuron that does not exist in its separate neuronal elements. This proposal fails to explain how all parts of the multineuron are influenced by every feature of the EIs characteristic of the participating subgroups of neurons. Yet integration cannot be achieved by action at a distance unless it uses some intervening physical process.

The third explanation is that as these patterns of firing occur in densely packed resonating cell ensembles, shifts of potassium and calcium ions produce complex patterns of charge. Ionic binding to mucopolysaccharide filaments and surfaces of glial cells, as well as on neurons, creates charge sinks, sources, and gradients. A three-dimensional charge volume thus arises, with a topography that is *supracellular*. The EIs in many neural ensembles have been focused and integrated as the improbable distribution of charge density in space and time within this supracellular domain, which could be called a *hyperneuron*. Consciousness may emerge from negative entropy, when the matter of the brain achieves highly improbable levels of organization of energy. Densely packed neurons produce charge distributions that are extremely improbable from a thermodynamic view. These improbable charge distributions may act back upon neurons to sustain improbable firing patterns.

The postulated hyperneuron is a complex electric field. This field comes from lightly bound ions rather than free charges. Since the critical locus of the hyperneuron is postulated to be in regions of closely packed neurons such as a resonating cortical-thalamic-RF system, it is not relevant that previous attempts to disrupt transcortical fields by metallic insertions were unsuccessful. First, such experiments asked whether transcortical fields were crucial to perception, but did not rule out the possibility that consciousness involves corticofugal inputs to a thalamic-RF field. Furthermore, as is known from physical chemistry, unless the voltage exceeds the ionization constant of the ionic species plus the work function of the conductor surface, electrons cannot flow into a conductor to move between fluid regions of different ionic density. Thus, a metallic insertion into neural tissue may distort but will not disrupt an ionic field. Finally, the use of external fields in attempts to alter consciousness is not an adequate test of the hyperneuron concept. Any conceivable externally applied field would establish voltage gradients on the cellular scale that would be negligible in contrast to the enormous local gradients that are found in neural tissue. The flaw in this hyperneuron model is that we have no justification to expect that an energy distribution could produce consciousness.

There are surely other explanations of how EIs in different regions are transformed into integrated consciousness. For me, evaluation of these alternatives is the most intriguing problem of contemporary neuroscience.

REFERENCES

Bartlett, F., & John, E. R. Equipotentiality quantified: The anatomical distribution of the engram. *Science*, 1973, *181*, 764–767.

Bartlett, F., John, E. R., Shimokochi, M., & Kleinman, D. Electrophysiological signs of readout from memory. II. Computer classification of single evoked potential waveshapes. *Behavioral Biology*, 1975, *14*, 409–449.

Begleiter, H., Porjesz, B., Yerre, C., & Kissin, B. Evoked potential correlates of expected stimulus intensity. *Science*, 1973, *179*, 814–816.

Brown, W. S., Marsh, J. T., & Smith, J. C. Contextual meaning effects on speech-evoked potentials. *Behavioral Biology*, 1973, *9*, 755–761.

Cooper, R., McCallum, W. C., Newton, P., Papakostopoulos, D., Pocock, P. V., & Warren, W. J. Cortical potentials associated with the detection of visual events. *Science*, 1977, *196*, 74–77.

Clynes, M., Kohn, M., & Gradijan, J. Computer recognition of the brain's visual perception through learning the brain's physiologic language. *Institute of Electrical and Electronics Engineers, International Conference Record, Pt. 9*, 1967, 125–142.

Doty, R. W. The ablation of visual areas in the central nervous system. *Handbook of Sensory Physiology*, 1973, *VII/3*, 483–502.

Fishman, M. C., & Michael, C. R. Integration of auditory information in the cat's visual cortex. *Vision Research*, 1973, *13*, 1415–1419.

Granit, R., *The Purposive Brain*, Cambridge, Mass.: MIT Press, 1977.

Grinberg, J. & John, E. R. Unpublished results, 1974.

Herrington, R. N., & Schneidau, P. The effects of imagery on the visual evoked response. *Experientia*, 1968, *24*, 1136–1137.

Hoeppner, T. J. Stimulus analyzing mechanisms in the cat's visual cortex, *Experimental Neurology*, 1974, *45*, 257–267.

Horn, G., & Hill, R. M. Modifications of receptive fields of cells in the visual cortex occurring spontaneously and associated with bodily tilt. *Nature*, 1969, *221*, 186–188.

Hubel, D. H., & Wiesel, T. N. Receptive fields, binocular interaction and functional architecture in the cat's visual cortex. *Journal of Physiology*, 1962, *160*, 106–154.

John, E. R. Higher nervous functions: brain functions and learning. *Annual Review of Physiology*, 1961, *23*, 451–484.

John, E. R. *Mechanisms of memory*. New York: Academic Press, 1967.

John, E. R. Switchboard versus statistical theories of learning and memory. *Science*, 1972, *177*, 850–864.

John, E. R. A model of consciousness. In G. Schwartz and D. Shapiro (Eds.), *Consciousness and self-regulation: Advances in research* (Vol. 1). New York: Plenum Press, 1976.

John, E. R., Bartlett, F., Shimokochi, M., & Kleinman, D. Neural readout from memory. *Journal of Neurophysiology*, 1973, *36*, 893–924.

John, E. R., Bartlett, F., Shimokochi, M., & Kleinman, D. Electrophysiological signs of the readout from memory. *Behavioral Biology*, 1975, *14*, 247–282.

John, E. R., Herrington, R. N., & Sutton, S. Effects of visual form on the evoked response. *Science*, 1967, *155*, 1439–1442.

John, E. R., & Killam, K. F. Electrophysiological correlates of differential approach-avoidance conditioning in the cat. *Journal of Nervous Mental Disease*, 1960, *131*, 183.

John, E. R., & Kleinman, D. Stimulus generalization between differentiated visual, auditory and central stimuli. *Journal of Neurophysiology*, 1975, *38*, 1015–1034.

John, E. R., & Morgades, P. P. Neural correlates of conditioned responses studied with multiple chronically implanted moving microelectrodes. *Experimental Neurology*, 1969a, *23*, 412–425.

John, E. R., & Morgades, P. P. The pattern and anatomical distribution of evoked potentials and multiple unit activity elicited by conditioned stimuli in trained cats. *Communications in Behavioral Biology*, 1969b, Pt. A, *3*, 181–207.

John, E. R., Ruchkin, D. S., Leiman, A., Sachs, E., & Ahn, H. Electrophysiological studies of generalization using both peripheral and central conditioned stimuli. *23rd International Congress of the Physiological Sciences* (Tokyo), 1965, International Congress Series No. 87, 618–627.

John, E. R., & Schwartz, E. L. The neurophysiology of information processing and cognition. *Annual Review of Psychology*, 1978, *29*, 1–29.

John, E. R., Shimokochi, M., & Bartlett, F. Neural readout from memory during generalization. *Science*, 1969, *164*, 1519–1521.

Johnston, V. L., & Chesney, G. L. Electrophysiological correlates of meaning. *Science*, 1974, *186*, 944–946.

Kleinman, D., & John, E. R. Contradiction of auditory and visual information by brain stimulation. *Science*, 1975, *187*, 271–272.

Klinke, R., Fruhstorfer, H., & Finkenzeller, P. Evoked responses as a function of external and stored information. *Electroencephalography and Clinical Neurophysiology*, 1968, *26*, 216–219.

Libet, B. In J. C. Eccles (Ed.), *Brain and conscious experience*. Heidelberg and New York: Springer-Verlag, 1966.

Livanov, M. N., & Poliakov, K. L. The electrical reactions of the cerebral cortex of a rabbit during the formation of a conditioned defense reflex by means of rhythmic stimulation. *Izvestiya Akademya Nauk, USSR Series Biology*, 1945, *3*, 286.

Majkowski, J. The electroencephalogram and electromyogram of motor conditioned reflexes after paralysis with curare. *Electroencephalography and Clinical Neurophysiology*, 1958, *10*, 503–514.

Majkowski, J. Electrophysiological studies of learning in split-brain cats. *Electroencephalography and Clinical Neurophysiology*, 1967, *30*, 482–493.

Morrell, F. Visual system's view of acoustic space. *Nature*, 1972, *238*, 44–46.

Pasik, T., & Pasik, P. Extrageniculostriate vision in the monkey. IV. Critical structures for light vs no-light discrimination. *Brain Research*, 1973a, *56*, 165–182.

Pasik, T., & Pasik, P. Extrageniculostriate vision in the monkey. V. Role of accessory optic system. *Journal of Neurophysiology*, 1973b, *36*, 450–457.

Picton, T. W., Hillyard, S. A., & Galambos, R. Cortical evoked responses to omitted stimuli. In M. N. Livanov (Ed.), *Major problems of brain electrophysiology*. Moscow: Union of Soviet Socialist Republics Academy of Sciences, 1973.

Rakic, Pasko, *Local circuit neurons*. Cambridge, Mass.: MIT Press, 1976.

Ramos, A., & Schwartz, E. L. Observation of frequency specific discharges at the unit level in conditioned cats. *Physiology and Behavior*, 1976, *16*, 649–652.

Ramos, A., Schwartz, E. L., & John, E. R. Stable and plastic unit discharge patterns during behavioral generalization. *Science*, 1976a, *192*, 393–396.

Ramos, A., Schwartz, E. L., & John, E. R. Evoked potential-unit relationships in behaving cats. *Brain Research Bulletin*, 1976b, *1*, 69–75.

Ramos, A., Schwartz, E. L., & John, E. R. Examination of the participation of neurons in readout from memory. *Brain Research Bulletin*, 1976c, *1*, 77–86.

Rose, D. The hypercomplex cell classification in the cat's striate cortex. *Journal of Physiology*, 1974, *242*, 123P–125P.

Ruchkin, D. S., & John, E. R. Evoked potential correlates of generalization. *Science*, 1966, *153*, 209–211.

Schwartz, E. L., Ramos, A., & John, E. R. Single cell activity in chronic unit recording: A quantitative study of the unit amplitude spectrum. *Brain Research Bulletin*, 1976, *1*, 57–68.

Sherrington, C. S. *Integrative Activity of the Nervous System*. New Haven, Conn.: Yale Univ. Press, 1906.

Sprague, J. M., Berlucchi, G., & Rizzolatti, G. The role of the superior colliculus and pretectum in vision and visually guided behavior. *Handbook of Sensory Physiology*, 1973, *VII/3*, 27–102.

Stein, D. G., Rosen, J. J., & Butters, N. (Eds.). *Plasticity and recovery of function in the central nervous system*. New York: Academic Press, 1974.

Sutton, S., Braren, M., Zubin, J., & John, E. R. Evoked potential correlates of stimulus uncertainty. *Science*, 1965, *150*, 1187–1188.

Sutton, S., Tueting, P., Zubin, J., & John, E. R. Information delivery and the sensory evoked potential. *Science*, 1967, *155*, 1436–1439.

Tepas, D. I., Guiteras, V. L., & Klingaman, R. Variability of the human average evoked response to visual stimulation: A warning. *Electroencephalography and Clinical Neurophysiology*, 1974, *36*, 533–537.

Thatcher, R. W. Evoked potential correlates of semantic information processing. In S. Harnard (Ed.), *Lateralization in the nervous system*. New York: Academic Press, 1976.

Thatcher, R. W., & John, E. R. *Functional neuroscience* (Vol. I). Hillsdale, N.J.: L. Erlbaum Associates, 1977.

Ukhtomski, A. A. Essays on the physiology of the nervous system. In *Collected Works* (Vol. 4). Leningrad, 1945.

Walter, W. G., Cooper, R., McCallum, C., & Cohen, J. The origin and significance of the contingent negative variation or "expectancy wave." *Electroencephalography and Clinical Neurophysiology*, 1965, *18*, 720.

Weinberg, H., Walter, W. G., Cooper, R., & Aldridge, V. J. Emitted cerebral events. *Electroencephalography and Clinical Neurophysiology*, 1974, *36*, 449–456.

Weinberg, W. G., Walter, W., & Crow, H. H. Intracerebral events in humans related to real and imaginary stimuli. *Electroencephalography and Clinical Neurophysiology*, 1970, *29*, 1–9.

Weiskrantz, L. Behavioural analysis of the monkey's visual nervous system. *Proceedings of the Royal Society*, 1972, *182* (Series B).

Wiener, N. *Cybernetics or control and communication in the animal and the machine*. Cambridge, Mass.: MIT Press, 1961.

Motivation and Recovery of Function

PART II

ALAN N. EPSTEIN

A Comparison of Instinct and Motivation with Emphasis on Their Differences[1]

7

I will not describe my current research. Instead, I will do something more hazardous. I will discuss motivated and instinctive behaviors—a pair of old concepts that have been vexed with controversy, but ones from which I think new lessons can be learned. I will recall their similarities because they are prominent, so much so that they have led others (Eibl-Eibesfeldt, 1970; Lashley, 1938; McDougall, 1914; Stellar, 1960; Tinbergen, 1974) to consider them identical. But they are not. There are important differences between them, differences both in the kinds of animals in which they occur and in the characteristics of the behavior that are described by each concept. I am convinced that motivated behavior is biologically less common and psychologically more complex than instinctive behavior. And I believe that the differences deserve emphasis, because they are diagnostic of differences in underlying neurological structure which must be respected when we choose species for studying the brain mechanisms of motivated behavior.

First, consider the similarities. The concepts of motivation and instinct arose as explanations for the spontaneity of behavior and for its inconstancy in frequency and form in response to constant stimulus conditions. Both of these phenomena require a concept of changeable internal states that generate the readiness for behavior. Behavior is strikingly spontaneous. Unlike reflexive behavior that will not occur without adequate

---

[1] The personal research mentioned here and the writing of this essay were supported by Grant 03469 from the NINCDS.

stimulation (for example, Sherrington's myotatic reflex), much of what animals do is done without immediately preceding stimulation. This is true from protozoa to primates. For examples of what I have in mind recall Jennings' (1962) descriptions of the behavior of paramecia and the others of what he called the "lower organisms." Recall, also, Richter's studies of the periodic behaviors of the undisturbed rat (1927), and any of the several detailed descriptions that we now have of the behavior of apes in the wild (Goodall, 1965; Schaller, 1964). The phenomenon of spontaneity requires that behavior be conceived as the outcome of states within animals that directly generate or suppress action. Second, and also unlike reflexive behavior that is qualitatively unchanged across episodes of elicitation, instinctive and motivated behaviors may or may not be emitted in response to constant stimuli, and when they do occur they may take an entirely different form from stimulus episode to stimulus episode (for example, the behavior of a female rat in response to a male conspecific when she is or is not in heat). These inconstancies again require a concept of behavior as the outcome of endogenous states, but in this instance the state must be thought of as subject to changes that result in altered responsiveness to unaltered stimulus conditions. Instinctive and motivated behaviors are therefore not thought of as mere reactions to concurrent stimulation; instead they are both attributed to the operations of changeable internal states. These are variously referred to as tendencies, moods, or urges, but most universally as drives.

There are other prominent similarities. Instinctive and motivated behaviors are divisible into appetitive and consummatory phases (Craig, 1918). Each includes a phase of wandering or of search and approach that precedes a phase of ultimate and terminating behavior. Instinctive and motivated behaviors often serve the same functions for the animal, that is, locomotion, communication, reproduction, nutrition, predation, and defense. They often have similar physiological mechanisms that employ deficit signals, hormones, pheromones, and endogenous oscillators, which are utilized by innate servomechanisms and lead in many instances to the achievement of homeostasis. In both instinct and motivation the behavior itself may be compounded of innate and acquired responses [the maternal behavior of the digger wasps is *not* entirely innate (from Baerends, 1941, as cited by Tinbergen), and the maternal behavior of the rat is clearly *not* entirely acquired (Rosenblatt & Lehrman, 1963)]. And in animals with sufficiently complex nervous systems, the mechanisms underlying both instinctive and motivated behaviors reside within the brain. With the similarities enumerated it is not surprising that instinct and motivation have so often been used as names for the same thing.

But, now, consider the differences. Instinctive behavior is by far the more common of the two. Remember, I am including all animals in this discussion, not just the domestic pets and pests and the handful of primates that we psychologists refer to rather grandly as "organisms." Animal-kind is divisible into between 15 and 20 major phyla (different kinds of animals that have been isolated from each other in evolution for 400 million years or more, the vertebrates having arisen in the Ordovician Period). Remember also that the birds and mammals, which tend to preoccupy our thinking, are only the top layers of one of the smaller phyla (the Chordata) in both size and diversity of species. The entire phylum Chordata is only half the size of the Mollusca in number of species, and is an order of magnitude smaller than the Arthropoda. Even among our own kind the evidence of our diversity is modest. Fish are the most common vertebrates (twice as many species as birds and mammals combined), and the rodents are the most common mammals. The point here is twofold. First, there are different kinds of animals. They are great in diversity, each kind is an ancient lineage, and, with very few exceptions, they all have a complex repertory of behaviors. But the overwhelming majority of them are different from us and our most common subjects in design and ancestry. Second, animals of the other phyla, even those in our own that do not have hair or feathers, are not only different but also simpler than us, and simpler animals are likely to execute complex behaviors in simpler ways than do we and our favorite subjects.

It is my contention that most animals do employ the simpler mechanism. They make their livings with instinctive behavior that has the following distinguishing characteristics. It is largely heritable phenotype, that is, it is the outcome of genome realized in development, and is therefore species-specific both in action and in releasing and guiding stimuli (see Fraenkel & Gunn, 1961, for many examples). It occurs typically in small-brained, short-lived animals who often live in isolation and exhibit the behavior infrequently (often only once) in crucial situations in which it must be performed flawlessly. It is not entirely innate being subject to habituation and sensitization and often incorporates elements acquired in associative learning, but these are what I would call contextural and prescribed habits. Contextural habits are episodes of learning that are inserted into a sequence of innate behaviors (Tinbergen has referred to these as "localized" learning); this is exemplified by the place-learning of the digger wasp that has just deposited her egg in a newly dug nursery chamber and, before flying off to seek prey, memorizes its location (Tinbergen & Kruyt, 1938, as cited by Tinbergen). By prescribed habits I mean those that are constrained by what the particular kind of animal can or cannot learn and by the time in the animal's life history in which the opportunity for learning occurs. The ideal example here is the learning of species song in finches and sparrows (Marler & Hamilton, 1966). Contextural and prescribed learning do not diversify the behavior of individuals of the same species; on the contrary, in cooperation with genetic endowment and developmental history they reduce diversity and assure the species uniformity of instinctive behavior. Finally, instinctive behavior is displayed by animals that are blind to its ends. This must be true whenever the behavior is performed only once, and is illustrated by the so-called physiological traps in which many animals can be imprisoned whenever an innate requirement of their behavior is exploited so that it becomes self-desctructive or absurd.

Except for the absence of a demonstrated role for learning, the cocoon spinning of the silkmoth larva (Van der Kloot and Williams, 1953) is an ideal illustration of what I mean by instinctive behavior. Having reached an appropriate stage of development and being prepared by appropriate changes in hormonal state, the mature larva evacuates its gut and wanders away from the leaves it has been eating, settling after some hours in a suitable site for spinning. It then does this for half a day or more by executing an elaborate sequence of specialized movements that draw silk from its glands and form it into a double-layered envelope within which the larva pupates, overwinters, and ultimately emerges as an adult. Like the webs of spiders the cocoons are records of the species-specificity of the behavior. The kind of larva that is about to spin predicts the details of the behavior it will display, and these details of behavior produce a cocoon whose size and shape are as diagnostic of the species of moth that produced it as is the animal itself. The emergence of the adult depends, in the species studied by Van der Kloot, on the mutual alignment of the loosely woven upper poles (called valves) of the inner and outer envelopes of the cocoon. The correct alignment of the valves is assured by a negative geotaxis that keeps the animal head-up throughout the spinning episode. If the animal and the outer envelope it has just spun are turned 180 degrees just before it proceeds to the spinning of the inner envelope, the larva rights itself within the outer envelope, leaving the valve at what is now its lower pole. The animal then proceeds to spin its inner envelope, with its valve normally upright and thereby entombs itself within it own cocoon. This, then, is instinctive behavior and it is in my view the dominant, even exclusive, form of complex behavior among animals of all phyla including the simpler species of vertebrates. See Wigglesworth (1964), and Wells (1968) for reviews of the ubiquity and variety of instinctive behavior, and Evans (1973) for an affectionate description of the elaborate behaviors of one family of insects.

Motivated behavior, on the other hand, is more complex and more rare. In addition to being drive-activated and to sharing the characteristics previously described, it is more complex than instinctive behavior in the following respects. It occurs in big-brained, long-lived animals that typically live in social groups. It is exhibited repeatedly after an ontogeny during which its performance is improved. It is subject to habituation and sensitization, but is dominated by associative learning particularly in its appetitive phase. Animals displaying it do so while anticipating its ends and it is therefore truly goal-directed. And, finally, animals displaying it behave with affect. It therefore has characteristics that it does not share with instinctive behavior and that make it psychologically more complex.

Two of these characteristics are outcomes of learning. The first is *individuation* of the appetitive phase, that is, individual differences of approach, search, and selection are acquired in prior episodes of the behavior and are employed as operants thereafter (see Teitelbaum, 1977, for a full discussion of this important point), thus leading to diversification of the behavior among individuals of the same species and liberating them from the stereotype of species-specific performance [contrast the feeding behavior of the blowfly that has no options in the behavior it must employ to ingest food (Dethier, 1978) with the feeding behavior of the rat that can gain access to food by performing learned behaviors that are not normally part of its food-getting repertory].

The second learned characteristic of motivated behavior is *expectancy*. As the result of prior experience with goals, that is, with objects or situations that permit the performance of the appropriate consummatory behavior, the appetitive phase of motivated behavior is altered by the nature (kind or quantity) of the expected goal. Two very good examples of this are Crespi's old experiments (1944) showing that the speed with which hungry rats run to the end of an alleyway is predictable from the amount of food that they have received there in prior trials. These experiments show that some representation of the expected value of the goal objects must be operating to adjust the intensity of behavior that is begun at a distance from that goal and is executed before the goal is reached. A second example comes from work that is closer to my own interests (Kriekhaus & Wolf, 1968). Rats that have had prior experience operating levers that deliver plain or salty water will, when made salt deficient for the first time, approach the "salty" bar from a distance in preference to the "water" bar, and will work at it more vigorously even when both bars deliver no fluid. That is, having learned where salt is and how to get it, the rat expects to find it there and will work for it when made salt deficient. This incidentally provides an elegant example of the combination of inborn and learned elements in motivated behavior because the arousal of the appetite for salt, the prior condition for the heightened expectancy, is innate (Epstein & Stellar, 1955; Nachman, 1962).

The third distinctive characteristic of motivated behavior is *affect*. Motivated behavior is laden with affect and its performance is accompanied by overt expressions of internal affective states. Affect is expressed by very young animals, is often full-fledged when first exhibited, and is typically species-specific. It is therefore a characteristic of motivated behavior that does not depend for its initial expression on learning. Please appreciate that I am not lapsing here into nineteenth century mentalism. I am referring to behavior that is overtly expressed and observable, not to mere assertions about an animal's mood or to claims of knowledge of what it may be feeling. I am arguing that the performance of motivated behavior includes necessarily the exhibition of qualitatively different behavioral displays that are signs of changing internal states. These are not simple changes in limb movement or in intensity of locomotion, but are true displays, organized into recognizable patterns and sufficiently diversified to express a variety of internal states.

Solomon and Corbit (1974) clearly appreciate the role of affect in complex behavior. They have made affective expression the basis of a very attractive theory of motivations that range from hunger to addiction. I know of no better descriptions of these than the illustrations from Darwin's book on the expression of the emotions (1965). His talented illustrator depicts cats and dogs in various states of emotional expression. To understand why I consider them so important for this discussion, contrast these graphic portraits with the behavior of a cockroach. If I showed you a film clip of the running behavior of first a dog and then of a roach but omitted from each film the conditions that triggered the locomotion and the behavior with which it terminated, I believe you would have no trouble distinguishing from the style of the dog's behavior whether it was escaping a threat or approaching a safe shelter. But could you do the same for the roach? Could you tell from the way it ran whether it was threatened or eager? If I showed you a film of the entire behavioral episode and you knew what had initiated the running and what the animal did at the end of the sequence, you might be tempted, as Darwin was and as many still are, to infer a mood or emotion underlying the roach's running behavior. But this would be pointless anthropomorphism unless the differences in underlying state were plain in the animal's behavior as they are in the behavior of the dog. I believe, in other words, that motivated behavior is hedonic (see Stellar, 1974, for another discussion of this issue). It arises from mood, is performed with feeling, and results in pleasure or the escape from pain, and although the moods, feelings, and satisfactions themselves are private and beyond our reach as scientists, their overt expression is a necessary characteristic of motivation.

The distinction I have drawn should not be read to mean that I believe that animals display *either* instinctive behavior *or* motivated behavior. There are two classes of animals, in my view, but not these two. There are animals that display only instinct and there are those, a smaller number but more complex and big brained, that display both. Birds, for example, provide some of the classic examples of instinctive behavior (imprinting, egg-rolling in the grayleg goose), but my concept does not deny them the capacity for motivation.

Now, a last word about research strategy. Having emphasized the distinctions between the two forms of complex behavior, I think they should be respected when we choose species for the study of motivated behavior. Animals that cannot individuate their appetitive behavior, that show no evidence in it of expectancy, and that do not express affect are, in my view, not performing motivated behavior and are poor choices for its study. This excludes most animals, which may be unfortunate but should not be surprising. We should expect to find more complex neural and behavioral function in more complex animals. But it leaves us with a rich supply of mammals, birds, reptiles, and very likely with all of the remaining vertebrates except the simplest fishes. It does exclude most of the invertebrates either because they do not have brains or do not have the neural equipment for learning. It also excludes the arthropods, who may be brainy enough but who seem to have sacrificed the advantages of affective expression for those of a rigid exoskeleton. The molluscs are a likely group, particularly because they include the brainiest of the invertebrates and have the advantage for affective expression of fleshy, mobile bodies. But even here there are suggestions that they are limited to instinct and have not evolved the additional mechanisms necessary for motivation. First, like all other invertebrates they do not have an autonomic nervous system with which so much of affective expression is achieved. Second, although eating behavior (orientation to food odors, food grasping, and ingestion) has been studied extensively in the seahare (Kupferman, 1974) and its relatives, no one has yet succeeded in showing that these animals can learn where to find food (seaweed) when they are separated from it. Even the octopus, the champion among big-brained invertebrates, may be no more than the most complex instrument for cool, instinctive behavior. Wells (1977) reported that no changes in heart rate or heartbeat amplitude could be detected in the male octopus during sexual arousal or even in the midst of copulation!

Because changing states within the brain are characteristics of both instinct and motivation, the investigation of the neurological bases of drive can be pursued in a great variety of animals with some hope of finding generalizations that will be useful across phyla. My own work on the induction of thirst by angiotensin is an example (Epstein, 1978). But the commonality of interest in drive and the other characteristics shared by instinct and motivation must not obscure the important differences that distinguish them. Full understanding of the brain mechanisms of motivated behavior will not come from the study of animals that exhibit only instinctive behavior. It will come from the study of animals whose behavior is characterized by drive, by appetitive behavior marked by individuation and expectancy, and by overt expressions of affect.

REFERENCES

Craig, W. Appetites and aversions as constituents of instincts. *Biology Bulletin*, 1918, *34*, 91–107.
Crespi, L. P. Amount of reinforcement and level of performance. *Psychology Review*, 1944, *51*, 341–357.

Darwin, C. *The expression of the emotions in man and animals*. Chicago: Univ. of Chicago Press, 1965. (First publication in 1872)
Dethier, V. G. *The hungry fly*. Cambridge, Mass.: Harvard Univ. Press, 1978.
Eibl-Eibesfeldt, I. *Ethology, the biology of behavior*. New York: Holt, 1970.
Epstein, A. N. The neuroendocrinology of thirst and salt appetite. In W. F. Ganong & L. Martini (Eds.), *Frontiers in neuroendocrinology* (Vol. 5). New York: Raven Press, 1978, pp. 101–134.
Epstein, A. N., & Stellar, E. The control of salt preference in the adrenalectomized rat. *Journal of Comparative and Physiological Psychology*, 1955, *48*, 167–172.
Evans, H. E. *Wasp farm*. New York: Anchor Press, 1973.
Fraenkel, G. S., & Gunn, D. L. *The orientation of animals*. New York: Dover Press, 1961.
Goodall, J. Chimpanzees of the Gombe stream reserve. In DeVore (Ed.), *Primate behavior*, I. New York: Holt, 1965, pp. 425–481.
Jennings, H. S. *Behavior of the lower organisms*. Bloomington: Indiana Univ. Press, 1962. (First published in 1905)
Kriekhaus, E. E., & Wolf, G. Acquisition of sodium by rats: interaction of innate mechanisms and latent learning. *Journal of Comparative and Physiological Psychology*, 1968, *65*, 197–201.
Kupferman, I. Feeding behavior in Aplysia: A simple system for the study of motivation. *Behavioral Biology*, 1974, *10*, 1–26.
Lashley, K. S. Experimental analysis of instinctive behavior. *Psychology Review*, 1938, *45*, 445–471.
Marler, P., & Hamilton, W. J. *Mechanisms of animal behavior*. New York: Wiley, 1966.
McDougall, W. *An introduction to social psychology*. Boston: Luce, 1914.
Nachman, M. Taste preferences for sodium salts by adrenalectomized rats. *Journal of Comparative and Physiological Psychology*, 1962, *56*, 343–349.
Richter, C. P. Animal behavior and internal drives. *Quarterly Review of Biology*, 1927, *2*, 307–342.
Rosenblatt, J. S., & Lehrman, D. F. Maternal behavior of the laboratory rat. In H. I. Rheingold (Ed.), *Maternal behavior in mammals*. New York: Wiley, pp. 8–57.
Schaller, G. B. *The year of the gorilla*. Chicago: Univ. of Chicago Press, 1964.
Solomon, R. L., & Corbit, J. D. An opponent-process theory of motivation: I. Temporal dynamics of affect. *Psychology Review*, 1974, *81*, 119–145.
Stellar, E. Drive and motivation. In H. W. Magoun (Ed.), *Handbook of physiology* (Section 1, Vol. III), Washington, D.C., Am. Physiol. Soc., 1960, pp. 1471–1527.
Stellar, E. Brain mechanisms in hunger and other hedonic experiences. *Proceedings of American Philosophy Society*, 1974, *118*(3), 276–282.
Teitelbaum, P. The physiological analysis of motivated behavior. In P. G. Zimbardo & F. L. Ruch (Eds.), *Psychology and life*. Glenview, Ill.: Scott, Foresman, 1977.
Tinbergen, N. *The study of instinct* (2nd ed.). New York: Oxford Univ. Press, 1974.
Van der Kloot, W. G., & Williams, C. M. Cocoon construction in the Cecropia silkworm. *Behavior*, 1953, *5*, 141–163.
Wells, M. J. *Lower animals*. New York: World Univ. Library, 1968.
Wells, M. J. *Octopus: Physiology and behaviour of an advanced invertebrate*. London: Chapman and Hall, 1977.
Wigglesworth, V. B. *The life of insects*. London: Weidenfeld and Nicolson, 1964.

PHILIP TEITELBAUM
TIMOTHY SCHALLERT
MARC DE RYCK
IAN Q. WHISHAW
ILAN GOLANI

Motor Subsystems in Motivated Behavior[1]

8

To improve our understanding of motivated behavior, we have found it useful to break it down into subsystems, not only because they are simpler, but also because they can be recombined to reconstitute the original behavior. Complete surgical transection is a classic method of simplifying the nervous system to analyze behavior (Sherrington, 1906). However, with this method, one can never observe again the resynthesis of behavior that depends on nervous connections that go across the transection. For instance, a decerebrate animal can never put its fragmented behavior back together again. Therefore, we use the method of partial transection by localized lesions (Teitelbaum, 1974). When a system of the brain is damaged, the behavior controlled by it may disappear completely or disintegrate into only a very simple fragmentary form. Then, as recovery from such partial transection takes place, the behavior gradually reintegrates itself.

Brain systems running through the hypothalamus have long been known to play an important role in motivated behavior (Hess, 1954). Animals will press a lever thousands of times to stimulate themselves electrically in the lateral hypothalamus (LH), indicating that they find it reinforcing (Olds, 1977). Sustained LH stimulation can elicit eating or drinking in satiated animals, or even mating or attack (Flynn, 1973; Hoebel,

[1] Supported by National Research Council of Canada Grant A8273 to Ian Q. Whishaw, by NIH Grant RO1 NS 11671, University of Illinois Biomedical Research grant, Sloan Foundation, and William T. Grant Foundation to Philip Teitelbaum, and by a grant to Ilan Golani from the U.S.-Israel Binational Science Foundation (BSF), Jerusalem, Israel.

1975). If we damage the LH on both sides, rats, cats, dogs, monkeys, and even people will stop eating and drinking. The refusal to eat (aphagia) and to drink (adipsia) may persist until the animal dies (Anand & Brobeck, 1951). However, if kept alive by force-feeding, it will eventually recover (Teitelbaum & Epstein, 1962; Teitelbaum & Stellar, 1954). The behavioral pattern of recovery in adult LH animals is summarized diagrammatically in Figure 8.1. Every LH animal shows the same basic sequence of recovery. That this reflects a fundamental process of reorganization is suggested by the fact that a close parallel has been demonstrated between the stages of adult recovery and the stages of infantile development of the regulation of food and water intake (Cheng, Rozin, & Teitelbaum, 1971; Teitelbaum, 1971; Teitelbaum, Cheng, & Rozin, 1969).

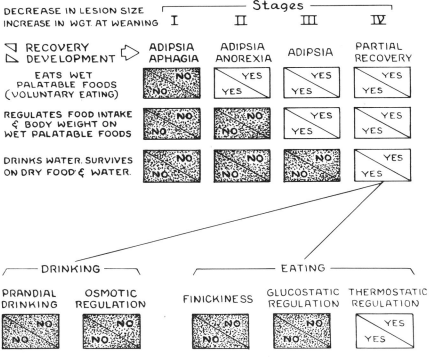

Figure 8.1. Comparison of the development of eating and drinking in infancy and its recovery after hypothalamic lesions in adults. The upper right half of each block represents the recovering lateral hypothalamic rat and the lower left, the growing infant rat (thyroidectomized or starvation-stunted to slow down development). Uniform coloring in each full block indicates similar responses in recovery and development (Teitelbaum et al., 1969).

Moreover, a parallel between development of the voluntary use of the hand and recovery of such use in brain-damaged adults has been shown independently in people (Seyffarth & Denny-Brown, 1948; Twitchell, 1951, 1965, 1970). More recently, the development-recovery parallel has proved fruitful in the study of sexual behavior (Nance, Phelps, Shryne, & Gorski, 1977; Twiggs, Popolow & Gerall, 1978), as well as catalepsy (Teitelbaum, Wolgin, De Ryck, & Marin, 1976). Although the underlying neural process may be different, the behavioral sequence is strikingly similar. In that sense, recovery recapitulates ontogeny.

If LH lesions are large enough, not only do the animals stop eating and drinking, but all forms of spontaneous motivated behavior appear to be abolished. The animal is akinetic and cataleptic (Levitt & Teitelbaum, 1975; Robinson & Whishaw, 1974). During recovery, however, "spontaneity" gradually reappears. To many workers, such states of unresponsiveness have seemed so global or bizarre that they have generally been considered as irrelevant, undesirable "side-effects" of excessively large lesions, presumably including neural systems unrelated to more "specific" forms of motivated behavior, such as eating, drinking, mating, and attack. However, it is often difficult to judge a priori which effects of a lesion are unrelated side-effects and which are relevant. One way of judging is: If the variables that control the apparent side-effect also control the behavior we are interested in, then despite its global or seemingly bizarre character, it is not unrelated, but truly involves neural systems that control the behavior in question (Wolgin & Teitelbaum, 1978).

In this chapter, we will show that this is true for catalepsy and akinesia, produced by LH damage and by many treatments that block catecholamine systems in the brain (Ungerstedt, 1971; Fog, 1972; Schallert, Whishaw, Ramirez, & Teitelbaum, 1978; Schallert & Whishaw, 1978). For instance, as shown in Figure 8.2, an LH cataleptic cat allows one foreleg to be placed up on its back at quite an extreme angle or both its forelimbs to be spread widely apart, and for a long period of time (minutes) does not replace the limb(s) to a more normal position (Wolgin & Teitelbaum, 1978). Similarly, a cataleptic rat may allow itself to remain in a quite awkward posture [Figure 8.2(d)].

These seemingly awake animals clearly look bizarre—what can they teach us about motivation? A painful stimulus, such as a tail-pinch (or other forms of stress; see Wagner & Woods, 1950), can cause a cataleptic cat that has been clinging unmoving to the back of a chair to climb up and then leap down to the floor (Teitelbaum & Wolgin, 1975). Similarly, an LH rat will sink to the bottom of a tank full of water ($33°C$), making little or no attempt to swim. However, when the water is colder ($23°C$), it will swim vigorously, and may even climb out, leap to the floor, and run away (Levitt & Teitelbaum, 1975; Robinson & Whishaw, 1974). In an otherwise totally aphagic rat or cat, a painful clamp on its tail can activate

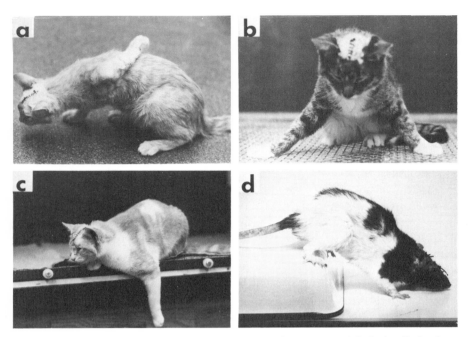

Figure 8.2. Cataleptic animals maintain awkward postures. (a) Left forelimb of an LH cat placed in extreme retroflexion by the experimenter. (b) Extreme abduction of the forelimbs. (c) One forelimb hangs down off a ledge. (From Wolgin and Teitelbaum, 1978.) (d) Cataleptic posture in a rat treated with 6-hydroxydopamine. (From Schallert et al., 1978.)

the animal to eat palatable food (Antelman, Rowland, & Fisher, 1976; Marshall, Levitan, & Stricker, 1976; Marshall, Richardson, & Teitelbaum, 1974a; Teitelbaum & Wolgin, 1975; Wolgin, Cytawa, & Teitelbaum, 1976; Wolgin & Teitelbaum, 1978). Indeed, sated normal rats can be induced to eat by tail-pinch (Antelman, Szechtman, Chin, & Fisher, 1975). Similarly, otherwise unresponsive rats will mate or carry out maternal behavior when painful stimuli are applied (Barfield & Sachs, 1968; Szechtman, Siegel, Rosenblatt, & Komisaruk, 1977). Clearly the variables that counteract catalepsy/akinesia also control normal motivated behavior.

It seems quite paradoxical that seemingly irrelevant stressful states (pain or cold) should facilitate normal motivation, let alone counteract catalepsy/akinesia. In the normal animal we assume that the urge to eat represents a central state that is quite separate from motivation to mate, and that both forms of appetitive behavior are incompatible with states of pain and stress. However, the phenomena of catalepsy/akinesia demonstrate that all behavior requires a background level of tonic general activation that can be increased by many avenues of stimulation, including stressful ones. In the presence of appropriate stimuli that can evoke and direct a particular form of motivated behavior, but which at that instant are not sufficient to do so, raising the level of tonic general activation by seemingly inappropriate stressful stimuli can be sufficient to make the customary stimuli effective once again.

From this standpoint, catalepsy/akinesia merely represents an exaggerated form of "motivational" unresponsiveness. Clearly, the animals are not paralyzed—complex behavior patterns are intact, but unresponsive to the usual internal and external stimuli that activate them in the normal (intact or undrugged) animal.

Sources of "Spontaneity" in Catalepsy/Akinesia

Let us consider the cataleptic-akinetic rat, lying prone, or standing crouched and immobile, neglecting all stimuli (Marshall & Teitelbaum, 1974; Marshall, Turner, & Teitelbaum, 1971) 24 hr after large, bilateral LH lesions. If such an animal is dropped from a supine position, it instantly comes to life in the air and rights itself so that it lands in a stable position on the ground. Similarly, a cataleptic cat, standing on three legs with its left foreleg twisted up onto its back [Figure 8.2(a)], if its center of gravity is only slightly displaced by a gentle push laterally toward the unstably supported left side, instantly replaces the left foreleg on the ground, thus regaining stable equilibrium and support. Any outside action that tends to displace the animal's center of gravity will be actively resisted by bracing reactions, yielding a form of "negativism," which can be understood as a release of the positive supporting reflexes (Schallert, Whishaw, De Ryck, & Teitelbaum, 1978; Van Harreveld & Kok, 1935). In other words, catalepsy/akinesia in animals is a state in which the major normal controls over "spontaneity" are those which maintain stable equilibrium and support. Vestibular or proprioceptive stimuli readily activate such an animal by acting on response subsystems of righting, bracing, clinging, and standing, all of which maintain stable support. Other response subsystems, involved in orienting, locomotion, and ingestion, are unresponsive to the stimuli that normally activate them. However, as we have seen earlier, they can be activated by strong stressful stimuli, such as pain and cold.

Sources of Spontaneity Revealed in Aberrations in Locomotion

An akinetic LH animal eventually begins to orient, eat, and walk around again. However, just as its food intake is achieved by only a fractional form of eating (i.e., reflexive nibbling at palatable foods without caloric regulation) (Teitelbaum & Epstein, 1962), so too is its locomotion only a fractional form of normal exploration. Such an animal seems relatively normal as it walks around, exploring an open field. However, its movements are extraordinarily stereotyped, consisting of relatively few movement patterns. This becomes clear if the animal should chance to walk into a corner. A normal animal would simply rear up, turn, and walk away. In contrast, the LH rat may be trapped in the corner for long periods, performing a repetitive series of head-scanning movements and stepping patterns so stereotyped as to resemble a waltz (Levitt & Teitelbaum, 1975). Which components of normal locomotion does such an animal lack, and how does it recover them?

To study this, Golani, Wolgin, and Teitelbaum (1979) used the Eshkol-Wachmann movement-notation system, originally developed for ballet choreography and analysis (Eshkol & Wachmann, 1958; Eshkol et al., 1968). Films were made of the movements of each animal throughout the process of recovery from total akinesia to relatively normal movement. These films were analyzed frame-by-frame, and written "musical scores" of movement sequences were prepared. In this way, it was possible to analyze the dimensions of movement along which recovery occurs. These are illustrated in tracings from representative film frames in Figures 8.3 to 8.7.

The akinetic rat lies prone, without antigravity support, locomotion, head-scanning, or head orienting. Long before forward locomotion appears, static postural antigravity support typically recovers (Figure 8.3), allowing the animal to crouch or stand. At about the same time, scanning movements of the head appear along the floor, first lateral (Figure 8.4) and then longitudinal (Figure 8.5). Recovery within each movement subsystem is cephalo-caudal: First the head is raised slightly and small lateral and longitudinal head scans appear along the floor; then more caudal limb and body segments are recruited into larger amplitude scanning movements. At first, only the front legs are recruited in such scanning movements, with the hindlegs remaining immobile, rooted to the ground (Figures 8.4 and 8.5). Then the hindlegs join in, allowing the animal to circle laterally, but forward locomotion with all four legs is still absent (Figure 8.4). Somewhat later, forward locomotion returns, but the head scans are still restricted to horizontal surfaces (the floor), so that when a vertical obstruction is encountered, the head does not scan upward (Figure 8.6). This is the stage of recovery in which the animal gets trapped in corners because it lacks upward tactile scans along vertical surfaces, and therefore cannot rear up along the vertical walls and turn around.

We should not make the mistake of thinking that such animals are normally motivated, merely lacking the motor capacity to make a particular movement. On the contrary, even when all the normal response patterns have reappeared in the animal's repertoire, and it can rear up and turn around in the corner or alley, it is just as likely to turn back into the dead end as it is to proceed out of the partial enclosure. It is, in effect, a little robot, lacking all goal-directedness, responding reflexively with a stereo-

8. Motor Subsystems in Motivated Behavior

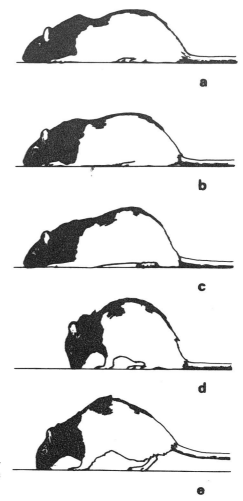

Figure 8.3. *Five phases in cephalocaudal recovery of postural support. (From Golani et al., 1979.)*

typed response pattern to each configuration of surfaces it happens to encounter.

If that is so, how come the animals usually manage to escape from the corner? Is not the increasing amplitude of their movements (see Figure 8.7) evidence of a struggle to surmount a barrier or to extricate itself from a trap? Notation analysis (Figure 8.8) reveals that after any prolonged arrest, each successive movement along a given dimension will be larger in amplitude than the one preceding it (Golani, Wolgin, & Teitelbaum, 1979). This occurs even without any trap, barrier, or restraint. Thus, even long after it has recovered virtually all normal movements, whenever such an LH animal is placed on the ground in an open field, it will freeze into immo-

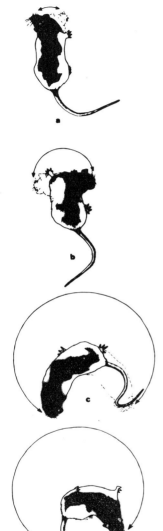

Figure 8.4. Top view of a rat performing increasingly larger amplitude horizontal lateral movements during four successive phases of recovery. Dashed line and solid line drawings indicate the extreme positions that the rat assumes during each phase. The arrows indicate the amplitude of the movements. The plus sign indicates the root of the movement, beyond which there is practically no recruitment of limb and body segments for movement. During increasingly larger lateral movements (b, c, d), the limb and body segments are recruited in a cephalocaudal order. (From Golani et al., 1979.)

bility for a while. Then it will begin to move, reactivating itself in a sequence (taking minutes) that recapitulates the pattern of recovery from akinesia (which had taken weeks). This "warm-up" phenomenon illustrates that the animal's own movements provide an important source of "spontaneous" self-activation. Similar warm-up can account for the spread of allied reflexes—for example, from grooming, to licking, to eating—seen

8. Motor Subsystems in Motivated Behavior

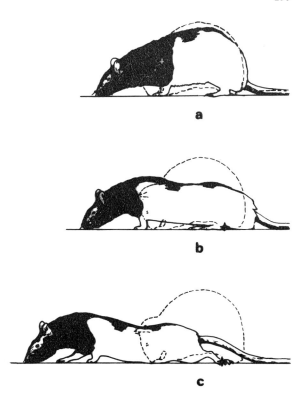

Figure 8.5. Side views of a rat performing increasingly larger amplitude longitudinal snout movements during three successive phases of recovery. Dashed line and solid line drawings indicate the extreme positions that the rat assumes during each phase. The plus sign indicates the root of the movement as in Figure 8.4.

in otherwise aphagic rats induced to eat by dropping water on the snout (Wolgin, Cytawa, & Teitelbaum, 1976). Such positive reafferent activation may also be important in states of extreme fatigue or cold, where there is danger of falling asleep if one does not keep moving.

Sources of Spontaneity in Obstinate Festinating Progression

In an attempt to resynthesize the LH syndrome in the rat by destroying, not the lateral hypothalamus, but the systems that course through it, we made a combination of small bilateral brainstem lesions—for example, in substantia nigra, locus coeruleus, midline raphe nuclei, and in the ventral pons (Schallert, Whishaw, & Teitelbaum, 1977). We did suceed in producing a form of aphagia, but instead of being cataleptic-akinetic, such an animal displayed an extremely interesting and informative form of forward locomotion. When placed on the ground, it was unable to stand, but fell over on its side. However, it got its legs underneath it again and kept its balance by beginning to walk forward [Figure 8.9(b)]. First it walked then began to run [Figure 8.9(c)], and then gallop, racing at full

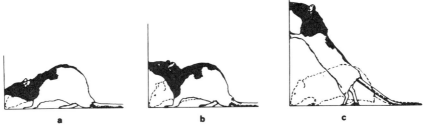

Figure 8.6. Three phases in the recovery of movement along vertical surfaces.

speed without being able to stop [Figure 8.9(d)]. It would gallop off the edge of a table or smash head-on into a wall or some other obstruction. The animal lacked all lateral head-scanning or head-orienting movements, moving straight ahead like a torpedo. When dropped supine in the air, it did not right itself, but merely tucked its head forward to its chest as it fell like a stone to earth. In a sense, with respect to posture and locomotion, this animal displays the inverse of catalepsy/akinesia. It lacks the subsystems of righting and static postural support that enable the cataleptic-akinetic animal to stand, crouch, or right itself in the air. It lacks the head-scanning movements seen so prominently throughout the course of recovery from catalepsy/akinesia. What it does retain, however, is pure forward locomotion, the subsystem that recovers only very late in catalepsy/akinesia in the rat. With only the locomotion subsystem operating, it is able to maintain a dynamic phasic equilibrium by walking and running. In the locomotion subsystem, as in the "warm-up" of head-scanning, reafferent self-activation appears in the positive feedback cycle of walk–run–gallop. This bears an interesting similarity to one form of "festination" in forward locomotion seen in some people with Parkinson's disease. When they begin to walk, they must run faster and faster; and without being able to stop, they are in great danger of injury by colliding with people or walls around them (Martin, 1967; Sacks, 1976). If such an animal is placed on a wire-mesh grid surface (that elicits grasping with the toes) instead of on the floor, it can walk without breaking into a run. However, as soon as it steps from the grid to the smooth surface of the floor, it begins to run again. This indicates that the clinging reaction elicited by the interrupted surface of the grid activates the static postural support system (the one used in cataleptic clinging) that inhibits the system involved in forward locomotion.

Similarly, as head-orientation and head-scanning reappear in recovery, the animal becomes able to inhibit its headlong forward acceleration by stopping to nibble at food or to investigate objects on the ground. In other words, not only are the response subsystems of head-scanning and head-orientation independent of the subsystem involved in forward locomotion (that in turn is independent of the subsystem involved in static postural support), but they are mutually antagonistic to each other. The "warm-up"

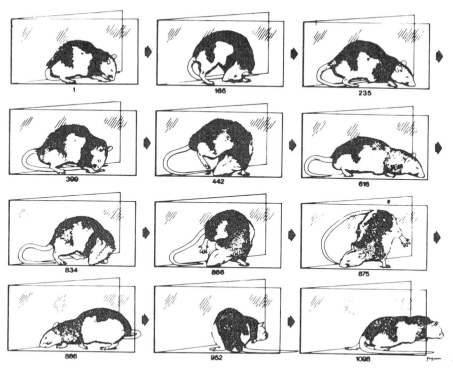

Figure 8.7. Behavioral trap of an LH rat in a corner, 4 days postoperatively. Tracings were made from one continuous film sequence, lasting more than 2 min. Only the first 34 sec are presented. (A movement-notation motor score of the first 7 sec is presented in Figure 8.8.) Each tracing shows the position reached at the maximal extent of the movement. After assuming each of these positions, the rat changed direction and moved along another dimension. The figure should be read the same way as the frame numbers run, that is, from left to right in each row, then to the row below. The rat performs only lateral and longitudinal movements with increasing amplitudes, ending in swiveling 360 degrees, and stretching as far as its rooted hindlegs allow. Thus, a small lateral movement (frames 2–165) is followed by a small longitudinal movement (166–235). Due to warm-up, the succeeding ventral flexion in the midline (236–399), lateral movement (400–442), and longitudinal movements (443–618) increase in amplitude until the rat can swivel (834–952) and stretch (952–1098) as far as anatomy permits. Lacking vertical movements along surfaces, the rat's snout must remain in contact with the ground throughout. During the last part of the sequence, not illustrated, the rat kept swiveling and stretching in and out of the corner. (From Golani et al., 1979.)

of head-scanning inhibits crouching and standing, thus freeing the akinetic animal to circle, and eventually to walk. Furthermore, a normal rat does not seem able to walk or run forward and scan or orient at the same time. If forced to run on a treadmill, a normal rat always breaks stride and stops momentarily if induced to orient its head laterally (by touching its skin with a hair, for instance). In switching from one behavior pattern

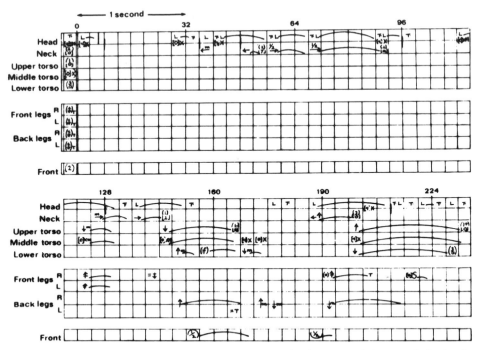

Figure 8.8. A movement-notation motor score of the first 7 sec of the same behavioral sequence that is depicted in Figure 8.7. The numerals on top indicate frame number on film. Film was taken at 32 frames per second. For information on what each symbol means, and how the score is read, see Golani et al., 1979.

to another, the normal animal is able to switch from one response subsystem to another so quickly that it all appears to be one continuous flow of behavior. However, by studying the abnormalities of catalepsy/akinesia, or its converse, festinating locomotion, we become aware more readily of such behavioral switching from one response subsystem to another.

While the festinating animal is galloping, if it is plucked from the ground and held in the air, it becomes completely motionless. For as long as 30 min (the limit of our patience), it remained completely immobile, showing no "spontaneous" movements of head, trunk, or legs (see Figure 8.9(a)). In the air, the animal seemed to be nothing more than a simple heart-lung machine: Its heart beat and it breathed, but no other behavior could be seen. When put back on the ground, however, it instantly began to walk, run, and gallop again. In other words, this animal's "spontaneity" comes from the ground.

Finally, by making large, lateral hypothalamic lesions more anteriorly, we can produce a paradoxical form of catalepsy/akinesia (Schallert *et al.*, 1977). As shown in Figure 8.10(a), if such an animal is placed over a bar, it seems cataleptic, that is, it remains unmoving in bizarre postures

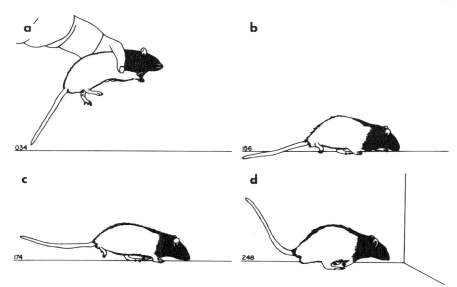

Figure 8.9. Subsystem for forward locomotion isolated in rat with combined midbrain lesions. (Pictures traced from a motion film; at the lower left of each panel are the film frame numbers for this behavioral sequence, which was shot at 64 frames per second.) (a) In the air, the animal is completely motionless. (b) On the ground, the animal begins to walk. (c) Then it runs. (d) Then it gallops at full speed, without being able to stop. (From Schallert et al., 1977.)

for long periods. It does not show "spontaneous" head scans, and does not climb down. However, if we move a visual or a smell stimulus toward it, or if we touch its snout or its flank, it instantly orients [Figure 8.10 (b)–(d)], follows the stimulus, and bites it. In such an animal, we have depressed the subsystems involved in "spontaneous" locomotion and head-scanning, but we have not depressed the head-orientation subsystem. Therefore, stimuli that can elicit orienting provide an independent source of "spontaneity" in such animals.

In summary, by analyzing the dimensions of movement in recovery from akinesia and in other aberrations in locomotion we have revealed independent behavioral subsystems that may form the substrate for exploration and orienting to goal objects. By various forms of localized brain damage, or with the use of drugs, we have developed simplified animal preparations in which the response subsystems of locomotion, postural support, head-scanning, and head-orienting are physically isolated from each other. In such simplified, isolated subsystems we see clearly the role of reafferent activation in the walk–run–gallop cycle of festinating locomotion, and in the warm-up of head-scanning during recovery from akinesia. Their selective damage or incomplete recovery reveals seemingly paradoxical phenomena that now make sense: In catalepsy/akinesia, an animal that rights

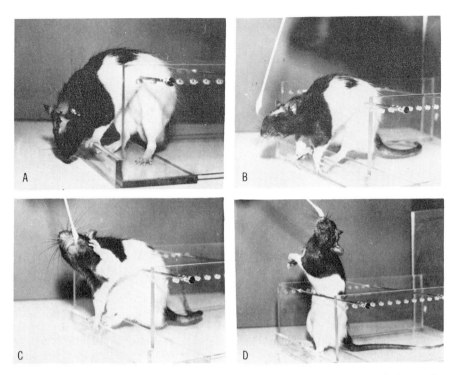

Figure 8.10. *With large anterolateral hypothalamic lesions, "paradoxical" catalepsy is produced. (a) The animal hangs cataleptically, unmoving for long periods. (b, c, and d) Such a rat comes to life, if a stimulus to which it can orient is moved toward it (From Schallert et al., 1977.)*

itself beautifully in free fall, but will stand with bizarre limb postures (as long as its center of gravity is stable); an animal that walks forward relatively normally, but gets trapped in corners; or an animal that seems cataleptic-akinetic when clinging in an awkward posture, but which instantly comes to life if approached by a stimulus that elicits orienting.

Behavioral "traps" in partial enclosures reveal that much of what seems "spontaneous" in motivated behavior is in fact not so: The variety of actions that yield the illusion of spontaneity is caused by the variety of environmental stimuli successively encountered as a result of each automatic response to the previous stimulus.

This leaves us with an essential paradox: We firmly believe from the awareness that accompanies our own acts that we have freedom to control our own actions. In animals, as well as people, operant behavior seems to embody such freedom; for example, the arbitrary operant appears to be the antithesis of the involuntary, automatic reflex (Teitelbaum, 1967, 1976). However, when we break down the systems involved in motivated, operant behavior—by partial destruction of the lateral hypothalamus, for example—then in every instance, we reveal reflexive behavioral subcomponents. Apparent "spontaneity" becomes, when analyzed, mere reflexes. These are compounded in an amazing manner, but particular stimuli can be determined for each source of "spontaneity."

We are faced with a fundamental problem of emergence. Somehow, with the addition of processes of learning and memory, the activation of simple reflexes in combination yields behavior that seems directed toward a goal that exists only in memory, that is, nonreflexive. To understand how such emergent properties arise, we must continue to discover how reflexes are transformed into operants.

ACKNOWLEDGMENTS

We thank Nancy Peshkin for technical assistance and Debie Kassner-Whelchel for typing the manuscript.

REFERENCES

Anand, B. K., & Brobeck, J. R. Hypothalamic control of food intake. *Yale Journal of Biology and Medicine*, 1951, *24*, 123–140.

Antelman, S. M., Szechtman, H., Chin, P., & Fisher, A. E. Tail-pinch-induced eating, gnawing, and licking behavior in rats: Dependence on the nigrostriatal dopamine system. *Brain Research*, 1975, *99*, 319–337.

Antelman, S. M., Rowland, N. E., & Fisher, A. E. Stress related recovery from lateral hypothalamic aphagia. *Brain Research*, 1976, *102*, 346–350.

Barfield, R. J., & Sachs, B. D. Sexual behavior: Stimulation by painful electric shock to the skin of male rats. *Science*, 1968, *161*, 392–394.

Cheng, M. F., Rozin, P., & Teitelbaum, P. Semi-starvation retards the development of food and water regulations in infant rats. *Journal of Comparative and Physiological Psychology*, 1971, *76*, 206–218.

Eshkol, N., & Wachmann, A. *Movement notation*. London: Weidenfeld and Nicolson, 1958.

Eshkol, N. et al. *Classical ballet*. Holon, Israel: Movement Notation Society, 1968.

Flynn, J. P. Patterning mechanisms, patterned reflexes, and attack behavior in cats. In J. K. Cole & D. D. Jensen (Eds.), *Nebraska symposium on motivation*. Lincoln, Nebraska: Univ. of Nebraska Press, 1973, *20*, 125–153.

Fog, R. *On stereotypy and catalepsy*. Copenhagen: Munksgaard, 1972.

Golani, I., Wolgin, D. L., & Teitelbaum, P. A proposed natural geometry of recovery from akinesia in the lateral hypothalamic rat. *Brain Research*, 1979, *164*, 237–267.

Hess, W. R. *Diencephalon: Autonomic and extrapyramidal functions*. New York: Grune & Stratton, 1954.

Hoebel, B. G. Brain reward and aversion systems in the control of feeding and sexual behavior. In J. K. Cole & T. B. Sonderegger (Eds.), *Nebraska Symposium on Motivation*. Lincoln, Nebraska: Univ. of Nebraska Press, 1975, *22*, 49–112.

Levitt, D. R., & Teitelbaum, P. Somnolence, akinesia, and sensory activation of motivated behavior in the lateral hypothalamic syndrome. *Proceedings of National Academy of Sciences*, 1975, *72*, 2819–2823.

Marshall, J. F., Turner, B. H., & Teitelbaum, P. Sensory neglect produced by lateral hypothalamic damage. *Science,* 1971, *174,* 523–525.

Marshall, J. F., Richardson, J. S., & Teitelbaum, P. Nigrostriatal bundle damage and the lateral hypothalamic syndrome. *Journal of Comparative and Physiological Psychology,* 1974, *87,* 808–830.

Marshall, J. F., & Teitelbaum, P. Further analysis of sensory inattention following lateral hypothalamic damage in rats. *Journal of Comparative and Physiological Psychology,* 1974, *86,* 375–395.

Marshall, J. F., Levitan, D., & Stricker, E. M. Activation-induced restoration of sensorimotor functions in rats with dopamine-depleting brain lesions. *Journal of Comparative and Physiological Psychology,* 1976, *90,* 536–546.

Martin, J. P. *The basal ganglia and posture.* Philadelphia: Lippincott, 1967.

Nance, D. M., Phelps, C., Shryne, J. E., & Gorski, R. A. Alterations by estrogen and hypothyroidism in the effects of septal lesions on lordosis behavior of male rats. *Brain Research Bulletin,* 1977, *2,* 49–53.

Olds, J. *Drives and reinforcements: Behavioral studies of hypothalamic functions.* New York: Raven Press, 1977.

Robinson, T. E., & Whishaw, I. Q. Effects of posterior hypothalamic lesions on voluntary behavior and hippocampal electroencephalograms in the rat. *Journal of Comparative and Physiological Psychology,* 1974, *86,* 768–786.

Sacks, O. W. *Awakenings.* New York: Vintage Books, 1976.

Schallert, T., & Whishaw, I. Q. Two types of aphagia and two types of sensorimotor impairment after lateral hypothalamic lesions: Observations in normal-weight, dieted, and fattened rats. *Journal of Comparative and Physiological Psychology,* 1978, *92,* 720–741.

Schallert, T., Whishaw, I. Q., De Ryck, M., & Teitelbaum, P. The postures of catecholamine-depletion catalepsy: Their possible adaptive value in thermoregulation. *Physiology and Behavior,* 1978, *21,* 817–820.

Schallert, T., Whishaw, I. Q., Ramirez, V. D., & Teitelbaum, P. Compulsive, abnormal walking caused by anticholinergics in akinetic 6-hydroxydopamine-treated rats. *Science,* 1978, *199,* 1461–1463.

Schallert, T., Whishaw, I. Q., & Teitelbaum, P. Sources of spontaneity in the behavior of aphagic rats. International Conference on Physiology of Food and Fluid Intake. Jouy-en-Josas, France, 1977. (Abstract)

Seyffarth, H., & Denny-Brown, D. The grasp reflex and the instinctive grasp reaction. *Brain,* 1948, *71,* 109–183.

Sherrington, C. S. *The integrative action of the nervous system.* Forge Village, Mass.: Murray Printing Company, 1906. (Issued as a Yale paperbound, 1961)

Szechtman, H., Siegel, H. I., Rosenblatt, J. S., & Komisaruk, B. R. Tail-pinch facilitates onset of maternal behavior in rats. *Physiology and Behavior,* 1977, *19,* 807–809.

Teitelbaum, P. The encephalization of hunger. In E. Stellar & J. M. Sprague (Eds.), *Progress in Physiological Psychology.* Vol. 4. New York: Academic Press, 1971.

Teitelbaum, P. The use of recovery of function to analyze the organization of motivated behavior in the nervous system. *Neuroscience Research Program Bulletin,* 1974, *12,* 255–260.

Teitelbaum, P., Cheng, M. F., & Rozin, P. Development of feeding parallels its recovery after hypothalamic damage. *Journal of Comparative and Physiological Psychology,* 1969, *67,* 430–441.

Teitelbaum, P., & Epstein, A. N. The lateral hypothalamic syndrome: Recovery of feeding and drinking after lateral hypothalamic lesions. *Psychological Review,* 1962, *69,* 74–90.

Teitelbaum, P., Schallert, T., Whishaw, I. Q., & Golani, I. Sources of "spontaneity" in motivated behavior. In E. Satinoff & P. Teitelbaum (Eds.), *Behavioral neurobiology.* New York: Plenum, in press.

Teitelbaum, P., & Stellar, E. Recovery from the failure to eat produced by hypothalamic lesions. *Science*, 1954, *120*, 894–895.

Teitelbaum, P., & Wolgin, D. L. Neurotransmitters and the regulation of food intake. *Progress in Brain Research*, 1975, *42*, 235–249.

Teitelbaum, P., Wolgin, D. L., De Ryck, M., & Marin, O. S. M. Bandage-backfall reaction: Occurs in infancy, hypothalamic damage, and catalepsy. *Proceedings of the National Academy of Sciences, U.S.A.*, 1976, *73*, 3311–3314.

Twiggs, D. G., Popolow, H. B., & Gerall, A. A. Medial preoptic lesions and male sexual behavior: Age and environmental interactions. *Science*, 1978, *200*, 1414–1415.

Twitchell, T. E. The restoration of motor function following hemiplegia in man. *Brain*, 1951, *74*, 443–480.

Twitchell, T. E. The automatic grasping responses of infants. *Neuropsychologia*, 1965, *3*, 247–259.

Twitchell, T. E. Reflex mechanisms and the development of prehension. In K. J. Connolly (Ed.), *Mechanisms of motor skill development*. New York: Academic Press, 1970, 25–45.

Ungerstedt, U. Adipsia and aphagia after 6-hydroxydopamine induced degeneration of the nigrostriatal dopamine system. *Acta Physiologica Scandinavica*, 1971, *Supplementum 367*, 95–122.

Van Harreveld, A., & Kok, D. J. À propos de la nature de la catalepsie experimentale. *Archives Néerlandaises de Physiologie de l'homme et des animaux*, 1935, *20*, 411–429.

Wagner, H. N., & Woods, J. W. Interruption of bulbocapnine catalepsy in rats by environmental stress. *Archives of Neurology and Psychiatry*, 1950, *64*, 720–725.

Wolgin, D. L., Cytawa, J., & Teitelbaum, P. The role of activation in the regulation of food intake. In D. Novin, W. Wyrwicka, & G. Bray (Eds.), *Hunger: Basic mechanisms and clinical implications*. New York: Raven Press, 1976, 179–191.

Wolgin, D. L., & Teitelbaum, P. The role of activation and sensory stimuli in recovery from lateral hypothalamic damage in the cat. *Journal of Comparative and Physiological Psychology*, 1978, *92*, 474–500.

CARL W. COTMAN

Synaptic Growth as a Plasticity Mechanism in the Brain

9

Introduction

Frequently, neural circuitry in adults is thought of as being fixed and unchangeable. However, it is now clear that neural circuitry is changeable even at the level of synaptic growth. Some years ago it was discovered in the peripheral nervous system that when a fraction of an input to a neuron or group of neurons is lost, the axons of the residual undamaged inputs sprout and form new connections in place of those lost. This phenomenon, termed "axon sprouting," has now been described in a number of places in the peripheral and central nervous systems (Cotman, 1978).

In this chapter, I will briefly describe our studies on the dentate gyrus of the rat hippocampal formation. Many of the conclusions on the nature and mechanism of the response to partial denervation have been elucidated using this system. A comprehensive review has been published (Cotman & Nadler, 1978).

Studies on the reorganization of neural circuitry are important for two reasons. First, they tell us how the brain may repair itself after injury. This ability has obvious clinical significance, but it is also relevant to physiological psychology, where lesions are commonly used to investigate the role of brain areas in behavior. After such lesions, the rewiring of neural circuitry must be taken into account. Second, studies on axon sprouting tell us about the fundamental capacities of the brain to grow new synapses and adjust its circuitry. Perhaps such responses are restricted to repair, but it is likely they can be induced in other ways not requiring lesions.

The Response to Denervation

In the dentate gyrus, the neurons are almost entirely granule cells (Figure 9.1). These cells are organized in a layer with their dendrites arborizing in a zone called the molecular layer. Synapses are strictly ordered on these cells. The bottom quarter of the dendritic field receives a projection called the commissural-associational projection from contralateral and ipsilateral hippocampus, respectively. In the outer three-quarters of the molecular layer there is a massive cortical projection from the ipsilateral entorhinal area and a minor projection from the septum and contralateral entorhinal cortex.

Studies have been carried out on the responses to a variety of lesions which deprive the granule cells of different portions of their input. I will describe only briefly the results following a unilateral entorhinal lesion.

We have used electron microscopy to follow the status of synapses at various times after the lesion. Initially, there is a rapid and massive loss of terminals from the outer three-fourths of the molecular layer in which the entorhinal input terminates. This, however, is only the beginning. At approximately one week after the lesion there is a rapid reappearance of synapses that continues for 200 or more days (Figure 9.2). The recovery of synapses is nearly complete. Thus, what is lost is regained.

Where do the new synapses come from? As mentioned previously, the remaining (contralateral) entorhinal cortex has a minor input into this area that appears to originate from cells of the same cortical layer as the ipsilateral projection. Interestingly the remaining entorhinal input does not capture all of the territory of the lost one. Only a few percent of the

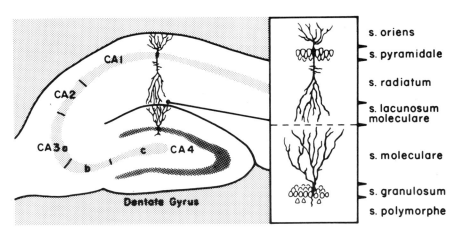

Figure 9.1. Diagrammatic cross section of the rat hippocampal formation showing the denate gyrus. The granule cell bodies are found in a layer (s. granulosum) as are their dendrites (s. moleculare). (From Cotman and Nadler, 1978.)

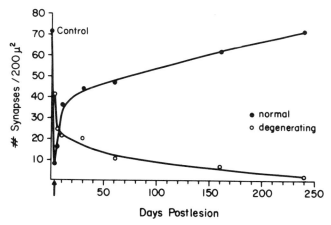

Figure 9.2. *Time course of degeneration and reinnervation in the perforant path terminal zone after an ipsilateral entorhinal lesion. (From Cotman and Nadler, 1978.)*

new terminals are of entorhinal origin (Cotman et al., 1977). Therefore, the most closely related input does not have a preference for making new synapses. The septal input in the outer zone also appears to reorganize its connections. Moreover, commissural-associational synapses that are normally restricted to the inner zone spread out and expand into the denervated area. However, they do not move all the way out to the tips of the dendrites. Instead they form a new boundary about halfway out.

Generalizations

Based on these and many other studies in the hippocampus and other brain areas, several generalizations or principles about axon sprouting can be drawn (Cotman & Nadler, 1978).

1. New functional synapses are formed by about one week. However, the process may continue for several weeks.
2. The quantity of synapses lost are nearly all replaced.
3. No new pathways are created if the lesion is done in the mature nervous system. If the lesion is created in a developing animal, there are instances of new pathways created. In adults, however, the rule is that existing inputs to neurons are rearranged.
4. Growth is selective. Some populations will grow; others will not. There are a few constraints that partially predict when a population may grow.
5. In order to get growth, it is necessary that the fiber system be im-

mediately adjacent to the denervated zone. Neighbors have a chance, but distal neighbors do not. Proximity to a denervated zone is a necessary but insufficient condition for growth.

Mechanisms

We can consider the mechanisms in two stages: (a) an initiation stage and (b) the actual stage of synapse formation. Growth must start and, in order to be significant, must result in synapses. The critical question in the initiation stage is whether or not it is possible to promote growth without actually causing extensive damage. Indications are that trophic materials that are synthesized in the cell body and flow to the terminals may regulate the growth of terminal fields. Researchers have suggested, for example, that each neuron gives off a growth-retarding factor that prevents the processes of other neurons from growing further. In the peripheral nervous system of the salamander, treatment of axons with colchicine (an inhibitor of axoplasmic flow) allows neighboring axons to form new terminals (Diamond, Cooper, Turner, & MacIntyre, 1976). We carried out similar experiments in the mammalian CNS.

We applied colchicine to the fimbria (that carries the commissural axons) in order to block the axoplasmic flow of materials transported in this tract. The treatment does not disturb significantly the conduction properties of these fibers, but it blocks the flow of proteins from one hippocampus to the other. A number of days after we applied the drug to axons in the fimbria, we examined the inner part of the molecular layer in the dentate gyrus in which commissural fibers terminate. A lesion of the fimbria results in a 20–25% decrease in the number of synapses 4 days after the operation in this zone as seen in the electron microscope. Numerous degenerating synapses are seen. In contrast, in colchicine-treated animals 4 days after treatment, there is no net loss of synapses and there are no degenerating synapses. Degeneration seems minimal, if at all present. By 11 days, in these animals, the number of synapses increases and by 60–70 days there is a further increase. The number of synapses in the inner molecular layer increases about 20% in colchicine-treated animals. Thus, a disturbance in the metabolism of a group of fibers can trigger synaptic growth. This then raises the possibility that altered metabolic states might induce synapse formation. Certainly, denervation of the type associated with cell loss is unnecessary.

The rate of reactive growth is not always the same. It is possible to speed up or slow down the rate of synapse formation after lesions. One of the conditions we have used is to make a partial entorhinal lesion and follow this within a few days by a complete entorhinal lesion. In a typical experiment the interlesion interval is four days and the response is measured two days after the second lesion is completed. Normally a growth response requires about six days and two days would be an insufficient survival time to detect such a response. However, with a conditioning or priming lesion, the total response will occur in a period of two days. Hence, there is a remarkable acceleration of the growth, thus indicating that with progressive damage the time course of recovery is not the same as for a single lesion. Other data indicate that not only does the rate increase, but also the extent of the fiber growth may increase.

New synapse formation involves the construction of a new synaptic junction. As shown in Figure 9.3, one possible mechanism is the reuse of old synaptic sites. When a terminal dies, it leaves a vacant place on the membrane. A new fiber may grow and simply reoccupy this place. On the other hand, the sites may be lost when the terminals die and new fibers may induce a new synaptic junction. Each mechanism has its own inplications. The first mechanism does not give the target cell a chance to control which fibers it takes nor a chance to reorder them on the dendrite. It also makes the process of synapse formation distinct from a developmental process in which the sites do not, as a rule, exist. The second mechanism is similar to a developmental one.

These two possible mechanisms can be distinguished as follows: The simple prediction is that if there is reutilization of sites (the first mechanism), the same total number should exist all the way through the degeneration–reinnervation process. This is not the case. We are able to show conclusively that the total number of the synaptic sites is reduced shortly after the operation and subsequently recovers over time (Matthews *et al.*,

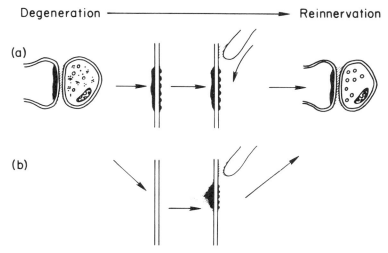

Figure 9.3. Schematic diagram of reinnervation by (a) occupation of a vacant (old) postsynaptic site or (b) induction of a new postsynaptic site. (From Cotman, 1976.)

1976). When it has been possible to test the new synapses by neurophysiological methods, they are functional (Cotman & Nadler, 1978). Thus the entire process for synaptogenesis can take place in the adult brain within a few days.

Conclusion and Summary

We might ask what is the significance of axon sprouting or reactive synaptogenesis? In all systems that have been studied a clearly aberrant pattern of connections is formed and in some cases this produces abnormal behavior (Schneider, 1973). In other cases, however, the regrowth of new connections aids in the recovery from damage (Goldberger & Murray, 1978). Each system has to be dealt with in the context of its own changes. I have given some of the rules that provide us some predictive power over the types of changes that may occur. Whatever the outcome of the particular changes, it is clear that the adult brain has the capacity to reorganize dynamically its circuitry and that after brain damage this plasticity must be taken into account.

It would be surprising if this extraordinary ability were used only for the purposes of repair. Indeed, the finding that synaptogenesis can be induced without denervation, such as shown in the colchicine experiments, supports the contention that perhaps neuronal circuitry continuously remodels itself according to a balance of trophic materials in the environment. It seems that studies on the reaction of brain damage reveal the inherent plastic properties of the adult brain. For this reason I prefer the term reactive synaptogenesis over axon sprouting to describe the reorganization of neuronal circuitry. Reactive synaptogenesis, as the term implies, provides insights into the types of changes that can occur in the wiring of the brain, the mechanisms that underlie these changes, and the perturbations, both normal and pathological that will elicit them. The challenge of the future is to narrow the gap between environmentally induced changes and those induced internally by changes in mental state.

REFERENCES

Cotman, C. W. Lesion-induced synaptogenesis in brain: A study of dynamic changes in neuronal membrane specializations. *Journal of Supramolecular Striate*, 1976, *4*, 319–327.
Cotman, C. W. (Ed.). *Neuronal plasticity*. New York: Raven Press, 1978.
Cotman, C. W., Gentry, C., & Steward, O. Synaptic replacement in the dentate gyrus after unilateral entorhinal lesion: Electron microscopic analysis of the extent of replacement of synapses by the remaining entorhinal cortex. *Journal of Neurocytology*, 1977, *6*, 455–464.
Cotman, C. W., & Nadler, J. V. Reactive synaptogenesis in the hippocampus, In C. W. Cotman (Ed.), *Neuronal plasticity*. New York: Raven Press, 1978, pp. 227–271.

Diamond, J., Cooper, E., Turner, C., & MacIntyre, L. Trophic regulation of nerve sprouting. *Science,* 1976, *193*, 371–377.

Goldberger, M. E., & Murray, M. Recovery of movement and axonal sprouting may obey some of the same laws. In C. W. Cotman (Ed.), *Neural plasticity*. New York: Raven Press, 1978, pp. 73–96.

Matthews, D. A., Cotman, C., & Lynch, G. An electron microscopic study of lesion-induced synaptogenesis in the dentate gyrus of the adult. II. Reappearance of morphologically normal synaptic contacts. *Brain Research,* 1976, *115*, 23–41.

Schneider, G. E. Early lesions of superior colliculus: Factors affecting the formation of abnormal retinal projections. *Brain Behavior and Evolution,* 1973, *8*, 73–109.

LESLIE H. HICKS

The Basal Ganglia and Psychomotor Behavior[1]

10

The problem with which we have been primarily interested is the relationship of basal ganglia structures to the speed and efficiency with which organisms respond to significant environmental stimuli. Clearly, detection of stimuli and responses to them depends broadly on sensory and motor nervous functions. But there are differences in psychomotor performance during the life-span of an organism (i.e., aging) and across different categories of organisms (for example, brain damage or psychoses) that cannot be related directly to differences in primary sensory or motor nervous structures.

It was the search for central nervous system correlates of the psychomotor slowing that accompanies normal aging that led us to focus on the basal ganglia as possible structures that are related to this process (Hicks & Birren, 1970). Rather than studying anatomical or electrophysiological changes in the basal ganglia in organisms of different ages, we chose to alter the basal ganglia of rats by electrolytic lesions and to study the behavioral consequences.

The term "basal ganglia" has been used to refer to a number of different subcortical nuclear masses, even in the fairly recent history of neuroanatomy. As Mettler (1968) points out, nearly all the large gray masses at the base of the brain, including the thalamus, have been historically considered as basal ganglia by some neuroanatomists. But the current consensus of what structures comprise the basal ganglia would list the caudate nucleus,

[1] Supported by PHS Grant NS09630 from the National Institute of Neurological Diseases and Stroke.

putamen, globus pallidus, and the claustrum. Most authors also consider the subthalamic nucleus and the substantia nigra as sufficiently closely related to the structures just listed to warrant their mention whenever the basal ganglia are discussed (see Figure 10.1).

Typically, the basal ganglia have been viewed as motor structures that were anatomically the major part of the extrapyramidal motor system. Their involvement in a number of dyskinesias in human patients, including athetosis and parkinsonism, has been considered prime evidence for their role as fundamentally movement-related structures.

However, the view that is informative of our approach to basal ganglia function is that of Brodal (1969), which deemphasizes the motor role of the basal ganglia. After remarking on the lack of nuclei in the basal ganglia that connect directly to the spinal cord, Brodal states that "the basal ganglia can scarcely be considered as important 'motor centers'" (Brodal, 1969, p. 188). He goes on to say that "the basal ganglia appear to be first and foremost concerned in the collaboration between the cerebral cortex and the thalamus" (p. 188). This collaborative or integrative role of the basal ganglia makes it likely that they are concerned with a number of behaviors in which sensory-motor integration is important.

The tasks we used to study psychomotor behavior were active avoidance learning in a shuttle box and learning of a simple maze. Our first experiments involved bilateral lesions of the globus pallidus. This structure, which is phylogenetically older than the other structures of the corpus striatum (namely, the caudate and the putamen), receives a heavy inflow

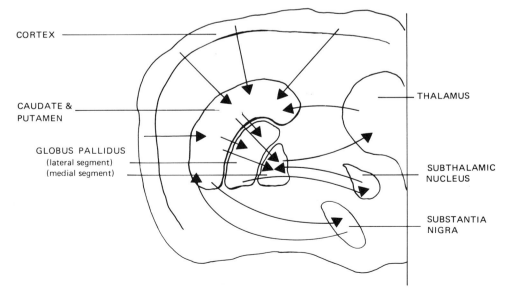

Figure 10.1. A diagram showing the major fiber connections of the basal ganglia.

of fibers from other basal ganglia. Anatomically, therefore, it is a key nucleus in the basal ganglia.

The pathophysiology of the globus pallidus offers some evidence of its function in various organisms. For example, in human patients who inhale sufficient carbon monoxide to induce coma, there is symmetrical damage to the globus pallidus with a sequel of disorders in movement (Denny-Brown, 1962). These patients are unable to speak, show rigidity of arms and legs, or flexion of the arms accompanied by extension of the legs. Denny-Brown also states that in human patients, "small pallidal lesions have not been associated with extrapyramidal signs other than slowness in initiation of movement" (Denny-Brown, 1962, p. 61). Our avoidance experiments were designed to investigate that proposition.

After training on a two-way shuttle avoidance task, 16 rats received bilateral electrolytic lesions in the globus pallidus. The performance of this operated group was compared with that of an unoperated control group.

The task consisted of an acquisition phase that lasted ten days. On days 1 through 5, the rats received 25 trials per day on a two-way shuttle avoidance task. The task required the subject to respond within a 5-sec period during which warning signals consisting of a buzzer and a light were on. If the rat did not cross a barrier to the other side of the cage during this period, an electric shock was delivered through the grid bars on the floor of the cage. On days 6–8, the warning signal duration was reduced to 3.5 sec, and on days 9 and 10, the signal was reduced to 2.5 sec. The procedure was designed to press the subjects into faster responding. Twenty-five trials were given on each of the ten days. Performance on Days 9 and 10 was used as the standard against which postlesion avoidance was measured.

Ten days following the placement of bilateral lesions in the globus pallidus, the rats were given two successive retention testing sessions of 25 trials per day on the avoidance task with a warning signal duration of 2.5 sec. Additional retention sessions took place at 20 and 30 days after surgery.

The results showed that operated and control animals made a similar number of avoidances with similar latencies during the last two presurgery sessions. But on tests of retention at 10, 20, and 30 days following surgery, unoperated animals made more and faster avoidances than did the lesioned animals (Figure 10.2).

Performance of rats with bilateral lesions in the substantia nigra was compared with that of unoperated control animals at 10, 20, and 30 days after surgery. Figure 10.2 shows the mean number of avoidances for this comparison. On all the retention tests, substantia nigra animals made fewer avoidance responses and their avoidance latencies were longer than those of unoperated controls, but were not different from globus pallidus animals.

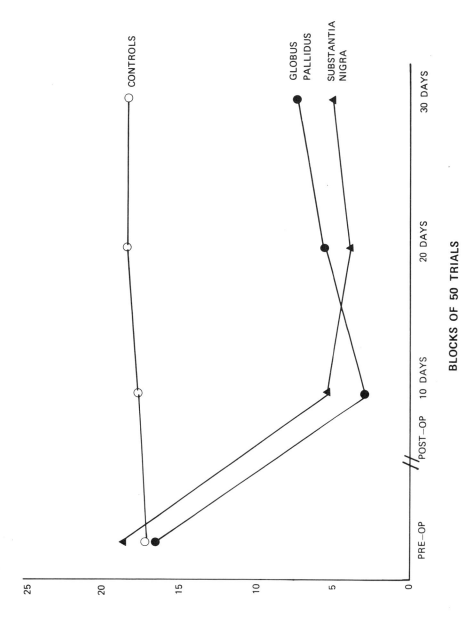

Figure 10.2. Mean number of avoidance responses for globus pallidus, substantia nigra, and unoperated controls before and after the operation.

In another experiment with a slightly different procedure, animals receiving bilateral lesions in the subthalamic nuclei were compared with sham-operated as well as unoperated controls on the acquisition of a two-way active avoidance task. Animals received 50 trials per day for four days. The warning signal duration, during which a successful avoidance response was possible, was 2.5 sec on all trials.

Subthalamic animals made significantly fewer avoidance responses during acquisition than did either sham-operated or unoperated controls. On a later test of retention by the groups, which now included an additional subthalamic group made by operating the former unoperated controls, both subthalamic groups performed significantly worse than did the sham-operated animals (Figures 10.3 and 10.4). The performance of none of these groups, including the controls, is particularly impressive, although it is characteristic of the Sprague–Dawley strain on this task.

Avoidance tasks were also used to evaluate the effects of caudate nucleus lesions. To examine the question of possible regional differences in functions within this nucleus, a claim made by some investigators (Neill & Grossman, 1970; Winocur, 1974), the caudate was divided into six segments and lesioned accordingly in six separate groups of animals. Table 10.1 gives information about the groups.

After 10 days of preoperative avoidance training, there were no differences among the caudate groups and the controls. But postoperative avoidance behavior, evaluated three times at 10-day intervals after surgery, demonstrated differences. All caudate groups were inferior to controls 10 days postoperatively. The differences were present but less pronounced 20 days postoperatively, and for all groups except the posterior-lesioned animals, the caudates were only slightly inferior to control animals at 30 days after the operation (Figure 10.5).

The results of the two-way avoidance task demonstrate clearly that rats receiving bilateral lesions in each of the basal ganglia structures that we tested, globus pallidus, substantia nigra, subthalamic nucleus, and caudate

Table 10.1
Caudate Group Designation and Lesion Sites

Group	Bregma coordinates		
	AP	RL	H
Anterodorsal	2.6	2.5	−4.0
Anteroventral	2.3	3.0	−5.5
Middorsal	1.4	2.9	−4.0
Midventral	1.4	3.8	−5.5
Posterodorsal	− .8	4.2	−4.0
Posteroventral	− .8	5.0	−5.5

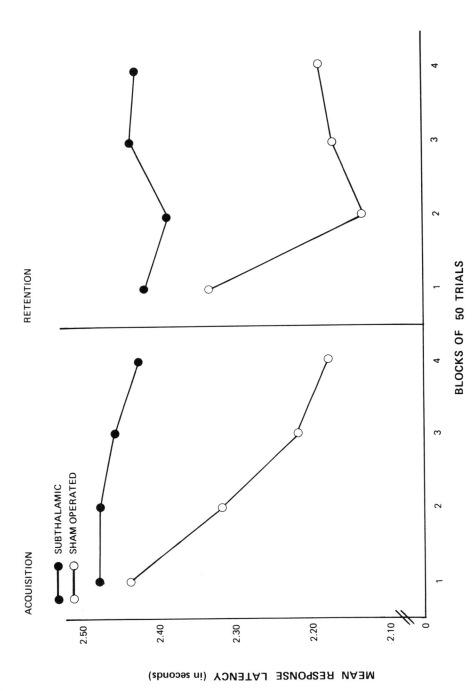

Figure 10.3. Mean response latencies on avoidance task during acquisition and retention for subthalamic and sham-operated animals.

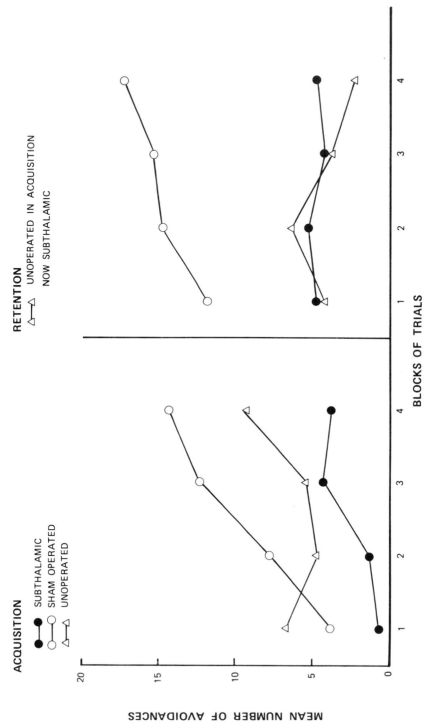

Figure 10.4. Mean number of avoidance responses during acquisition for subthalamic, sham-operated, and unoperated animals and during retention for subthalamic, sham, and subthalamic group from the formerly unoperated controls.

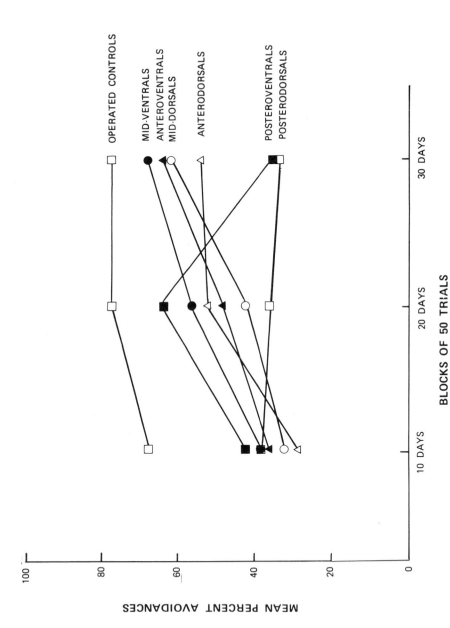

Figure 10.5. Mean percentage of avoidance responses for caudate-lesioned anterior, medial, posterior, and operated control animals at each of the postoperative retention sessions on two-way active avoidance.

nucleus, show performance deficits. On a task requiring the rapid initiation of a movement in response to a sensory stimulus, animals with basal ganglia lesions are impaired. But the task requirements do not permit an easy analysis of what specific components of behavior involved in this performance are most impaired. For example, is it motivation in the form of decreased arousal or increased fear? Is it sensory, that is, impaired attention or sensation itself? Or is it motor, the inability to prepare for and execute this quick movement?

Performance on another behavioral task, place, and response learning was also examined in these caudate-lesioned animals. Overtraining was introduced as a variable that might affect an animal's propensity for place or response habits. Animals were given 20 trials per day until they made 10 consecutive correct choices of the side on which the food was located. One-half the animals received 100 overtraining trials, the other half received no overtraining trials. To determine whether an animal was a place or response learner, a critical trial was run starting from the side opposite to that used in the original learning. On this trial, if the animal turned in the same direction as in original learning, it was designated a response learner; if it went to the same goal-box, necessitating a turn in the opposite direction from that of original learning, it was designated a place learner.

Based on an earlier study (Hicks, 1964) more response learners were expected in the overtrained group and more place learners in the non-overtrained group. Our additional interest in this study was to determine whether lesions in this extrapyramidal "motor" structure would lead to more place than response learners, and what effect overtraining would have. Of course, we were interested in the simple question of whether the lesioned animals showed a deficit compared to control animals.

The major conclusion from the results of the place and response task is that the anterodorsal and both posterior caudate groups showed increased trials to criterion on the reversal phase of the task compared to sham-operated controls (Figure 10.6).

A second finding was a relative increase in place learners in the operated overtrained posterior caudate groups. This latter finding is similar to some preliminary results that we have obtained with rats lesioned in the region of the claustrum and the caudate. These claustral-caudate animals were predominantly place learners. With increasing overtraining, they did not shift to motor or proprioceptive cues, but maintained performance based on external cues.

We have also carried out studies of place and response learning on animals with lesions in the substantia nigra, the subthalamic nucleus, and the nucleus center median of the thalamus. These results can be reported simply: There were no differences between operated and sham-operated animals in performance on this task.

Basal ganglia structures have become immensely popular with brain

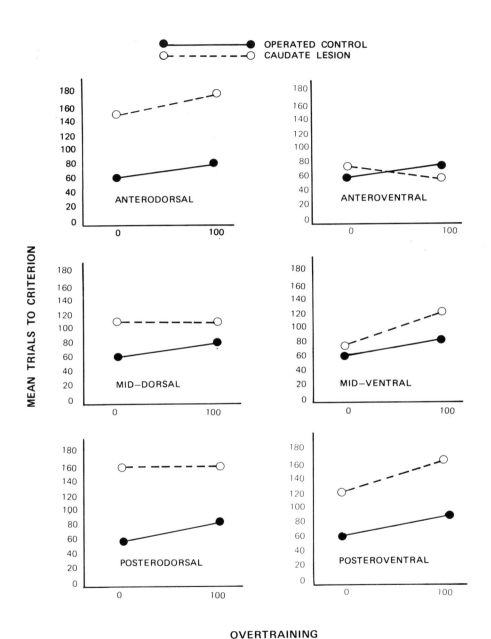

Figure 10.6. Mean trials to criterion on place and response task reversal as a function of degree of overtraining.

researchers. A wide array of the latest techniques in neuroanatomy, neurophysiology, neurochemistry, and neuropsychopharmacology has been used to study these nuclei. What have these studies told us about the role of the basal ganglia in behavior? We can attempt a brief separate summary for each of the nuclei.

First, we shall discuss the globus pallidus. Anatomically, it receives fibers from the striatum (most of the striatal efferents) in a topically organized fashion. The pallidum has massive efferent bundles of fibers (the major one being the *ansa lenticularis*) that project to the thalamus (ventral lateral and ventral anterior), the subthalamic nucleus, the substantia nigra, the red nucleus, the reticular formation, the hypothalamus, and the habenula. And although some investigators (Denny-Brown, 1962) consider it to be the single most important structure in the motor system of primates, the role of this well-connected nucleus in behavior is not clearly understood.

Although there have been reports like ours of deficits in active avoidance learning for rats (Beatty & Siders, 1977) and for cats (Laursen, 1962), there have been other studies reporting no symptoms following bilateral lesions of the pallidum in monkeys (Kennard, 1944; Ranson & Berry, 1941).

Among the most striking behavioral symptoms in rats following bilateral lesions to the globus pallidus are profound aphagia and adipsia. This finding was first reported by Morgane (1961) and has been described subsequently by several investigators (e.g., Levine & Schwartzbaum, 1973; Marshall, Richardson, & Teitelbaum, 1974). Our pallidal animals also showed aphagia and adipsia. These results raise the possibility that the pallidum, in rats, is important in species-specific behavior associated with movements such as chewing and swallowing involved in feeding and drinking.

Following lesions to the subthalamic nucleus in rats, we found a deficit in active avoidance learning, but not on a T-maze learning task. But Thompson (1974) reported memory deficits for rats with subthalamic lesions on a maze discrimination-learning task. Although lesions in this structure produce ballistic movement disorders in human patients, we observed no gross motor disturbance in rats with bilateral subthalamic damage. However, Lindsley, Ranf, Fernandez, & Wyrwicka (1975) found marked hypokinesia in cats with bilateral subthalamic lesions.

As in our experiment, lesions of the substantia nigra in rats seem consistently to produce deficits on avoidance tasks (Delacour, Echavarria, Senault, & Houcine, 1977; Mitcham & Thomas, 1972; Fibiger, Phillips, & Zis, 1974). However, there seem to be species differences in nigral lesion effects. Laursen (1962) reported that his cats with nigral lesions were unfit for behavioral studies because of a severe generalized muscular weakness.

Typically, investigators stress the correlation between dopamine depletion following nigral lesions and the behavioral effects. Although dopamine is depleted in human Parkinson patients, not all investigators agree that lesions in the substantia nigra are basic to Parkinsonism. And, again, it is not possible to produce parkinsonlike movement disorders in animals with lesions to this structure alone.

Lesions of the caudate nucleus in the rat impair performance on tasks as diverse as passive avoidance learning (Kirkby & Kimble, 1968), one- and two-way active avoidance learning (Kirkby & Kimble, 1968), position reversal learning (Chorover & Gross, 1963; Divac, 1971), and brightness discrimination reversal (Kirkby, 1969). There are, however, contradictory findings. Some investigators report no deficits on active avoidance learning, and the location of the lesion within the caudate is said to be the determining factor. Indeed, we have preliminary data from two separate studies on rats in a T-maze in which there were either no differences between caudate and control animals or in which the caudate animals performed better than the controls.

Although it is true that cortical afferents to the caudate are topographically organized, it is not always easy to produce differential behavioral performance associated with differences in location of lesions in the rat caudate. For the monkey and the cat, in which the cortical-caudate projections are much heavier than in the rat, regional specialization of the caudate associated with cortical specialization has been demonstrated better (Divac, 1972; Rosvold, 1968).

These animal studies show that the caudate is involved in a variety of psychomotor tasks and that its functions, similar to the functions of its cortical fiber sources, are not narrowly motor. Sensory, perceptual, and emotional factors are also involved. The existence of high levels of neurotransmitter substances in the caudate has also contributed to a revival of speculation about its primary role in mental illness (Stevens, 1973). Mettler advanced an earlier view in 1955. Based on his studies of striatal anatomy, he stated that schizophrenia resulted from striatal dysfunction. Following striatal damage, organisms experience difficulty in starting and stopping their behavior. Although schizophrenia is considered by most observers as primarily a disorder of thought processes, this disordered thought must be finally expressed in verbal or bodily movement. And therefore there is no reason to discard Mettler's hypothesis as rather far-fetched. Although, in fact, until the recent neurochemical discoveries, it never caught on.

Electrolytic lesions are often criticized for the nonspecificity of their effects. Fibers of passage, as well as cell bodies, and fibers containing different neurotransmitters at a particular brain site are indiscriminately destroyed. Chemical lesioning techniques with their selective destruction of specific neurochemical pathways or, in some cases, their sparing of axons of passage are preferred for their increased precision. But Delacour (1977) points out that in order to produce a deficit in active avoidance behavior in rats by injection of 6–OHDA into the substantia nigra, the dose level must be sufficiently high so that the lesion effect is nonspecific rather than confined to the dopaminergic nigrostriatal pathway. If this is so, conclusions that the active avoidance deficit following lesions to the substantia nigra or the caudate result primarily from interference with the dopaminergic nigrostriatal pathway cannot be made confidently.

The basal ganglia seem to have a greater involvement in disorders of movement in human beings than in infra-human animals. It is difficult to reproduce Parkinsonism in animals by destroying the basal ganglia. Can symptoms resembling those of Huntington's chorea be produced reliably in animals by damaging the striatum? Originally, we were interested in the basal ganglia's role in the speed of initiating a movement. Because this behavior is in response to an external stimulus, it is clear that sensory factors are involved, in addition to whatever "integrative activity" goes on between the detection of a stimulus and the motor response to it. Lesions in the basal ganglia structures we have examined do produce a deficit in active avoidance performance. Perhaps this is evidence for their function in the initiation of movement. As has been stated earlier elsewhere (Hicks & Birren, 1970), the diffuse connections of the basal ganglia with other important sensory (reticular formation), motor (cerebral motor cortex, cerebellum), and association (association cortex) structures make it possible that they are implicated in the psychomotor slowing that occurs in brain-damaged and older human beings. But more and better evidence is needed before a firm conclusion can be made.

REFERENCES

Beatty, W. W., & Siders, W. A. Effects of small lesions in the globus pallidus on open-field and avoidance behavior in male and female rats. *Bulletin of the Psychonomic Society*, 1977, *10*, 98–100.

Brodal, A. *Neurological anatomy* (2nd ed.). New York: Oxford Univ. Press, 1969.

Chorover, S., & Gross, C. Caudate nucleus lesions: Behavioral effects in the rat. *Science*, 1963, *141*, 826–827.

Delacour, J., Echavarria, M. T., Senault, B., & Houcine, D. Specificity of avoidance deficits produced by 6-hydroxydopamine lesions of the nigrostriatal system of the rat. *Journal of Comparative and Physiological Psychology*, 1977, *91*, 875–885.

Denny-Brown, D. *The basal ganglia*. London: Oxford Univ. Press, 1962.

Divac, I. Frontal lobe system and spatial reversal in the rat. *Neuropsychologia*, 1971, *9*, 175–183.

Divac, I. Neostriatum and functions of prefrontal cortex. *Acta Neurobiologiae Experimentalis*, 1972, *32*, 461–477.

Fibiger, H. C., Phillips, A. G., & Zis, A. P. Deficits in instrumental responding after 6-hydroxydopamine lesions of the nigro-neostriatal dopaminergic projection. *Pharmacology Biochemistry and Behavior*, 1974, *2*, 87–96.

Hicks, L. H. Effects of overtraining on acquisition and reversal of place and response learning. *Psychological Reports*, 1964, *15*, 459–462.

Hicks, L. H., & Birren, J. E. Aging, brain damage, and psychomotor slowing. *Psychological Bulletin*, 1970, *74*, 377–396.

Kennard, M. A. Experimental analysis of the functions of the basal ganglia in monkeys and chimpanzees. *Journal of Neurophysiology*, 1944, *7*, 127–150.

Kirkby, R. J. Caudate nucleus lesions and perseverative behavior. *Physiology and Behavior*, 1969, *4*, 451–454.

Kirkby, R. J., & Kimble, D. P. Avoidance and escape behavior following striatal lesions in the rat. *Experimental Neurology*, 1968, *20*, 215–227.

Laursen, A. M. Conditioned avoidance behavior of cats with lesions in globus pallidus. *Acta Physiologica Scandinavica*, 1962, *55*, 1–9.

Levine, M. S., & Schwartzbaum, J. S. Sensorimotor functions of the striatopallidal system and lateral hypothalamus and consummatory behavior in rats. *Journal of Comparative and Physiological Psychology*, 1973, *85*, 615–635.

Lindsley, D. F., Ranf, S. K., Fernandez, F. C., & Wyrwicka, W. Effects of anti-Parkinsonian drugs on the motor activity and EEG of cats with subthalamic lesions. *Experimental Neurology*, 1975, *47*, 404–418.

Marshall, J. F., Richardson, J. S., & Teitelbaum, P. Nigrostriatal bundle damage and the lateral hypothalamic syndrome. *Journal of Comparative and Physiological Psychology*, 1974, *87*, 808–830.

Mettler, F. A. Perceptual capacity functions of the corpus striatum and schizophrenia. *Psychiatric Quarterly*, 1955, *29*, 89–111.

Mettler, F. A. Anatomy of the basal ganglia. In P. J. Vinken & G. W. Bruyn (Eds.), *Handbook of clinical neurology* (Vol. 6). New York: Wiley, 1968.

Mitcham, J. C., & Thomas, R. K., Jr. Effects of substantia nigra and caudate nucleus lesions on avoidance learning in rats. *Journal of Comparative and Physiological Psychology*, 1972, *81*, 101–107.

Morgane, P. J. Alterations in feeding and drinking behavior of rats with lesions in globi pallidi. *American Journal of Physiology*, 1961, *201*, 420–428.

Neill, D. B., & Grossman, S. P. Behavioral effects of lesions or cholinergic blockade of the dorsal and ventral caudate of rats. *Journal of Comparative and Physiological Psychology*, 1970, *71*, 311–317.

Ranson, S. W., & Berry, C. Observation on monkeys with bilateral lesions of the globus pallidus. *Archives of Neurology and Psychiatry*, 1941, *46*, 504–508.

Rosvold, H. E. The prefrontal cortex and caudate nucleus: A system for effecting correction in response mechanisms. In C. Rupp (Ed.), *Mind as a tissue*. New York: Harper & Row, 1968.

Stevens, J. R. An anatomy of schizophrenia. *Archives of General Psychiatry*, 1973, *29*, 177–189.

Thompson, R. Localization of the "maze memory system" in the white rat. *Physiological Psychology*, 1974, *2*, 1–17.

Winocur, G. Functional dissociation within the caudate nucleus of rats. *Journal of Comparative and Physiological Psychology*, 1974, *86*, 432–439.

JOHN F. MARSHALL

Basal Ganglia Dopaminergic Control of Sensorimotor Functions Related to Motivated Behavior[1]

11

Tissue in and around the region of the lateral hypothalamus has long been thought to play a critical role in motivated behaviors, especially in the control of feeding and drinking (Anand & Brobeck, 1951; Morgane, 1961; Teitelbaum & Epstein, 1962). However, the importance of this tissue in more fundamental processes of sensorimotor integration has been recognized only recently. Unilateral electrical stimulation of this region in cats creates contralateral sensory fields for tactually and visually elicited reflex components of "stimulation-bound" attack on a rat (see review by Flynn, Edwards, & Bandler, 1971). A similar enhancement of contralateral sensory fields has been seen during lateral hypothalamic stimulation in rats (Beagley & Holley, 1977; Smith, 1972). In contrast, unilateral electrolytic lesions of the lateral hypothalamic region of rats result in deficits in head orientation to touch, odors, and visual stimuli on the contralateral body side (Figure 11.1). Following bilateral damage to this structure, rats are inattentive to stimuli emanating from either body side, and show catalepsy, akinesia, deficits in limb use, aphagia, and adipsia (Levitt & Teitelbaum, 1975; Marshall & Teitelbaum, 1974; Marshall, Turner, & Teitelbaum, 1971; Turner, 1973). A similar syndrome of sensorimotor impairments has been observed in cats with such lesions (Teitelbaum & Wolgin, 1975).

[1] The work was supported by Public Health Service research Grant MH 31426.

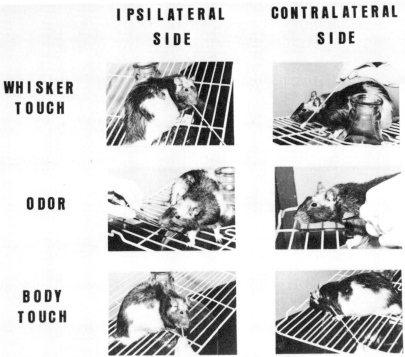

Figure 11.1. A rat with unilateral (right) lateral hypothalamic damage shows precise head orientation and biting to various kinds of stimuli (whisker touch, odor, body touch) on the ipsilateral side while neglecting the same stimuli presented contralaterally. This rat, as others, also failed to orient to contralateral visual stimuli. (From Marshall et al., 1971.)

Brain Dopamine and Sensory Inattention

Although the inattention syndrome observed after lateral hypothalamic damage might be due to the destruction of cells located within this region, such lesions would also be expected to interrupt axons traversing this structure; for example, ascending dopamine (DA)-containing axons whose cell bodies are located in the ventral midbrain course through the lateral hypothalamus. Ungerstedt (1971b) demonstrated that the severe ingestive impairments of rats with lateral hypothalamic lesions appeared attributable to severing these fibers, a finding that has since been confirmed and extended (see review by Stricker & Zigmond, 1976).

These findings suggested to us that the hypothalamic lesion might also produce sensorimotor impairments by interrupting these dopaminergic neurons. To examine this possibility, the ascending DA-containing neurons were destroyed using intracerebral injection of the catecholamine neurotoxin 6-hydroxydopamine (6–OH–DA) along their course. Rats with extensive damage to these cells of one hemisphere showed impairments in orientation to contralateral touch, odors, or visual stimuli, whereas orientation to ipsilateral stimuli was unaffected. Bilateral damage to these neurons resulted in bilateral sensorimotor deficits and aphagia and adipsia, a syndrome remarkably similar to that observed after lateral hypothalamic lesions (Ljungberg and Ungerstedt, 1976; Marshall, Richardson, & Teitelbaum, 1974).

Recovery From Inattention

One of the most remarkable aspects of the sensory inattention seen after either type of brain damage is its spontaneous recovery. In the weeks following 6–OH–DA injections or hypothalamic lesions, most rats regain their ability to orient toward tactile, olfactory, and visual stimuli on the affected side (Marshall et al., 1971, 1974; Marshall & Teitelbaum, 1974). After bilateral damage, this recovery of sensorimotor capacities accompanies the well-known restitution of ingestive behaviors (Teitelbaum & Epstein, 1962).

The spontaneous recovery of somatosensory orientation has been of particular interest. When the rat with lateral hypothalamic damage regains the ability to orient to touch of the body surface, it does so first to contact with the snout. As recovery proceeds, the body regions from which orientation can be elicited spread along the affected body side in a rostral to caudal direction (Marshall et al., 1971; Marshall & Teitelbaum, 1974) (Figure 11.2). I have found recently that the recovery of somatosensory orientation in rats given unilateral DA-depleting 6–OH–DA injections shows the same pattern. Approximately 50% of the 6–OH–DA-treated rats that show an initial somatosensory inattention recover spontaneously. Orientation occurs first to touch of rostral body regions and spreads progressively caudally during the postoperative period (Figure 11.3; Marshall, 1979).

Mechanisms of Recovery From Inattention

How do rats that have suffered apparently irreversible damage to the ascending DA-containing neurons recover from their initial sensorimotor deficits? Two classes of explanations exist. First, the behavioral recovery may depend upon an increased capacity of nondopaminergic neurons to substitute for the functions of lost nerve cells. For example, brain regions outside of the projection of central dopaminerigc fibers (e.g., superior colliculus) are known to contribute to rats' ability to orient to impinging visual, tactile, and olfactory stimuli (reviewed by Marshall, 1978; Marshall

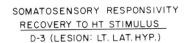
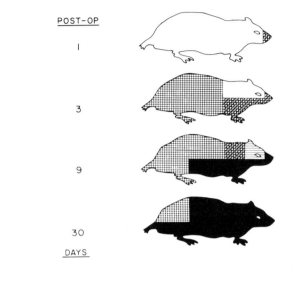

Figure 11.2. Pattern of recovery (after left lateral hypothalamic damage) of orientation to touch of various regions of the contralateral body surface. (HT stimulus indicates light touch of the hairs along the body surface with a cotton swab.) (From Marshall et al., 1971.)

& Teitelbaum, 1974). Such brain regions might vicariously assume the functions of lost DA-containing neurons.

Alternatively, the recovery of function may depend upon a restoration of DA receptor activity within the damaged fiber system. Assuming that some DA-containing neurons survive the injury, DA receptor activity might gradually increase toward normal during the postoperative period because of an enhanced capacity for synthesis and release of DA within remaining nerve terminals or an increased postsynaptic response to released DA ("postsynaptic supersensitivity"). Such pre- and postsynaptic changes have been shown to occur following damage to ascending dopaminergic fibers (Agid *et al.*, 1973; Creese, Burt, & Snyder, 1977; Marshall & Ungerstedt, 1977; Schoenfeld & Uretsky, 1972; Ungerstedt, 1971a).

Stricker and Zigmond (1976) have presented an elegant model to account for the recovery of ingestive behavior after DA-depleting brain injury. Their model hypothesizes that plastic changes occurring at unin-

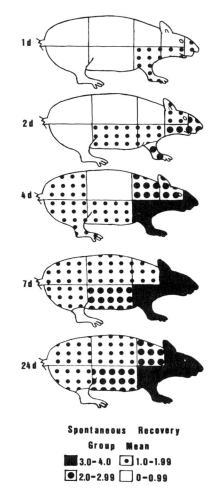

Figure 11.3. Mean orientation scores (N = 8) to touch (4 grams pressure) of the contralateral (right) body surface at 1, 2, 4, 7, and 24 days after left hemisphere 6-OH-DA injection (8 μg in 4 μl). Scores are based upon the following scale: 4 = precise head orientation and biting; 3 = precise head orientation without biting; 2 = turning of head approximately halfway toward touch; 1 = turning of head less than 1 cm in direction of touch; 0 = no head orientation toward touch. (From Marshall, 1979.)

jured dopaminergic synapses account for the restoration of eating and drinking. Can such changes account for the recovery of sensorimotor functions? Two experiments suggest that they may.

In the first experiment, rats were given bilateral injections of 6–OH–DA along the course of the ascending dopaminergic fibers and were tested for their ability to orient toward touch of the body surface, odors, and moving visual stimuli. Only those rats that displayed sensory inattention were used. Groups of 6 to 12 rats each were given injections of the specific DA receptor stimulating agent apomorphine (.05, .10, or .20 mg/kg) or its vehicle (.1% ascorbic acid, .9% NaCl; all i.p.) 2, 3, 5, and 8 days postoperatively. The rats were tested for orientation to stimuli before and at 5 min intervals after apomorphine administration for the duration of drug action (20 min).

Apomorphine administration reversed the sensory inattention of these

6–OH–DA-treated rats (F = 12.85, p < .01). The improvement was most marked at .10 mg/kg at 2 to 5 days postoperatively (Figure 11.4). The highest dose of apomorphine was ineffective in restoring orientation, as was the .10 mg/kg dose on Day 8, because of the appearance of stereotyped behaviors (sniffing, licking) that prevented the animal from orienting to applied stimuli. The restoration was not limited to the somatosensory modality: .10 mg/kg apomorphine reinstated orientation to moving visual stimuli and odors in 10 of 12 rats that were tested on the third day postoperatively. The restorative action of apomorphine lasted approximately 20 min. These findings parallel closely the improvement in sensorimotor functions of Parkinsonian patients given L-dopa.

Administration of a potent and specific DA receptor blocking agent spiroperidol (.05 mg/kg, i.p.) 2 hr before apomorphine could abolish completely the restorative effect of the DA receptor stimulating agent. These

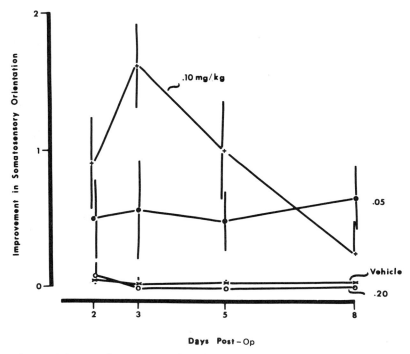

Figure 11.4. Mean improvement in orientation to touch (2 grams pressure) of both body surfaces following administration of apomorphine or its vehicle to rats with bilateral 6-OH-DA injections (16 µg in 8 µl). Scores are based upon the following scale: 3 = precise head orientation with or without biting; 2 = turning of head approximately halfway toward touch; 1 = turning of head less than 1 cm in direction of touch; 0 = no head orientation toward touch. Vertical lines represent one standard error of mean. (From Marshall and Gotthelf, 1979.)

findings strongly suggest that in rats that have had their presynaptic stores of DA largely destroyed using 6–OH–DA injections, the pharmacological restoration of DA receptor activity is highly effective in reinstating orientation to impinging stimuli.

A second line of evidence links dopaminergic mechanisms to the recovery of orientational capacities. Rats were administered 6–OH–DA along the ascending dopaminergic neurons of the left hemisphere. All showed an initial inattention followed by substantial recovery during the first month postoperatively, after which their neurological status remained stable. Two to three months postoperatively, the rats were given injections of the catecholamine synthesis inhibitor alpha-methyl-para-tyrosine (AMPT; 70 or 100 mg/kg of the salt) or the dopamine receptor blocking agent spiroperidol (.05 mg/kg, both i.p.). All animals were tested for their ability to orient toward touch of each body side before and after drug administration.

Both doses of AMPT resulted in a reversal of the spontaneous recovery process that had occurred on the contralateral (right) body surface (Figure 11.5). At these doses, however, the drug had no significant effect on orientation to ipsilateral touch. Of particular interest was the finding that the drug preferentially interfered with orientation to touch of the caudal body regions. As the behavioral effect of AMPT increased (from 0 to 4.5 hr after administration), the area of somatosensory loss spread progressively in a caudal to rostral direction. As the drug effect dissipated, somatosensory orientation was reinstated in a rostral to caudal direction, thus recapitulating in an abbreviated time frame the spontaneous recovery that had occurred during the first postoperative month.

Quite similar results were obtained using spiroperidol, thus suggesting that the recovery of somatosensory orientation depends specifically upon the maintenance of DA receptor activity in the hemisphere of the 6–OH–DA injection. These findings also suggest that orientation to the most caudal body regions is the most susceptible to disruption by pharmacological interference with DA receptor activity.

Conclusions

The findings reviewed in this chapter indicate that DA-containing neurons of the rat brain are critical to certain sensorimotor capacities, most notably to its ability to orient toward tactile, visual, and olfactory stimuli. The sensorimotor deficits of rats with damage to these neurons appear to contribute to the initial aphagia (Marshall & Teitelbaum, 1974; Marshall, Turner, & Teitelbaum, 1971) and may contribute as well to the later stages of the syndrome of ingestive deficits.

The remarkable capacity of these rats to recover from sensorimotor

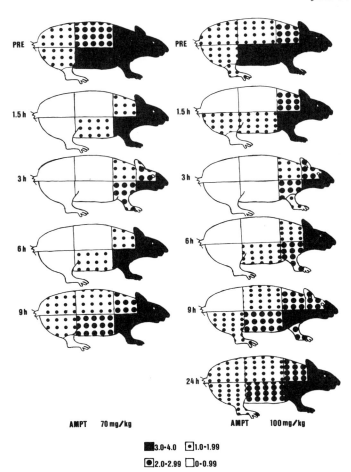

Figure 11.5. Mean orientation scores (N = 8) to touch (4 grams pressure) or contralateral (right) body surface before and at various times after alpha-methyl-para-tyrosine (AMPT) in rats with prior left hemisphere 6-OH-DA injection (8 µg in 4 µl). Scale is same as in Figure 11.3. Administrations of the two doses of AMPT were separated by 48 hrs. (From Marshall, 1979.)

deficits appears related to compensatory neurochemical events occurring at remaining dopaminergic synapses. First, administration of a DA receptor stimulant to animals with severe DA depletions can restore orientation capacities, suggesting that a restitution of DA receptor activity is *sufficient* for behavioral recovery to occur. Second, administration of drugs that interfere with DA neurotransmission to animals that have spontaneously recovered from their somatosensory inattention reinstates sensorimotor dysfunctions, suggesting that a restitution of DA receptor activation may be *necessary* for behavioral recovery. The mechanisms underlying the rostral to caudal recovery progression of somatosensory orientation are under investigation.

REFERENCES

Agid, Y., Javoy, F., & Glowinski, J. Hyperactivity of remaining dopaminergic neurons after partial destruction of nigro-striatal dopaminergic system in the rat. *Nature New Biology*, 1973, *245*, 150–151.

Anand, B. K., & Brobeck, J. R. Hypothalamic control of food intake in rats and cats. *Yale Journal of Biology and Medicine*, 1951, *24*, 123–140.

Beagley, W. K., & Holley, T. L. Hypothalamic stimulation facilitates contralateral visual control of a learned response. *Science*, 1977, *196*, 321–322.

Creese, I., Burt, D. R., & Snyder, S. H. Dopamine receptor binding enhancement accompanies lesion-induced behavioral supersensitivity. *Science*, 1977, *197*, 596–598.

Flynn, J. P., Edwards, S. B., & Bandler, R. J. Changes in sensory and motor systems during centrally elicited attack. *Behavioral Science*, 1971, *16*, 1–19.

Levitt, D. R., & Teitelbaum, P. Somnolence, akinesia, and sensory activation of motivated behavior in the lateral hypothalamic syndrome. *Proceedings of the National Academy of Sciences, U.S.A.*, 1975, *72*, 2819–2823.

Ljungberg, T., & Ungerstedt, U. Sensory inattention produced by 6-hydroxydopamine-induced degeneration of ascending dopamine neurons in the brain. *Experimental Neurology*, 1976, *53*, 585–600.

Marshall, J. F. Comparison of the sensorimotor dysfunctions produced by damage to lateral hypothalamus or superior colliculus in the rat. *Experimental Neurology*, 1978, *58*, 203–217.

Marshall, J. F., & Gotthelf, T. Sensory inattention in rats with 6-hydroxydopamine-induced degeneration of ascending dopaminergic neurons: Apomorphine-induced reversal of deficits. *Experimental Neurology*, 1979, *65*, 398–411.

Marshall, J. F. Somatosensory inattention after dopamine-depleting intracerebral 6-OHDA injections: Spontaneous recovery and pharmacological control. *Brain Research*, 1979, *177*, 311–324.

Marshall, J. F., & Teitelbaum, P. Further analysis of sensory inattention following lateral hypothalamic damage in rats. *Journal of Comparative and Physiological Psychology*, 1974, *86*, 375–395.

Marshall, J. F., & Ungerstedt, U. Supersensitivity to apomorphine following destruction of the ascending dopamine neurons: Quantification using the rotational model. *European Journal of Pharmacology*, 1977, *41*, 361–367.

Marshall, J. F., Turner, B. H., & Teitelbaum, P. Sensory neglect produced by lateral hypothalamic damage. *Science*, 1971, *174*, 523–525.

Marshall, J. F., Richardson, J. S., & Teitelbaum, P. Nigrostriatal bundle damage and the lateral hypothalamic syndrome. *Journal of Comparative and Physiological Psychology*, 1974, *87*, 808–830.

Morgane, P. J. Alterations in feeding and drinking behavior of rats with lesions of globi pallidi. *American Journal of Physiology*, 1961, *201*, 420–428.

Schoenfeld, R., & Uretsky, N. Altered response to apomorphine in 6-hydroxydopamine-treated rats. *European Journal of Pharmacology*, 1972, *19*, 115–118.

Smith, D. A. Increased perioral responsiveness: A possible explanation for the switching of behavior observed during lateral hypothalamic stimulation. *Physiology and Behavior,* 1972, *8,* 617–621.

Stricker, E. M., & Zigmond, M. J. Recovery of function following damage to central catecholamine-containing neurons: A neurochemical model for the lateral hypothalamic syndrome. In J. M. Sprague & A. N. Epstein (Eds.), *Progress in physiological psychology and psychobiology* (Vol. 6). New York: Academic Press, 1976.

Teitelbaum, P., & Epstein, A. N. The lateral hypothalamic syndrome: Recovery of feeding and drinking after lateral hypothalamic lesions. *Psychological Review,* 1962, *69,* 74–90.

Teitelbaum, P., & Wolgin, D. L. Neurotransmitters and the regulation of food intake. In W. H. Gispen *et al.* (Eds.), *Progress in brain research* (Vol. 42). Amsterdam: Elsevier, 1975.

Turner, B. H. A sensorimotor syndrome produced by lesions of the amygdala and lateral hypothalamus. *Journal of Comparative and Physiological Psychology,* 1973, *82,* 37–47.

Ungerstedt, U. Postsynaptic supersensitivity after 6-hydroxydopamine induced degeneration of the nigro-striatal dopamine system. *Acta Physiologica Scandinavica,* 1971a (Suppl. 367), 69–93.

Ungerstedt, U. Adipsia and aphagia after 6-hydroxydopamine induced degeneration of the nigro-striatal dopamine system. *Acta Physiologica Scandinavica,* 1971b (Suppl. 367), 95–122.

YU. I. ALEKSANDROV
YU. V. GRINCHENKO

Hierarchical Organization of Physiological Subsystems in Elementary Food Acquisition Behavior

12

The problem of physiological mechanisms of behavior is essentially a problem of organizing the physiological functions of many elements into a singly integrated functional system of the behavioral act. A functional system of any behavioral act constitutes the integration of many subsystems, and at the same time is part of a system on a higher level. An analysis of the hierarchical organization of systems (Anokhin, 1973) is therefore of fundamental importance to a study of behavioral mechanisms. Our task was to study experimentally the hierarchical organization of subsystems of separate movements, muscle activity, and neuron firing in actuating mechanisms of the elementary behavioral act of taking food that is common to food acquisition behavior of variable complexity. The functioning of motor cortex neurons, spinal motor neurons, and the muscle proprioceptive apparatus is associated with the realization of the actuating mechanisms of behavior. Therefore, a solution to the aforementioned problem necessitates a study of the correlation between the neuron activity of the so-called motor system and the subsystem hierarchy of the actuating mechanisms in the functional system of the food-taking behavioral act. Such a study requires a preliminary explanation of the relationship between the subsystems of individual movements that are identifiable by the results they are supposed to achieve and the relationship between individual movements and muscle activity.

Method

Large (10 × 10 × 10 mm) and small (2–3 mm thick flakes) pieces of carrot pinned to the stem of the feeder device (Figure 12.1[a,1]) were presented to rabbits loosely bound by the feet. The end result of the food acquisition act (seizing the carrot by the teeth) was determined by a contact

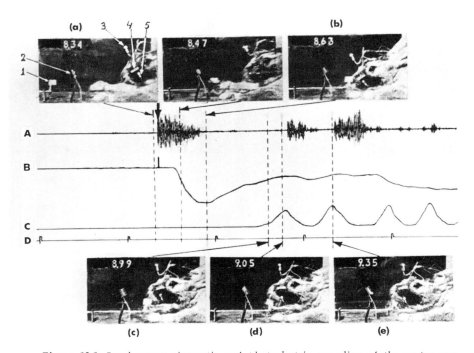

Figure 12.1. Synchronous cinematic and photoelectric recording of the motor components in a single food-acquisition act. (A) Recording of noises during presentation of food (left) and taking of food (right). When the food presentation was turned on, an accompanying clicking sound served as the start stimulus for the behavioral act. From now on this is shown by an arrow. (B) Recording of head movement/downward deflection of the pen—rapid phase of movement; upward deflection—slow phase. Between those phases is a horizontal "plateau" section of the curve that corresponds to the cessation of head movement. (C) Recording of the vertical component of lower jaw movement (lowering the jaw corresponds to an upward shift of the pen). (D) 500 msec time marker; (a)–(e), individual movie frames. Numbers in the frame corners—time in seconds. The dotted lines indicate the times in the recording that correspond to frame time of the arrow. Designations: (a, 1) movable pin with the piece of carrot; (a, 2) photoelectric plate; (a, 3) cranial-fastened light source with the distance between the cranium and plate (a, 2) indicated with respect to changes in the plate's photo EMF; (a, 4) photoelectric plate fastened to the nasal bone; (a, 5) light source fastened by an implanted wood screw to the lower jaw. The photo-EMF of the plate (a, 4) changes with movements of the lower jaw. The representation in (b) is blurred because of the high speed of movement. The same is true for (c) which corresponds to the "plateau," a clear representation. The maximum opening of the mouth is achieved directly after (d) and the taking of food occurs after (e), which corresponds to the second taking of food, at which time gnawing occurs.

microphone recording of the sounds made by the rabbit's taking the carrot with its teeth. The lower jaw movements were recorded by a photoelectric method that we devised. The head movements were also recorded photoelectrically. Synchronous film recordings of the movements were made in some acts (Figure 12.1[a–e]). Intramuscularly implanted bipolar wire electrodes recorded the electrical activity of the m. splenius and the masticatory muscles mm. masseter p. prof., mylohyoideus, digastricus, and pterygoideus lateralis. An eight-channel "Nikhon-Koden" polygraph was used to record electrical muscle activity, head and lower jaw movements, and the achievement of the end result.

As an illustration of "motor system"-related cells, recordings were made of the anterior portion of the motor cortex whose stimulation produced well-defined masticatory movements (Sumi, 1969). Also recorded were the neuron activity of the trigeminal mesencephalic nucleus (trigeminal mesencephalic neurons) that are the first-order sensory neurons that send out the peripheral process to the proprioreceptors of the masticatory muscles, and the firing activity of a number of mesencephalic neurons. The cells that were recorded in corresponding coordinates (P—12,0; H—13; OL—1,5–2,5 of the stereotaxic atlas) (McBride & Klemm, 1968) were identified as TMS (trigeminal mesencephalic) neurons because of the relationship between their activity and the masticatory cycles and because of responses to palpation of the masticatory muscles and to direct electric stimulation of the m. masseter. The position of the microelectrode track was controlled morphologically. The experimental data were processed on a laboratory minicomputer.

Experimental Results and Discussion

An analysis of the cinematographs indicates that the first component of the rabbit's movement toward the carrot was a rapid lowering and forward extension of its head (Figure 12.1[b])—the "rapid phase." The lower jaw did not leave the rest position during this phase of the movement with the exception of the development of "microchews" (see Figures 12.2[1,A], 12.3[1,A]) that we identified as the initial consummatory act according to Craig (1918). The lower jaw's fixation was due to the stress created by the masticatory muscles during the head's rapid movement, at which time the muscles exhibited low-amplitude tonic activity. When the minimum distance was reached between the food and the head, the latter's movement slowed markedly; after that movement ceased, the "slow phase" of the movement evolved for a 30–80 msec "plateau" that consisted of the coordinated opening of the mouth and the movement of the head. Consequently, the buccal orifice was placed over the carrot (Figure 12.1[c]), and the food was taken (Figure 12.1[d,e]—secondary food take).

A significant degree of variability was found in the means of achieving

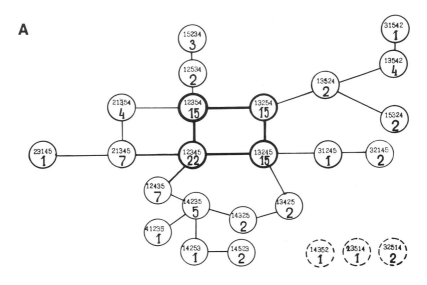

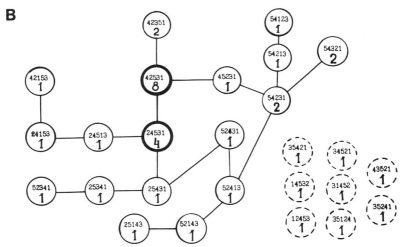

Figure 12.2. *The association between the neuron activity of the anterior-lateral cortex (I) and the midbrain (II) and the implementation of the entire behavioral act (A) and with a separate phase of movement (B). Designations: (I, B) neuron activation of the anteriolateral cortex occurs when the head is lowered only in (a) the "standard" behavioral act, but not when (b) the food is taken from the experimenter's hand. The time marker in (A) is 100 msec; in (B) is 250 msec. (II, B) Activation of the mesencephalic neuron occurs only when the head is lowered in (a) the "standard" behavioral act, in (b) "background" lowering, and even (c) during defensive integration; that is, during forcible lowering of the head. Time marker is 100 msec. Designations 1, 2, 3, and 4, are the same as Figure 12.5.*

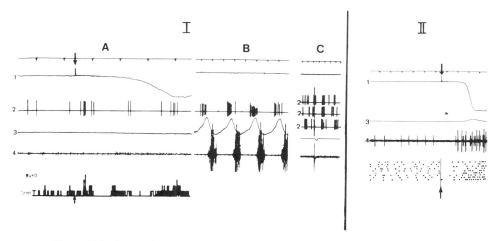

Figure 12.3. Comparison of the link between the activity of a TMS neuron (I) and anteriolateral cortex neuron (II) and the mouth's opening and closing during mastication and when these movements are involved in the taking of food. Designations: (I) similarity in TMS neuron activation during movements of the lower jaw in (A) the "standard" act of taking food, and (B) in chewing, (C) activation recording of the same neuron during intramuscular stimulation of the m. masseter, and (D) during palpation of the m. masseter. The time marker is 100 msec. (II) The activity of the anteriolateral cortex neuron is exclusively timed with the first opening of the mouth for taking food in a "standard" behavioral act (A, B, C). (C) represents the taking of food with subsequent gnawing and regular chewing. (D) The rabbit leans toward the food without taking it—activity is absent. Of course, the relation of a neuron's activity to a specific level of organization does not mean that there is no link with another level. Thus, the neuron activity that is being compared with an entire behavioral act (see text) also depends on the structure of the movement with which it coincides in time. See the change in the neuron activity during the "extended" opening of the mouth as shown by the arrow in (C). The jaw movement and EMF activity, characteristic of food taking, was observed only after the interval indicated by the arrow. The time marker is 250 msec. The designations 1, 2, 3, and 4 are the same as in Figure 12.5.

the end result of the rapid phase—maximum approach of the head to the carrot—and the end result of the slow phase—coincidence of the buccal orifice and food. We mean by the means of achieving the end result that activity "which a given system manifests externally and which is formed in the course of selecting this system from among many other possible activities" (Anokhin, 1973, p. 80), or the selection of the system's "degree of freedom." The variability in the degree of freedom was not chaotic or a simple manifestation of stochasticity. Concise regular characteristics were identified in their distribution.

After discovering that the sequence in which the muscles are involved in the rapid phase of a movement changes from act to act, for the sake of a more convenient analysis of the order of their involvement, we assigned the numbers 1, 2, 3, 4, 5 to mm. masseter, digastricus, mylohyoideus, splenius, and pterygoideus lateralis, respectively. Thus, the order of muscle involvement in each given act—"ranking" (Vartazarov et al., 1975)—was presented as a combination of five numbers. The reliable ($p < .02$) connection between the order of muscle involvement and achievement time of the end result (the time from the initial head movement to the actual taking of food) indicates that the ranking indicator is not a random one, but actually linked to the characteristics of the behavioral act. In the rabbits that exhibited a stable automated act, the rankings that differed from each other by not more than one reciprocal permutation of a pair of neighboring elements (by one "inversion") comprised the majority of rankings and formed a "coordinated" group in which the "center" of the most probable rankings was identified (Figure 12.4[A]). In the situation represented in this figure, only three rankings out of 119 observations "fall out." They cannot be approached through a continuous inversion chain. One can see fundamental differences in a comparison of this characteristic distribution to those identified in the case of the uncoordinated behavioral act (Figure 12.4[B]). That is, the 'fallout" rankings here include 8 out of 39 observations. This is a significant difference, particularly if we consider that an increase in the number of observations in an '"unbalanced" process leads to an increase in data straggling. Thus, the degree of freedom distribution turns out to be linked to the degree of the behavioral act's automation and the extent of its coordination.

An analysis of the variability of the means of achieving the result of the slow phase indicated that the stabilization of the head's position before the food was taken occurs at a rather fixed time interval after the maximum opening of the mouth—0–50 msec in 80–90% of the cases for various rabbits. The link between the onset of the head's slow movement following the "plateau" and the onset of the mouth's opening was significantly less rigid—the relative variation interval of these factors exceeded the aforementioned interval by three to seven times. The examined indices, like the rankings, are variable but are not random. The correlation between the onsets of the head's movement and the mouth's opening is reliably ($p < .05$) linked to the size of the carrot. When a small piece of carrot was presented, the head movement began later than it did when a larger piece was presented.

The appetent phase can be singled out in a "microethological" approach to analyzing the taking of food—movement towards the food and the consummatory phase—the eating of the food following the taking of food. Thus, the decrease we noted in the variability of the link between the subsystems of head and lower jaw movements during the approaching result of the act, that is, the taking of food, turns out to be comparable to the decrease observed by ethologists in the variability of appetent behavior as the consummatory act draws near (Tinbergen, 1955). Consequently, the variability in the degree of freedom is an index that is characteristically

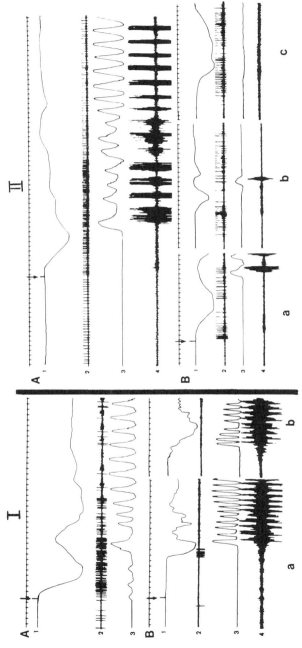

Figure 12.4. Distribution of rankings in the coordinated (A) and uncoordinated (B) acts. The small numbers in circles correspond to the numbers of the mice (see text). The large numbers inside the circles signify the number of observations for a given ranking. The boldface circles signify the center of the coordinated group. The dotted circles on the lower right are the "fall-out" rankings. See text for details.

183

linked to the whole behavioral act and changes in time as the behavior evolves.

The preceding analysis indicates that a variability in the means of achieving the end result of a system on any level is observed in the course of examining various organizational levels of a functional system. But when a system is examined as a subsystem, it turns out to be invariable with respect to the place of its subresult in the hierarchy of the "big" system of which it has become a part. In other words, all of the various suborganizational forms of elements that play an identical role in a big system, that is, enhance the achievement of the same subresult, act as a subsystem in a system of a higher organizational level. This has determined our approach to analyzing neuronal activity from the viewpoint of its link to various organizational levels of an elementary behavioral act.

From the viewpoints of Anokhin's functional system theory, the organizational processes of a behavioral act—afferent synthesis and decision making—occur in the latent period of the actuating mechanism's involvement, that is, in the latent period of EMG activation and movement whose development corresponds to the realization processes—the action program Shvyrkov. An analysis of the activity of 53 cortical and 50 mesencephalic neurons disclosed changes in that activity, not only in connection with the functioning of the actuating mechanisms but also in accordance with organizational processes. The early activations that coincide in time with the development of organizational processes were observed in 8 cortical and 22 mesencephalic neurons. The latent period of the early activations of the motor cortex neurons was not less than 40–50 msec. The activations varied with respect to the number of impulses and the latent period. The identification of those activations necessitated the construction of poststimulus histograms. The early activations of the mesencephalic neurons were marked by a high degree of stability, and their latent period was significantly less than the 7 msec minima. Warranting special attention is the fact that short latent activations of 16 to 32 msec were observed in 3 out of 16 TMS neurons (identified proprioceptive elements). Figure 12.5 illus-

Figure 12.5. Changes in the activity of the trigeminal-mesencephalic neuron (I) and an individual motor unit of m. masseter (II) in the latent period of a behavioral act. (I, A) top: Recording of a separate act; bottom: histogram of a TGM neuron activity, constructed from the time of the appearance of the feeder's starter click. The channel width is 1.25 msec. (B) Connection between the neuron activity and individual chewing cycles; (C) TGM neuron activity during the intramuscular stimulation. An inhibitory pause with postinhibitory activation, characteristic of spindle afferents, is observed. (II) top: Recording of the activity of two motor units of m. masseter in the act of taking food. bottom: rasters of impulse activity of a motor unit with low-amplitude potentials in sequential behavior acts. The rasters were constructed from the starting click of the feeder device. Designations: (1) head movement recording; (2) neuronogram; (3) recording of lower jaw movements; (4) m. masseter EMF. The time marker is 100 msec.

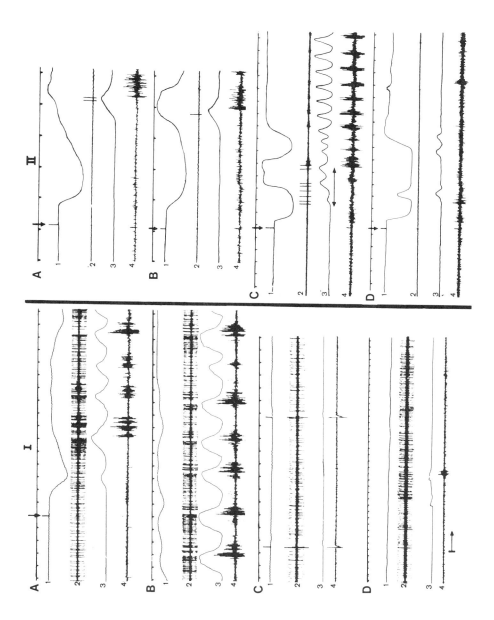

trates the activity of a TMS neuron that exhibited activation with a 32 msec latent period after the click and before the beginning of movement and EMG activation, and exhibited the properties of an afferent spindle during the tests. This is important in connection with the fact that the commonly accepted concept on proprioceptor function presumes that their activity is analyzed only in connection with the functioning of the actuating mechanisms. The latent period of the behavioral act, that is, the time required by the processes of afferent synthesis and decision making, has not been recorded by even the investigators studying not cyclic and imposed movements but singular arbitrary ones. Moreover, activity of the proprioceptors in this interval has not been analyzed.

The presence of an early activation in spindles is in agreement with data on the possibility that skeletal motor activity is surpassed significantly by spindle activity (Matsunami, 1972), and with the data on the activation of spindles outside their "own" muscle (Burg, Szumski, & Strupler, 1974), as well as the data on the appearance of gamma-activation with a latent period of 18 msec in response to a conditional defensive click-stimulus (Buchwald, Beatty, & Eldred, 1961). The appearance of early activations of TMS-neurons in the interval corresponding to the development of the behavioral act's organizational processes confirms Ganit's (1975) hypothesis on the connection between spindle activity that occurs before an arbitrary movement and the arbitrary act's preparatory processes. The changes that occur in the alpha-skeletal motor activity that were identified in an analysis of individual motor unit activity (Figure 12.5[II]) apparently should be associated with those preparatory processes.

An analysis of the firing of cortical and mesencephalic neurons that are related to actuating mechanisms indicated that the neuron activation of both those structures can be time-lined to both the entire act (Figure 12.2 [I,A]; [II,A]) and to a separate movement (Figure 12.2[I,B]; [II,B]). A study of the activation properties related in time to individual movements disclosed significant differences between the cortical and mesencephalic neurons. The cortical cells were activated in relation to a given movement, as a rule, only when it was involved in strictly defined behavior (Figure 12.2 [I,B]). In the case where an activation was extinguished when there was a change in behavior, it underwent a fundamental reorganization. On the other hand, mesencephalic neurons were activated in a similar fashion in connection with a given movement in different behavioral situations (Figure 12.2[II,B]).

In order to study that difference in detail, a special comparison was made between the neuron activity of both structures during the opening and closing of the mouth in the chewing and gnawing processes, and when these movements were involved in the behavioral act of taking the food. Out of the 17 mesencephalic neurons examined for this purpose, the relationships that were exhibited between neuron activity and the opening and closing of the mouth in chewing were maintained during the taking of food in 13 neurons, although there may have been some modification of activity in the latter case that is related to changes in the characteristics of the behavior itself (Figure 12.3[I]). Out of nine cortical neurons in which a link with individual chewing cycles was observed, in only three neurons did that link remain more or less constant when the jaw's movement was involved in the behavioral act. The activity of the remaining neurons changed fundamentally. Moreover, the activations of seven cortical neurons, consisting of one or several spikes, were related exclusively to the opening of the mouth in order to take the food, and were absent in all of the other animal's movements, including the gnawing and chewing of food, at which time the mouth was already opening and closing (Figure 12.3[II]). In spite of cellular differences of the examined regions, there were elements in each of them whose properties corresponded to the total cellular characteristics of another region.

Thus, most neurons in the cortex whose activity was time-linked to even the implementation of a separate movement exhibited a dependence upon the entire act. This dependence is the link that was identified during the involvement of a given movement in strictly defined behavior. Characteristic property of the mesencephalic neurons that were firing in connection with a specific movement was the retention of that bond as the utilization of a given movement was affected in different ways. The data we have obtained give us grounds for saying that activity of the cortical neurons is chiefly related to the organizational level of a behavioral act, and that the activity of mesencephalic neurons is primarily related to the organization of individual subsystems.

Conclusions

1. The functional system of any organizational level is variable in analyzing the means for achieving its result—degrees of freedom. The degrees of freedom distribution is connected with the characteristics of a behavioral act. The number of a system's degrees of freedom decreases by the time the system's result is achieved. A system cannot be distinguished through morphology. A system can be identified only "from the top," that is, by determining the position of its result in the hierarchy of the "big" system's results, into which system it has been included as a subsystem.
2. Motor cortex activity and proprioceptive afferentation which, according to traditional concepts, are linked to the actuating mechanisms of behavior, are also essential to the organizational processes of a behavioral act, that is, afferent synthesis and decision-making.
3. The activity of different neurons that fire at the same time of a be-

havioral act (e.g., during the first opening of the mouth for taking food) can be related to various hierarchical organizational levels of a behavioral act's actuating mechanisms.

ACKNOWLEDGMENT

The authors are grateful to N. G. Gladkovich for his assistance in morphological control.

REFERENCES

Aleksandrov, Yu. I., & Grinchenko, Yu. V. Method for the photoelectric recording of lower jaw individual components of masticatory movements. *USSR Physiology Journal,* 63 (No. 7), 1062. (In Russian)
Anokhin, P. K. Fundamental problems in the general theory of functional systems. In *Principles of the systems organization of functions,* 1973, p. 5. (In Russian)
Anokhin, P. K. The Problem of decision-making in psychology and physiology. *Problems of psychology,* 1974, (No. 4), 21. (In Russian)
Buchwald, J. S., Beatty, D., Eldred E. Conditional responses of gamma and alpha notoneurons in the cat trained to conditioned avoidance. *Experimental Neurology,* 1961, 4, 91.
Burg, D., Szumski, A. J., Strupler, A., & Velho F. Assessment of fusimotor contribution to reflex reinforcement in humans. *Journal of Neurology, Neurosurgery, and Psychology,* 1974 (37), p. 1012.
Craig, W. Appetites and aversions as constituents of instincts. *Biological Bulletin of Marine Biology Laboratory,* 1918 (34), 91.
Granit, R. The functional role of muscle spindles—facts and hypotheses. *Brain,* 1975, 98, 53.
Matsunami, K., & Kubota, K. Muscle afferents of trigeminal mesencephalic tract nucleus and mastication in chronic monkey. *Japanese Journal of Physiology,* 1972, 22, 545.
McBride, R. L., & Klemm, N. R. Stereotaxic atlas of rabbit brain. *Communication in Bahavioral Biology,* 1968, 2 (Pt. A), 779.
Shvyrkov, V. V. On the relationship between physiological and psychological processes in the functional system of the behaviour act." *Studia Psychology,* 1977, 19, 82.
Sumi, T. Some properties of cortically evoked swallowing and chewing in rabbits. *Brain Research,* 1969, 15, 107.
Tinbergen, N. *The study of instinct.* New York: Oxford Univ. Press, 1955.
Vartazarov, I. S., Zharomskii, V. S., Gorlov, I. G., Sobinyanov, V. A., Khvastunov, R. M. *Methods of investigatory analysis.* 1975. (In Russian)

EVELYN SATINOFF

Independence of Behavioral and Autonomic Thermoregulatory Responses[1]

13

Behavior occupies a peculiar position in the study of thermoregulation. To most psychologists, the constancy of internal body temperature is primarily an autonomic achievement; the primitive reflexive responses an animal uses are the most important, with behavioral responses a sophisticated, later developed ability. To many physiologists, autonomic responses are also paramount, and for them, in any case, there is no real need to study behavior as opposed to autonomic thermoregulation because "Except for the effector actions which close the regulatory loops, the two systems are alike" (Hardy, Stolwijk, & Gagge, 1971).

In this chapter I will try to document several points that do not fit with either of the preceding views. First, I will give evidence that the two systems are dissimilar not only in effector action, but also at the integrative level. Behavioral and autonomic responses are not controlled by a single central integrator—one thermostat—but in fact are neuroanatomically separate from one another. Second, behavioral thermoregulatory responses are more primitive than autonomic responses in the sense that they are phylogenetically older, that is, they are well developed in species that either have not developed or have only rudimentary autonomic mechanisms of temperature control. Third, in mammalian ontogeny, behavioral responses are not only more effective than autonomic responses, but also may be crucial for maintaining a constant body temperature.

[1] The preparation of this paper and some of the work reported in it was supported by U.S. Navy Grant N00014-77-C-0465 and National Science Foundation Grant BNS77-03151.

At first it was not necessary to postulate neuroanatomical separation; it appeared that behavioral and autonomic thermoregulatory responses were both localized in the preoptic/anterior hypothalamic area. It was well known that if that area of the brain was heated, animals would pant and vasodilate and their body temperatures would fall (Fusco, Hardy, & Hammel, 1961; Magoun, Harrison, Brobeck, & Ranson, 1938). If the preoptic area was locally cooled, animals would shiver and vasoconstrict and their body temperatures would rise (Hammel, Hardy, & Fusco, 1960; also see Satinoff, 1974, for review). Then, in 1964, Satinoff showed that local cooling of the preoptic area also elicited motivated responses designed to increase heat conservation. Not only did rats shiver and increase their body temperature when the preoptic area was cooled at ambient temperatures of 5° and 24°C, but further, if given an opportunity to press a bar to turn a heat lamp on, the animals pressed much more for heat when their brains were being cooled than when they were not (Figure 13.1). Later, Carlisle (1966) demonstrated the opposite effect—in the cold, rats worked less for external heat when their preoptic areas were heated. These effects have been replicated many times in several different species (e.g., baboons, Gale, Matthews, & Young, 1970; squirrel monkeys, Adair, Casby, & Stolwijk, 1970; pigs, Baldwin & Ingram, 1967). Thus, it appeared that the neural control of thermoregulation could be described adequately by a single integrator, localized in the preoptic area, which, when thermally stimulated, elicited both autonomic and behavioral thermoregulatory responses.

It had also been known for some time that if lesions were made in the preoptic area, animals could not maintain their body temperatures by

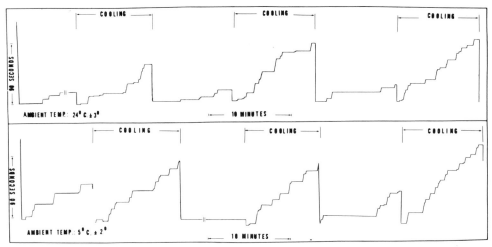

Figure 13.1. Cumulative record of a typical rat showing amount of bar pressing for heat during successive intervals of preoptic cooling and no cooling at ambient temperature of 5 and 24°C. (From Satinoff, 1964. American Physiological Society Reprinted by permission.)

autonomic means in thermally extreme environments (Andersson, Gale, Hokfelt, & Larsson, 1965; Frazier, Alpers, & Lewy, 1936). It was therefore reasonable to assume, because of both the thermal stimulation results and the prevailing model of a single integrator, that lesions of the preoptic area should also impair behavioral responding for heat. However, when the experiments were done, this did not turn out to be the case. Preoptic lesions did indeed impair autonomic responses in rats. When they were placed at an ambient temperature of 5°C for 1 hr a week, their body temperatures dropped as much as 6.8°C. However, if the rats were allowed to press a bar to keep a heat lamp on, in 2-hr tests in the cold they kept the bar depressed for a mean of 32% of the time (Figure 13.2) and maintained their body temperatures within .5°C of normal (Satinoff & Rutstein, 1970). Carlisle (1969) reported similar results and Lipton (1968) showed the other side of the picture. Rats with preoptic lesions that would surely have died of hyperthermia in the heat pressed a bar to turn the heat lamp off and a cooling fan on and thus survived the periods of heat stress. The results of the lesion experiments do not support an identity of function between behavioral and autonomic responses. Operant responses are not integrated solely in the preoptic area, because they continue to function efficiently when that region is largely destroyed. Additional evidence for this separation comes from the demonstration that lateral hypothalamic lesions disrupt thermoregulatory operants while leaving autonomic responses largely intact. Well-trained rats that had pressed a lever for heat in the cold no longer did so after small lateral hypothalamic lesions. Nevertheless, the animals were able to maintain normal body temperatures reflexively (Satinoff & Shan, 1971). In adult mammals, then, autonomic and operant responses to thermal

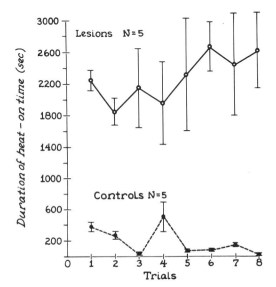

Figure 13.2. Duration of heat on time for eight weekly 2-hr tests in the cold of rats with preoptic lesions and controls maintained at 80% normal body weight. (From Satinoff and Rutstein, 1970. American Psychological Association. Reprinted by permission.)

stresses are functionally and neuroanatomically separate, and animals can compensate for deficits in one system by using the mechanisms of the other system.

Phylogenetically these two systems are also separable. Most fish, amphibia, and reptiles have no reflexive thermoregulatory responses at all; yet in their natural environments they regulate their body temperatures within well-defined limits using a variety of heat-seeking and heat-escaping behaviors (Whittow, 1970). In the laboratory, all ectotherms examined so far show clear thermoregulatory preferences in choice situations (Satinoff & Hendersen, 1977). Thermoregulatory behaviors in ectotherms appear to be controlled by a combination of skin, brain, and body temperatures, just as they are in mammals. Heating or cooling the brains of lizards or fish alters the time they spend in hot or cold environments the way similar thermal stimulation would do in mammals (Hammel, Caldwell, & Abrams, 1967; Hammel, Stromme, & Myhre, 1969).

Because ectotherms regulate their body temperatures behaviorally, we can ask the interesting question of whether they develop fevers. When iguanas were injected with a bacterial pyrogen (one that causes fever in mammals) and maintained at an ambient temperature of 30°C, no fever developed. This is because iguanas do not have any means of increasing internal heat production. However, when the lizards were allowed to shuttle back and forth between a 50 and 30°C environment, the iguanas injected with pyrogen preferred to remain in the warmer chamber long enough to elevate their body temperatures by over 2°C (Vaughn, Bernheim, & Kluger, 1974). Similar selection of warmer temperatures after bacterial injections has been reported for teleost fish (Reynolds, Casterlin, & Covert, 1976), amphibia (Casterlin & Reynolds, 1977a; Myhre, Cabanac, & Myhre, 1977), and crayfish (Casterlin & Reynolds, 1977b). Thus, ectotherms that do not have autonomic thermoregulatory mechanisms nevertheless develop a fever by using behavioral responses in a manner very similar to adult mammals.

For some time investigators thought that infant mammals were similar to ectotherms in having no autonomic means of thermoregulation, because they could not maintain their body temperatures in cool or cold environments. However, there is considerable evidence now demonstrating that infants of several mammalian species increase their metabolic rate at ambient temperatures below their thermoneutral zones (Dawes & Mestyán, 1963; Hull, 1965). In fact, infant rats have a maximum capacity for heat production that exceeds that of adults (Conklin & Heggeness, 1971). The major problem of infant mammals is lack of control over heat loss—they have very poor insulation and an unfavorable surface-to-volume ratio. These physical factors will mature, but early in development even a very high heat production cannot make up for excessive heat loss. Even though heat production, in the form of nonshivering thermogenesis, can be well developed at birth, it is an extremely costly use of metabolic energy that could better be used for growth and development. One might think that infants would do better to drift into hypothermia in the cold, but hypothermic rats develop much more slowly than normothermic pups (Stone, Bonnet, & Hofer, 1976) and premature human infants maintained in incubators that support slightly low body temperatures have higher mortality rates and more metabolic disturbances than do euthermic infants (Buetow & Klein, 1964; Silverman, Fertig, & Berger, 1958). Therefore, it is desirable for newborn mammals to maintain a relatively high body temperature, and the major responses used to achieve this are behavioral. All infant mammals examined so far seek heat as early as one day after birth and prefer warm fur or even warm metallic coils to cold fur (puppies, Jeddi, 1970; rabbits, Jeddi, 1970; monkeys, Harlow, 1971). In thermally graded alleys neonatal mice (Ogilvie & Stinson, 1966), pigs (Mount, 1963), and rabbits (Baccino, 1935; Satinoff, McEwen, & Williams, 1976) all choose to remain at warmer temperatures than do older animals and their body temperatures do not fall. In nature, the major method of thermoregulatory behavior in social animals is huddling. In an elegant series of studies, Alberts (1978) has shown that when nest temperatures were low, the direction of movement of rat pups in the huddle was downward, into the warm center of the pile. When nest temperatures were raised, the direction of pup flow reversed to the surface of the pile. Thus, huddling by litters of rat pups is an active process serving to regulate the body temperature of individual members of the huddle. In fact, one might argue that one of the reasons that species whose young are small and uninsulated have large litters is for thermoregulatory purposes.

Behavior is clearly the most efficient means by which infant mammals maintain normal body temperatures. It is also far more sensitive to thermal disturbances than are autonomic mechanisms. Satinoff, McEwen, & Williams (1976) demonstrated that when newborn rabbits were injected with a bacterial pyrogen (one that caused autonomic fever in adult rabbits) and incubated for 2 hr at their thermoneutral temperature, they did not develop a fever. However, when they were allowed to select an ambient temperature in a thermally graded alleyway, pups injected with the pyrogen (Piromen, 500 μg/kg) chose to remain at significantly warmer temperatures than did the saline-injected controls (Figure 13.3). In this experiment the pups were removed from the alleyway after they had remained in the same place for 5 successive mins. In later studies, Kleitman and Satinoff (1980) allowed the pups to stay in the gradient for 30 mins. At the end of that time the body temperatures of the pyrogen-injected pups had risen an average of 1°C, which is significantly different from the mean rise of .5°C of the saline-injected pups ($p < .001$). Thus, in infant rabbits, the behavioral threshold of response to a pyrogen is lower than the autonomic threshold. Thus, behavior is separated from reflexes in ontogeny, and is the more efficient of the two classes of responses in regulating a newborn's

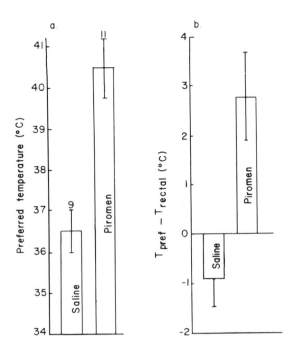

Figure 13.3. (a) Preferred gradient temperature (°C) as a function of saline or pyrogen injection. Numbers over bars denote number of rabbit pups in each group. Bars are ± 1 sem. (b) Preferred gradient temperature (T) minus pretest core temperature as a function of saline or pyrogen injection. (From Satinoff, McEwen, and Williams, 1976. American Association for the Advancement of Science. Reprinted by permission.)

body temperature. So far we can say nothing about how sophisticated the infantile heat seeking is. It may be a learned response or it may be a thermokinesis, that is, infants may simply be more active in cold or hot environments and become quiescent when the ambient temperature is optimal. This is a matter for experimental determination. Nevertheless, whatever it is, it is clearly the major method of achieving a nearly normal body temperature.

In summary, thermoregulatory behavioral responses appear to be phylogenetically older and ontogenetically more sensitive than autonomic responses. In adult mammals, which can use both types of responses, behavior and reflexes are generally activated simultaneously, and this may give the illusion of a single integrator for all thermoregulatory responses. In reality, the neural networks for the two classes of responses are separate. The reason that they occur together in the same situations is that they are oriented toward the same goal—thermal comfort.

REFERENCES

Adair, E. R., Casby, J. U., & Stolwijk, J. A. J. Behavioral temperature regulation in the squirrel monkey: Changes induced by shifts in hypothalamic temperature. *Journal of Comparative and Physiological Psychology*, 1970, 72, 17–27.

Alberts, J. R. Huddling by rat pups: Group behavioral mechanisms of temperature regu-

lation and energy conservation. *Journal of Comparative and Physiological Psychology,* 1978, *92,* 231–245.
Andersson, B., Gale, C., Hokfelt, B., & Larsson, B. Acute and chronic effects of preoptic lesions. *Acta Physiologica Scandinavica,* 1965, *65,* 45–60.
Baccino, M. La The optimum temperature of growth of young homeotherms. Various methods of determination. *Comptes Rendus de la Société de Biologie,* 1935, *119,* 1246–1248.
Baldwin, B. A., & Ingram, D. L. The effect of heating and cooling the hypothalamus on behavioral thermoregulation in the pig. *Journal of Physiology* (London): 1967, *191,* 375–392.
Buetow, K. C., & Klein, S. W. Effect of maintenance of "normal" skin temperature on survival of infants of low birth weight. *Pediatrics,* 1964, *34,* 163–170.
Carlisle, H. J. Behavioural significance of hypothalamic temperature-sensitive cells. *Nature,* 1966, *209,* 1324–1325.
Carlisle, H. J. The effects of preoptic and anterior hypothalamic lesions on behavioral thermoregulation in the cold. *Journal of Comparative and Physiological Psychology,* 1969, *69,* 391–402.
Casterlin, M. E., & Reynolds, W. W. Behavioral fever in anuran amphibian larvae. *Life Science,* 1977a, *20,* 593–596.
Casterlin, M. E., & Reynolds, W. W. Behavioral fever in crayfish. *Hydrobiologia,* 1977b, *56,* 99–101.
Conklin, P., & Heggeness, F. W. Maturation of temperature homeostasis in the rat. *American Journal of Physiology,* 1971, *220,* 333–336.
Dawes, G. S., & Mestyán, G. Changes in the oxygen consumption of newborn guinea pigs and rabbits on exposure to cold. *Journal of Physiology* (London), 1963, *168,* 22–42.
Frazier, C. H., Alpers, B. J., & Lewy, F. H. The anatomical localization of the hypothalamic center for the regulation of temperature. *Brain,* 1936, *59,* 122–129.
Fusco, M. M., Hardy, J. D., & Hammel, H. T. Interaction of central and peripheral factors in physiological temperature regulation. *American Journal of Physiology,* 1961, *200,* 572–580.
Gale, C. C., Mathews, M., & Young, J. Behavioral thermoregulatory responses to hypothalamic cooling and warming in baboons. *Physiology and Behavior,* 1970, *5,* 1–6.
Hammel, H. T., Caldwell, F. T., Jr., & Abrams, R. M. Regulation of body temperature in the blue-tongued lizard. *Science,* 1967, *156,* 1260–1262.
Hammel, H. T., Hardy, J. D., & Fusco, M. M. Thermoregulatory responses to hypothalamic cooling in unanesthetized dogs. *American Journal of Physiology,* 1960, *198,* 481–486.
Hammel, H. T., Stromme, S. B., & Myhre, K. Forebrain temperature activates behavioral thermoregulatory responses in Arctic sculpins. *Science,* 1969, *165,* 83–85.
Hardy, J. D., Stolwijk, J., & Gagge, A. P. Man. In G. C. Whittow (Ed.), *Comparative physiology of thermoregulation* (Vol. 2). New York: Academic Press, 1971, pp. 327–380.
Harlow, H. *Learning to love.* San Francisco: Albion, 1971.
Hull, D. Oxygen consumption and body temperature of newborn rabbits and kittens exposed to cold. *Journal of Physiology* (London), 1965, *177,* 192–202.
Jeddi, E. Contact comfort and behavioral thermoregulation. *Physiology and Behavior,* 1970, *5,* 1487–1493.
Jeddi, E. Thermoregulatory efficiency of neonatal rabbit search for fur comfort contact. *International Journal of Biometeorology,* 1971, *15,* 337–341.
Kleitman, N. & Satinoff, E. Fever in normal and maternally neglected newborn rabbits. In J. M. Lipton (Ed.), *FEVER.* New York: Raven Press, 1980, pp. 197–205.
Lipton, J. M. Effects of preoptic lesions on heat-escape responding and colonic temperature in the rat. *Physiology and Behavior,* 1968, *3,* 165–169.

Magoun, H. W., Harrison, F., Brobeck, J., & Ranson, S. W. Activation of heat loss mechanisms by local heating of the brain. *Journal of Neurophysiology*, 1938, *1*, 101–114.

Mount, L. E. Environmental temperature preferred by the young pig. *Nature*, 1963, *199*, 1212–1213.

Myhre, K., Cabanac, M., & Myhre, G. Fever and behavioural temperature regulation in the frog *Rana esculenta*. *Acta Physiologica Scandinavica*, 1977, *101*, 219–229.

Ogilvie, D. M., & Stinson, R. H. The effect of age on temperature selection by laboratory mice (*Mus musculus*). *Canadian Journal of Zoology*, 1966, *44*, 511–517.

Reynolds, W. W., Casterlin, M. E., & Covert, J. B. Behavioural fever in teleost fishes. *Nature*, 1976, *259*, 41–42.

Satinoff, E. Behavioral thermoregulation in response to local cooling of the rat brain. *American Journal of Physiology*, 1964, *206*, 1389–1394.

Satinoff, E. Neural integration of thermoregulatory responses. In L. V. DiCara (Ed.), *Limbic and autonomic nervous system: Advances in research*. New York: Plenum Press, 1974, pp. 41–83.

Satinoff, E., & Hendersen, R. Thermoregulatory behavior. In W. K. Honig and J. Staddon (Eds.), *Handbook of operant behavior*. Englewood Cliffs, N.J.: Prentice-Hall, 1977, pp. 153–173.

Satinoff, E., McEwen, G. N., Jr., & Williams, B. A. Behavioral fever in newborn rabbits. *Science*, 1976, *193*, 1139–1140.

Satinoff, E., & Rutstein, J. Behavioral thermoregulation in rats with anterior hypothalamic lesions. *Journal of Comparative and Physiological Psychology*, 1970, *71*, 77–82.

Satinoff, E., & Shan, S. Loss of behavioral thermoregulation after lateral hypothalamic lesions in rats. *Journal of Comparative and Physiological Psychology*, 1971, *77*, 302–312.

Silverman, W. A., Fertig, J. W., & Berger, A. P. The influence of the thermal environment upon the survival of newly born premature infants. *Pediatrics*, 1958, *22*, 876–886.

Stone, E. A., Bonnet, K. A., & Hofer, M. A. Survival and development of maternally deprived rats: Role of body temperature. *Psychosomatic Medicine*, 1976, *38*, 242–249.

Vaughn, L. K., Bernheim, H. A., & Kluger, M. J. Fever in the lizard *Dipsosaurus dorsalis*. *Nature*, 1974, *252*, 473–474.

Whittow, G. C. (Ed.). *Comparative physiology of thermoregulation* (Vol. 1): Invertebrates and nonmammalian vertebrates. New York: Academic Press, 1970.

Neuronal Processes of Learning

PART III

V. B. SHVYRKOV

Goal as a System-Forming Factor in Behavior and Learning

14

Behavior constitutes a complex phenomenon that assumes both physiological and psychological aspects of investigation. At the present time, there is apparently little hope of understanding the mechanisms of goal-directed behavior within the framework of the "stimulus-reaction" pattern, from whose viewpoint both the integral behavioral act and the activity of individual neurons in behavior have simple cause–effect relationships to environmental factors or stimuli. Nevertheless, until now most investigations of neuronal behavior mechanisms are based on the "stimulus–reaction" pattern, and this determines both the experimental method employed and the interpretation of results. For example, neuron activity of sensory structures is analyzed in terms of a "reaction" to stimuli with particular physical properties, whereas the activity of motor areas is compared to the motor reactions of a particular muscle or joint. The psychological aspect of behavior in such an approach turns out to be quite superfluous, but the apparent goal-directiveness of behavior remains unexplained.

This chapter attempts to analyze neuron activity in goal-directed behavior from the viewpoint of the functional system theory, first formulated by P. K. Anokhin (1935–1974). We believe that this theory creates a new basis for understanding all biological processes and thereby raises new problems and introduces specific methods of studying goal-directed behavior.

In the broadest sense, behavior can be defined as the relationship between an organism and its environment in which both the organism and the environment are integral entities. Only the experience of a species and of an individual organism makes it possible to identify the particular objects in a "continuous" environment that guide the animal in its search for food, shelter, or a sex partner. And it is only that experience that makes it possible to create from the activities of the organism's various elements, behavior that is directed toward the attainment of these goals.

The determination of environmental behavior is mediated by internal informational or psychological models of environmental objects that correspond to organizations of environmental elements. In a certain aspect, psychological processes (the "mind") represent a system of informational models of different environmental objects and means of transferring one environmental organization to another. In psychology, these models are called *images* and *actions*. The structure of links between individual models reflects the linkage structure of environmental objects and comprises the memory storage or experience of the organism. This store expands in the process of learning. In definitive behavior, any behavioral act can be derived only from memory storage.

Behavior determination by psychological images is accomplished in qualitatively specific systems processes or "organizational processes of physiological processes," which take place only in the natural behavior of an integral organism, and are absent in the presence of narcotics, muscle relaxants or other preparations that disrupt the integrity of the CNS (Shvyrkov, 1978b). Even as a phenomenon, a behavioral act exists only as the organization of specific physiological processes throughout the entire brain and organism. The organization of individual physiological functions into the unified functional system of an integral behavioral act, even in the presence of a drug, has specific developmental phases (Anokhin, 1973a, b), occurs simultaneously in various regions of the brain (Shvyrkov, 1977), and generally brings about the organizational determination of physiological functions by organizing the environmental elements (Shvyrkov, 1978a, b). The determination is of an informational character and is associated with the double nature of the systems processes: their substrate—the activity of the organism's elements, and their informational content—the properties and relationships of the environmental elements. In other words, the organization of physiological functions in the functional system of a behavioral act reflects the organization of the environmental elements. Inasmuch as the organization of physiological functions is a behavioral act, behavior determination by psychological processes is the organizational determination of *specific* physiological functions by memory-derived images of *specific* objects.

The image of any particular object is maintained only in the organization of total brain processes. It can be fixed and retrieved from memory only by a specific organization or a system of neuron activities. An individual neuron in such a system receives the specific organization of a synaptic afferent transmission that constitutes a part of all interneural influences in the entire system. Therefore, a certain part of the integral image can be fixed and reproduced in the memory of a single neuron. This reproduction is probably accomplished as a potentiation of a specific synaptic organization (Anokhin, 1974; Sokolov, 1969), which is mediated by intraneural molecular memory mechanisms (Anokhin, 1974; Matties, 1973). Therefore, the memory of a neuron may be characterized not only as a part of a specific image, but also in terms of the potentiation of synapses mediating the effects of a specific combination of other cells. The image of a specific external object, derived from memory during a change in intraneural metabolic processes, is per se, which is a condition for the activation of *each* individual neuron and at the same time causing a specific organization of *all* interneural links.

The impulse activity of an individual neuron represents its physiological functioning, which is expressed in terms of the neuron's effects upon other cells in accordance with the constant distributional topography of its axon collaterals. The appearance of impulse activity in the selective aggregate of neurons signifies their involvement in the system, which, on the other hand, determines the organizational specificity of selected peripheral processes in a specific behavioral act.

Such a solution to a psychophysiological problem that is based on the functional system theory opens up completely new areas of research in neurophysiology and behavior and raises a plethora of problems that require solutions through experimentation. This chapter is concerned with one of these problems—that of explaining the role played by the various images that comprise experience in determining individual neuron activity in goal-directed behavior.

Method

This task requires a knowledge of the memory structure or experience of the organism under study that, naturally, is self-contained and that is usually not taken into account. The control portion of the experience experiment can probably be produced rather simply in rabbits, because they are kept in solitary cages in the vivarium under deprivation conditions and have a rather limited initial experience. The control series of behavioral acts in the experimental rabbits was produced in preliminary learning over a period of one to two weeks. The rabbits received no food in the vivarium during this entire time.

A hungry rabbit was placed into an experimental chamber 50×50 cm, with a food box and four levers placed along its perimeter (Figure 14.1). In the process of exploring the chamber, the rabbit pressed one of the levers that triggered a flash of light (the photoflash lamp was placed on the chamber's ceiling 90 cm from the floor). One second after the flash of light the food box was filled with 10–20 grams of cabbage or carrots. In the first stage of the experiment, the light and food box appeared when any of the

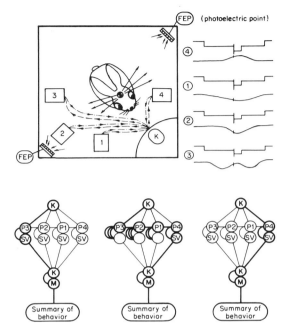

Figure 14.1. Top Left: *Arrangement of food box (K), levers (3, 2, 1, 4), and photoelectric plates in the experimental chamber. On the right are characteristic recordings corresponding to approaches to various levers.* Bottom: *Arrangement of links between behavioral acts in the cycle with operative pedal No. 3 (left), in "emergency" learning (center), and in a cycle with operative lever No. 4 (right).*

four levers was pressed. Later, only one of the levers would produce the light and food. Consequently, the pressing of inoperative levers was quickly extinguished and behavior became cyclic. The rabbit went directly from the food box to the operative lever and went back to the food box after the operative lever was pressed and the light was produced. After 20 to 40 of these cycles, the lever was made inoperative. The rabbit consequently exhibited all of the stages of learning that we have called "emergency" learning, because it is not formation in the full sense of a new skill, but rather a derivation from the memory of one of the previously learned skills. During the discord stage, the rabbit continued to press the inoperative lever three to ten times, but, as no light was produced, it would not go from the lever to the food box, although the animal was sometimes observed to have probed the empty food box with its snout. In the trial-and-error stage, the rabbit pressed the other levers in various sequences and various frequencies, until it pressed the lever the experimenter made operative. After producing the light, the rabbit went immediately to the food box. The stage of new skill formation was then observed: In the course of three to five cycles, after obtaining the carrots, the rabbit pressed not only the operative lever, but the others as well. In the stage of a completely consolidated cycle, the rabbit moved again from the food box to only the new operative lever, and from that lever to the food box. After 20 to 40 cycles of this behavior, one of the other levers (in random order) was made operative, and all the stages of "emergency" learning were observed repeatedly.

The pattern of behavioral acts that comprise the rabbit's experience is presented at the bottom of Figure 14.1, where environmental objects are designated as circles and the lines connecting them represent the actions leading the rabbit from one object to another. In cyclic behavior, the rabbit goes from the food box up to the lever, presses the lever to produce the light, goes up to the food box, lowers its head, and takes the food. This is followed by consummatory behavior, including chewing, swallowing, and so forth, that almost continuously accompanies cyclic appetitive behavior. Motivation in this pattern reproduces memory material from bottom to top and external events from top to bottom. Images of identical objects in the experience structure perform various roles. They act as goals prior to the implementation of corresponding actions, and act as results following those actions. The set of images and the bonds between them that comprise this controlled experience is sufficiently varied. Therefore, the individual images in one cycle must be connected in sequence. In different cycles they are linked alternately. A shift from one system of reproducible images to another takes place in the learning stages.

Following the formation of this experience, basic experiments were conducted to record the impulse activity of individual neurons. We used glass microelectrodes with a tip diameter of about 1 μ and filled with 3 m KCl solution, and a micromanipulator that was built in the laboratory (Grinchenko & Shvyrkov, 1974). The experiment recorded the rabbit's pressing of the levers, its probing into the food box, and its movements around the cage. A dimly lit incandescent miniature bulb was fastened to the rabbit's head, and highly sensitive photoelectric plates, as shown in Figure 14.1, were placed at the corners of the experimental chamber. Electrical recording from the plates gave not only the onset of movements, but also the times at which the rabbit passed by the individual levers. Characteristic recordings are presented in the top right of Figure 14.1.

All indicators were tape recorded on a four-channel recorder. Following the experiment, the recordings were reproduced on a printout autorecorder (at one-tenth of the tape speed) and were processed on a laboratory minicomputer.

In order to explain the connection between individual neuron activity and individual behavioral acts, rasters and histograms were constructed from the onset of lever pressing and appearance of light, as well as from the moment the rabbit inserted or withdrew its snout from the food box. In order to determine the images that bring about neuron activity, the activity of the same neuron was compared in different cycles and behavioral acts. Finally, in order to explain informational restructuring in the neural activity that occurred in "emergency" learning, the dynamics of neuron activity was analyzed in various stages of learning. Several recordings were made of neuron impulse activity in the same rabbit; recordings, were first made of the visual cortical region and then in the motor region. The activity of 70 neurons in the cortical motor region and 45 neurons of the visual region were recorded in five rabbits. The activity of 36 neurons of the motor cortex and 21 neurons of the visual cortex were analyzed in two or more cycles.

Results and Discussion

Of the 115 neurons that were analyzed, the activity of only seven cells was found to be unassociated with any of the factors being recorded in the experiment. The activity of the remaining neurons both in the visual and motor regions of the cortex was found to be structured in accordance with specific behavioral stages. Thus, 34 neurons were activated in only a single behavioral act, whereas 74 cells exhibited several activations corresponding to all or some of the behavioral acts in a cycle. Any combination of activations was possible, and could be different in cycles in which there were different operative levers. A specific time structure of firings could also be identified within the activations. In accordance with the concepts on the hierarchical organization of functional systems (Anokhin, 1973), the reciprocal grouping of several activations reflects the form of neuron participation in the functional system of an entire cycle. A single activation reflects that participation—in the system of an act and the structure of each activation does this—in the physiological subsystems forming a behavioral act. In this chapter we have limited ourselves to analyzing only individual activations identified as an increase in the firing frequency of a neuron at a specific stage of behavior.

In constructing rasters and histograms that reflect the onset of various events (pressing of lever, appearance of light, snout lowered into food box), we found that neuron activations of both the visual and motor cortex usually precede the future results and cease with the appearance of those results, as is shown, for example, in Figure 14.2. In other words, activations in behavioral acts are not reactions to preceding external events, but cause the appearance of events that are future ones relative to the activations. Moreover, the results terminate the activity of the individual neurons directed towards the attainment of those results which also terminate the whole organism's corresponding activity in behavior (Shvyrkov & Grinchenko, 1972). Thus, the involvement of individual neurons in the functional system of a behavioral act does not depend on the images of the results already achieved, but is entirely determined only by the images that comprise the goals.

The fact that the activities of both the visual and motor neurons are identically associated with the structure of a behavioral act indicates that

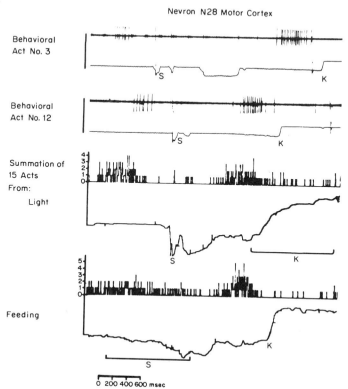

Figure 14.2. *Preactivations of a motor cortex neuron.* Top: *Neurograms in two separate cycles completed by the rabbit in different rhythms. (C) Flash of light when lever is pressed; (K) lowering of snout into food box.* Bottom: *Histograms and cumulative markings of lever pressings and snout probes into the food box (below). Horizontal lines indicate the time spreads of snout probes (K) and the appearance of light (C). Bin is equal to 5 μsec in the histograms.*

neuron activations of various morphological brain structures in behavior signify not only the functioning of physiological "sensory" or "motor" mechanisms, but also reflect the involvement of neurons that have various physiological functions in a unified functional system of a particular behavioral act. The association between neurons of other brain structures and behavioral stages has also been described (Andrianov & Fadeev, 1976; Miller, Satton, Pfingest, Ryan, & Gourevich, 1972; Ranck, 1976; Reymann, Shvyrkov, & Grinchenko, 1977; Sparks & Travis, 1968; Travis & Sparks, 1967).

The activations of any individual neurons reflect their involvement in the actuating mechanisms of behavioral acts; that is, in an action, and they correlate with other coordinated and goal-directed processes during the

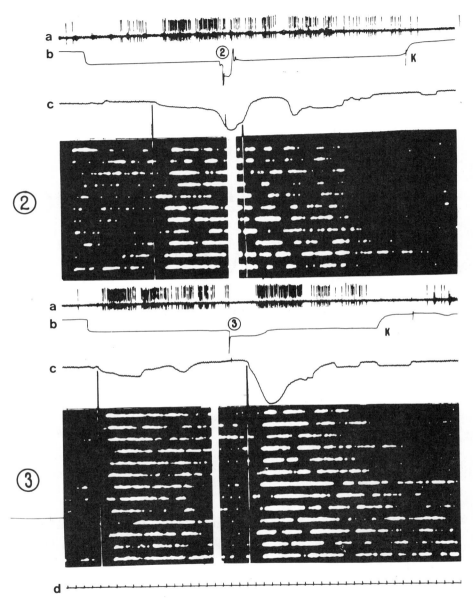

Figure 14.3. The association between activations of a visual cortex neuron and movements in cycles with operative levers Nos. 2 and 3. Designations: (a) neurogram, (b) markings of lever pressings (2) and snout lowering into the food box (K); (c) markings for rabbit movements. The rasters were constructed from the onset of movements from the food box to the lever (left) and from the lever to the food box (right); (d) 100 μsec marker.

action. The image of an external object that serves as the goal determines the activity of all neurons in behavior, and the action is made possible not only by "motor" neurons but by all the remaining neurons. Therefore, neuron activity in behavior not only precedes the result, but also correlates to external movements. An example of the correlation of a visual neuron's activity to movements is shown in Figure 14.3.

Activations of the same neuron in stereotype behavior appear in the same behavioral acts. Consequently, the activations are informationally determined by the same goal. The constancy with which the activations of individual neurons are associated with a specific goal was particularly clear when the neuron participated in only one behavioral act out of the entire cycle. Examples of such neurons are shown in Figures 14.4 and 14.5. Figure 14.4 illustrates a visual cortex neuron that was activated by pressing lever No. 3 only. When the lever became inoperative, activation lasted somewhat longer, probably because no light was produced when the lever was pressed. When either of the other operative or inoperative levers was pressed, the neuron was not activated. Figure 14.5 shows a motor cortex neuron that

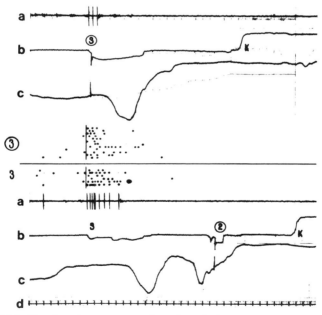

Figure 14.4. Visual cortex neuron activated exclusively by pressing lever No. 3; (a), (b), (c), (d), are same as in Figure 14.3. Top: Activations when operative lever No. 3 was pressed (3). Bottom: Neuron activations when inoperative lever No. 3 was pressed (3) and the absence of activations when operative lever No. 2 is pressed (2). (3): Raster of activations when operative lever No. 3 is pressed, 3: designation activation raster when inoperative lever No. 3 is pressed. Rasters were constructed from the time the lever was pressed.

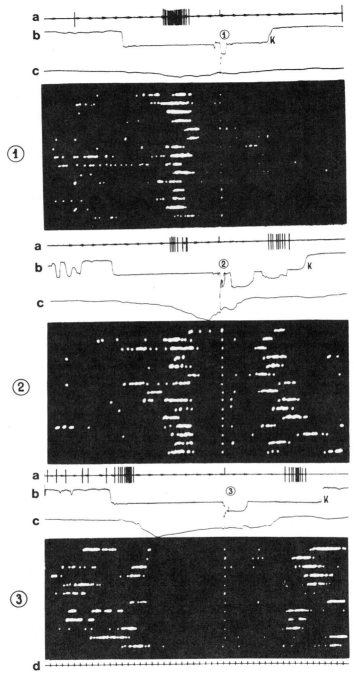

Figure 14.5. *Motor cortex neuron activated upon approach to lever No. 1; (a), (b), (c), (d) are same as in Figure 14.3. (1) shows activation raster upon approach to pedal No. 1; (2): activations roster upon movement from food box to lever No. 2 and from lever No. 2 to food box; (3): activation raster upon approach to lever No. 3 from food box and from the lever to the food box.*

was activated in all cycles as lever No. 1 was being approached, regardless of whether or not the rabbit proceeded to the food box from some other lever or levers.

Thus, certain neurons are activated by a specific goal under any conditions. In this case, this association has a clearly causal character, but not the "probability" or "stochastic" nature that is characteristic of activation associated with preceding events. The clear and constant association between individual neuron activations and specific future results allows one to presume that in stereotype behavior the same goal causes activations of the same neuron system.

The images that appear as goals can be derived from memory only through the internal links between all images in the experience structure. It is possible that a goal derived from memory by motivation and the result which acts directly from the environment are represented in individual neurons by complementary organizations of potentiated synapses. Although a neuron receives different synaptic transmissions in activations "from within" and "from without," one organization of a synaptic transmission assures the activation of the neuron while the other organization eliminates this activation and causes the following goals to be retrieved from memory.

The processes of goal change and behavioral acts in sequential behavior are very complex and include stages of result recognition, afferent synthesis, decision making, formation of the acceptor of action results, and action programs. These processes correspond to stages in the formation of a behavioral act system and last about 100 msec from the time the result of the past act is achieved to the beginning of the next act. We have analyzed neuronal mechanisms of these processes in other works (Shvyrkov, 1978a, b).

All "reactions" to "stimuli" in behavior actually constitute an activity determined by goals, as is easily demonstrated by future-related events relative to the "stimulus" (Aleksandrov, 1975; Shvyrkov & Shvyrkov, 1975).

The activations that we analyzed were determined by images of integral environmental objects such as levers, the food box, or portions of food. The so-called receptive neuron fields identified in analytical experiments probably reflect the fact that in the absence of behavior only very elementary subsystems on the physiological level can be organized, where goals and results appear as information about very detailed environmental elements. These are identified, however, in accordance with innate or acquired experience, but not in accordance with the "physical properties of a stimulus" that are synthetically identified by the experimenter. A change in goals also takes place on the hierarchical level of these subsystems in accordance with the goal change of the behavioral level. The appearance of "reactions" to specific stimuli probably means that specific goals are derived from memory only after specific results are achieved in accordance with the structure of image associations in experience.

The activations both in acts preceding the flash of light and in those that were closer to consummatory behavior were observed in a single stereotype cycle in the 36 motor neurons of the cortex and 21 visual neurons that were examined during the cycle changes. These activations were not extinguished in eight visual neurons and eight motor neurons in the discord stage when the lever became inoperative. This indicates the degree to which such activations are determined by images corresponding to the lever and the flash of light. Activations in acts more removed from consummatory behavior were extinguished in twelve motor and four visual neurons after two to three unsuccessful lever pressings, whereas those activations were maintained when the animal probed his snout into the food box (see Figure 14.6). Inasmuch as the approaches to, and pressings of, the inoperative lever were made without the participation of these neurons, their physiological functions are probably really essential only in subsequent acts. One can therefore assume that the participation of neurons in acts further removed from consummatory behavior is determined by the goals of subsequent acts of the cycle.

The goals of all of the cycle's subsequent acts in stereotype cyclic behavior are probably reproduced from memory even in the first act; that is, in the approach to the lever. Thus, the goal of the first act causes the formation of an integral functional system, whereas the goals of subsequent acts that do not disrupt the integral system of current behavior cause the partial formation of a subsystem. The fact that "anticipatory" activations were observed in only 32 of the 46 neurons activated in near-consummatory behavioral acts supports the hypothesis about the partial formation of subsequent act systems. The physiological functioning of neurons that yield anticipatory activations can be manifested, for example, in the preceding heightened tonicity of certain muscles or in the appearance of salivary flow prior to the appearance of food, as has been demonstrated by classical conditioned reflex methods.

A comparison of neuron activity in different cycles has shown that 9 neurons in each region were activated at a specific stage of behavior, regardless of the cycle. This also applied to the anticipatory activations of these neurons (Figures 14.7 and 14.8). In one cycle, the other 24 neurons (14 motor and 10 visual) were activated in a single behavioral fragment, whereas in the other cycle the neurons were included into another system (Figure 14.9) or yielded activations, including anticipatory ones, in one cycle only (Figure 14.10).

These observations probably indicate that the goals of separate acts in a cycle's functional system are in specific antagonistic or synergistic logical relationships and cause one another's inhibition or appearance. The structural differences in the activity of individual neurons in different cycles are probably tied to the restructuring of logical links between the goals of individual acts in the transition from one cycle to another.

Note must be taken of three more motor and two visual cortical

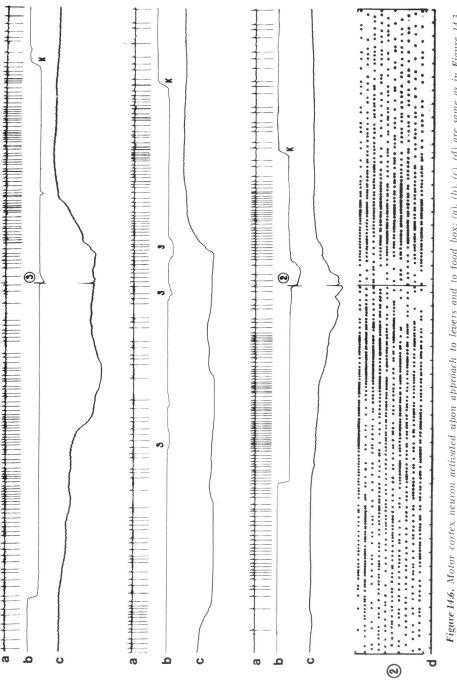

Figure 14.6. Motor cortex neuron activated upon approach to levers and to food box: (a), (b), (c), (d) are same as in Figure 14.3. Top: Activation upon approaches to operative lever No. 3 and food box; below: absence of activation upon approach to inoperative lever No. 3 and activation maintenance/retention/upon approach to empty food box; lower: activations upon approaches to operative lever No. 2 and to food box; below: activation raster in cycles with operative lever No. 2, constructed from the onset of flash of light.

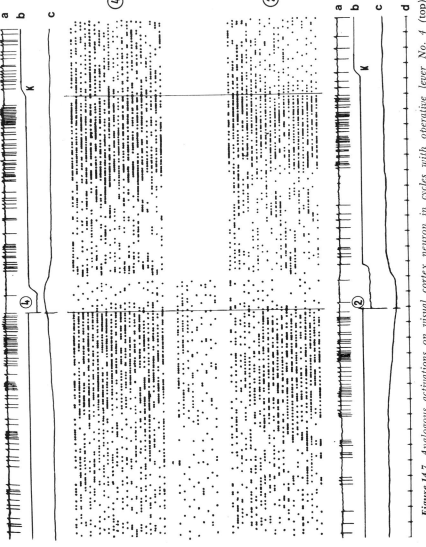

Figure 14.7. Analogous activations on visual cortex neuron in cycles with operative lever No. 4 (top) and No. 2 (bottom); (a), (b), (c), (d) are as in Figure 14.3. (4) shows activation raster in cycles with operative lever No. 4. Designation 4: activation raster in the discord stage. (2) shows activation raster in the cyclic formation stage with operative lever No. 2; the vertical lines signify the starting times at which the rasters were constructed (left: the light's appearance; right: the time the snout was lowered into the food box).

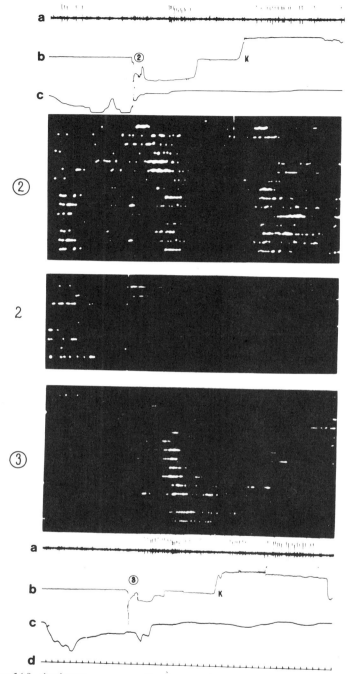

Figure 14.8. Analogous neuron activations of the motor cortex in cycles with operative lever No. 2 (top) and the operative lever No. 3 (bottom); (a) (b), (c), (d), are as in Figure 14.3. (2) shows activation raster during approach to the food box and taking of carrots with operative lever No. 2, constructed from the time the snout was lowered into the food box. Designation 2 shows raster in the discord stage, constructed from the time the nonoperative lever No. 2 was pressed; (3) shows activation raster during approach to the food box and the taking of carrots in the cycle with operative lever No. 3.

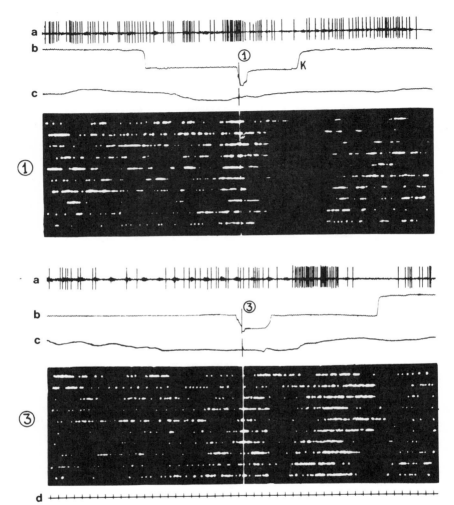

Figure 14.9. *Various activation in a motor cortex neuron in cycles with operative lever No. 1 (top) and No. 3 (bottom); (a), (b), (c), (d) are as in Figure 14.3. (1) shows activation raster when operative lever No. 1 is pressed; (3) is the activation raster in approach to food box in the cycle with operative lever No. 3. Rasters were constructed from the start of the flash of light.*

neurons that were activated in lever approach and pressing acts only during the discord and probe pressing stages. As soon as the lever became inoperative, the activations of these neurons were extinguished (Figure 14.2). One can also probably associate such activations with the restructuring of goals derived from memory in the learning stages.

All of the phenomena which we observed in the learning stages might be summarized as follows.

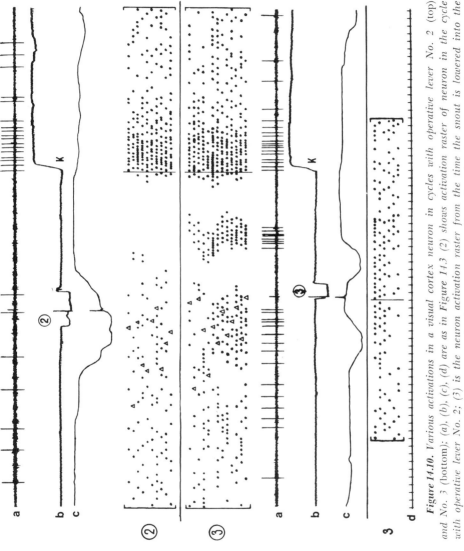

Figure 14.10. Various activations in a visual cortex neuron in cycles with operative lever No. 2 (top) and No. 3 (bottom): (a), (b), (c), (d) are as in Figure 14.3 (2) shows activation raster of neuron in the cycle with operative lever No. 2; (3) is the neuron activation raster from the time the snout is lowered into the food box. Triangles indicate the times the levers were pressed. Designation 3: neuron activation raster in the discord stage, constructed from the time the nonoperative lever No. 3 was pressed.

1. The activations in neurons implicated in the system by the approach or pressing goal in the discord stage are maintained (Figures 14.4 and 14.5).
2. The activations in neurons implicated by the goals of the cycle's future acts are gradually extinguished (Figures 14.6, 14.7, 14.8, and 14.10) and appear in supplemental neurons that are activated only in the discord and trial pressing stages (Figure 14.11).
3. Activations determined by the image of the previously used lever in the trial pressing stage disappear.
4. The activations determined by images of the tested levers appear, and the activations in neurons connected with the learning stages only are maintained.
5. The activations corresponding to the approaches and to pressing of the new operative lever are maintained in the new cycle's formation stage, and the anticipatory activations are developed gradually as the activations in neurons associated with learning stages are extinguished.

In the learning process, there is a restructuring of the intergoal links that correspond to one cycle, in response to the new link structure that corresponds to another cycle. This restructuring probably begins with memory materials closest to consummatory behavior. In addition to goals of one cycle, the goals of other cycles that are in alternate relationships are activated during the discord stage. This is also probably a manifestation of activations in previously nonactive neurons and the elimination of anticipatory activations. Trial pressings are initiated by goals of different cycles in sequence. Finally, during the formation of a new cycle, there is a gradual elimination of alternative goals corresponding to different cycles, and only one cycle is realized. This is probably a manifestation of anticipatory activations.

Phenomenologically, the dynamics of anticipatory activations are not distinguishable from the dynamics of the so-called conditional reactions that are observed in neurons of different brain regions when external and electrocutaneous stimuli are combined, for example (Rabinovich, 1975; Shul'gina, 1969; Shvyrkov, 1969; Vasilevskii, 1968). Our data indicates informationally, that anticipatory activations are determined by the goals of future acts. This conclusion can also be applied to cases of "conditional reactions." This is particularly true if one considers the latter's dependence upon future events (Aleksandrov, 1975; Shvyrkov & Shvyrkov, 1975).

It seems to us that all of the phenomenology accumulated in the neurophysiology of behavior can be explained by approaching the analysis of the neuronal mechanisms of goal-directed behavior from the viewpoint of the functional system theory. At the same time, that approach raises

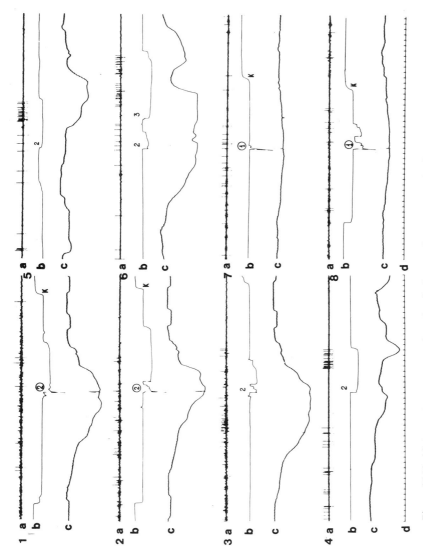

Figure 14.11. Motor cortex neuron activated only by pressing nonoperative levers; (a), (b), (c), (d) are as in Figure 14.3. Designations 1 and 2: no activations in cycles with operative lever No. 2; 3, 4, 5, 6: gradual development of activations when nonoperative levers were pressed; 7, 8: extinction of activations when operative lever No. 1 was pressed.

217

many new problems, the most important of which, we believe, is the hierarchical organization of systems and goals of different levels in behavior.

Although many of the preceding hypotheses require further confirmation, we believe that we have proven experimentally that goals informationally determine the activations of individual neurons, both in behavior and in "emergency" learning, and they constitute a system-forming factor.

Conclusions

1. In behavior, activations of neurons in different brain regions reflect their involvement in integral functional systems of behavioral acts that are determined by psychological images of environmental objects. The activations are initiated by goals and are terminated by the achieved results of the behavioral acts.
2. In stereotype behavior, the same goal, as a rule, activates the same neurons.
3. In stereotype cyclic behavior, neuron activations are determined both by goals of current behavioral acts and by the goals of subsequent acts.
4. Neuron activation in different cycles are determined by goals that are in different logical relationships to each other.
5. In "emergency" learning stages, neurons are activated by goals that are derived from memory in the processes of restructuring the past experience.

REFERENCES

Aleksandrov, Yu. I. Changes in the configuration of conditional reactions in the visual region of a rabbit's brain when reinforcement parameters are changed. *Journal of Higher Nervous Activity*, 1975, 25(No. 4), 760. (In Russian)

Andrianov, V. V., & Fadeev, Yu. A. Impulse activity of visual cortex neurons in successive stages of instrumental behavior. *Journal of Higher Nervous Activity*, 1976, 26(No. 5) 916.

Anokhin, P. K. *The problem of center and periphery in modern neurophysiology.* Gor'kii 1935. (In Russian)

Anokhin, P. K. Fundamental problems in the general theory of functional systems. In *Fundamentals of the systems organization of functions.* Moscow, 1973a, p. 5. (In Russian)

Anokhin, P. K. Systems analysis of the conditioned reflex. *Journal of Higher Nervous Activity*, 1973b, 23(No. 2), 229.

Anokhin, P. K. Systems analysis of a neuron's integrative activity. *Progress in the Physiological Sciences*, 1974, (No. 2), 5. (In Russian)

Grinchenko, Yu. V., & Shvyrkov, V. B. A simple micromanipulator for studying neuro activity in rabbits in free behavior. *Journal of Higher Nervous Activity*, 1974, 24(No. 4), 870.

Matties, H. The biochemical basis of learning and memory. *Life Science*, 1974, *15*(No. 12), 2017.

Miller, I. M., Satton, D., Pfingest, B., Ryan, A., & Gourevich, C. Single cell activity in the auditory cortex of rhesus monkeys: behavioral dependency. *Science*, 1972, *177*, 449.

Rabinovich, M. Ya. *Switching function of the brain. Neuron mechanisms.* Moscow, 1975. (In Russian)

Ranck, J. B. Behavioral correlates and firing repertoires of neurons in the dorsal hippocampal formation and septum of unrestrained rats. In *The Hippocampus* (Vol. 2), New York: Plenum, 1976.

Reymann, K., Shvyrkov, B., & Grinchenko, Yu. V. Über die Beteiligung hippokampaler Neuren an den Teilschritten eines Nahrungsvehaltens. *Acta Biol. Med. Ger.*, 1977, *36*, 1107.

Shul'gina, G. I. Reactions of cortical neurons in the rabbit brain to the early developmental stage of the defense conditional response to rhythmic light. *Journal of Higher Nervous Activity*, 1969, *19*(No. 5), 778.

Shvyrkov, V. B. Comparative characteristics of anticipatory and unconditional stimuli in a rabbit's somatosensory cortex in the development of the conditional defense reflex. *Journal of Higher Nervous Activity*, 1969, *19*(No. 1), 3.

Shvyrkov, V. B. Electrophysiological correlates of sytems processes in the elementary behavioral act. In *Functional importance of electrical brain processes*. Moscow, 1977, p. 95. (In Russian)

Shvyrkov, V. B. Theory of the functional system as a methodological basis of behavioral neurophysiology. *Progress in the Physiological Sciences*, 1978a, *9*(No. 1), 81.

Shvyrkov, V. B. Theory of functional systems—the basis of psychological and physiological synthesis. In *Functional system theory in physiology and pychology*. Moscow, 1978b, p. 11. (In Russian)

Shvyrkov, V. B., & Grinchenko, Yu. V. Electrophysiological study of an action results acceptor in instrumental behavior. *Journal of Higher Nervous Activity*, 1972, *22*(No. 4), 792.

Shvyrkov, N. A., & Shvyrkov, V. B. Activity of the visual cortex neurons in feeding and defense behavior. *Neurophysiology*, 1975, *7*(No. 1), 100.

Sokolov, E. N. *Memory mechanisms*. Moscow, 1969. (In Russian)

Sparks, D. L., & Travis, R. P. Patterns of reticular unit activity observed during the performance of discriminative task. *Physiology and Behavior*, 1968, *3*, 961.

Travis, R. P., & Sparks, D. L. Change in unit activity during stimuli associated with food and shock reinforcement. *Physiology and Behavior*, 1967, *2*, 171.

Vasilevskii, N. N. *Neuronal mechanisms of the cerebral cortex.* Leningrad, 1968. (In Russian)

RICHARD F. THOMPSON
THEODORE W. BERGER
STEPHEN D. BERRY

Brain Mechanisms of Learning[1]

15

Modern analysis of brain mechanisms of learning and memory began with Pavlov's formulation of cortical representation (see Asratyan, Chapter 2 this volume; Pavlov, 1927) and Lashley's concept of the localized memory trace or "engram" (see Lashley, 1929). A very large number of brain lesion studies of learning yielded negative results, in the sense that spatially localized memory traces could not be demonstrated in the brain (see Lashley, 1950).

This outcome has forced most workers to abandon the notion of the localized engram; indeed, Lashley had abandoned it by 1929 as a result of his own work. The logic underlying the notion of the localized engram seems to derive from an oversimplified conception of causality. It is an example of what the Soviet psychologist Boris Lomov (Chapter 1, this volume) has termed *linear causality*. The basic idea is a linear or series chain of events from stimulus inputs to final motor output, with a critical change developing at some point in the sequence in the brain. Hence, there would be a direct linear causal chain from this change, the engram, to the learned behavior. Except in very simple systems, it is doubtful if such elementary linear causality ever obtains in the central nervous system.

---

[1] Supported by research grants from the National Institute of Mental Health (MH26530), National Institutes of Health (NS12268), National Science Foundation (BMS 75-00453), and the McKnight Foundation, with the support of the Center for Advanced Study in the Behavioral Sciences, Stanford, California, the National Institute of Mental Health Grant 5T 32MH14581-03 and National Science Foundation Grant BNS 76-22943 A02.

Given that a localized engram does not exist and hence that brain mechanisms of learning cannot be accounted for in terms of linear causality, what are the alternatives? We wish to suggest that the memory system of higher animals—mammals—consists of a number of brain systems that play various roles during learning. These systems can be defined, or at least characterized, by anatomical and physiological criteria. Various systems may exist more or less separately, or overlap, or merge. They can have hierarchical organization, as in a sensory "system," a partly temporal organization, as in certain motor systems, or alternative organizations that have not yet been characterized. The roles these various hypothetical brain systems play in learning and memory may or may not correspond to conceptual categories or terms that now exist. Note that such a multiple systems theory can also account for results of the lesion studies that give rise to Lashley's concepts of equipotentiality and mass action. A discrete lesion might interrupt only a part of one or more systems—they could still function, although perhaps not as well. A system, almost by definition, is not localized to one anatomical place. The larger the lesion, the more systems that are damaged, and the greater is the damage to some, thus yielding greater impairment.

Note that this neural "systems" approach to learning has certain similarities to Pavlov's hierarchical systems, particularly as developed by his student Asratyan (Chapter 2, this volume) and his associates, for example, Gasanov (Chapter 22, this volume); to the functional systems approach developed by another of Pavlov's students, Anokhin (1968) and his associates, for example, Sudakov (1965) and Shvyrkov (1977; Chapter 11, this volume); and to the general approach developed by Sokolov and his associates in their work on the orienting reflex (e.g., Sokolov & Vinogradova, 1975).

Analysis of brain systems and mechanisms underlying learning and memory has many parallels with analysis of motor systems. Both are concerned with relating brain events to behavior. The obvious difference is, of course, that learning involves change in behavior as a result of experience. We wish to find the neuronal substrate of the *change* in behavior as opposed to the *substrate* of the behavior, per se. This is perhaps just another way of stating the learning versus performance distinction that is emphasized in psychology. However, it is a critical distinction from an experimental viewpoint. The paradigms used must permit one to distinguish between neurophysiological substrates of learning and behavioral performance. There must be changes that develop within and/or among the various brain systems involved in learning and memory. This, then, is the engram; it is not a localized change at one place in the brain nor a diffuse net, but rather sets of changes in sets of definable interacting brain systems.

Given the preceding rationale, we have adopted the general approach of recording neuronal unit activity during the course of learning in a simple and discrete Pavlovian conditioned response situation. The goal is to characterize the activity of various brain systems in learning and memory. Once this is accomplished, the structures and systems that exhibit altered activity with learning will have been identified and analysis of synaptic mechanisms will be feasible. We began by identifying the immediate neuronal substrate of the behavioral conditioned response—the motoneurons—and characterizing their activity during learning. Results of these studies will be indicated briefly below. Having defined the pattern of change of neuronal activity during learning at the final common path, it can be used as a neural performance measure against which to compare activity of higher brain structures and systems.

To date, we have focused on the limbic system—the hippocampus and related structures. The mammalian hippocampus has been implicated in learning and memory in a wide variety of experimental and clinical conditions (Isaacson & Pribram, 1976; Olds, Disterhoft, Segal, Kornblith, & Hirsh, 1972; Scoville & Milner, 1957; Sokolov, 1977; Sokolov & Vinogradova, 1975; and others). However, the precise role of the hippocampus in learning has not been clear. Our results indicate that the system seems to play a very specific role in learning, at least in the simple classical conditioning situation we use: Neurons of the hippocampus rapidly develop a *temporal* model of the behavioral response to be learned, but they develop this model only under conditions in which behavioral learning will subsequently occur.

We selected a preparation developed by Gormezano (1966, 1972)—a classical conditioning of the rabbit nictitating membrane response to a tone-conditioned stimulus (CS) using a corneal airpuff unconditioned stimulus (UCS)—as a simple and discrete model of mammalian learning. This system has several advantages, which have been detailed elsewhere (Thompson, *et al.*, 1976). These are both practical and conceptual. They are practical in that the animal is held motionless but not drugged or paralyzed, significant learning occurs within a single 2-hr training session, and the airpuff UCS (as opposed to shock) does not give recording artifact. They are conceptual in that, thanks to the extensive studies of Gormezano and associates, the learned response is very well characterized—it is an extremely well-behaved Pavlovian response and shows virtually no pseudoconditioning or sensitization; learning versus performance substrates can be distinguished at the neuronal level, and the actual amplitude-time course of the behavioral response is easily measured and quantified.

Methods

The details of our procedures are given elsewhere and will be indicated only briefly here (Berger, Alger, & Thompson, 1976; Berger & Thompson, 1978a, b, c; Berry & Thompson, 1978). We have adopted the behavioral procedures developed by Gormezano (1966). Animals are given a tone CS (1 KHz, 85 db, 350 msec duration) and a corneal airpuff UCS (210 grams/ cm pressure source, duration 100 msec, occurring during the last msec of the tone CS—they terminate simultaneously). A mean intertrial interval of 60 sec is used and varied from 50–70 sec to eliminate possible temporal conditioning. Animals are given eight paired trials and one tone alone test trial per block, and are typically given 13 such blocks in a day, that is, training session. Unpaired control animals are given a pseudorandom sequence of unpaired CS and UCS presentations (explicitly unpaired procedure) with a mean interval of 30 sec (varied from 20–40) for approximately the same total number of stimulus presentations per session as conditioning animals. The exact amplitude-time course of the NM extension response is measured by a micropotentiometer, recorded on tape, digitized at 3 msec intervals and stored in the computer. Later analysis involves computation of onset latencies, eight-trial averaged responses and measurement of the area under the NM response curve. This latter measure provides a useful index of the "amount" of the response in terms of both amplitude and time (Cegavske, Patterson, & Thompson, 1979).

Unit spike discharges of neurons (either multiple unit clusters or isolated single unit potentials) are recorded using metal microelectrodes and stored on tape. The unit discharges are picked off by a discriminator, converted to standard pulses and fed into the computer. The basic data collection program counts the number of unit discharges in each 3 msec time bin. Data collection begins 250 msec prior to tone CS onset (the pre-CS period), continues through the 250 msec of tone (the CS period), and then for an additional 250 msec beginning with airpuff onset (the UCS period). Airpuff "onset" time is the time at which the airpuff actually arrives at the cornea. Unit counts are cumulated for display, for example, in eight-paired-trial frequency histograms; all histogram data shown here are cumulated and displayed in 15 msec time bins. Cumulated eight-trial unit count data are also converted to standard scores, relative to background (pre-CS) activity; for example, for the CS period the standard score is the mean CS counts minus the mean pre-CS counts divided by the standard deviation of the pre-CS activity, the latter being computed on an entire day's session. The unit standard score measure for an eight-trial block for a given time period (e.g., the CS period or the UCS period) can be compared with the area under the averaged NM response curve for that same block of trials.

The multiple-unit microelectrode, of insulated stainless steel with a $5-7\mu$ tip diameter and a $40-50\mu$ shaft exposed, is permanently implanted (while monitoring unit activity for localization) using halothane anesthesia in the structure to be studied. For single unit recording, a small chronic microdrive system is implanted in the skull overlying the target structure, and single unit microelectrodes—$3-5\mu$ tip diameter, insulated to the tip, 500 K–1 M ohm resistance—are inserted for each recording session. At least 1 week is allowed between surgical implantation procedures and the beginning of training–recording sessions.

The Final Common Path

The highest correlation possible between neuronal and behavioral events should hold for the behavior and its immediate neuronal precedent, the activity of motoneurons in the final common path controlling the behavioral response. In initial studies, we identified the abducens (6th nerve) motoneurons as the final common path for the NM extension response. Nictitating membrane extension is a largely passive consequence of eyeball retraction via the retractor bulbus muscle, innervated by the 6th nerve (Cegavske, Thompson, Patterson, & Gormezano, 1976; Young, Cegavske, & Thompson, 1976). We completed a study comparing eight-paired conditioning and eight-unpaired control animals with multiple-unit recording electrodes implanted in the abducens nucleus ipsilateral to the eye being conditioned (Cegavske, Patterson, & Thompson, 1979). Examples of eight-trial average NM responses and histograms of abducens unit activity are shown in Figure 15.1 for a conditioning animal before (A) and after (B) learning and for a control animal to airpuff (C) and to tone (D). Results are clear; there is a very close coupling between abducens unit activity and the behavioral response, independent of whether the animal has learned or not, whether it is a conditioning or control animal, and whether it is a conditioned response or a reflex response. Whatever the abducens neurons do, so does the nictitating membrane.

This close correspondence of the amplitude-time course of the NM response and the histogram of unit activity from the motor nucleus is extremely useful. It means that the easily recorded NM response actually portrays a temporal course of the histogram of unit activity in the motor nucleus. It is particularly helpful when studying changes in neuronal activity in higher brain structures during learning. It is necessary to compare the temporal patterns of neural activity in such structures against the pattern of activity in the final common path during acquisition of the conditioned response. Given the present findings (e.g., Figure 15.1), it is not

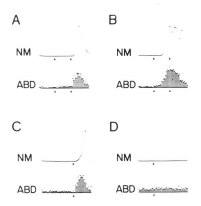

Figure 15.1. Examples of eight-trial averaged behavioral nictitating membrane (NM) responses and associated multiple unit histograms of abducens nucleus activity (15 msec time bins) for a conditioning animal at the beginning and end of training (A, B) and a control animal at the beginning of training for the airpuff UCS (C) and the tone CS (D). Note the close correspondence between the histogram of unit activity recorded from the final common path and the temporal form of the behavioral NM response in all cases. (Cegavske, Patterson & Thompson, 1979. Reproduced by permission.)

necessary to record activity of abducens motoneurons during acquisition; measurement of the NM response form will suffice.

The Hippocampal System

As previously noted, we selected the hippocampus as an initial brain system in which to explore possible learning-related changes in this clear-cut classical conditioning paradigm.

HIPPOCAMPAL EEG

A very simple study of hippocampal EEG indicated that neuronal activity in the hippocampus is indeed related to learning in this paradigm (Berry & Thompson, 1978, 1979). In brief, 2 min time samples of spontaneous hippocampal EEG were recorded at the beginning and end of each training day in 16 animals. In the rabbit, the hippocampal EEG is dominated by rhythmic slow activity—a large amplitude, an almost sinusoidal waveform of approximately 3–8 Hz (so-called theta activity), which occurs in the waking state in response to many forms of stimulation and during paradoxical sleep (Green & Arduini, 1954; Winson, 1972). It is generally believed to be a good index of behavioral "state"—prominent theta indicates arousal (e.g., Lindsley & Wilson, 1975).

In order to characterize overall (2–22 Hz) EEG in terms of frequency, a low/high dichotomy ratio was computed: The percentage of 8–22 Hz activity divided by the percentage of 2–8 Hz (theta) activity. The correlation between this measure and trials to criterion was highly significant ($r = +.72$, $df = 14$, $p < .01$). Note the clear linear trend of the correlation illustrated by the scatter plot and best-fitting linear regression line in Figure 15.2.

Thus, a brief time sample of hippocampal EEG taken prior to the onset of training is highly predictive of subsequent learning rate, even over a period of days. A higher proportion of hippocampal theta (2–8 Hz) predicts faster rates of learning. To our knowledge, this is the first demonstration that a purely *neurophysiological* measure taken prior to the beginning of training can predict the subsequent *behavioral* rate of learning. This result is nicely consistent with consolidation studies showing a positive relationship between amount of theta in the posttraining EEG and subsequent retention performance (Landfield, McGaugh, & Tusa, 1972) and with studies reporting change in hippocampal EEG frequency and phase relations during training (Adey, 1966; Coleman & Lindsley, 1977; Grastyan, Lissack, Madarasz, & Donhoffer, 1959).

Prokasy (1972) has developed a most interesting mathematical model of behavioral learning for this particular paradigm, which indicates that

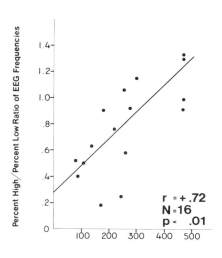

Figure 15.2. Scatter plot and best-fitting regression line for the relationship between trials to criterion and the EEG frequency ratio (percent 8–22 Hz activity divided by percent 2–8 Hz activity) computed on the 2 min time sample of EEG taken just prior to training. Data are from 16 rabbits. (From Berry & Thompson, 1978. Reproduced by permission.)

learning occurs in two phases: (a) an initial phase that extends from the beginning of training until the animal begins to give conditioned responses and (b) a second phase that extends from this point until the response is well learned. The initial phase is more variable and more likely to be influenced by "motivation," "arousal," and other conditions.

Partly to determine if Prokasy's model could be extended to physiological measures, we compared the number of trials of training required to give the 5th conditioned response against the amount of *change* in the low/high EEG ratio over training for each of the 16 animals previously described. The result, shown in Figure 15.3, is striking and would seem to provide "physiological" substantiation of Prokasy's model. The correlation between the amount of change in the EEG ratio and the number of trials to the 5th CR is −.93, a highly significant value.

These results are in close agreement with the data of Coleman and Lindsley (1977) in their analysis of hippocampal EEG during lever press learning for reward in cats. The data provide further support for the general notion of an inverted "U" function relating alerting and arousal to learning (Duffy, 1962; Hebb, 1955; Lindsley, 1951; Malmo, 1959). Finally, our data indicate that hippocampal activity is closely related to learning in the rabbit NM paradigm.

HIPPOCAMPAL UNIT ACTIVITY

Our unit analysis of hippocampal activity during learning began with multiple unit recordings (typically 4–12 units were recorded) from the pyramidal cell layer of CA1–2 and CA3–4, and from the granule cell layer

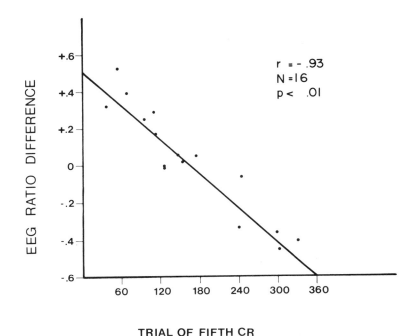

Figure 15.3. Scatter plot and regression line showing the relationship between the amount of change in the EEG frequency ratio (see text) and the number of trials to the fifth conditioned response for the 16 animals of Figure 15.2. (From Berry & Thompson, 1979. Reproduced by permission.)

of the dentate gyrus (Berger, Alger, & Thompson, 1976; Berger & Thompson, 1978a). Essentially, the same result was obtained from all these regions. An example is shown in Figure 15.4. The hippocampal unit poststimulus histogram and averaged NM response for the first block of eight trials are given for one animal in Figure 15.4A and for the same animal after learning criterion was reached in Figure 15.4B; over the first block of eight trials there is a large increase in unit activity in the UCS period that precedes and closely parallels the behavioral NM response form. Over training, this hippocampal unit response increases and moves into the CS period as behavioral learning develops. Indeed, Figures 15.4A and B closely resemble unit activity from the motor nucleus (see Figure 15.1). Actually, the average latency of the hippocampal response is shorter than that for motorneurons (42 msec less than NM onset for hippocampal units and 17 msec less than NM onset for motor units).

In marked contrast, the control animal hippocampal data are completely different from motoneuron activity. The eight-trial hippocampal unit activity and averaged NM are shown for airpuff-alone trials at the beginning and end of unpaired training for a control animal in Figures

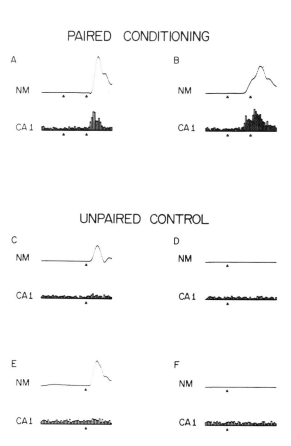

Figure 15.4. Examples of eight-trial averaged behavioral NM responses and associated multiple unit histograms of hippocampal activity for a conditioning (A, B) and a control (C–F) animal at the beginning and end of training. Note the very large increase in hippocampal unit activity that develops in the conditioning animal. Upper trace: Average nictitating membrane response for one block of eight trials. Lower trace: Hippocampal unit poststimulus histogram (15 msec time bins) for one block of eight trials. (A) First block of eight-paired conditioning trials, Day 1. (B) Last block of eight-paired conditioning trials, Day 1, after conditioning has occurred. (C) First block of eight-unpaired UCS-alone trials, Day 1. (E) Last block of eight-unpaired UCS-alone trials, Day 2. (D) First block of eight unpaired CS-alone trials, Day 1. (F) Last block of eight unpaired CS-alone trials, Day 2. (From Berger & Thompson, 1978a. Reproduced by permission.)

15.4C and E. Although there is a clear reflex NM response, there is little associated unit activity in the hippocampus. There is essentially no NM response or evoked hippocampal activity in tone-alone trials (Figures 15.4D and F).

The hippocampal unit responses illustrated in Figure 15.4 are closely paralleled for all animals in both conditioning and control groups. In Figure 15.5 are shown the standard scores of unit activity for both paired ($N = 21$) and unpaired ($N = 12$) groups across all blocks of training trials. For both UCS and CS periods, unit activity in the hippocampus for conditioned animals increases and remains high over all 26 blocks of paired trials (334 trials total, solid lines). In contrast, standard scores for animals given control training remain low across blocks of unpaired trials (broken lines) in both the CS and UCS periods. Although the unpaired standard scores for the UCS period are positive, the differences between paired and unpaired groups are quite dramatic and statistically highly significant.

Behavioral learning closely parallels the development of the hippo-

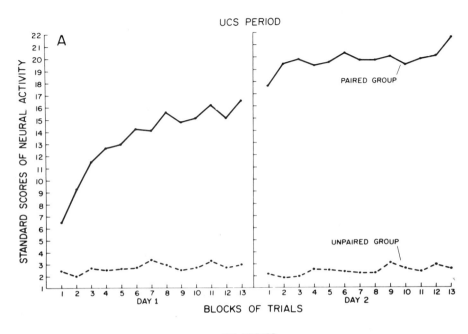
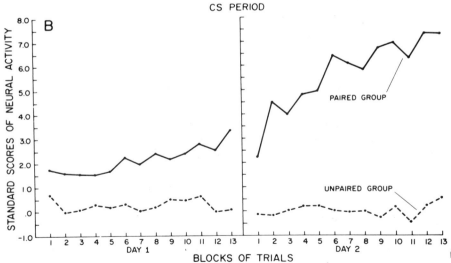

Figure 15.5. Group curves of standard scores of hippocampal neural activity throughout training. (A) Standard scores of unit activity for UCS period, day 1 and day 2. (B) Standard scores of unit activity for CS period, Days 1 and 2. Paired conditioning group (N = 21, Day 1; N = 14, Day 2): solid lines. Unpaired control group (N = 12, Days 1 and 2): broken lines. Note expanded y-axis and different zero point for CS period graph. (From Berger & Thompson, 1978a. Reproduced by permission.)

campal unit response in the *CS* period. For the average data, this occurred on about block 6 (see Figure 15.5). On the average, behavioral conditioned responses began to occur when the hippocampal unit activity in the *UCS* period had increased to about 12–13 standard scores (see Figure 15.5). This activity increases linearly over initial blocks of training and begins to decrease its rate of growth at about the time behavioral learning begins to occur.

Because hippocampal unit activity in the UCS period appeared so highly developed at the end of the first block of paired trials, an individual trial analysis for the first eight pairings was completed for all animals and control groups. In brief, the two groups begin with the same low levels of hippocampal activity, but diverge significantly by the end of the first eight-trial block.

Data acquired from animals that occasionally give spontaneous NM responses during trial periods provide further evidence against "motor" or "sensory" interpretation. Examples are shown in Figure 15.6A–C. In this example (A) the animal has not yet learned behaviorally. These are individual paired trials, hence the variability in the NM and the low levels of the hippocampal unit histograms. Note that although there is little or no hippocampal unit activity associated with even large spontaneous NM responses (e.g., Figure 15.6C), there is clear hippocampal unit activity associated with the NM response to the *paired* stimuli. Figure 15.6D shows an individual test trial for an animal that has learned behaviorally. Al-

Figure 15.6. Spontaneous, reflex, and conditioned nictitating membrane responses and associated hippocampal unit activity. Upper trace: *Individual nictitating membrane response from a single trial.* Lower trace: *Hippocampal unit poststimulus histogram from a single trial. (A), (B), and (C) Paired conditioning trial; (D) test trial. See text for explanation.* Vertical bar in C equals 16 unit counts per 15 msec time bin. Note the virtual absence of hippocampal unit activity associated with spontaneous NM responses and the much larger activity associated with the NM response to paired stimulation. (From Berger & Thompson, 1978a. Reproduced by permission.)

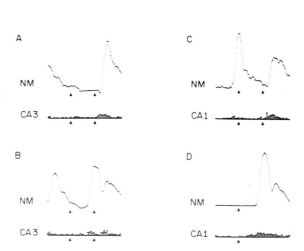

though there is no airpuff, there is a clear hippocampal response associated with learned NM response.

To summarize briefly, under conditions of paired training where behavioral learning will occur, unit activity in the hippocampus increases rapidly, initially in the UCS period, forms a temporal "model" of the behavioral response, and precedes it in time. As hippocampal activity begins to occur in the CS period, behavioral learning begins to occur. This increased unit activity in the hippocampus does not develop in unpaired control animals. In paired animals the hippocampal activity begins to develop in the first few trials of training.

In the context of "goal-directed behavior," we wish to suggest that unit activity in the hippocampus may in fact code the goal-directed aspects of behavior in learning. There is little increase in hippocampal unit activity during unsignaled *reflex* responses to airpuff (Figure 15.4) or during spontaneous responses (Figure 15.6). However, in the case of a signaled response (tone—airpuff pairing; see Figure 15.4) there is a marked increase in hippocampal activity that develops after only a few such trials, long before behavioral learning. This hippocampal response may well be the earliest sign of goal-directed learning to develop in the brain.

This argument implies that unsignaled reflex responses are *not* goal directed. They are, of course, biologically adaptive, but they are not goal directed or "purposive" in Tolman's sense. In classical conditioning, the CS serves as a signal for the occurrence of the UCS. Under these conditions, it is possible for the animal's behavior to become goal directed. If this argument is correct, the hippocampus may then form the basic substrate for goal-directed behavior in the brain. This usage of the term "goal directed," incidentally, would seem to correspond to Epstein's (1979) use of the term "motivation," and seems to us to be at least close to the meaning employed by Shvyrkov at this conference.

HIPPOCAMPAL PROJECTIONS

If the large, learning-dependent response that develops in the hippocampus is to exert an influence on other brain structures and systems and ultimately on the activity of motoneurons in the final common path for the behavioral conditioned response—the abducens nucleus—then it must be projected out of the hippocampus to other structures.

Recent anatomical data indicate that the majority of fibers projecting from the hippocampus to the septal nuclei project to the lateral septal nucleus (Meibach & Siegel, 1977; Nauta, 1956; Raisman, Cowan, & Powell, 1966; Swanson & Cowan, 1977). The medial septal nucleus, in contrast, is predominantly a source of fibers projecting to the hippocampus (Anderson, Bruland, & Kaada, 1961; Guillery, 1956; Mosko, Lynch, & Cotman, 1973; Segal & Landis, 1974; Storm-Mathisen, 1970).

15. Brain Mechanisms of Learning

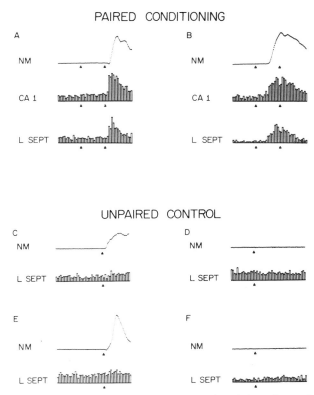

Figure 15.7. Simultaneously recorded hippocampal and lateral septal unit responses from a paired conditioning animal. Upper trace: Average nictitating membrane response for one block of eight trials. Middle trace: Hippocampal unit poststimulus histogram for once block of eight trials. Lower trace: Lateral septal unit poststimulus histogram for one block of eight trials. (A) and (B) show results for one paired conditioning animal. (A) One block of paired trials early on day 1; (B) one block of conditioning trials late on Day 1. (C)–(F) shows results for one unpaired control animal. (C) and (E) UCS-alone trials from early and late in unpaired training, respectively. (D) and (F) CS-alone trials from early and late in unpaired training, respectively. Vertical bar in (A) equals 46 unit counts per 15 msec time bin. (From Berger & Thompson, 1978c. Reproduced by permission.)

Examples of simultaneous recordings from the lateral septal nucleus and the hippocampus are shown in Figure 15.7. As is clearly seen, the same growth in unit activity occurs in the lateral septum as in the hippocampus, and only under paired training, not in unpaired controls. Group data comparing unit standard scores for the lateral septum and hippocampus recorded simultaneously during the course of training (seven animals) are shown in Figure 15.8. Note that initial growth of activity is faster in the hippocampus than in the lateral septal nucleus in the UCS period but that lateral septal activity eventually catches up. The growth of activity during

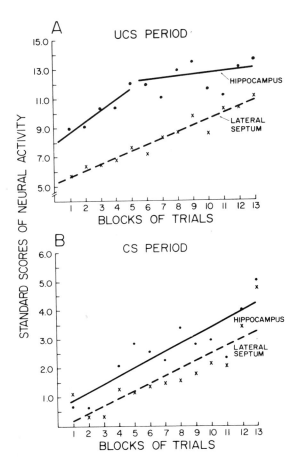

Figure 15.8. Linear regression analysis of unit responses recorded simultaneously from hippocampus and lateral septum during paired conditioning. (A) UCS period standard scores of unit activity. (B) CS period standard scores of unit activity. Solid lines: *Hippocampus;* broken lines: *lateral septum.* Note expanded y-axis and different zero point for CS period graph. (From Berger & Thompson, 1978c. Reproduced by permission.)

the CS period is the same in both structures. It is as though the increasing activity in the hippocampus induces a similar plasticity in the lateral septal nucleus, but it takes a bit more time to develop initially.

Results from the medial septal nucleus are quite different. An example is shown in Figure 15.9. Here there is evoked unit activity to the onsets of the stimuli. However, this activity does not grow over training. Instead, it decreases or habituates, at the same time that unit activity is growing in the hippocampus and the lateral septum. The medial septum does not exhibit the learning-dependent plasticity—the increase in unit activity that models the behavioral response. Instead, it appears to be providing the hippocampus with information about the occurence of stimuli. Results for both lateral and medial septal nuclei are in complete accord with current anatomical data and appear to show a functional projection of learning-dependent increases in unit activity over a defined anatomical pathway (hippocampus-lateral septum).

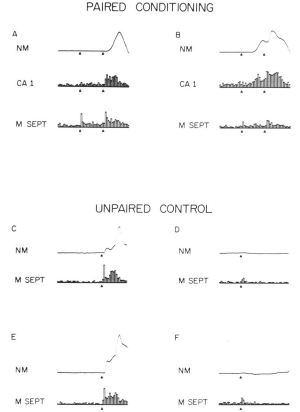

Figure 15.9. Simultaneously recorded hippocampal and medial septal unit responses from a paired conditioning animal. Upper trace: *Average nictitating membrane response for one block of eight trials.* Middle trace: *Hippocampal unit poststimulus histogram for one block of eight trials.* Lower trace: *Medial septal unit poststimulus histogram for one block of eight trials.* (A) and (B) show results for one paired conditioning animal. (A) One block of paired trials early on Day 1; (B) one block of conditioning trials late on Day 1. (C)–(F) show results for one unpaired control animal; (C) and (E) UCS-alone trials from early and late in unpaired training, respectively; (D) and (F) CS-alone trials from early and late in unpaired training, respectively. Vertical bar in (A) equals 46 unit counts per 15 msec time bin. (From Berger & Thompson, 1978c. Reproduced by permission.)

SINGLE UNIT ANALYSIS

All of the hippocampal data previously discussed involved multiple unit recording—measurement of the activity of small clusters of units. The fact that this unit response grows to such a large extent over the course of training (an average increase of 20 standard scores—see Figure 15.5) implies that a substantial number of units in the hippocampus are involved. However, more detailed information about unit activity requires isolated single unit recording. A related issue of great importance concerns analysis of the synaptic mechanisms underlying the increase in neuronal activity. It is first necessary to identify the classes of neurons involved.

The anatomy of the hippocampus provides a convenient method for identifying at least one class of neurons, the pyramidal cells. As Spencer & Kandel (1961) showed, pyramidal cells can be antidromically activated by electrical stimulation of the fornix. We utilized this method with a chronically implanted bipolar stimulating electrode in the fornix. Actually, the fimbriafornix system has both efferent axons from pyramidal cells and afferent fibers projecting to the hippocampus. Electrical stimulation of the fornix can produce antidromic (backward nonsynaptic) firing of pyramidal cells (determined by short latency, low variability of latency, and ability to follow at high frequencies) or orthodromic mono- or polysynaptic activation of hippocampal neurons. The latter could be interneurons or pyramidal cells.

We recently completed an initial study of 36 hippocampal neurons using this identification technique (Berger & Thompson, 1978b). The procedure involves lowering the microelectrode with the chronically implanted microdrive until a spontaneously active neuron was isolated. It was then tested with fornix stimulation and then studied in the conditioning paradigm.

There were three categories of units in terms of response to fornix stimulation: (a) antidromically activated pyramidal neurons; (b) orthodromically activated neurons (otherwise unidentifiable); and (c) some neurons that could not be activated at all by fornix stimulation. This last class of neurons tended to have very low spontaneous activity rates. In terms of the patterns of unit activity related to learning, results were unexpectedly clear (see Figure 15.10). The majority of neurons identified as

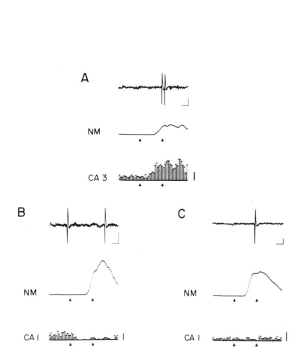

Figure 15.10. Middle traces show average NM response and bottom traces show poststimulus histograms generated by isolated single hippocampal units recorded during paired conditioning. Top traces show examples of spontaneous activity from the single cells that generated the respective poststimulus histograms. Calibrations for upper raw data trace equal 5 µV and 5 msec. (A) Data collected from an antidromically activated unit (pyramidal cell). Vertical histogram calibration is equivalent to 25 unit counts per 15 msec time bin. (B) Data collected from an orthodromically activated unit. Calibration equals 21 counts per 15 msec time. (C) Data collected from a hippocampal cell not activated by fornix stimulation. Calibration equals 28 counts per 15 msec time bin. (From Berger & Thompson, 1978b. Reproduced by permission.)

pyramidal cells (16 of the 20) generated the typical "multiple unit" histogram (see Figure 15.10A). The majority of orthodromically activated neurons showed inhibition during the trial period (see Figure 15.10B). The third class—cells that did not respond to fornix stimulation—showed no apparent changes in activity over the trial periods (Figure 15.10C).

Consequently, it seems that the growth in unit activity in the hippocampus during learning is due primarily to increased activity of pyramidal cells. Having identified the pyramidal cells as the major class of neurons generating the learning-dependent hippocampal response, it now becomes possible to analyze the underlying synaptic mechanisms.

All of our results can be summed up in the following rather simple statement: In the learning paradigm we employ, *the growth of the hippocampal unit response is completely predictive of subsequent behavioral learning.* If the hippocampal response does not develop, the animal will not learn. If it develops rapidly, the animal will learn rapidly. If it develops slowly, the animal will learn slowly. The hippocampal response develops within the first few trials of training and is projected to other brain structures (e.g., lateral septum). The response appears to be generated primarily by pyramidal neurons. Finally, the temporal form of the hippocampus unit response predicts the temporal form of the behavioral response; it forms a *temporal model* of the learned behavioral response.

In the context of goal-directed behavior, the hippocampus may function as a portion of the basic brain system that seems to code the goal-directed or "salient" aspects of the stimulus world in relation to adaptive behavior.

ACKNOWLEDGMENTS

We thank Fe Glanzman and Carol Cooper for the histology, Kathy Berry for preparing the illustrations, and Carol Hibbert and Sharon Phillips for their contributions to the manuscript.

REFERENCES

Adey, W. R. Neurophysiological correlates of information transaction and storage in brain tissue. In E. Stellar & J. M. Sprague (Eds.), *Progress in physiological psychology* (Vol. 1). New York: Academic Press, 1966. Pp. 3–43.

Andersen, P., Bruland, P. H., & Kaada, B. R. Activation of field CA1 of the hippcoampus by septal simulation. *Acta Physiologica Scandinavica*, 1961, *51*, 29–40.

Anokhin, P. K. *Biology and neurophysiology of the conditioned reflex and its role in adaptive behavior.* New York: Pergamon Press, 1974. (Russian edition published by Meditsina, Moscow, 1968.)

Berry, S. D., & Thompson, R. F. Prediction of learning rate from the hippocampal electroencephalogram. *Science*, 1978, *200*, 1298–1300.

Berry, S. D., & Thompson, R. F. EEG-multiple unit relationships during classical conditioning of the NM response in rabbits. In preparation, 1979.

Berger, T. W., & Thompson, R. F. Neuronal plasticity in the limbic system during classical conditioning of the rabbit nictitating membrane response. I. The hippocampus. *Brain Research*, 1978a, *145*, 323–346.

Berger, T. W., & Thompson, R. F. Identification of pyramidal cells as the critical elements in hippocampal neuronal plasticity during learning. *Proceedings of the National Academy of Sciences*, 1978b, *75*(3), 1572–1576.

Berger, T. W., & Thompson, R. F. Neuronal plasticity in the limbic system during classical conditioning of the rabbit nictitating membrane response. II. Septum and mammillary bodies. *Brain Research*, 1978, *156*, 293–314.

Berger, T. W., Alger, B. E., & Thompson, R. F. Neuronal substrates of classical conditioning in the hippocampus. *Science*, 1976, *192*, 438–485.

Cegavske, C. F., Patterson, M. M., & Thompson, R. F. Activity of units in the abducens nucleus (the final common path) during classical conditioning of the rabbit nictitating membrane response. In preparation, 1979.

Cegavske, C. F., Thompson, R. F., Patterson, M. M., & Gormezano, I. Mechanisms of efferent neuronal control of the reflex nictitating membrane response in the rabbit. *Journal of Comparative and Physiological Psychology*, 1976, *90*, 411–423.

Coleman, J. R., & Lindsley, D. B. Behavioral and hippocampal electrical changes during operant learning in cats and effects of stimulating two hypothalamic-hippocampal systems. *Electroencephalography and Clinical Neurophysiology*, 1977, *42*, 309–331.

Duffy, E. *Activation and behavior*. New York: Wiley, 1962.

Gormezano, I. Classical conditioning. In J. B. Sidowski (Ed.), *Experimental methods and instrumentation in psychology*. New York: McGraw-Hill, 1966.

Gormezano, I. Investigations of defense and reward conditioning in the rabbit. In A. B. Black & W. F. Prokasy (Eds.), *Classical conditioning II: Current research and theory*. New York: Appleton-Century-Crofts, 1972, 151–181.

Grastyan, E., Lissak, K., Madarasz, L., & Donhoffer, H. Hippocampal electrical activity during the development of conditioned reflexes. *Electroencephalograph and Clinical Neurophysiology*, 1959, *11*, 409–430.

Green, J. D., & Arduini, A. Hippocampal electrical activity and arousal. *Journal of Neurophysiology*, 1954, *17*, 533–557.

Guillery, R. W. Degeneration in the postcommissural fornix and mammillary peduncle of the rat. *Journal of Anatomy*, 1956, *90*, 350–370.

Hebb, D. O. Drives and the c.n.s. (conceptual nervous system). *Psychological Review*, 1955, *62*, 243–254.

Isaacson, R. L., & Pribram, K. H. *The hippocampus* (Vols. 1 and 2) New York: Plenum, 1975.

Landfield, P. W., McGaugh, J. L., & Tusa, R. J. Theta rhythm: a temporal correlate of memory storage processes in the rat. *Science*, 1972, *175*, 87–89.

Lashley, K. S. *Brain mechanisms and intelligence*. Chicago: Univ. of Chicago Press, 1929.

Lashley, K. S. In search of the engram. In *Symposium of the society for experimental biology* (Vol. 4). London: Cambridge Univ. Press, 1950.

Lindsley, D. B. Emotion. In S. S. Stevens (Ed.), *Handbook of experimental psychology*. New York: Wiley, 1951, 473–516.

Lindsley, D. M., & Wilson, C. L. In R. L. Isaacson & K. H. Pribram (Eds.), *The hippocampus* (Vol. 2). New York: Plenum Press, 1975. Pp. 247–278.

Malmo, R. B. Activation: A neuropsychological dimension. *Psychological Review*, 1959, *66*, 367–386.

Meibach, R. C., & Siegel, A. Efferent connections of the septal area of the rat: An analysis

utilizing retrograde and anterograde transport methods. *Brain Research*, 1977, *119*, 1–20.
Mosko, S., Lynch, G., & Cotman, C. W. The distribution of septal projections to the hippocampus of the rat. *Journal of Comparative Neurology*, 1973, *152*, 163–174.
Nauta, W. J. H. An experimental study of the fornix in the rat. *Journal of Comparative Neurology*, 1956, *104*, 247–272.
Olds, J., Disterhoft, J. F. Segal, M., Kornblith, C. L., & Hirsh, R. Learning centers of rat brain mapped by measuring latencies of conditioned unit responses. *Journal of Neurophysiology*, 1972, *35*, 202–219.
Pavlov, I. *Conditioned reflexes*. New York: Oxford University Press, 1927.
Prokasy, W. F. Developments with the two-phase model applied to human eyelid conditioning. In A. H. Black & W. F. Prokasy (Eds.), *Classical conditioning II: Current research and theory*. New York: Appleton-Century-Crofts, 1972, 119–147.
Raisman, G., Cowan, W. M., & Powell, T. P. S. An experimental analysis of the efferent projection of the hippocampus. *Brain*, 1966, *89*, 83–108.
Scoville, W. B., & Milner, B. Loss of recent memory after bilateral hippocampal lesions. *Journal of Neurology and Psychiatry*, 1957, *20*, 11–21.
Segal, M., & Landis, S. Afferents to the hippocampus of the rat studied with the method of retrograde transport of horseradish peroxidase. *Brain Research*, 1974, *78*, 1–15.
Shvyrkov, V. B. On the relationship between physiological and psychological processes in the functional system of the behaviour act. *Studia Psychologica (Bratislava)*, 1977, *19*(2), 82–96.
Sokolov, E. N. Brain functions: Neuronal mechanisms of learning and memory. *Annual Review of Psychology*, 1977, *28*, 85–112.
Sokolov, E. N., & Vinogradova, O. S. *Neuronal mechanisms of the orienting reflex*. (Weinberger, N. M., Ed. English Edition). Hillsdale, N.J.: Lawrence Erlbaum Associates, 1975.
Spencer, W. A., & Kandel, E. R. Hippocampal neuron responses to selective activation of recurrent collaterals of hippocampal axons. *Experimental Neurology*, 1961, *4*, 149–161.
Storm-Mathisen, J. Quantitative histochemistry of acetylcholinesterase in rat hippocampal region correlated to histochemical staining. *Journal of Neurochemistry*, 1970, *17*, 739–750.
Sudakov, K. V. The interaction of the hypothalamus, midbrain reticular formation, and thalamus in the mechanism of selective ascending cortical activation during physiological hunger. *Fiziol • zh • SSSR*, 1965, *51*(4), 449–456. (In Russian)
Swanson, L. W., & Cowan, W. M. An autoradiographic study of the organization of efferent connections of the hippocampal formation in the rat. *Journal of Comparative Neurology*, 1977, *172*, 49–84.
Thompson, R. F., Berger, T. W., Cegavske, C. F., Patterson, M. M., Roemer, R. A., Teyler, T. J., & Young, R. A. The search for the engram. *American Psychologist*, 1976, *31*, 209–227.
Winson, J. Interspecies differences in the occurrence of theta. *Behavioral Biology*, 1972, *7*, 479–487.
Young, R. A., Cegavske, C. F., & Thompson, R. F. Tone-induced changes in excitability of abducens motoneurons in the reflex path of the rabbit nictitating membrane response. *Journal of Comparative and Physiological Psychology*, 1976, *90*, 424–434.

NORMAN M. WEINBERGER

Neurophysiological Studies of Learning in Association with the Pupillary Dilation Conditioned Reflex[1]

16

Introduction

The goal of our research is to determine if there exist general neural principles that underlie the functional plasticity of neurons that develops during the formation of conditioned reflexes, specifically the pupillary dilation conditioned reflex. This research is presently conducted at the cellular or multicellular level with the intention that these studies will one day prove helpful in the investigation and the discovery of molecular mechanisms of conditioning. It is by no means certain that a single set of general principles exists either at the intercellular or intracellular levels of brain organization.

The present investigations should be seen as complementary to experiments that attempt to determine the total neural circuit involved in the formation of a conditioned reflex (Cohen, 1974; Thompson, 1976; Woody, 1974). Although we do concern ourselves with simultaneous recording and analysis of both behavioral and neuronal data, as will be explained later, we are not attempting simply to define the neural circuit from the afferent through the efferent limbs of the pupillary dilation conditioned reflex.

---

[1] This research was supported in part by Research Grants MH 11250 from the National Institute of Mental Health and BNS 76-81924 from the National Science Foundation to N. M. Weinberger, Predoctoral Fellowship MH 51342 and Training Grant MH 11095 to T. D. Oleson, Postdoctoral Fellowship MH 11095-08 from the National Institute of Mental Health to J. M. Cassady, and Training Grant MH 05440 and Predoctoral Fellowship MH 14599 from the National Institute of Mental Health to J. H. Ashe.

Rather, we "use" the development of a behavioral conditioned reflex as a control, whose nature will be explained later. As a subsidiary interest, we have in fact investigated the neural circuitry of the efferent limb of this conditioned reflex, but space limitations permit only a brief recapitulation that will be provided later. However, I ask your indulgence in thinking about the data and ideas in terms of how they may advance progress in realizing the goal of finding out whether general neuronal principles of cellular conditioning exist and, if so, of determining their nature. Incidentally, I will take the liberty of using the phrase "neural" or "neuronal learning" to refer to systematic changes in the discharge characteristics of cells, which are due to the association between a conditional and an unconditional stimulus. Progress to date has been slow, owing to the great difficulty of the problem and the need to invoke tedious but necessary experimental steps and controls.

Concerning Neural Principles

We owe to Pavlov the discovery of principles at the behavioral level that are responsible for the elaboration of conditioned reflexes. These are so well known that they need not be restated here. Numerous investigators, particularly those using electrophysiological approaches, have studied the distribution of functional changes throughout the nervous system during conditioning. There is yet another approach to the neurophysiology of conditioning, which is the search for those characteristics of cellular activity or response to stimuli that are necessary and sufficient for the development of cellular conditioned responses, that is, systematic change in the activity of single neurons. This line of study was apparently initiated by Yoshii and Ogura in 1960, who reported that during the pairing of a conditioned and an unconditioned stimulus, many neurons in the brain stem reticular formation exhibited the development of change in their response to the conditioned stimulus, mainly an increase in discharge rate. These authors also examined the initial response of the cells to the CS and US that was tested before the stimuli were paired, and they reported that neurons that were excited by these stimuli were more likely to develop neuronal learning than were cells that responded to either the CS or the US, but not both, or did not respond initially to either stimulus. From this and subsequent studies by other workers (among them Ben-Ari, 1972; Ben-Ari & LaSalle, 1972; Bures & Buresova, 1965, 1967, 1970; Chow, Lindsley, & Gollender, 1968; Kotlyar & Frolov, 1971; O'Brien & Fox, 1969a,b; O'Brien, Packham, & Brunnhoelzl, 1973; Polonskaya, 1974; Voronin, Gerstein, Kudryashov, & Ioffe, 1975; Voronin & Ioffe, 1974; Voronin & Kozhedub, 1972), it has been believed that an important *neural rule* governing cellular learning is bimodal convergence upon a neuron *before* training is initiated. This, then, is an example of a proposed general neural principle governing learning in the brain. Another example is that attributed to Kotlyar and Mayorov (1971), who reported that plastic neurons were characterized by a spontaneous rate of approximately nine discharges per second; cells with a higher or lower rate of background activity were not changed as a result of training. This provisional principle might be stated as follows: Neurons that develop conditioned plasticity must have a certain restricted amount of background activity, neither too low nor too high.

In a larger context, the following possibilities suggest themselves: (a) all neurons in the nervous system are capable of "learning"; if true, then either (1) all neurons exhibit such plasticity on every learning occasion or (2) some neurons learn on every appropriate occasion. Possibility (1) is unlikely in view of the fact that all published studies report neurons that fail to develop physiological changes during learning. If alternative (2) is correct, then there can be no morphological principles regarding neuronal learning. On the other hand, it is possible that (b) there is a specialized population(s) of cells that learn; if true, then by definition there is also a specialized population(s) of cells that are not plastic in relation to Pavlovian conditioning.

Where might these populations be located within the neuraxis? Learning cells might be (a) embedded in a matrix of nonplastic cells at all levels and in all nuclei and structures throughout the neuraxis. On the other hand, neurons that learn may be (b) more or less segregated, such as comprising a subregion of a nucleus, or a particular population of cells within a large structure. Such learning cells could still be distributed throughout the neuraxis, but would not be located in all neural systems or subsystems. There is abundant electrophysiological evidence that conditioning involves many levels and regions of the brain, but this is not to say that it involves all regions. If populations of neurons that do learn can be segregated from populations of cells that do not learn, then it would be possible to compare the electrical and even morphological characteristics of the neurons. In the search for general principles of neuronal learning, similarities between populations of learning versus nonlearning cells could be eliminated from consideration. On the other hand, *differences* in predisposing discharge characteristics, in morphology (including cell type, nature of afferents and efferents) and perhaps even in histochemical and molecular characteristics, would suggest general principles underlying the neuronal plasticity caused by using Pavlovian training procedures.

The physiological and anatomical approaches are not mutually exclusive; rather, it may be that structure and function are complementary. The strategy we have used is to be concerned first with anatomical considerations with the goal of locating systems or regions that contain neurons that learn, and only then to begin a systematic study of the functional characteristics of single neurons. Studies completed to date have employed so-called multiple-unit recordings that we regard as a convenient method to provide general localization of plastic changes. Work in progress is confined to single cell extracellular approaches during the elaboration of the pupillary dilation conditioned reflex in the cat.

Experimental Approach

Before presenting findings, it is necessary to consider certain control procedures that have been instituted to permit clear interpretation of results. At the behavioral level, it is essential to control for nonassociative factors, such as general excitability of the preparation. Such procedures, which include the use of discrimination training, training periods in which the stimuli are not paired, and so forth, were of course devised by Pavlov and are too well known to require further comment here. However, the behavioral control procedures may not be sufficient for the investigation of neural conditioning.

Although various experimental designs do control for nonassociative factors, neural conditioning is subject to other confounding variables. For example, suppose that an auditory stimulus is used as the CS, and a conditioned response develops in the auditory system (e.g., primary auditory cortex, A1). This could reflect genuine association, or it could result from the animal changing its bodily orientation, particularly the head, toward the sound source (e.g., loudspeaker) during training. For example, the resultant potentiation of the auditory cortex response to the CS could be due not to association of the CS and US, but rather to an increase in effective CS intensity at the auditory receptor. The general problem, then, is insuring that the stimuli used during conditioning are constant at the receptor level. In the absence of such controls, it may be impossible to decide whether the changes in brain activity are due to the association between the CS and US or a change in the actual stimulus. A suitable control procedure is available. Control of acoustic stimulation must account for three sources of variability: (*a*) relationship of the ears to a sound source (Marsh, Worden, & Hicks, 1962; Wiener, Pfeiffer, & Backus, 1966), (*b*) action of the middle ear muscles (Galambos & Rupert, 1959; Carmel & Starr, 1963), and, (*c*) masking noise produced by the subject's own movements (Imig & Weinberger, 1970; Irvine & Webster, 1972). All three can be controlled by neuromuscular paralysis. The use of a muscle-relaxed preparation has certain other advantages, including a mechanically stable system that is useful for extracellular recording and essential for intracellular recording, and the opportunity to record pupillary dilation in the absence of eye movements. The pupillary dilation conditioned response is elaborated very quickly using defensive conditioning procedures (tactile or acoustic CS and electrocutaneous US); this permits one to record from a single neuron during the whole course of Pavlovian training.

Details about the preparation and data analysis have been presented elsewhere (Oleson, Ashe, & Weinberger, 1975; Oleson, Westenberg, & Weinberger, 1972; Ryugo & Weinberger, 1978). The salient features of the preparation are that cats are prepared under general anesthesia with either indwelling electrodes or a chamber on the skull through which electrodes can be positioned, and are fitted with a skull pedestal that allows for later nontraumatic fixation of the head during training. Actual training takes place several days later under neuromuscular paralysis, and intubation of the trachea for artificial respiration is accomplished without discomfort or surgery. Conditioned stimuli are acoustic, white noise or pure tone (occasionally tactile), and the unconditioned stimulus is electrocutaneous stimulation of a forepaw. Pupillary size is sensed by an infrared pupillometer positioned directly in front of one eye. The animal is enclosed in an acoustic chamber to control for incidental stimuli because pupillary dilation is easily evoked by any stimulus.

Results

To begin with, it was necessary to establish that the pupillary dilation reflex exhibited characteristics of Pavlovian conditioning. Figure 16.1 provides typical records from different stages of training. Initial presentation of white noise evokes a large dilation (Figure 16.1A), whereas continuous aperiodic presentation of this stimulus (range of 30–90 sec) leads to a decrement in this response (Figure 16.1B). The response decrement is related to stimulus repetition rather than to inability of the pupil to dilate, as witnessed by the spontaneous dilation following the trial (Figure 16.1B). (See also Cooper, Ashe, & Weinberger, 1978; Weinberger, Oleson, & Ashe,

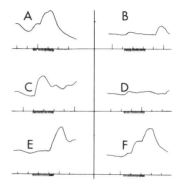

Figure 16.1. Pupillary dilation responses to white noise pips for a single cat before (A) habituation, (B) after habituation, (C) during early sensitization trials, (D) after repeated sensitization trials, (E) early in the conditioning series, and (F) after acquisition. Upward vertical lines indicate one second marker, downward vertical lines indicate auditory pips and shock. A rise in pupillometer writeout indicates increased dilation. That the pupil response may appear to precede the stimulus marker is attributable to the curvilinear coordinates of the polygraph paper. Note in particular the conditioned dilation reflex in (F). (Oleson et al., 1972. Reprinted with permission.)

1975.) Introduction of pawshock, intermixed in a random fashion among presentations of the acoustic stimulus, often produces a short-lived restoration of the habituated dilation, that is, dishabituation (Figure 16.1C), but continuation of random presentations of the acoustic and shock stimuli soon result in only a minimal dilation to white noise (Figure 16.1D), thus providing a baseline upon which the effects of stimulus pairing can be assessed. The first such trial is shown in Figure 16.1E, where it can be seen that the acoustic stimulus has a minimal effect, whereas the unconditioned response to shock is pronounced. After several paired trials, the conditioned stimulus evokes a conditioned pupillary dilation response (Figure 16.1F), which exhibits a systematic growth in amplitude as a function of repeated trials (Oleson, Westenberg, & Weinberger, 1972).

Differentiation can be established easily. Figure 16.2 shows records from an experiment in which two acoustic stimuli and pawshock were presented. During the control period, all three stimuli were given in random sequence. As it happens, the dilation to a pure tone stimulus was less potent than that to white noise (Figure 16.2, Sens.). The tone stimulus was then paired with shock, whereas the white noise was never followed by shock, with the result that the tone (CS) came to elicit a large dilation, whereas the white noise (DS) did not (Figure 16.2, Cond.). When the contingencies are reversed, reversal of this differentiation is also established (i.e., "discrimination reversal"); repeated reversals can also be established (Oleson, Westenberg, & Weinberger, 1972). In other experiments, inhibition of delay has been successfully investigated (Oleson, Vododnick, & Weinberger, 1973). Figure 16.3 presents a series of records in which the interval between onset of the CS and onset of the US was gradually increased from 3–16 secs. The latency to peak of the dilation conditioned response increased as a direct function of the CS–US interval. Furthermore, inhibition of the pupil is observed immediately following CS onset, prior to initiation of the dilation conditioned reflex (Figure 16.3D, E, F).

Also, the pupillary dilation conditioned reflex can be under the in-

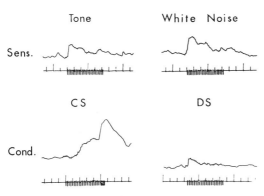

Figure 16.2. Examples of pupillary responses to tone and white noise stimuli from an individual cat during sensitization and differential conditioning procedures. Stimulus markers are as in Figure 16.1. (Oleson et al., 1972. Reproduced with permission.)

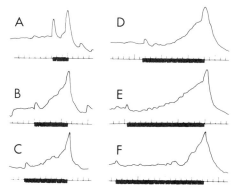

Figure 16.3. Pupillary dilation responses to white noise pips for a single cat as a function of increasing the CS–US interval from 2.5 sec (A) to 16 sec (F). Stimulus markers are as in Figures 16.1 and 16.2 (Oleson et al., 1975. Reprinted with permission.)

hibitory control of an additional specifically nonreinforced stimulus (Weinberger, Oleson, & Haste, 1973).

These findings indicate that the pupillary dilation reflex is subject to laws of Pavlovian conditioning and therefore that this preparation can be used to investigate principles underlying neuronal plasticity that result from the procedures of classical conditioning.

We have studied the efferent limb of this conditional reflex, but limitations of space preclude a detailed exposition. Briefly, we have found that inhibition of the parasympathetic supply to the iris (short ciliary nerve) seems to play a prime role in the conditioned pupillary dilation reflex. Thus, this conditioned reflex develops in the sympathectomized cat (Ashe, Cooper, & Weinberger, 1978a), inhibition in the short-ciliary nerve precedes excitation in the long-ciliary (sympathetic) nerve (Ashe & Cooper, 1978) and neurons in the accessory oculomotor nuclei of the midbrain develop systematic changes in discharge rate to the conditioned stimulus that parallel the development of the pupillary conditioned reflex (Ashe, Cooper, & Weinberger, 1978b).

Major experiments addressed the issue of the location of at least some plastic changes, beginning with the system of the modality of the conditioned stimulus, that is, the auditory system. Recordings of multiple-unit discharges were obtained simultaneously from the bulbar and cortical levels of the auditory system, that is, the cochlear nucleus and primary auditory cortex, respectively. The training paradigm included both differentiation and differentiation reversal. A training session consisted of three sequential phases. The first "sensitization" control period consisted of the presentation of white noise, tone and shock in a random and intermixed sequence. The second phase, "conditioning," consisted of the presentation of the acoustic stimulus designated as the CS, followed by shock (US) on every trial. In the third phase, "discrimination," CS–US trials were intermixed with presentations of the nonreinforced, DS. Thus, the probability of a CS, DS, or US was the same in the first (sensitization) and third (discrimination) phases, the only difference being in the contingencies of the stimuli. On the first day of training, the conditioned stimulus was white noise and the differential stimulus (DS) was pure tone (2 kHz).[2] The roles of these stimuli were reversed during the second training session which took place one week later. Appropriate pupillary behavior developed in all subjects, that is, a large conditioned reflex was elaborated to the conditioned stimulus but not the differential stimulus of both the initial and later training sessions. Having established a behavioral differential conditioned response and reversed this response when stimulus contingencies are reversed, it is of interest to consider the neural accompaniments.

Representative records for the primary auditory cortex are given in Figure 16.4. Note, relative to the control sensitization period (white noise, tone, and shock randomly intermixed), that stimulus pairing results in an augmentation of the response of auditory cortex neurons to the conditioned stimulus (compare Figures 16.4A and C, WN). Furthermore, this augmentation is also evident during the following discrimination testing period (Figure 16.4D). The initial response to the tone (DS) was an excitation followed by inhibition (Figure 16.4A, Tn). Discrimination training resulted in a slight augmentation of the excitation and elimination of the inhibition, that is, the response to the DS was one of increased excitation. However, this degree of excitation was significantly less than that that developed to the white noise (CS), so that neural differentiation did in fact develop in conjunction with pupillary differentiation (for statistical details, see Oleson, Ashe, & Weinberger, 1975). Reversal of these neural effects also was established on the second training session, one week later (Figure 16.4, Day 2).

Multiple-unit activity from the anteroventral cochlear nucleus is illustrated in Figure 16.5. As in the case of primary auditory cortex, the response to the conditioned stimulus was augmented significantly during conditioning training and discrimination training on Day 1 (Compare Figures 16.5A and C, D). Interestingly, the response to the differential stimulus (Tn) was significantly depressed during discrimination training. One week later, during reversal training, neural responses to the tone (CS) were augmented relative to the sensitization control period (Figures 16.5A and D, Day 2). Responses to the DS (WN) were reduced from those of the sensitization period.

In summary, we conclude that neurons within the pathway of the conditioned stimulus increase their activity to the conditioned stimulus

---

[2] Tactile stimulation was also presented continuously and recordings from the cuneate nucleus and primary somatosensory-motor cortex were obtained and analyzed. Limitations of space preclude considerations of these findings. The reader is referred to Oleson, Westenberg, and Weinberger (1972) for details. Also, an extensive analysis of the results of the initial control sensitization period (stimuli randomly intermixed) has been presented elsewhere (Weinberger, Oleson, & Ashe, 1975).

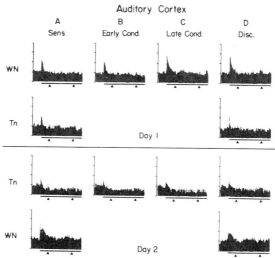

Figure 16.4. Histograms of multiple-unit activity in the auditory cortex of animal T22 during (A) sensitization, (B) early conditioning (trials 1–10), (C) late conditioning (trials 46–55), and (D) discrimination for initial training (Day 1) and reversal training (Day 2). Response to the acoustic stimuli during sensitization on Day 1 consisted of a brief increase in activity followed by sustained discharges below pretrial levels ("habituation"). Note the increase in the initial discharge and the reduction or abolition of the subsequent inhibition during conditioning and discrimination. On Day 2, the enhanced response to the white noise stimulus, which developed on Day 1, is still evident (Day 2, A). The inhibition to the tone is systematically reduced during conditioning (B, C) and discrimination (D). In this and Figure 16.5, each histogram represents a block of 10 trials. Calibrations are 96 spikes per division.

during the elaboration of the pupillary conditioned reflex and that this change is either not as great to the differential stimulus (auditory cortex) or is in the reverse direction (i.e., depressed, cochlear nucleus). Incidentally, analysis of the temporal course of these changes on a trial-by-trial basis revealed that the augmentation of neural responses (neural learning) within each subject occurred first in the cortex and only later in the cochlear nucleus. Furthermore, the cortical learning occurred at the same time as the development of the pupillary conditioned response.

In light of the neuronal changes in the auditory system, it is important to determine whether they were endogenous or reflected alterations in the receptor. Although this possibility is unlikely, particularly since neuronal conditioning developed in the auditory cortex prior to appearing in the cochlear nucleus, more direct data were obtained by recording the cochlear microphonic response from the round window during training. In brief, although the cochlear microphonic was not constant during training, neither did it exhibit systematic changes that could account for systematic alterations in the central auditory system (Ashe, Cassady, & Weinberger, 1976).

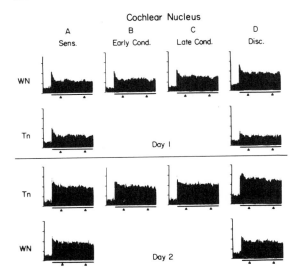

Figure 16.5. Histograms of multiple-unit activity in the anteroventral cochlear nucleus of animal T18 during (A) sensitization, (B) early conditioning (trials 1–10), (C) late conditioning (trials 41–50), and (D) discrimination (trials 71–80), for Day 1 and 2. Note the increase in onset discharge and higher level of sustained firing during CS+ presentation, whereas responses to the CS− show a marked reduction (A versus D). Calibrations are 96 spikes per division. (Oleson et al., 1973. Reproduced with permission.)

Thus, the observed neural learning is due neither to changes in effective stimulus intensity nor in all probability to the olivocochlear bundle.

It was now possible to address directly the problems stated at the beginning of this report. Given the fact that there are neurons in the auditory system that exhibit functional plasticity during conditioning, what are their physiological properties and, to begin with, are they distributed throughout the system or segregated in discrete loci? This issue could be approached best by investigating nuclei within the auditory system that are characterized by having morphologically and physiologically distinct subregions. We have selected the medial geniculate nucleus—the thalamic portion of the auditory system. Examination of the morphology of this nucleus, first by Cajal and more recently by Morest (1964, 1965a,b), indicates that it is comprised of discrete subdivisions, the major regions being the ventral, dorsal, and medial (also called magnocellular) divisions. While each region is responsive to ascending auditory volleys and while each projects to auditory responsive cortex, there are important differences. Thus, the ventral division contains small cells that are organized in concentric lamina and have restricted polar dendritic fields, subject to synaptic input from the inferior colliculus in a topographic fashion. The medial and dorsal regions contain larger multipolar, isodendritic neurons with larger radiate dendritic fields, subject to input from less restricted regions of the midbrain. In particular, the medial division contains cells whose dendrites extend over greater distances than cells of either the ventral or dorsal divisions, and is also the most sparsely populated part of the medial geniculate. Figure 16.6 presents our Golgi reconstruction through these subdivisions, in agreement with the findings of Morest (1964).

Two microelectrodes were placed simultaneously in two of these three

Figure 16.6. Camera lucida reconstruction of the medial geniculate body (MGB) from Golgi material. Typical distribution of neuronal types through the middle of the MGB of the adult cat (coronal section, Golgi-Cox). Abbreviations: D, dorsal division; M, medial division; VL, ventral division, pars lateralis; VO, ventral division, pars ovoida.

subdivisions. Records were obtained from clusters of cells ("multiple-unit recording") in order to facilitate experiments; this type of recording should be sufficiently discrete to permit characterization of small regions of the brain. Records obtained simultaneously from the medial and ventral sub-

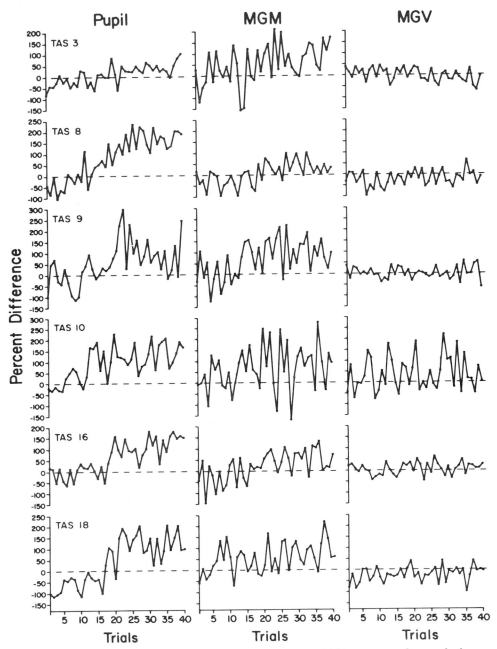

Figure 16.7. Trial-by-trial plots of pupillary, MGm, and MGv response changes during conditioning. Each point represents the normalized response to the CS+ during its 1.0 sec presentation, expressed as a percent difference score relative to its mean sensitization value (dashed line). Both pupil and MGm neuronal activity exhibit a systematic growth of responsivity during conditioning. MGv neuronal activity fails to demonstrate such conditioned response enhancement. Data are from all animals that develop pupillary conditioning and had placements in both the MGm and MGv of the same medial geniculate body. (Oleson et al., 1973. Reproduced with permission.)

divisions in six cats are given in Figure 16.7, together with the pupillary record. These trial-by-trial graphs show in each case that (a) a pupillary dilation conditioned reflex was established, (b) that within the medial division, neural activity was significantly increased during training, and (c) no neural conditioning occurred within the ventral division. In fact, in a total of 14 ventral placements, there were no instances of significant changes in neural responses to the conditioned stimulus. Interestingly, the same uniformly negative outcome was obtained in the dorsal division ($N = 9$). Neural learning was found in six of eight medial placements. Additionally, differential training was also instituted, with the result that neurons in the medial division that exhibited conditioning also developed differential responding between the white noise CS and the pure tone DS; differentiation was not obtained in the ventral or dorsal divisions.

As previously noted, neural learning was obtained only within the medial division in six of eight cases. The two failures may be interpreted as examples of nonplasticity. However, there is an alternative possibility that these two failures were obtained from subjects that were in some way "substandard." Physiologists are familiar with "poor" preparations, that is, those that yield negative results despite meeting appropriate physiological criteria for health. Likewise, psychologists sometimes find that they have "stupid" subjects, that is, those that fail to learn a task for no obvious reason. Presumably, some critical factors are absent in such "negative" cases; a discussion of these factors is beyond the scope of this chapter. However, although it may not be possible to identify the causes of failure to obtain a positive result, that is, neural conditioning, it may be possible to decide between the alternatives of (a) nonplasticity and (b) substandard subject. This issue is related directly to the rationale for simultaneously recording a behavioral measure, that is, pupillary dilation, during the recording of neural data, the behavioral control procedure that was outlined in the Introduction. If a negative neural finding is obtained in a subject that develops a behavioral conditioned response, then one can reject the alternative that the subject was substandard, stupid, and so on, and at the very least conclude that the factors necessary for neural conditioning were present somewhere in the brain. However, if a behavioral conditioned response is not elaborated, then one may reasonably conclude that the preparation was inadequate, and consequently that a failure to obtain neural conditioning cannot be interpreted simply as evidence that the brain site in question is nonplastic. In fact, negative neural findings from a negative behavioral subject, or from a subject lacking behavioral controls, cannot be clearly interpreted and must fall into the category of "nebulous" findings.

Given these considerations, the behavioral findings of the two subjects that failed to develop neural learning in the medial division are of critical interest. These data are presented in Figure 16.8, and they reveal that the two cases in which neurons in the medial region failed to learn were ob-

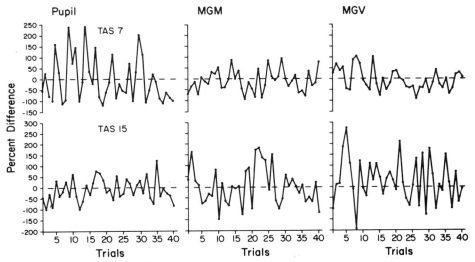

Figure 16.8. Trial-by-trial plots of pupillary, MGm, and MGv response changes during conditioning in both cats that failed to attain acquisition criteria. Each point represents the normalized response to the CS+, expressed as a percent difference score relative to its mean senstization value (dashed line). Neither pupil nor neuronal activity exhibits any systematic change in responsiveness during conditioning. (Ryugo et al., 1978. Reproduced with permission.)

tained from subjects in which a pupillary dilation conditioned response also failed to be elaborated. In short, these negative neural findings cannot be attributed to nonplasticity in the medial division.

Overall, the results of this study demonstrate differential plasticity during conditioning in morphologically distinct regions of the medial geniculate nucleus such that cells in the medial division are consistently

---

Figure 16.9. Top: Histograms of single neuron discharges for successive blocks of five trials during sensitization (SENS. 1, SENS. 2) and conditioning (COND. 1–4) for a unit in the medial region of the medial geniculate body. The horizontal bar designates the presentation of the white noise stimulus, 1 sec in duration. Vertical calibration, six spikes per division. Note the decrement in CS evoked discharges from SENS. 1 to SENS. 2, and the subsequent increase in response to the CS during the four blocks of conditioning (20 trials total). Bottom: Mean spikes per second for the evoked discharges for the five trial blocks whose histograms appear in the upper part of the figure. In this and Figure 16.10, the evoked discharge is computed as the number of spikes during presentation of the white noise CS minus the number of spikes during the period immediately preceding CS presentation. Additionally, representative single trial records of pupillary responses are presented, each record corresponding roughly to that portion of the conditioning session denoted on the abscissa. An upward deflection of the stimulus marker indicates presentation of white noise (CS), a downward deflection indicates presentation of shock (US). Note the decrease in pupillary response during sensitization and the subsequent growth of this response during conditioning. (Ryugo et al., 1978. Reproduced with permission.)

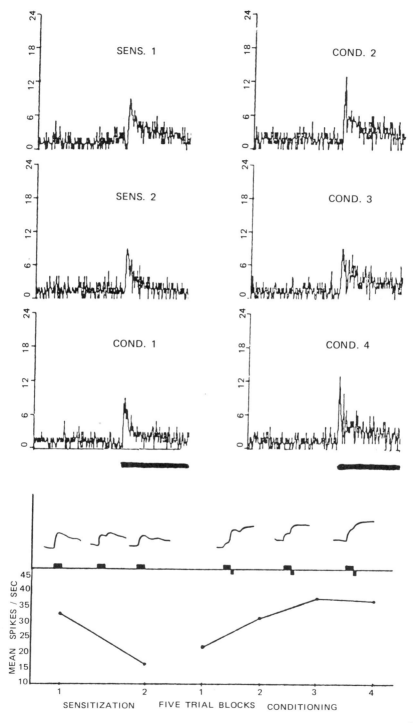

conditionable in contrast to the lack of plasticity of cells in the ventral and dorsal regions.

Having located areas of differential plasticity (for the particular experimental conditions employed), we now turned to an analysis of the physiological characteristics of isolated single neurons in the subdivisions of the medial geniculate body. The issue is whether the discharge characteristics of these cells, prior to the institution of training procedures, are predictive of plasticity during actual training. The experiment involves a single training session in which data are obtained from only one cell per cat. Although it would be possible to record from other cells following the training period, such data would not be from "naive" cells, and would relate only to retention, not to initial acquisition. Because of the requirement that only one cell be analyzed from each subject, progress is slow and this experiment is still in progress. Therefore, findings relevant to this issue will not be presented at this time. However, in the interest of furthering communication, I would like to provide two examples of single unit data and behavior that underscore the idea that investigations into the characteristics of plasticity of neurons, in distinction to the tracing of neural circuitry involved in a particular conditioned reflex, should include the simultaneous study of behavioral plasticity.

Histograms of a cell during two periods of training, sensitization (random presentations of CS and US), and conditioning (paired presentation) are given in the top part of Figure 16.9. These reveal that the neuron exhibited a decrease in discharge to the acoustic stimulus during random presentations, and a systematic increase in response to the conditioned stimulus during paired-presentation. The lower part of Figure 16.9 provides a summary of these data showing the evoked discharge. Also shown are representation records of pupillary behavior, indicating that a pupillary dilation conditioned reflex was acquired simultaneously. Figure 16.10 presents the

---

Figure 16.10. Top: *Histograms of single neuron discharges for successive blocks of five trials during sensitization (SENS. 1, SENS. 2) and conditioning (COND. 1–8) for a unit in the medial region of the medial geniculate body. The horizontal bar designates the presentation of the white noise stimulus, 1 sec in duration. Vertical calibration, three spikes per division. Note the decrement in CS evoked discharges from SENS. 1 to SENS. 2, and, in contrast to Figure 16.9, the failure to develop an augmented response to the stimulus during conditioning (40 trials total). Bottom: Mean spikes per second for the five trial blocks whose histograms appear in the upper part of the figure. Additionally, representative single trial records of pupillary response are presented, each record corresponding roughly to that portion of the conditioning session denoted on the abscissa. Stimulus markers are the same as in Figure 16.9. As for the subject whose data are shown in Figure 16.9, there is a decrement in pupillary response to the white noise during the sensitization period. However, in contrast to the data from Figure 16.9, this subject failed to develop a pupillary dilation conditioned reflex; the large upward deflections of the pupillary record during conditioning are unconditioned responses to the unconditioned stimulus. (Ryugo et al., 1978. Reproduced with permission.)*

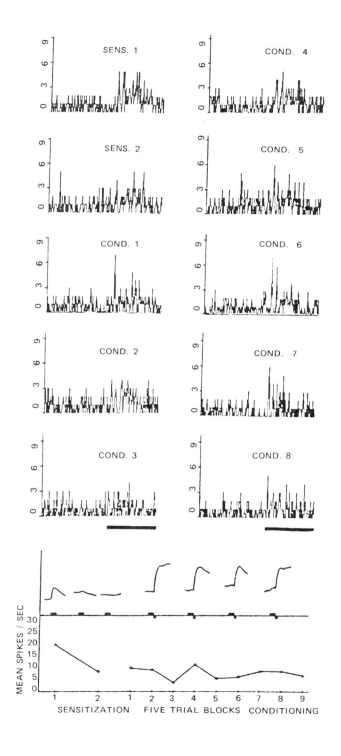

same type of data from another neuron recorded from another naive cat. This cell also exhibited a decrement in response during sensitization, but unlike the cell whose data are given in Figure 16.9, this neuron failed to develop neural conditioning, even though many paired trials were given. Samples of pupillary data are presented in the lower part of Figure 16.10; this animal failed to elaborate a pupillary dilation conditioned reflex. Thus, this neuron, and the specific cell population to which it belongs, cannot be classified as nonconditionable because the preparation was inadequate. Of interest, it appears that neurons exhibiting a decrement in response during random pairing ("habituation") do not necessarily develop a neural conditioned response. In other words, it may be that plasticity as evidenced by both neural and behavioral habituation is not necessarily predictive of plasticity of the conditioning type.

Concluding Comments

In summary, there is evidence that conditionable neurons are located throughout the sensory system of the conditioned stimulus, in this case, the auditory system. The physiological plasticity obtained is not due to artifacts such as change in effective CS intensity at the receptor and feedback from movements, for the subjects were maintained under neuromuscular paralysis. Furthermore, although conditionable cells are found at loci along the auditory neuraxis, they are segregated within the thalamic level at least. Because the segregation corresponds to morphologically distinct subpopulations, it is possible that conditionable cells may be identified morphologically.

It should be understood that this state of affairs leaves open the issue of *situational plasticity,* by which is meant that cells may exhibit plasticity under one set of conditions and not under other circumstances. Situational plasticity is almost certainly the case. However, this does not mean that *all* neurons are inherently plastic. Indeed, it would be difficult to prove this assertion because in theory one would have to test a set of presumptive "nonplastic" cells under an almost infinitely large number of circumstances and conditions before concluding that indeed they are not conditionable. All that can be done reasonably is to examine neurons under a wide variety of systematically controlled situations; experiments presented here constitute work in this direction. Finally, if indeed all neurons are conditionable, given the "correct" circumstances, then it would be important to determine the critical variables that render a neuron amenable to physiological change under "correct" circumstances. The present work hopefully contributes to this line of thought as well.

ACKNOWLEDGMENTS

I wish to acknowledge with pleasure the colleagues who have made major contributions to these studies: Drs. John H. Ashe, J. Michael Cassady, Terrence D. Oleson, David K. Ryugo, Ms. Carol Cooper, Mr. Simon Boughey, Ms. Lola Moffitt, and Mrs. Elaine Hackelman.

REFERENCES

Ashe, J. H., Cassady, J. M., & Weinberger, N. M. The relationship of the cochlear microphonic potential to the acquisition of a classically conditioned pupillary dilation response. *Behavioral Biology*, 1976, *16*, 45–62.

Ashe, J. H., & Cooper, C. L. Multifiber efferent activity in postganglionic sympathetic and parasympathetic nerves related to the latency of spontaneous and evoked pupillary dilation. *Experimental Neurology*, 1978, *59*, 413–434.

Ashe, J. H., Cooper, C. L., & Weinberger, N. M. Mesencephalic multiple-unit activity during acquisition of conditioned pupillary dilation. *Brain Research Bulletin*, 1978a, *3*, 143–154.

Ashe, J. H., Cooper, C. L., & Weinberger, N. M. Role of the parasympathetic pupillomotor system in classically conditioned pupillary dilation in the cat. *Behavioral Biology*, 1978b, *23*, 1–13.

Ben-Ari, Y. Plasticity at unitary level. I. An experimental design. *Electroncephalography and Clinical Neurophysiology*, 1972, *32*, 655–665.

Ben-Ari, Y., & LaSalle, G. E. G. Plasticity at unitary level. II. Modifications during sensory-sensory association procedures. *Electroencephalography and Clinical Neurophysiology*, 1972, *32*, 667–679.

Bures, J., & Buresova, O. Relationship between spontaneous and evoked unit activity in the inferior colliculus of rats. *Journal of Neurophysiology*, 1965, *28*, 641–644.

Bures, J., & Buresova, O. Plastic changes of unit activity based on reinforcing properties of extracellular stimulation of single neurons. *Journal of Neurophysiology*, 1967, *30*, 98–113.

Bures, J., & Buresova, O. Plasticity in single neurones and neural populations. In G. Horn and R. A. Hinde (Eds.), *Short-term changes in neural activity and behavior*. London: Cambridge Univ. Press. 1970, pp. 363–403.

Carmel, P. W., & Starr, A. Acoustical and nonacoustical factors modifying middle ear muscle activity in waking cats. *Journal of Neurophysiology*, 1963, *26*, 598–616.

Chow, K. L., Lindsley, D. F., & Gollender, M. Modification of response patterns of lateral geniculate neurons after paired stimulation of contralateral and ipsilateral eyes. *Journal of Neurophysiology*, 1968, *31*, 729–739.

Cohen, D. H. The neural pathways and informational flow mediating a conditioned autonomic response. In L. V. DiCara (Ed.), *Limbic and autonomic nervous systems research*. New York: Plenum Press, 1974.

Cooper, C. L., Ashe, J. H., & Weinberger, N. M. Effects of stimulus omission during habituation of the pupillary dilation reflex. *Physiological Psychology*, 1978, *6*, 1–6.

Galambos, R., & Rupert, A. Action of middle ear muscles in normal cats. *Journal of Acoustical Society of America*, 1959, *31*, 349–355.

Imig, T. J., & Weinberger, N. M. Auditory system multi-unit activity and behavior in the rat. *Psychonomic Science*, 1970, *18*, 164–165.

Irvine, D. R. F., & Webster, W. R. Studies of peripheral gating in the auditory system of cats. *Electroencephalography and Clinical Neurophysiology*, 1972, *32*, 545–556.

Kotlyar, B. I., & Frolov, A. G. Reorganization of unit activity in the lateral geniculate body during sound-light association. *Journal of Higher Nervous Activity*, 1971, *21*, 827–835. (In Russian)

Kotlyar, B. I., & Mayorov, V. I. Activity of the visual cortex units in rabbits in the course of association of sound with rhythmic light. *Journal of Higher Nervous Activity*, 1971, *21*, 157–163. (In Russian)

Marsh, J. T., Worden, F. G., & Hicks, L. Some effects of room acoustics on evoked auditory potentials. *Science*, 1962, *137*, 280–282.

Morest, D. K. The neuronal architecture of the medial geniculate body of the cat. *Journal of Anatomy* (London), 1964, *98*, 611–630.

Morest, D. K. The laminar structure of the medial geniculate body of the cat. *Journal of Anatomy* (London), 1965a, *99*, 143–160.

Morest, D. K. The lateral tegmental system of the midbrain and the medial geniculate body: Study with Golgi and Nauta methods in cat. *Journal of Anatomy* (London), 1965b, *99*, 611–634.

O'Brien, J. H., & Fox, S. S. Single-cell activity in cat motor cortex. I. Modifications during classical conditioning procedures. *Journal of Neurophysiology*, 1969a, *32*, 267–284.

O'Brien, J. H., & Fox, S. S. Single-cell activity in cat motor cortex. II. Functional characteristics of the cell related to conditioning changes. *Journal of Neurophysiology*, 1969b, *32*, 285–296.

O'Brien, J. H., Packham, S. C., & Brunnhoelzl, W. W. Features of spike train related to learning. *Journal of Neurophysiology*, 1973, *36*, 1051–1061.

Oleson, T. D., Ashe, J. H., & Weinberger, N. M. Modification of auditory and somatosensory system activity during pupillary conditioning in the paralyzed cat. *Journal of Neurophysiology*, 1975, *38*, 1114–1139.

Oleson, T. D., Vododnick, D. S., & Weinberger, N. M. Pupillary inhibition of delay during Pavlovian conditioning in paralyzed cat. *Behavioral Biology*, 1973, *8*, 337–346.

Oleson, T. D., Westenberg, I. S., & Weinberger, N. M. Characteristics of the pupillary dilation response during Pavlovian conditioning in paralyzed cats. *Behavioral Biology*, 1972, *7*, 829–840.

Polonskaya, E. L. Conditioned reactions of neurones in the magnocellular part of the medial geniculate body. *Journal of Higher Nervous Activity*, 1974, *24*, 986–995. (In Russian)

Ryugo, D. K., & Weinberger, N. M. Differential plasticity of morphologically distinct neuron populations in the medial geniculate body of the cat during classical conditioning. *Behavioral Biology*, 1978, *22*, 275–301.

Thompson, R. F. The search for the engram. *American Psychologist*, 1976, *31*, 209–225.

Voronin, L. L., Gerstein, G. L., Kudryashov, I. E., & Ioffe, S. V. Elaboration of a conditioned reflex in a single experiment with simultaneous recording of neural activity. *Brain Research*, 1975, *92*, 385–403.

Voronin, L. L., & Ioffe, S. V. Changes in unit postsynaptic responses at sensorimotor cortex with conditioning in rabbits. *Acta Neurobiologiae Experimentalis*, 1974, *34*, 505–513.

Voronin, L. L., & Kozhedub, R. G. Cell analog of the conditioned reflex to cortical electrical stimulation. *Neuroscience and Behavioral Physiology*, 1972, *5*, 339–346.

Weinberger, N. M., Oleson, T. D., & Ashe, J. H. Sensory system neural activity during habituation of the pupillary orienting reflex. *Behavioral Biology*, 1975, *15*, 283–301.

Weinberger, N. M., Oleson, T. D., & Haste, D. Inhibitory control of conditional pupillary dilation response in the paralyzed cat. *Behavioral Biology*, 1973, *9*, 307–316.

Wiener, J. M., Pfeiffer, R. R., & Backus, A. S. M. On the sound pressure transformation by the head and auditory meatus of the cat. *Acta Otolaryngology* 1966, *61*, 255–269.

Woody, C. D. Aspects of the electrophysiology of cortical processes related to the development and performance of learned motor responses. *The Physiologist,* 1974, *17,* 49–69.

Yoshii, N., & Ogura, H. Studies on the unit discharge of brainstem reticular formation in the cat. I. Changes of reticular unit discharge following conditioning procedure. *Medical Journal of Osaka University,* 1960, *11,* 1–17.

MICHAEL M. PATTERSON

Mechanisms of Classical Conditioning of Spinal Reflexes[1]

17

Introduction

The mammalian spinal cord and the isolated reflex arcs of the spinalized preparation have long been a favorite substrate for studies of basic nervous system properties, including neuroanatomical investigations, studies of reflex mechanisms, and sensitization-habituation processes (e.g., Groves & Thompson, 1970). In these areas, the mammalian spinal cord has been utilized to provide much of the information on neural process presently available. Indeed, the cord has many features that are similar to that of the brain itself, including an "isodendritic core" (Ramon-Moliner & Nauta, 1966) and reflex circuits that range from simple monosynaptic arcs to complex functional units capable of organizing an integrated sexual response. Thus, studies of the spinal cord have traditionally served two roles in neural studies: to determine the basic structure and function of the cord itself and to utilize the neural properties of the cord as a model for brain function. We have reviewed these areas of research in relation to the issue of spinal reflex alterability (Patterson, 1976).

Although studies of neuroanatomy and simple forms of response plasticity such as sensitization and habituation have been widely accepted from the spinal model, the effects of applying a classical conditioning paradigm

[1] The author gratefully acknowledges the support of the National Institute of Neurological and Communicative Disorders and Stroke (Grants 1-RO1-NS10647 and 1-RO1-NS14545) and the Research Bureau of the American Osteopathic Association (Grants T-7274, T-7374, T-7474) in the research and preparation of the paper.

to spinal reflex pathways have long been surrounded by controversy. The basis of the dispute lies in two areas. The first is in a philosophical-physiological view that holds the cerebral hemispheres to be the seat of all learning. This particular view was held by Pavlov (1927) and even earlier. Here, lower brain areas were believed to be mechanically "hard-wired" and to carry on the basic aspects of vegetative existence, whereas only the cerebral hemispheres were capable of performing the adaptive functions of learning. However, there seems little reason to hold a priori that neural circuits below the cortical level cannot alter input–output relationships in much the same way as cerebral circuits. Indeed, the recent studies of Norman, Villablanca, Brown, Schwafel, and Buchwald (1974) and Oakley and Russell (1975) strongly suggest learning in decorticate animals that is much like that observed in normal animals. The second area is the question of whether changes in spinal response properties are due to associative (hence learned) processes or can be better classified under nonassociative categories. This argument is best approached by conducting appropriate experimental manipulations to determine the congruence of spinal processes with the results from intact subjects under associative and control procedures.

Early Studies

The initial attempts to apply the classical conditioning paradigm to spinal reflexes grew out of work by Culler and Mettler (1934). They attempted to condition a defensive response in the decorticate dog. Although reporting some success, they did not find the response specificity to the training situation that occurred in intact subjects. The success of these studies led to the attempts to determine the simplest neural arrangement necessary to sustain a learned response through using the spinal preparation. Culler presented his initial work with spinal dogs in 1937, with a full report in 1940 (Shurrager & Culler, 1940). Kellogg (Pronko & Kellogg, 1942) soon suggested that Culler's results were caused by increased excitability of the reflex paths due to the presentation of shock stimuli or other non-learned phenomena. The controversy was started. Several other studies, both those reporting positive results and those reporting negative results, have appeared since early work (see Patterson, 1976, for a complete review).

A long series of studies performed over many years by Nesmeyanova in her Russian laboratories has become available to American workers (Nesmeyanova, 1977). The studies reported span a large area of spinal cord work and many processes, including the spinal conditioning paradigm. In this part of her work, Nesmeyanova reports finding response alterations, but generally assigns them to factors other than the associative process of the classical conditioning paradigm. The studies producing this conclusion will be examined in relation to our studies in another section of this chapter.

The results of the earlier work in the area of spinal conditioning procedures are thus quite mixed, with most showing alteration or plasticity of spinal responses. The controversy is thus not over the existence of alterations in spinal reflex pathways, but the extent to which such changes can or should be classified as a learned or associative response change.

Recent Work

Our work in spinal conditioning procedures began in 1969 at Irvine, California, in the Department of Psychobiology. Fitzgerald and Thompson (1967) had strongly suggested response changes under classical conditioning procedures using an isolated cat spinal cord preparation similar to that used for sensitization and habituation studies (Groves & Thompson, 1970). With modification, this preparation was utilized for our early work. The preparation has been detailed elsewhere (e.g., Patterson, Cegavske, & Thompson, 1973). Briefly, the basic preparation and procedures are as follows. An adult cat is spinalized under ether anesthesia by visually controlled spinal cord transection at the T12-L1 junction. The animal is then paralyzed, respirated, and ether discontinued. The superficial peroneal sensory and deep peroneal motor nerves are dissected from the hind leg (usually the left) for a distance sufficient to allow the cut nerve ends to be tied over stimulating and recording electrodes, respectively. The bipolar electrodes with nerves tied onto the distal wire are placed in a mineral oil pool maintained at body temperature and formed by the cut skin edges. The animal is maintained at a 5% expired CO_2 level and constant, normal body temperature. For the classical conditioning situation, the conditioned stimulus (CS) is delivered to the sensory nerve and consists of a train of 1 msec pulses. The unconditioned stimulus (UCS) is delivered to the ankle skin of the same leg and is generally a standard 50 volt pulse train of 2 msec duration pulses. The CS intensity is set for each subject to give a small response along the motor nerve, while the UCS provides a maximal response in most subjects. Each animal is allowed to rest until the response to the CS is stable for at least 15 min.

The initial studies (Patterson, Cegavske, & Thompson, 1973) were performed to assess response alterations to CS-UCS pairings as compared to control conditions. The first study used a fixed 60 sec intertrial interval (ITI) and a 250 msec interstimulus interval (ISI), with the CS and UCS being, respectively, 750 and 500 msec in duration and ending together. The results of this study are shown in Figure 17.1. It is seen that the response to the CS increased over trials when the CS and UCS were paired (paired condition), but not when presented 30 sec apart (unpaired condition). Most of the increase occurred in the first 20–25 trials and decreased in extinction (CS-alone). In a second study (Figure 17.2) a randomized ITI (mean 60

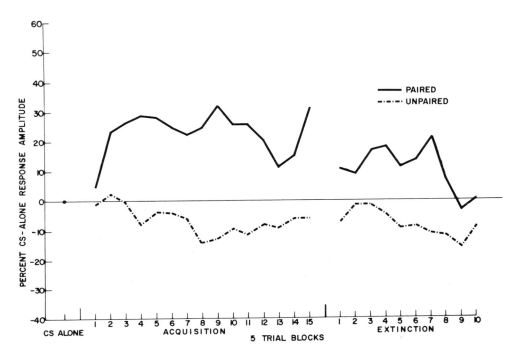

Figure 17.1. Mean response amplitudes in acquisition and extinction over five trial blocks as a percentage of CS-alone response amplitude for paired and unpaired groups (From Patterson, Cegavske, and Thompson, 1973, copyright 1973 by the American Psychological Association. Reprinted by permission.)

sec, paired; 30 sec, unpaired) was used and a CS-only control group added. Here, the pairing of CS and UCS produced response increases, with none in the unpaired and CS-only conditions. In a separate part of these studies, we presented the CS at various times following the UCS to estimate the amount and duration of response changes produced by the UCS alone. The results showed a marked response increase beginning almost immediately following the UCS, but disappearing before the 30 or 60 sec ITI of the unpaired or paired conditioning group. Thus the results of pairing were not due to a carryover sensitization from the UCS.

The unpaired condition, although controlling for pairing, does not control for stimulus overlap that occurs in the paired condition, possibly causing an additive effect not produced by the unpaired control. In another study (Patterson, 1975a) we utilized a backward paradigm to assess the overlap effect and to determine the effect of stimulus asynchrony on the spinal response. Figure 17.3 shows the effect of pairing the CS and UCS normally and at 100 msec and 250 msec backward intervals (UCS onset before CS onset). In most classical conditioning studies, such backward

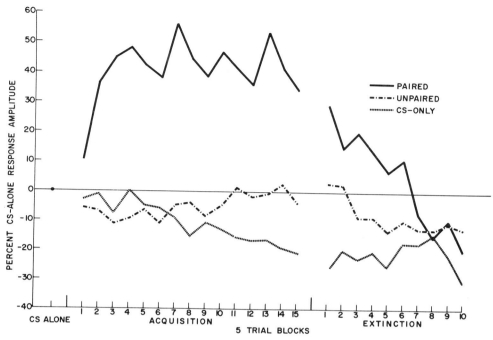

Figure 17.2. Mean response amplitudes in acquisition and extinction over five trial blocks as a percentage of CS-alone response amplitude for paired, unpaired and CS-only groups. (From Patterson, Cegavske, and Thompson, 1973, copyright 1973 by the American Psychological Association. Reprinted by permission.)

pairing produces no associative excitatory learning (e.g., Smith, Coleman, & Gormezano, 1969). The results indicated no increase with either backward interval but only with the forward pairing.

In a more recent study (Patterson, 1975b), we varied the forward stimulus asynchrony or ISI in the spinal conditioning preparation. Figure 17.4 shows the results of presenting the CS and UCS to different groups of animals at 0, 20, 60, 125, 250, 500, 1000, 2000, and 4000 msec ISIs. The greatest response increase is seen with the 250 msec ISI, lesser at 60, 125, 500, and 1000 msec intervals, and little at 20, 2000, and 4000, and also a slight decrease or inhibition with simultaneous stimulus onsets. These results parallel closely the results often obtained by ISI manipulations in classical conditioning studies in intact subjects (e.g., Smith, Coleman, & Gormezano, 1969), suggesting an optimal asynchrony for the response alteration in the spinal preparation.

In a separate unpublished study, we have examined the effect of extinction (CS-alone) trials on response retention and have found that the response increase is retained after either 20, 40, or 60 paired trials if a probe CS is then given once each 10 min for 2 hr. If, however, the CS

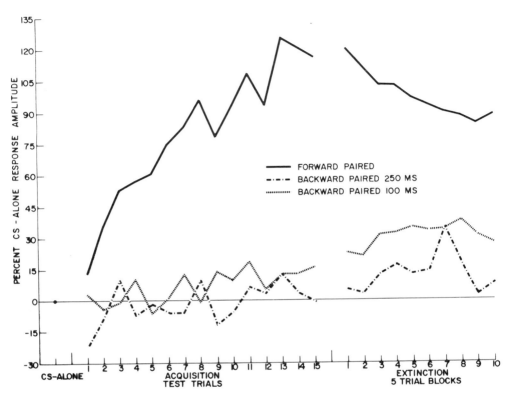

Figure 17.3. Mean response amplitude as a percentage of CS-alone amplitude for test trials in acquisition and over five trial blocks in extinction for the forward 250, backward 100, and backward 250 groups. (From Patterson, 1975a, copyright 1975 by the Psychonomic Society. Reprinted by permission.)

is given once each minute in regular extinction, the response decreases to base levels within 1 hr. In a separate, also unpublished, study, we have examined the effects of pairing and CS-only presentations to the left leg on response amplitude of the flexion response in the right leg. The preliminary results are shown in Figure 17.5, where it can be seen that the response increase produced by pairing in the left leg is seen later in the right leg, although not as strongly, thus suggesting a generalization effect. We have also found that a 45 sec ITI seems to produce the largest response increases.

The effects produced by stimulus pairing in this preparation are obviously not produced by muscle changes or receptor alterations. These effects are bypassed by the paralyzed preparation, stimulus placement, and response recording techniques. In addition, our preliminary data on the neural mechanisms suggest that the afferent terminals of the sensory paths and the motoneurons are not altered in excitability during the response

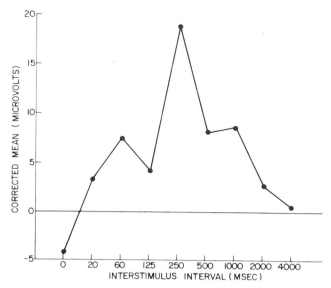

Figure 17.4. Mean corrected response amplitudes in acquisition as a function of ISI.

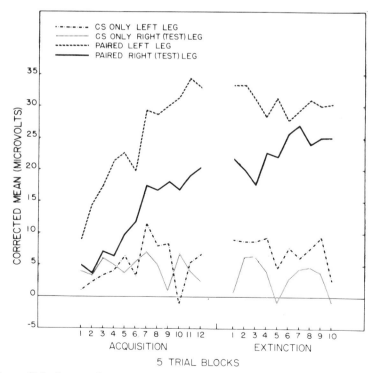

Figure 17.5. Corrected mean response amplitudes over five trial blocks in acquisition and extinction for experimental and control legs (paired left and right legs, respectively) and for a CS-alone control group (CS-only left and right legs, respectively).

changes, leaving the interneuron pools as the most likely source of the response alterations.

Recent elegant work by Durkovic (1975; Light & Durkovic, 1977) using a similar but unparalyzed preparation has confirmed our early findings and has extended the results to show that sensory afferents of the alpha class or smaller must be stimulated by the UCS to produce the response increase in spinal preparations. Durkovic has also shown that blood pressure changes have no effect on the response alterations. As in our work, Durkovic is now beginning to examine the reflex pathways for clues of the neural basis for the phenomenon.

The elgant work of Nesmeyanova, which suggests no associative pairing effects, differed substantially from our preparation. Her work generally involved stimulus application at receptive fields much farther separated from those stimulated in our studies, among other differences. She found threshold changes with 400–1000 trials in semichronic or acute preparations, whereas we found changes over a few trials. Thus, it may be that any one of several as yet unexplored differences may be responsible for the different interpretation of the results.

An Interpretation

This work and that of Durkovic show a definite response alteration with stimulus pairing in the flexor reflex of the acute spinal cat. In many respects so far tested, the response alterations parallel those seen in classical conditioning situations in intact animals, such as nictitating membrane (NM) conditioning of the rabbit (e.g., Gormezano, 1972) and cat (Patterson, Olah, & Clement, 1977), and in human work, such as eyelid conditioning (e.g., Ross & Ross, 1971). In other respects, however, it differs. In the spinal preparation, there is a predetermined alpha response (reflex to the CS) that does not (or is not) habituated prior to pairing, and no latency shifts in the response during pairing have been seen in our work. Thus the spinal response change is basically an amplitude increase with pairing. It is clear that no new connections are formed that were not present prior to training. However, it has been shown (Young, Cegavske, & Thompson, 1976) that even in the rabbit NM preparation, the CS elicits an alpha excitability in the motoneurons, which produce the behavioral response, and suggests that existing pathways may be used in CR formation. Thus, the spinal cord may lack the elegant machinery for response selection and adaptation of the cortex or upper brain stem, but the results of our work and Durkovic's studies suggest that the machinery for the basic pairing or associative effects of classical conditioning is present. Particularly intriguing in this regard are the results of the ISI study that show an optimal interval close to that shown in intact animal studies.

These results suggest that the stimulus asynchrony necessary in conditioning may not be a property of vast neural networks of the brain, but of a few synapses, perhaps of even single cells of the pathway where CS and UCS inputs converge, and that the basic properties of response alteration through stimulus pairing may be the properties of synaptic interactions on cells rather than of large networks. Thus, it would appear from this work that the basic elements necessary for the very fundamental process of associative learning are present in the reflex paths of the mammalian spinal cord and that the work can provide clues to the basis of associative learning in the brain as well as shed light on the normal function of the intact spinal cord.

REFERENCES

Culler, E. Observations on the spinal dog. *Psychological Bulletin*, 1937, *34*, 742–743.

Culler, E., & Mettler, F. Conditioned behavior in a decorticate dog. *Journal of Comparative Psychology*, 1934, *18*, 291–303.

Durkovic, R. Classical conditioning, sensitization, and habituation of the flexion reflex of the spinal cat. *Physiology and Behavior*, 1975, *14*(3), 297–304.

Fitzgerald, L. A., & Thompson, R. F. Classical conditioning of the hindlimb flexion reflex in the acute spinal cat. *Psychonomic Science*, 1967, *8*, 213–214.

Gormezano, I. Investigations of defense and reward conditioning in the rabbit. In A. Black & W. Prokasy (Eds.), *Classical conditioning II: Current theory and research*. New York: Appleton-Century-Crofts, 1972.

Groves, P. M., and Thompson, R. F. Habituation: A dual-process theory. *Psychological Review*, 1970, *77*, 419–450.

Light, A., & Durkovic, R. G. US intensity and blood pressure effects on classical conditioning and sensitization in spinal cat. *Physiological Psychology*, 1977, *5*, 81–88.

Nesmeyanova, T. *Experimental studies in regeneration of spinal neurons*. New York: Wiley, 1977.

Norman, R. J., Villablanca, J. R., Brown, K. A., Schwafel, J. A., & Buchwald, J. S. Classical eyeblink conditioning in the bilaterally hemispherectomized cat. *Experimental Neurology*, 1974, *44*, 363–380.

Oakley, D. A., & Russell, I. S. Role of cortex in Pavlovian discrimination-learning. *Physiology and Behavior*, 1975, *15*, 315–321.

Patterson, M. M. Effects of forward and backward classical conditioning procedures on a spinal cat hind-limb flexor nerve response. *Physiological Psychology*, 1975a, *3*(1), 86–91.

Patterson, M. M. Interstimulus interval effects on a classical conditioning paradigm in the acute spinalized cat. Paper presented at the Society for Neuroscience, New York, November, 1975b.

Patterson, M. M. Mechanisms of classical conditioning and fixation in spinal mammals. In A. Riesen & R. Thompson (Eds.), *Advances in psychobiology* (Vol. III). New York: Wiley, 1976.

Patterson, M. M., Cegavske, C. F., & Thompson, R. F. Effects of a classical conditioning paradigm on hind-limb flexor nerve response in immobilized spinal cats. *Journal of Comparative and Physiological Psychology*, 1973, *84*, 88–97.

Patterson, M. M., Olah, J., & Clement, J. Classical nictitating membrane conditioning in the awake, normal, restrained cat. *Science*, 1977, *196*, 1124–1126.

Pavlov I. P. Conditioned reflexes (trans. by G. V. Anrep). London: Oxford University Press, 1927.

Pronko, N., & Kellogg, W. The phenomenon of the muscle twitch in flexion conditioning. *Journal of Experimental Psychology*, 1942, *31*, 232–238.

Ramon-Moliner, E., & Nauta, W. J. The isodendritic core of the brain stem. *Journal of Comparative Neurology*, 1966, *126*, 311–336.

Ross, S. M., & Ross, L. E. Comparison of trace and delay classical eyelid conditioning as a function of interstimulus interval. *Journal of Experimental Psychology*, 1971, *91*, 165–167.

Shurrager, P. S., & Culler, E. Conditioning in the spinal dog. *Journal of Experimental Psychology*, 1940, *26*, 133–159.

Smith, M., Coleman, S., & Gormezano, I. Classical conditioning of the rabbit's nictitating membrane response at backward, simultaneous, and forward CS-UCS intervals. *Journal of Comparative and Physiological Psychology*, 1969, *69*(2), 226–231.

Young, R. A., Cegavske, C. F., & Thompson, R. F. Tone-induced changes in excitability of abducens motoneurons in the reflex path of the rabbit nictitating membrane response. *Journal of Comparative and Physiological Psychology*, 1976, *90*, 424–434.

A. P. KARPOV

Analysis of Neuron Activity in the Rabbit's Olfactory Bulb during Food-Acquisition Behavior

18

The assumption that perception constitutes an active process that must include efferent links is commonly recognized in modern psychology. However, until now, investigators have been "charting stimuli" in the course of studying any sensory processes, recording changes in some particular factors, and then calling this a "response."

All the studies that have been concerned with the neurophysiological mechanisms of olfaction have employed the traditional method of investigation that calls for the recording of an olfactory stimulus by olfactometers of various design or even the electrical stimulation of olfactory structures, such as various sections of the olfactory mucosa or the olfactory bulb, as well as olfactory analyzer pathways, which is then followed by the recording of "reactions" in some particular structure.

Under natural conditions, an organism is not subject to the effects of any particular stimulant, but rather it fixes events that occur in the environment. All of the events that are unexpected at any given moment induce a behavioral response of an orientational-investigatory nature or are ignored altogether. As early as 1947, R. R. Woodworth wrote that "perception is always prompted by a direct motive that is inherent in a given situation and which might be called a desire to perceive."

As seen from the functional system theory of P. K. Anokhin, information perceived by an analyzer is utilized to secure the systems processes of any particular behavioral act. Hence each stimulus whose information is used in a behavioral continuum has a double role. On the one hand, the stimulus is expected to be within the acceptor of action results; that is, it is the goal and result of one functional system, and, on the other hand, this stimulus is the trigger for developing an ensuing functional system (Anokhin, 1968; Shvyrkov, 1973, 1978).

The study of the neurophysiological mechanisms of olfaction has many aspects, one of which is how to utilize information about the environment in the system's mechanisms of behavior. The task of this study is to try to explain how the neuron activity of the olfactory bulb is organized in food-acquisition behavior that is guided by olfaction.

The methods we used were determined by methodological prerequisites. The work was done on rabbits in free behavior. We selected a form of animal behavior so that a "scent stimulus" was not presented to the rabbit, but rather the rabbit itself would try to obtain the scent by its own actions. For this purpose we made a special chamber-cage, to whose anterior walls we attached a food box divided into three sections (Figure 18.1), each of which was a pencil box. Some kind of aromatic substance or food was placed into the extreme left or right boxes. The boxes were closed by rubber band tension. In order to prevent the rabbit from prematurely recognizing the food box's contents by its odor, air was constantly withdrawn

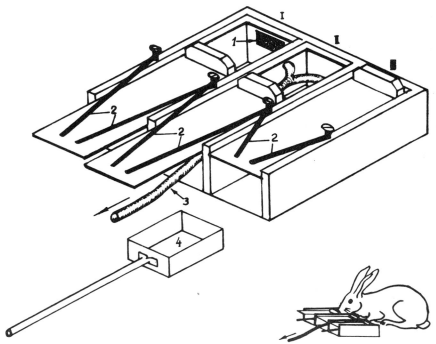

Figure 18.1. Design of food box for studying food-acquisition olfaction-directed behavior. Bottom right: *Rabbit examines one of the food box sections.* Designations: *(I, II, III) sections of the food box; (1) photoelectric plate; (2) elastic bands; (3) ust pipe of water-jet pump; (4) removable food box.*

from the boxes by a water-jet pump. The dynamics of the boxes' opening and closing was recorded photoelectrically.

As a preliminary step, the animals were trained to open the box covers and obtain food. To do this, the rabbit had to push open the box cover with its nose and take the food, if there were any there. When the box was empty or when it contained some substance with a repulsive odor (2% ammonia solution, 4.5% vinegar or urine of other rabbits), the rabbit was trained to release the cover and examine another section of the food box. The contents of each pencil box were changed by the experimenter in random order. The rabbit usually consistently examined the sections of the food box until it found the food (pieces of carrot or cabbage). When the rabbit examined the empty food box or the one containing substances with repulsive odors, it drew back the pencil box covers a total of .5–2 cm and then released them. This is proof of the fact that the rabbit is guided in his behavior only by olfaction when it cannot see the contents of the food box or examine it with its whiskers. When the rabbit detects food odor, it draws back the cover again and lowers its snout into the food box.

During the preceding described behavior, we recorded the overall electrical activity of the olfactory bulb, the activity of individual nerve cells of the olfactory bulb, as well as respiration. Recording respiration was essential to solving the problems at hand, because the respiratory movements accomplish the delivery of aromatic substance molecules to the olfactory mucosa. In addition, these movements apparently perform a function in olfaction that is analogous, for example, to eye movements in visual perception. Respiration was recorded by a photoelectric sensor sewn into the trachea.

The experimental data were processed by plotting histograms for neuron activity and averaging the overall electrical activity of the olfactory bulb and respiration relative to the noted behavioral instances: the opening of the food boxes, the closing of the boxes, the lowering of the animal's snout into the box to obtain food, as well as from the moment of inhalation, designated as "decisive." An accounting was made of the inhalation that was followed either by the food box's cover being drawn back or by the animal's lowering its snout into the food box. The data were processed on a laboratory minicomputer.

Before proceeding to a description and discussion of the results, we shall analyze in more detail the behavior already described. From the viewpoint of the functional system theory that behavior might be viewed as a large, complexly organized functional system whose useful adaptive result is the satisfaction of food motivation. To achieve this, the rabbit completes an entire cyclic behavior. Figure 18.2 shows three single acts in which the rabbit opened the food boxes containing vinegar, urine, and carrots. Apparently, the goal of any opening act is either to detect food or detect its absence. This information is essential to the completion of the subsequent

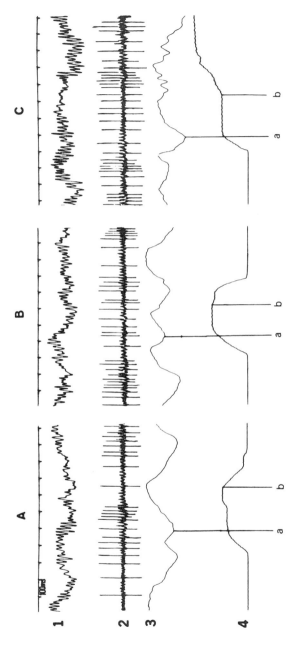

Figure 18.2. Overall electrical activity of the olfactory bulb. (1) Activity of olfactory bulb neuron; (2) and activity of respiration neuron; (3) during examination of food box with vinegar (A); food box with urine (B), and food box with cabbage (C). (4) Dynamics of food box opening. (a) is onset of "decisive inhalation," (b) is onset of releasing the food box covers (A) and (B) and the beginning of lowering snout into the food box (C).

acts—shifting to the neighboring box and opening it or lowering the snout into the box to get food. Thus, each sequential act of examining the food boxes also constitutes a complex functional system with its stagewise result.

Evidently, under the conditions of the described behavior, possible contact with a scent lasts from the onset of the decisive inhalation (point a in Figure 18.2) to the closing of the food box or the lowering of the snout into the food box (point b in Figure 18.2). Apparently, all the processes associated with making a decision about any particular scent occur during the time of one respiratory cycle.

From the viewpoint of the functional system theory, a respiratory act constitutes a functional system at the physiological level, whose useful adaptive result is the passage of a specific volume of air into the lungs. At the same time, each respiratory act is accompanied by the passage of aromatic substances contained in the surrounding air to the olfactory mucosa.

When the rabbit is resting in the cage, the olfactory information is probably an inevitable consequence or an incidental result of an inhalation's physiological system and is used as a component of situation afferentation in the afferent synthesis of current behavioral acts. The functional inhalation systems in current behavior are developed relatively independent of behavior, so that the same behavior act, such as the opening of the food boxes, for example, could begin in any phase of the respiratory cycle, as was observed in the experiments we conducted.

The time at which the food box is opened in food-acquisition behavior corresponds to the functional onset of the actuating mechanisms of the system of opening. The possibility of inhaling the air over the aromatic substance placed in the food boxes should be seen as the result of that system. One of the odors in that air is a result parameter of the functional system of inhalation that follows the functional system of opening. The functional system of inhalation at this time already acts as a functional system at the behavioral level, since the result of this system becomes a part of afferent synthesis, no longer as a situation component but as a trigger for the next behavioral act.

The experiments have shown that the time required for recognizing odors and for making a decision about a particular odor is about 250 msec on the average (the interval a–b in Figure 18.2). These figures are comparable to those indicated by other investigators as so-called reaction times to stimuli of other sensory modalities.

A study of the overall electrical activity of the olfactory bulb in the previously described behavior indicates that upon averaging various behavioral factors in an electrogram, one often detects a fluctuation in potential that is similar in configuration to the traditional evoked potential recorded by many investigators in various regions of the brain in response to stimuli from other sensory modalities. The upper curves 1 and 2 in Figure 18.3 (A, B, C) represent averaged recordings of the overall electrical

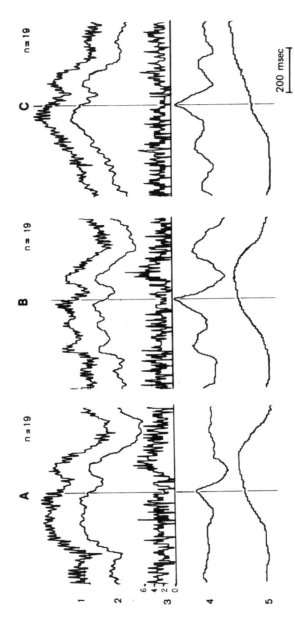

Figure 183. Coincidence of neuron (3) activation to negative fluctuation in an overall electrogram of an olfactory bulb (1, 2) and to respiration (4) during examination of the food boxes with vinegar (A) and with urine (B). Neuron activation was absent during examination of food boxes containing cabbage (C). Averaging of the electrogram, respiration and markings, and the plotting of neuron activity histograms were heavily relative to the end of the "decisive" inhalation.

activity of the olfactory bulb, from the end of the "decisive" inhalation to the beginning of exhalation, during the rabbit's examination of the food boxes containing vinegar, urine, and cabbage (first curve is the bandpass from the top 200 Hz and the second curve is 36 Hz). One can see that the "evoked-potentials" type fluctuations in the olfactory bulb electrogram correlate with the respiratory cycle, and are apparently associated with the detection of olfactory signals for accomplishing food-acquisition behavior.

There have been quite a few studies of the so-called background activity of olfactory bulb neurons, as well as activity associated with the effect of olfactory stimuli, and there are presently a number of works in which recordings have been made of olfactory bulb neurons whose activity is associated with various phases of the respiratory cycle (Eksler & Safonov, 1975; Macrides & Chorover, 1971; Walsh, 1956; and others). Fifty-seven out of the 96 neurons that were recorded in our experiments exhibited that association. The remaining 39 neurons were fired with relative regularity, without any connection to the respiratory cycle. It is essential to note that we did not detect any primary association between the activity of "respiratory" neurons and a particular phase of the respiratory cycle. Besides, the correlation between neuron activity and any particular phase of the respiratory cycle did not remain constant. As can be seen from Figure 18.4, neuron activation in the animal's quiescent state, when the rabbit does "not behave" visually, is associated with the exhalation–inhalation period and a reduction in activity to the end of the inhalation and the beginning of exhalation. During the rabbit's exploratory behavior, when it is sniffing the chamber's walls and its respiratory movements are frequent and superficial, the firing peak of the same neuron is observed in the inhalation period with a maximum on the inhalation peak and with inhibition during exhalation (Figure 18.4B).

A change in activity in connection with the stages of food-acquisition behavior was recorded in neurons that fired in connection with respiration and in neurons that were fired in relatively regular single impulses. Changes in the activity of ten olfactory neurons were observed in the period preceding the opening of the food boxes, as well as after the covers were released or the snout was lowered into the food box. The changes in activity that occurred during these periods could often be associated with a change in respiration and did not depend on the aromatic substance. During the same time, recordings were also made of neurons whose activity during these periods changed specifically in relation to the substance. Thus, Figure 18.5 shows a well-defined activation after the food boxes with urine were closed (Figure 18.5B), whereas no such activation was detected after the empty food boxes and the food boxes with cabbage were closed (Figures 18.5A and C). Evidently, these changes in the activity were not associated with odor differentiation, since no decision was taken with respect to odor,

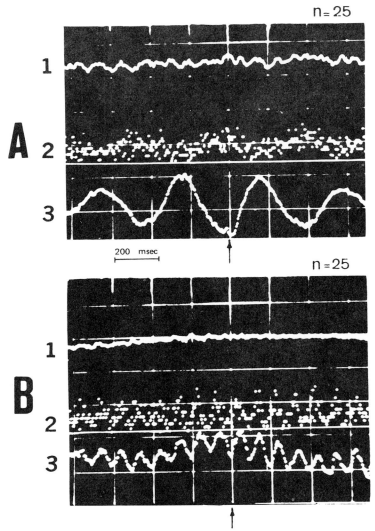

Figure 18.4. Change in the correlation of neuron activation during various aspects of behavior. (A) During the resting state of an alert rabbit; (B) during exploratory behavior-sniffing. Designations: (1) averaged EG recording of the olfactory bulb; (2) neuron activity histograms; (3) averaged recording of the animal's respiration.

but they were associated with the involvement of such neurons in the actuating mechanisms of olfaction-directed behavior.

When odor identification did occur and a decision was made with respect to that odor, an activity change was observed in seven neurons. Note that we did not find any relationship between the nature of these changes and any particular substance or substances of various groups (rejected sub-

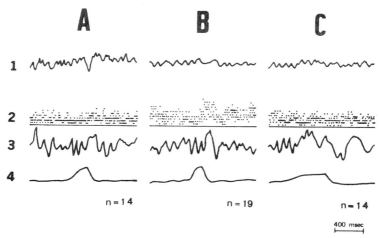

Figure 18.5. Association between olfactory bulb neuron activity and behavior directed by odor perception. Neuron activation after the closing of food box lid in box with urine (B). During the examinations of food boxes with cabbage (C), or examination of empty food boxes (A); neuron did not change its activity (A and C).

stances or food). Various neurons exhibited different forms of activity while the rabbit was examining food boxes with identical aromatic substances, and at the same time, different odors could change the activity of various neurons in an identical fashion. Thus, the neuron shown in Figure 18.3 (3) related its activity to the animal's respiration when it was examining the food boxes containing vinegar (Figure 18.3A) and when it opened the food boxes with urine (B). Thus, the neuron's activation corresponds to the end of exhalation and to the beginning of inhalation. Such activity was not observed when the animal examined the food boxes containing cabbage (Figure 18.3C).

Thus, the activity of olfactory bulb neurons in behavior does not constitute a "response reaction" to olfactory stimulation, but characterizes the utilization of olfactory information in various mechanisms of the behavioral act's functional system. Apparently, the activity of olfactory neurons is determined by their simultaneous involvement in the functional system of respiration and in the system's processes of behavior. One would think, then, that the processes of odor recognition are not conditioned by the activity of specific detectors, but by a specific organization of the activity of many olfactory bulb neurons.

REFERENCES

Anokhin, P. K. *Biology and the neurophysiology of the conditioned reflex.* Moscow, 1968. (In Russian)

Eksler, N. D., & Safonov, V. A. The activity of olfactory bulb neurons, synchronized with respiration. *Scientific Papers of Higher Institutions. Biological Sciences,* 1975 (No. 4), p. 39. (In Russian)

Macrides, F., & Chorover, S. L. Olfactory bulb units. Activity correlated with inhalation cycles and odor quality. *Science,* 1971, *175,* p. 84.

Shvyrkov, V. B. Neuronal mechanisms of learning as a component of the functional system of a behavioral act. In *Principles of the systems organization of functions.* Moscow, 1973, p. 156. (In Russian)

Shvyrkov, V. B. The functional system theory as a methodological basis of behavioral neurophysiology. *Progress in Physiology,* 1978, *9* (No. 1). (In Russian).

Walsh, R. R. Single cell spike activity in the olfactory bulb. *American Journal of Physiology,* 1956, *186,* p. 255.

Woodworth, R. S. Reinforcement of perception. *American Journal of Psychology,* 1974, *LX,* p. 119.

DAVID H. COHEN

The Functional Neuroanatomy of a Conditioned Response[1]

19

Introduction

Although the literature is replete with hypotheses of how the brain stores and retrieves information, experimental efforts to discover the actual cellular mechanisms of information storage have largely met with frustration. One could suggest that this reflects primarily a lack of appropriate experimental techniques, an explanation that may well have been valid in the earlier history of the field (Cohen, in press). However, this is no longer fully convincing, and in recent years it has been proposed that the major obstacle may in fact be a lack of effective model systems (Cohen, 1969; Kandel & Spencer, 1968). Most investigators accepting this premise adopted an approach of developing "simple" systems, primarily involving invertebrate (e.g., Kandel, 1976) and to a lesser extent simplified vertebrate (e.g., Groves & Thompson, 1970) preparations. Perhaps the most exciting progress in the field has derived from this approach. However, such progress has largely centered around nonassociative learning and frequently shorter-term effects. These "simple" systems have been selected primarily on anatomical criteria such as neuronal number, size and identifiability, rather than on the basis of behavioral capability, and they have not readily demonstrated long-term, associative learning.

[1] The Research described here was supported by National Science Foundation Grants GB-13816, GB-35204, BMS-74-22258, BNS-75-20537, and the Scottish Rite Foundation. Part of this research was conducted during the tenure of Research Career Development Award HL-16579 from the National Heart and Lung Institute.

This motivated our efforts to develop a vertebrate model system that would ultimately allow the cellular analysis of long-term associative learning (Cohen, 1969, 1974a, in press). Important in guiding this effort were certain general criteria proposed by Kandel and Spencer (1968) and Cohen (1969). First, the system must have a quantifiable behavioral response whose probability of occurrence can be modified on a long-term basis by appropriate training conditions, and this modification must be consistent with established principles of associative learning. Second, and of paramount importance, the neuroanatomical pathways mediating the development of the learned response must be specified in detail, because without such information it is difficult to make statements regarding the meaning, or even the relevancy, of changes in neuronal activity during training.

The Behavioral Model

Thus, our initial task was to develop an effective behavioral system, and for previously described reasons we selected visually conditioned heart rate change in the pigeon as our model (Cohen, 1969, 1974a). In developing this system over the years we have applied a standardized paradigm where a 6 sec pulse of whole field retinal illumination (conditioned stimulus) is immediately followed by a .5 sec footshock (unconditioned stimulus). The footshock elicits cardioacceleration (unconditioned response), and after a sufficient number of light-shock pairings retinal illumination itself reliably elicits a heart rate change of predictable dynamics (conditioned response). Cohen and Durkovic (1966) first showed that in this system stable conditioning develops within 20 pairings and asymptotic performance within 30–50 pairings—findings that have been described in considerable detail in a recent analysis of more than 450 animals (Cohen & Goff, 1978a) (Figure 19.1). Moreover, these acquisition characteristics occur with intertrial intervals as short as 3.5 min (Cohen & Macdonald, 1971), such that stable

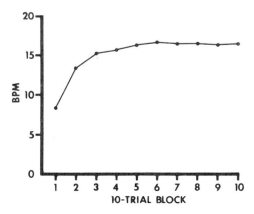

Figure 19.1. Acquisition of conditioned heart rate change in the pigeon. The curve represents mean heart rate changes between the 6 sec conditioned stimulus and the immediately preceding 6 sec control periods. Each point represents a group mean for a block of ten training trials. Abbreviation: BPM = beats per minute.

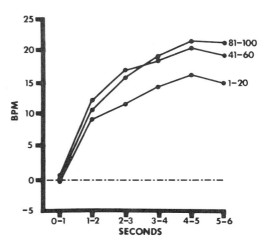

Figure 19.2. Conditioned response dynamics at various points in training. The curves show mean heart rate changes from baseline for each succeeding 1 sec interval of the conditioned stimulus period, averaged for trial blocks 1–20, 41–60, and 81–100. Note that stable conditioning occurs in block 1–20 and asymptotic performance by block 41–60. Abbreviation: BPM = beats per minute. (From Cohen, 1974a.)

conditioning obtains in $\cong$ 1 hr. and asymptotic performance in $\cong$ 2 hr. Regarding conditioned response dynamics, these too have been described in detail (Cohen, 1974a; Cohen & Goff, 1978a). Briefly, the response is a monotonic cardioacceleration with a latency of $\cong$ 1 sec and with maximal values in the fifth or sixth seconds of the conditioned stimulus period (Figure 19.2). It might also be noted that once established, this conditioned response is highly resistant to extinction (Cohen, 1969; Cohen & Durkovic, 1966) and that it is not secondary to concomitantly developing conditioned rseponses (Cohen, 1969, 1974a; Cohen & Goff, 1978a).

As our knowledge of the behavior of this system advanced, it became essential to assure that its attractive properties were retained in a preparation suitable for cellular neurophysiological investigations. This was accomplished by developing procedures for pharmacological immobilization that had minimal cardiovascular effects and allowed conditioning performance equivalent to that of nonimmobilized animals (Gold & Cohen, in preparation; Migani-Wall, Cabot, & Cohen, 1977). Furthermore, these preparations permit precise control of the visual stimulus (Duff & Cohen, 1975a, b), such that we now have available an immobilized conditioning preparation allowing precise stimulus control and training to asymptotic performance in a single session of $\cong$ 2 hr.

A final feature of the system to be considered here is its capacity for nonassociative learning within the context of our standardized paradigm. When the light is initially presented, prior to any experience with footshock, it elicits a slight cardioacceleration—the orienting response (Cohen & Goff, 1978a; Cohen & Macdonald, 1971). The dynamics of this response resemble those of the conditioned response, and it habituates in $\cong$ 10 stimulus presentations with the parameters of our paradigm (Cohen & Macdonald, 1971) (Figure 19.3). If random footshocks are then introduced,

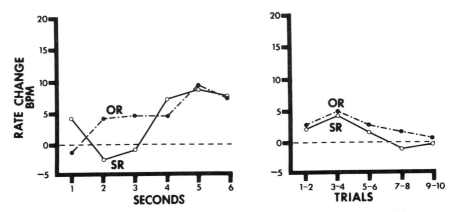

Figure 19.3. Panel A illustrates the dynamics of orienting and sensitization responses. The curves show mean heart rate changes from baseline for each succeeding 1 sec interval of the light period. The orienting response (OR) curve is an average of the responses of over 100 animals to the initial light presentation. The sensitization response (SR) curve is an averaged response to the light of 20 birds over the first four unpaired light-shock presentations. Panel B describes the habituation of the OR and SR shown in panel A. The curves show mean heart rate changes between the stimulus period and a preceding 6 sec control period averaged over two-trial blocks. Abbreviation: BPM = beats per minute. (From Cohen, 1974a.)

sensitization develops, and this response has distinctly different dynamics from either the orienting or conditioned responses (Figure 19.3). This sensitization responding also extinguishes within $\simeq$ 10 stimulus presentations, and thus orienting and sensitization contribute little to the conditioned response magnitudes. However, the magnitudes of these nonassociative responses can be systematically manipulated (Cohen, 1974b), thereby providing ample opportunity to investigate the mechanisms of both habituation and sensitization in a system demonstrating conditioning.

In summary of the behavioral model, it is a robust, effective, and technically attractive system for cellular neurophysiological analysis of long-term associative learning. It has the general advantages of providing a well-defined, quantifiable behavioral output and of potentially allowing study of many classes of questions, including analysis of nonassociative learning. Moreover, it has many specific technical advantages, such as pharmacological immobilization, rapid acquisition, and consistent response dynamics among animals (Cohen, 1974a; in press).

The Relevant Anatomical Pathways

Given an effective behavioral model, identifying the pathways mediating development of the conditioned heart rate change is the most essential and challenging criterion the model system must satisfy. Thus, specifying these pathways has constituted a major phase of our effort, and to implement this we began by defining four major segments of the system: (a) the visual pathways transmitting the conditioned stimulus information; (b) the somatosensory pathways transmitting the unconditioned stimulus information; (c) the descending pathways mediating expression of the conditioned response; and (d) the efferent pathways mediating the unconditioned response. Our approach has entailed beginning at the periphery of each of these segments and tracing them systematically centrally. Thus, for the conditioned and unconditioned stimulus pathways analysis was initiated at the sensory periphery, whereas for efferent segments it began at the extrinsic cardiac nerves. The assumption is that a systematic analysis from the periphery centrally along the input and output segments of the system will ultimately identify the sites of convergence of the conditioned and unconditioned stimulus pathways, as well as the sites of their coupling to the conditioned response pathways.

PATHWAYS TRANSMITTING THE CONDITIONED STIMULUS INFORMATION

Regarding the visual periphery, we have shown that bilateral enucleation prevents conditioned response development, excluding participation of non-retinal photoreceptors (Wild, Cohen, & Migani-Wall, in preparation). We next explored structures receiving retinal projections for their possible involvement in transmitting the conditioned stimulus information. The experimental approach was to destroy bilaterally a given optic tract terminal field in a group of untrained animals and after 10–14 days training them in our standardized paradigm, along with appropriate control animals. The results indicated that none of these lesions affects conditioned response development (Cohen, 1974a; in preparation). Perhaps this should not be surprising, since each of these optic tract projections is responsive to whole field illumination, therefore providing considerable opportunity for parallel transmission of the stimulus information.

An experimental series was then initiated to evaluate possible involvement of the various central visual pathways. Since in mammals severe deficits in various visual learning tasks occur following interruption of the visual pathways ascending to the cortex, we began by examining the homologous pathways in the avian brain. As a brief background, two such pathways have been described: the thalamofugal and tectofugal systems (Figure 19.4), which are respectively homologous to the mammalian geniculostriate and tecto-thalamo-extrastriate pathways (see Cohen & Karten, 1974). The thalamofugal system includes a crossed retinal projection to the principal optic nucleus of the thalamus, which then projects bilaterally upon the visual Wulst of the telencephalon. The tectofugal system involves a crossed retinal projection, at least one tectal interneuron, and a massive tectal projection via the brachium of the colliculus upon the ipsilateral nucleus

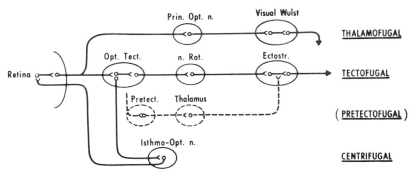

Figure 19.4. Schematic illustration of the major ascending visual pathways in the bird and of the centrifugal system to the retina. The dashed lines indicating the pretectofugal pathway reflect its speculative nature, and the hypothesized pretectofugal projection from the thalamus to the ectostriatum is intended to suggest a termination either upon or in the immediate region of the ectostriatum. Abbreviations: Ectostr. = ectostriatum; Isthmo-Opt. n. = isthmo-optic nucleus; n. Rot. = nucleus rotundus; Opt. Tect. = optic tectum; Pretect. = pretectal region; Prin. Opt. n. = principal optic nucleus of the thalamus. (From Cohen, 1974a.)

rotundus of the thalamus. Nucleus rotundus then projects upon a distinct cell mass of the ipsilateral telencephalon, the ectostriatum.

Our initial question was whether interrupting either of these systems would affect conditioned response development. With respect to the thalamofugal pathway, interruption at either thalamic or telencephalic levels had minimal effect; analogous findings followed interruption of the tectofugal pathway (see Cohen, 1974a). In contrast, combined destruction of these two systems by lesions of both the visual Wulst and ectostriatum prevented conditioned response development (Cohen, 1974a; in preparation) (Figure 19.5), thus clearly implicating these pathways in the transmission of the conditioned stimulus information. However, in attempting to cross-validate this finding by combined interruption of the two pathways at the thalamic level, only a transient deficit in conditioned response development was found. This, in conjunction with various electrophysiological observations (see Cohen, 1974a), suggested the existence of a third ascending visual pathway that either converged with the tectofugal pathway upon the ectostriatum or terminated in the immediate vicinity of the ectostriatum. Given the similarity of avian and mammalian visual pathways, we hypothesized that this pathway is of pretectal origin (Figure 19.4) and we thus conducted a study in which the principal optic nucleus, nucleus rotundus, and the pretectal terminal field of the optic tract were destroyed. This resulted in acquisition deficits comparable with those following the combined lesion of the visual Wulst and ectostriatum (Figure 19.5); lesions of the pretectal region alone produced no deficit (Cohen, 1974a; in preparation).

Thus, each of the ascending pathways is capable of transmitting con-

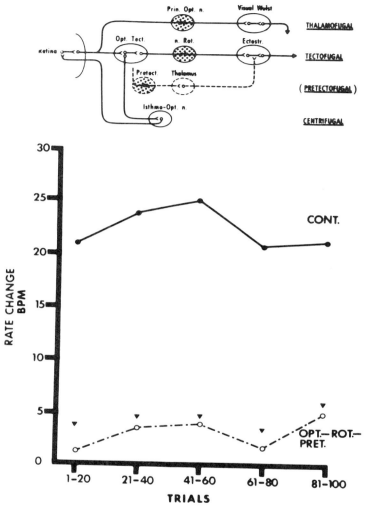

Figure 19.5. The effects of combined interruption of the principal optic nucleus, nucleus rotundus, and the pretectal terminal field of the optic tract. The upper part of the figure repeats Figure 19.4 with the lesioned structures stippled. The curves represent mean heart rate changes between conditioned stimulus and preceding control periods. Each point represents a group mean for a block of 20 training trials. The solid inverted triangles indicate the response levels of birds with a combined lesion of the ectostriatum and visual Wulst. Abbreviations: CONT. = control curve; OPT.-ROT.-PRET. = subtelencephalic lesion group; BPM = Beats per minute. See the caption to Figure 19.4 for abbreviations for the upper part of the figure. (From Cohen, 1974a.)

ditioned stimulus information that is effective in establishing conditioned heart rate change, and it is only with their combined destruction that response development is precluded. It should be appreciated, however, that this does not necessarily imply the impossibility of establishing some conditioned heart rate change with either prolonged training or some modification of the training procedure. The critical point is that in our standardized paradigm near-total loss obtains for two to three times the number of trials required for asymptotic performance in the intact animal. Since we are interested in the pathways that normally mediate response development, we have not pursued aggressive training regimens to determine if such responses could be established following the combined lesion.

In summary, we now have a first-approximation to the pathways transmitting the conditioned stimulus information. Without the integrity of the visual pathways ascending to the telencephalon, conditioned response acquisition is compromised, and thus these pathways are clearly necessary. Although we cannot definitively argue that they are the only pathways transmitting effective conditioned stimulus information, the results are certainly suggestive of this.

DESCENDING PATHWAYS MEDIATING CONDITIONED RESPONSE EXPRESSION

Following the strategy of initiating analysis at the periphery of each segment of the system, our earliest efforts to describe the conditioned response pathways began at the final common path for the heart rate change (Cohen, 1974a,c, in press). A behavioral study involving various combinations of cardiac denervation and pharmacological blockade demonstrated that the response is mediated entirely by the extrinsic cardiac nerves, that both the vagi and sympathetics participate, and that the sympathetic innervation provides the major contribution to response magnitude (Cohen & Pitts, 1968). Subsequent anatomical study of the pre- and postganglionic cardiac innervation indicated that (*a*) the right sympathetic cardiac nerve exerts the primary chronotropic influence; (*b*) the cells of origin of this nerve are restricted to the sympathetic ganglia associated with the last three cervical segments (ganglia 12–14), with the majority in ganglion 14; (*c*) the preganglionic projections upon these cells arise from the last cervical and upper three thoracic segments; and (*d*) the preganglionic cells of origin are located in the column of Terni, a well-defined cell group just dorsal to the central canal (Figure 19.6) (Macdonald & Cohen, 1970). Since then, the fiber spectrum of the right sympathetic cardiac nerve has been quantitatively described with electron microscopy; the fiber group associated with cardioacceleration has been determined; and criteria have been established for electrophysiologically identifying postganglionic cardiac neurons (Cabot & Cohen, 1977).

Concomitant with these studies of the sympathetic innervation, the

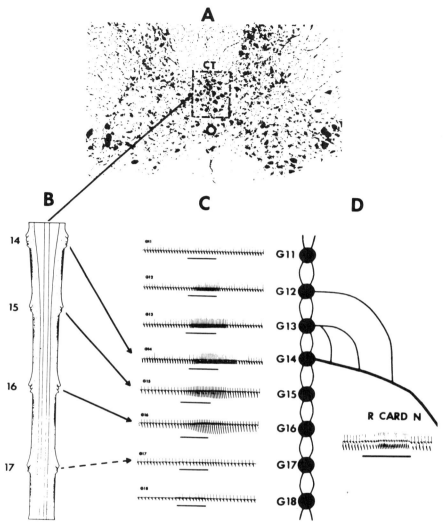

Figure 19.6. Schematic illustration of the sympathetic component of the final common path. Panel A, adapted from Macdonald and Cohen (1970), shows a photomicrograph of a transverse section through segment 15 of the pigeon spinal cord. The outlined cell group dorsal to the central canal is the sympathetic preganglionic column, the column of Terni (CT). Panel B illustrates schematically a horizontal view of the pigeon spinal cord from segments 14–17. Preganglionic fibers from the column of Terni that influence heart rate are shown emerging from segments 14–16 and sometimes 17. Panel C shows the heart rate change following stimulation of sympathetic ganglia G11–G18. (From Macdonald and Cohen, 1970.) Horizontal bars indicate stimulation periods. Panel D illustrates sympathetic ganglia 11–18 with the branches of the right cardioaccelerator nerve (R CARD N) emerging from ganglia 12–14. (From Macdonald and Cohen, 1970.) Also shown is the effect of stimulating the right cardioaccelerator nerve. (From Cohen, 1974a.)

291

parasympathetic cardiac innervation was also investigated. In this regard, the cells of origin of the vagal cardiac fibers were anatomically localized to a limited region of the dorsal motor nucleus (Cohen, Schnall, Macdonald, & Pitts, 1970), and we have since quantitatively described the fiber spectrum of the vagus and various of its branches, determined the vagal compound action potential components associated with bradycardia, and established criteria for electrophysiologically identifying vagal cardiac cells (Schwaber & Cohen, 1978a,b). Thus, the final common path of our model system is well described, and criteria for identifying these motoneurons in cellular neurophysiological conditioning studies have been established.

Given this description of the final common path, the next task is to delineate the central pathways converging upon these neurons and to evaluate their involvement in mediating expression of the conditioned response. The initial approach involved a systematic survey with electrical stimulation to describe those structures of the pigeon nervous system that influence heart rate and/or arterial blood pressure. With small electrodes and low stimulating currents, we sought to define functionally the pathways that could potentially mediate the conditioned response; we could then evaluate the actual involvement of each pathway by assessing the effect on conditioning performance of its interruption. This general approach was, in fact, successful (Macdonald & Cohen, 1973), and the resulting map of "cardioactive" regions was found to be remarkably similar to those described for mammalian brains (Cohen & Macdonald, 1974).

One such functionally defined pathway that produces striking pressor–accelerator responses can be followed from the hypothalamus through the ventromedial brainstem at mesencephalic and rostral pontine levels; it then shifts to a ventrolateral position that is maintained through the medulla (Figure 19.7). The dorsal motor nucleus is apparently accessed by fibers from this pathway that course dorsomedially in the plexus of Horsley, and the spinal continuation of this pressor–accelerator system appears to occupy a position in the lateral funiculus. Another important feature of this ventral brainstem system is that its most rostral component, the medial hypothalamus, receives a direct projection over a well-defined pathway from the most "cardioactive" structure of the telencephalon, the avian amygdalar homologue (Figure 19.7).

This particular system has been selected for mention because lesion studies indicate it mediates expression of the conditioned response. In this regard, we have shown that lesions of the avian amygdala or its hypothalamic projection prevent conditioned response development (Cohen, 1975). Equivalent conditioning deficits follow destruction of the hypothalamic terminal field of the amygdalar projection (Cohen & Macdonald, 1976) and the ventral area of Tsai (Cohen, 1974a; Durkovic & Cohen, 1969a). Ongoing lesion studies of this brainstem pathway are confirming its in-

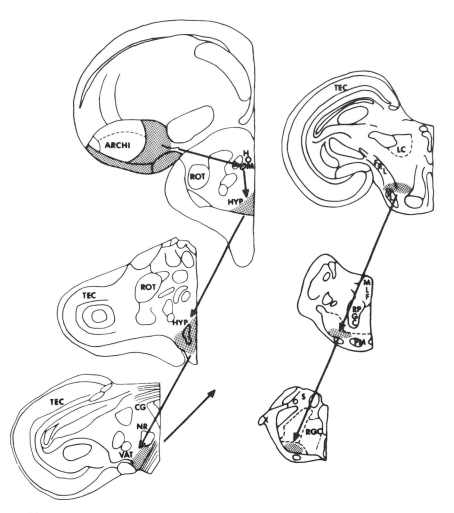

Figure 19.7. Schematic illustration of the descending system hypothesized to mediate expression of the conditioned response. This is shown on schematic sections by the stippled areas joined across successively caudal sections by arrows. Abbreviations: *ARCHI* = archistriatum; *CG* = central gray; *FRL* = lateral reticular formation; *HOM* = tractus occipitomesencephalicus, pars hypothalami; *HYP* = hypothalamus; *LC* = locus ceruleus; *MLF* = medial longitudinal fasciculus; *NR* = red nucleus; *PM* = medial pontine nucleus; *RGC* = nucleus reticularis gigantocellularis; *ROT* = nucleus rotundus; *RPGC* = nucleus reticularis pontis caudalis, pars gigantocellularis; *S* = solitary nucleus; *TEC* = optic tectum; *VAT* = ventral area of Tsai; *X* = vagal rootlets. (From Cohen, 1974a.)

volvement in the expression of the conditioned response, and an important aspect of this demonstration is that large brainstem lesions not involving the pathway have virtually no effect on conditioned response development (e.g., Durkovic & Cohen, 1969b). Analogously, destruction of other "cardioactive" regions, such as the septum (Cohen & Goff, 1978b), does not affect the conditioned heart rate change.

Thus, evidence is accumulating that the descending pathway for the conditioned response involves an amygdalar projection to the medial hypothalamus, which then influences the final common path neurons via a ventral brainstem pathway (Figure 19.7). However, beyond the amygdalo-hypothalamic projection, this pathway is largely defined functionally, and one of our principal ongoing efforts is to describe its constituent cell groups and their connections.

COUPLING OF THE CONDITIONED STIMULUS AND CONDITIONED RESPONSE PATHWAYS

Given that the visual conditioned stimulus information is transmitted over pathways ascending to the telencephalon and that the descending pathways for conditioned response expression include the avian amygdala, one can ask how the telencephalic visual cell groups influence the amygdala. Resolving this would then describe the coupling between the conditioned stimulus and the conditioned response pathways, thereby delineating a throughput pathway from the eye to the heart. A reasonable alternative is that the visual information reaches the amygdala by an intratelencephalic route, and we are realizing significant progress in describing the projections of the relevant visual structures of the telencephalon (Ritchie & Cohen, 1977). Similarly, we are beginning to identify telencephalic cell groups that project upon the amygdala and are exploring their connectional relationship to the intratelencephalic visual pathways.

SUMMARY COMMENTS ON IDENTIFICATION OF THE RELEVANT PATHWAYS

The phase of our model system to satisfy the pathway identification criterion requires considerable further effort. However, very encouraging progress has been realized, and we are rather close to describing the major throughput pathway from the retina to the final common path. (Space limitations preclude review of our progress in mapping the unconditioned stimulus and unconditioned response pathways; see Cohen, 1974a; in press.) A particularly promising feature of our findings has been the specificity of the identified pathways, and as detailed anatomical studies progress we anticipate the evolution of an increasingly precise description.

Cellular Neurophysiological Studies of Conditioning

Perhaps most exciting at this time is that sufficient information is now available on certain segments of the pathway to permit cellular neurophysiological studies during conditioning. Specifically, our descriptions of the relevant visual pathways and the final common path have allowed us to undertake cellular neurophysiological analyses of these structures during conditioned response development while efforts to map more central components of the system are continuing.

FINAL COMMON PATH

Although it is unlikely that training-induced change occurs at synapses on the motoneurons of the system, characterizing their behavior during conditioning establishes the response in a neurophysiological time domain; this is fundamental to the interpretation of the activity of more central neurons of the pathway. Moreover, describing the discharge characteristics of these neurons during conditioning could clarify a number of issues such as the relative sympathetic and vagal contributions to the response, the overlap of the conditioned and unconditioned response pathways at the level of the motoneurons, and the extent to which the pathways mediating habituation, sensitization, and conditioning converge at the final common path.

Using the criteria for identifying sympathetic cardiac neurons (Cabot & Cohen, 1977; in preparation), a number of these cells have been studied during conditioned responding in pretrained animals, and this has generated some important findings regarding the temporal properties of the system (Cabot & Cohen, in preparation). These sympathetic postganglionic neurons respond to the conditioned stimulus with a latency of $\cong 100$ msec. This response consists of a short train of spikes followed by a short depression of discharge and a subsequent return to maintained activity levels, or slightly higher (Figure 19.8). Most importantly, the short latency burst has a duration of only $\cong 300-400$ msec, and we find that simulating this discharge in electrical stimulation of the cardiac nerve will evoke a cardio-acceleratory response of 4–6 sec. Thus, it appears that the central processing time for the 6 sec behavioral response can be completed in $\cong 400$ msec of stimulus onset, and the maintenance of the conditioned response for many seconds is largely a function of the input-output relation between the cardiac nerve and the heart. We are pursuing these findings further with studies of sympathetic cardiac neurons during conditioned response development and, although preliminary, early results suggest that the short latency, short duration burst is common to orienting, sensitization, and conditioned responses. The primary effect of conditioning may then be to increase the probability of occurrence and magnitude of this burst in any given unit.

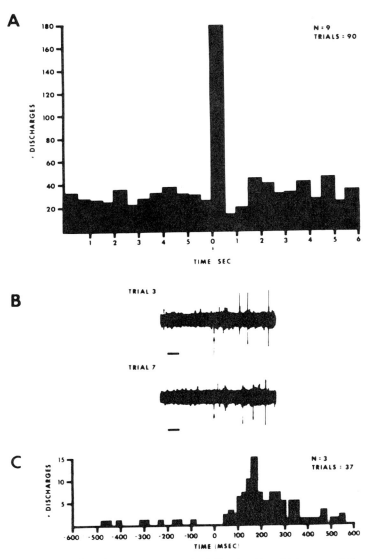

Figure 19.8. Panel A shows a summary peristimulus time histogram of the responses of nine sympathetic (cardiac) postganglionic neurons to 10 conditioned stimulus presentations. The animals from which these units were recorded had been pretrained to asymptotic conditioning performance. The arrow indicates onset of the 6 sec conditioned stimulus, and the bin width is 500 msec. Panel B shows representative responses of one of these units on two of the ten conditioned stimulus presentations. The arrows indicate stimulus onset, and the calibration bar represents 50 msec. Panel C shows a peristimulus time histogram of the discharges of three of these units for the 600 msec immediately before and after conditioned stimulus onset. (From Cabot and Cohen, unpublished observations.)

Recently, analogous studies have been initiated on the vagal cardioinhibitory neurons. In some respects, the responses of these neurons are more complex than those of the cardiac sympathetic cells, and they require considerable further study. However, preliminary results (Gold & Cohen, in preparation) do allow the conclusion that the conditioned stimulus elicits a short latency, variable duration decrease in discharge, such that early in the stimulus period the vagal and sympathetic components of the final common path are acting synergistically to produce the conditioned cardioacceleratory response (see Cohen & Pitts, 1968).

CONDITIONED STIMULUS PATHWAYS

The conditioned stimulus pathways are of particular interest, because, given their identification, one may investigate successively more central cell groups searching for the first neurons whose discharge patterns are modified by training. Along the motor pathways it is difficult to determine if changes represent local modification or merely reflect changes in more rostral structures. Thus, investigation of the conditioned stimulus pathways is perhaps the most promising with respect to identifying modified synaptic fields. If such changes do not occur along the relevant visual pathways, then one can conclude that they behave as input lines invariant with training and the focus would shift to structures upon which the telencephalic targets of these pathways project.

The first step in analyzing the conditioned stimulus pathways is to determine if the retinal output is modified by training. Although unlikely, this could occur as a consequence of (*a*) modified centrifugal control; (*b*) ancillary conditioned responses affecting the retina; (*c*) general systemic changes affecting the retina; and (*d*) actual changes of the retinal circuitry, perhaps the least likely alternative. Various indirect lines of evidence suggest no training-induced change in the retinal output. For example, interruption of the centrifugal innervation has no effect on conditioned response development, and no training-induced changes are evident in the electroretinogram (Wild, Cohen, & Migani-Wall, in preparation). However, the most definitive approach is to record the activity of single ganglion cells or their axons during conditioned response development. Based upon considerable background data from single optic tract axon recordings (Duff & Cohen, 1975a,b), a study of such single axons during conditioned response development is now near completion. The ongoing analyses suggest confirmation of the indirect evidence indicating no training-induced modification of either maintained or stimulus-evoked activity (Figure 19.9) (Wild, Cohen, & Migani-Wall, in preparation). This has now encouraged us to initiate similar neurophysiological conditioning studies of more central visual structures, studies that are now in progress.

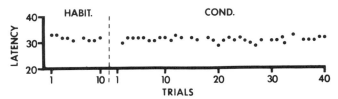

Figure 19.9. Illustration of the discharge of a single optic tract fiber over 10 habituation (HABIT.) and 40 conditioning (COND.) trials. This unit discharged with only a single spike at stimulus onset and had no off-response or maintained activity. Each solid circle indicates an on-discharge of a single spike, and the absence of a solid circle for any given trial indicates failure to respond on that trial. The latency of each discharge is shown on the ordinate. (From Wild, Cohen, and Migani-Wall, in preparation.)

Concluding Comments

To review the status of our model system, an effective behavioral preparation is well developed and has been adapted for cellular neurophysiological studies of conditioning. Thus, one of the major criteria for model system development has been satisfied. Regarding the criterion of identifying the relevant anatomical pathways, this has involved a substantial long-term effort that has realized considerable progress. The visual pathways transmitting the conditioned stimulus information have been largely identified. The descending pathways mediating conditioned response expression have been well described at the level of the final common path, and a functional description of the more central pathways has been obtained. Moreover, we are making significant progress in establishing the pathways by which the conditioned stimulus and conditioned response pathways are coupled. Although not reviewed here, advances have also been achieved in describing the unconditioned stimulus and unconditioned response pathways. Thus, although considerable effort is yet required in this mapping, the problem appears to be yielding.

Testimony to the success of the mapping program is that sufficient progress has been realized to permit initial cellular neurophysiological studies of conditioning, and these have been implemented for the final common path and for the periphery of the conditioned stimulus pathway. Although still at their inception, these single unit studies have already generated important information on the organization of the throughput pathway. The apparent lack of training-induced change in both maintained retinal activity and the retinal response to the conditioned stimulus gives us confidence in the stability of the system, excludes participation of centrifugal retinal inputs, and, most importantly, establishes a firm foundation for investigating more central visual structures involved in transmitting the conditioned stimulus information. Such investigations will be pivotal to identifying the most peripheral sites of training-induced modification.

The early results of cellular studies of the final common path have particularly important implications. Since the discharge of the postganglionic sympathetic neurons in response to the conditioned stimulus is only 300–400 msec in duration, the analysis of more central structures may be substantially simplified. Further, in describing the retinal response to the conditioned stimulus we find that the population response, as reconstructed from single unit data, has a duration of $\cong 60$ msec (Figure 19.10). This suggests that at conditioned stimulus onset a rather synchronous wave of activity is elicited, which in its transit along the throughput pathway undergoes a temporal dispersion of no more than a factor of five from the retinal gangion cells to the cardiac motoneurons (cf. Figures 19.8 and 19.10). In addition, the latency of the retinal ganglion cell response in our experimental situation is $\cong 30$ msec (Figures 19.9 and 19.10), and the sympathetic postganglionic latency is $\cong 100$ msec (Figure 19.8). Conse-

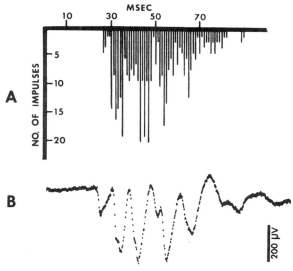

Figure 19.10. Panel A shows a poststimulus time histogram of 10 summed on-responses of 15 optic tract fibers randomly selected from a larger sample. Comparisons with other randomly selected samples indicate that this histogram is representative of the population response of the retinal ganglion cells in this stimulus situation. Panel B illustrates the optic tract field potential (16-sweep average) at stimulus onset. The time base is as in panel A; negativity is downward; and the stimulus intensity for both panels A and B is 600 ft-lamberts. Comparison of the histogram modes in panel A with the peaks of the negative field potential waves in panel B further suggests that the histogram is representative of the ganglion cell population response. (From Duff and Cohen, 1975a.)

quently, the central processing time for the shortest latency response component is $\cong$ 70 msec, remarkably shorter than one might have anticipated from a behavioral response persisting for many seconds.

The broader implication of these findings is that our system may be considerably more "analyzable" than our initial expectations. Although in no sense a "simple" system, our model is conceivably a "relatively simple" system, and there are promising possibilities for simplifying it yet further by studying subsystems of the identified circuitry. Yet, even in the context of the larger pathway, information is now emerging on such fundamental questions as the temporal properties of the informational flow during conditioning, the relationships among orienting, sensitization, and conditioned responses, and the sites at which training-induced modification might occur. In brief, the system has developed to a point where we are asking basic questions regarding the organization of the involved circuitry, and we may be on the threshold of identifying sites of storage—the prerequisite to studying the cellular mechanisms of long-term associative learning in a vertebrate brain.

ACKNOWLEDGMENT

I would like to acknowledge my gratitude to Nancy Richardson for her able secretarial assistance in the preparation of this chapter.

REFERENCES

Cabot, J. B., & Cohen, D. H. Avian sympathetic cardiac fibers and their cells of origin: Anatomical and electrophysiological characteristics. *Brain Research*, 1977, *131*, 73–87.

Cabot, J. B., & Cohen, D. H. Criteria for identifying and some response properties of the cells of origin of the cardiac sympathetic nerve of the pigeon. In preparation.

Cohen, D. H. Development of a vertebrate experimental model for cellular neurophysiologic studies of learning. *Conditional Reflex*, 1969, *4*, 61–80.

Cohen, D. H. The neural pathways and informational flow mediating a conditioned autonomic response. In L. V. DiCara (Ed.), *Limbic and autonomic nervous systems research*. New York: Plenum Press, 1974a.

Cohen, D. H. The effect of conditioned stimulus intensity on visually conditioned heart rate change in the pigeon: A sensitization mechanism. *Journal of Comparative and Physiological Psychology*, 1974b, *87*, 591–597.

Cohen, D. H. Analysis of the final common path for heart rate conditioning. In P. A. Obrist, A. H. Black, J. Brener, & L. V. DiCara (Eds.), *Cardiovascular psychophysiology*. Chicago: Aldine, 1974c.

Cohen, D. H. Involvement of the avian amygdalar homologue (archistriatum posterior and mediale) in defensively conditioned heart rate change. *Journal of Comparative Neurology*, 1975, *160*, 13–36.

Cohen, D. H. Current approaches to the mechanisms of memory. In J. McGaugh & R. F.

Thompson (Eds.), *Handbook of behavioral neurobiology: Learning and memory*. New York: Plenum Press, in press.

Cohen, D. H. Avian visual pathways involved in conditioned heart rate change to whole field illumination. In preparation.

Cohen, D. H., & Durkovic, R. G. Cardiac and respiratory conditioning, differentiation, and extinction in the pigeon. *Journal of the Experimental Analysis of Behavior*, 1966, *9*, 681–688.

Cohen, D. H., & Goff, D. M. Conditioned heart rate change in the pigeon: Analysis and prediction of acquisition patterns. *Physiological Psychology*, 1978a, *6*, 127–141.

Cohen, D. H., & Goff, D. M. Effect of avian basal forebrain lesions, including septum, on heart rate conditioning. *Brain Research Bulletin*, 1978b, *3*, 311–318.

Cohen, D. H., & Karten, H. J. The structural organization of avian brain: An overview. In I. J. Goodman and M. W. Schein (Eds.), *Birds: Brain and Behavior*. New York: Academic Press, 1974.

Cohen, D. H., & Macdonald, R. L. Some variables affecting orienting and conditioned heart-rate responses in the pigeon. *Journal of Comparative and Physiological Psychology*, 1971, *74*, 123–133.

Cohen, D. H., & Macdonald, R. L. A selective review of central neural pathways involved in cardiovascular control. In P. A. Obrist, A. H. Black, J. Brener, & L. V. DiCara (Eds.), *Cardiovascular psychophysiology*. Chicago: Aldine, 1974.

Cohen, D. H., & Macdonald, R. L. Involvement of the avian hypothalamus in defensively conditioned heart rate change. *Journal of Comparative Neurology*, 1976, *167*, 465–480.

Cohen, D. H., & Pitts, L. H. Vagal and sympathetic components of conditioned cardio-acceleration in the pigeon. *Brain Research*, 1968, *9*, 15–31.

Cohen, D. H., Schnall, A. M., Macdonald, R. L., & Pitts, L. H. Medullary cells of origin of vagal cardioinhibitory fibers in the pigeon. I. Anatomical studies of peripheral vagus nerve and the dorsal motor nucleus. *Journal of Comparative Neurology*, 1970, *140*, 299–320.

Duff, T. A., & Cohen, D. H. Retinal afferents to the pigeon optic tectum: Discharge characteristics in response to whole field illumination. *Brain Research*, 1975a, *92*, 1–19.

Duff, T. A., & Cohen, D. H. Optic chiasm fibers of the pigeon: discharge characteristics in response to whole field illumination. *Brain Research*, 1975b, *92*, 145–148.

Durkovic, R. G., & Cohen, D. H. Effects of rostral midbrain lesion on conditioning of heart and respiratory rate responses in the pigeon. *Journal of Comparative and Physiological Psychology*, 1969a, *68*, 184–192.

Durkovic, R. G., & Cohen, D. H. Effects of caudal midbrain lesions on conditioning of heart and respiratory rate responses in the pigeon. *Journal of Comparative and Physiological Psychology*, 1969b, *69*, 329–338.

Gold, M. R., & Cohen, D. H. Heart rate conditioning in pigeons immobilized with α-bungarotoxin. In preparation.

Groves, P. M., and Thompson, R. F. Habituation: A dual-process theory. *Psychological Review*, 1970, *77*, 419–450.

Kandel, E. R. *Cellular basis of behavior*. San Francisco: Freeman, 1976.

Kandel, E. R., & Spencer, W. A. Cellular neurophysiological approaches in the study of learning. *Physiological Reviews*, 1968, *48*, 65–134.

Macdonald, R. L., & Cohen, D. H. Cells of origin of sympathetic pre- and postganglionic cardioacceleratory fibers in the pigeon. *Journal of Comparative Neurology*, 1970, *140*, 343–358.

Macdonald, R. L., and Cohen, D. H. Heart rate and blood pressure responses to electrical stimulation of the central nervous system of the pigeon (*Columba livia*). *Journal of Comparative Neurology*, 1973, *150*, 109–136.

Migani-Wall, S. M., Cabot, J. B., & Cohen, D. H. Development of a neuromuscular blocking procedure for heart rate conditioning in immobilized pigeons. *Psychophysiology,* 1977, *14,* 499–506.

Ritchie, T. C., & Cohen, D. H. The avian tectofugal visual pathway: Projections of its telencephalic target, the ectostriatal complex. *Neuroscience Abstracts,* 1977, *3,* 94.

Schwaber, J. S., & Cohen, D. H. Electrophysiological and electron microscopic analysis of the vagus nerve of the pigeon, with particular reference to the cardiac innervation. *Brain Research,* 1978a, *147,* 65–78.

Schwaber, J. S., & Cohen, D. H. Field potential and single unit analysis of the dorsal motor nucleus of the pigeon: Criteria for identifying cells of origin of vagal cardiac fibers. *Brain Research,* 1978b, *147,* 79–90.

Wild, J. M., Cohen, D. H., & Migani-Wall, S. M. Demonstration that retinal output does not change during the development of a visually conditioned response. In preparation.

MICHAEL GABRIEL
KENT FOSTER
EDWARD ORONA

Unit Activity in Cingulate Cortex and Anteroventral Thalamus during Acquisition and Overtraining of Discriminative Avoidance Behavior in Rabbits

20

An important variety of goal-directed behavior is the avoidance of stimuli that are aversive. To accomplish avoidance an organism must learn to identify the signals that predict the occurrence of the aversive stimulus. The identification of predictive signals enables the organism to anticipate the aversive stimulus and perform the avoidance response.

The research in our laboratory is directed toward an understanding of neural mechanisms that underlie the learning and performance of avoidance behavior. To provide information on these mechanisms we have studied cellular (multiple-unit) activity of the central nervous system (CNS) during learning and performance of discriminative avoidance behavior in rabbits (Gabriel, Miller, & Saltwick, 1976, 1977a; Gabriel, Saltwick, & Miller, 1975, 1977b; Gabriel, Wheeler, & Thompson, 1973a,b). In our experiments the rabbits learn to avoid foot-shock by performing a locomotory response to an auditory signal, the CS^+, initiated 5 sec prior to the foot-shock. Pairing of the CS^+ with foot-shock establishes the CS^+ as a danger signal. A second signal, the CS^-, is randomly interspersed with the CS^+, but it is never followed by foot-shock. Thus, our procedure establishes the CS^- as a safety signal. The basic question that we have posed using these techniques is: Where in the central nervous system does one find unit activity that is discriminative in character? That is, what are the regions or systems of the CNS that acquire a unique neuronal response to the CS^+, relative to the CS^-, as the rabbit acquires discriminative behavior?

Our past studies have shown that acquisition of discriminative neuronal activity occurs during acquisition of discriminative behavior, within the anterior cingulate (AC) region of the rabbit cerebral cortex (Gabriel, et al., 1977a, 1977b). In the same studies, we observed acquisition of discriminative neuronal activity in the anteroventral nucleus of the thalamus (AV), a region having reciprocal interconnections with the AC (Berger, Milner, Swanson, Lynch, & Thompson, 1979; Domesick, 1969, 1972; Rose & Woolsey, 1948). Of special interest was the finding that the discriminative effect in AC occurred prior to the effect in AV. Specifically, the effect in AC was first observed at an early stage of training, when behavioral discrimination was incipient. The effect in AV occurred later in training when the behavioral discrimination was well learned. Thus, AC and AV participated in mediation of behavioral acquisition at different stages of the process. This finding provided evidence that our techniques were useful in analyzing distinctive neuroanatomical contributions to behavioral learning.

In our past studies, we presented neuronal data obtained in the beginning, middle, and end of behavioral acquisition. Data representing the middle of acquisition were obtained from the session of conditioning that occurred midway between the initial session and the session in which the rabbits first met the behavioral criterion. In this report we define the middle session as that session in which the rabbits first performed a significant behavioral discrimination, in order to observe more precisely the relation between incipient behavioral discrimination and neuronal activity in AC and AV.

We noted in our earlier findings that there was a small drop in the magnitude of neuronal response in AC in the criterial session relative to the middle session of acquisition. This suggested that the cortical response would show continuing decline and perhaps total disappearance with continuing training (overtraining) in the task. In the present study we sought to test this idea by examining whether or not neuronal responsiveness in AC and in AV would persist throughout overtraining in the task.

A third issue confronted by the present study concerned the neuroanatomical generality of our effects. Would electrodes located in regions of cingulate cortex other than AC yield discriminative effects with relations to behavioral acquisition similar to those observed in AC? To answer this question electrodes were implanted in the posterior as well as the anterior regions of cingulate cortex.

Method

SUBJECTS

The data of the present report were obtained from 50 rabbits. They weighed 1.5–2.0 kg at the time of their delivery to the laboratory and were maintained on *ad libitum* food and water throughout the experiments.

ELECTRODES AND SURGERY

Surgical implantation of electrodes was carried out after a minimum period of 48 hr of adaptation to laboratory cages. Stainless steel electrodes made from No. 00 insect pins were stereotaxically implanted while the rabbits were anesthetized. The electrode tips ranged in length from 20–60 μm. Multiple-unit activity was monitored continuously with an audio system and an oscilloscope as an aid to electrode placement during implantation. The target areas for implantation were AC, posterior cingulate cortex (PC), and AV. The AC is located in the medial wall of the cerebral hemisphere 3.5 mm posterior to bregma. It is in the anterior one-third of the region labeled area cingularis by J. E. Rose and Woolsey (1948). This corresponds to area retrosplenialis granularis (M. Rose, 1929), shown to have reciprocal fiber connections with the anteroventral nucleus of the thalamus. It is also the same as the area referred to as dorsal nonspecific cortex and shown photographically in a previous report (Gabriel et al., 1973a). The PC is the area between 8 and 14 mm posterior to bregma in the cortex of the medial wall. The subjects were given a minimum of 7 days to recover from surgery prior to behavioral conditioning.

BEHAVIORAL PROCEDURES

Differential conditioning was carried out in a wheel apparatus (Brogden & Culler, 1936), located in a shielding chamber, in a room adjacent to that containing the apparatus for controlling the experiment. The stimuli used as CSs were pure tones (8 kHz or 1 kHz), 85 db re 20 μN/m², with a rise time of 3 msec. The tones were played through a speaker located directly above the wheel. The unconditioned stimulus (UCS) was a constant-current shock (1.5 mA) delivered through the grid floor of the wheel, and a response was defined as any movement exceeding .96 cm on the circumference of the wheel. The CS⁺ was followed after 5 sec by UCS-onset and both stimuli were response-terminated. A response in the presence of the CS⁺ terminated it and prevented the UCS. The maximum duration of the simultaneous CS⁺ and UCS, given absence of response, was 1 sec. The CS⁻ was a tone of the frequency (8 kHz or 1 kHz) not chosen to be CS⁺. It was interspersed with the CS⁺ and it was response-terminated, but it was never followed by shock.

The intervals from the end of a trial to the onset of a new trial were 10, 15, 20, or 25 sec. These values occurred in a random sequence. Responses reset the interval. Sixty trials with each CS were given in each daily session. The order of the CSs was randomized with the restriction that each occurs equally often in each 60-trial block. Each rabbit was trained until its percentage of avoidance to the CS⁺ exceeded the percentage to the CS⁻ by at least 60 in any 60-trial block. This criterion had to be attained on two consecutive days. Following the attainment of criterion the rabbits were given either zero ($N = 20$), three ($N = 19$), or six ($N = 11$) additional days of conditioning, which shall be designated as overtraining.

Prior to conditioning, each rabbit received a session of pretraining, in which the tone signals to be CSs and the shock UCS were presented noncontingently to control for possible nonassociative effects of the stimuli. (See Gabriel and Saltwick [1977] for a complete description of the pretraining procedure).

RECORDING AND ANALYSIS OF ELECTROPHYSIOLOGICAL DATA

Throughout conditioning, neuronal activity was fed through an amplifier (bandwidth = 500–10,000 Hz) with a field-effect transistor input. The output of the amplifier was recorded on magnetic tape and simultaneously fed into Schmitt triggers, which produced a uniform discrete output pulse each time the input voltage exceeded a preset level. The trigger levels were set at a positive voltage independently for each record. A specific setting was adopted such that only five or six of the largest neuronal spikes on each record contributed to the analysis. Figure 20.1 shows representative neuronal records and trigger levels.

Output pulses from the Schmitt triggers were fed into a PDP8M digital computer that processed the neuronal records and controlled the behavioral experiment on-line. Computer analysis of the Schmitt trigger output yielded peristimulus histograms reflecting the frequency, summed over trials, of neuronal firing in consecutive 10 msec bins prior to and following the onsets of the tones. Separate histograms were constructed for the CS^+ and the CS^-. Thus, each histogram was based on a total of 60 trials. Standardized scores (z scores) normalized with respect to a pre-CS (baseline) period (300 msec) were computed for each of the first 20 bins after CS-onset. Also, the average frequency of neuronal firing was computed for each of the first six periods of 100 msec following CS-onset. A standard score (T-score) was obtained by normalizing each of these averages with respect to the pre-CS period. Further details of the procedures used to analyze the neuronal data are provided in Gabriel et al. (1976).

HISTOLOGY

Following completion of behavioral testing, each rabbit was given an overdose of barbiturate and perfused with saline followed by 10% formalin. The brains were frozen and sectioned at 40 μm, and the sections containing the electrode tracks were photographed while still wet (Fox & Eichman, 1959). There were 36 placements in AC, 12 in PC, and 13 in AV. Thirteen of the placements in AC and seven of those in AV had

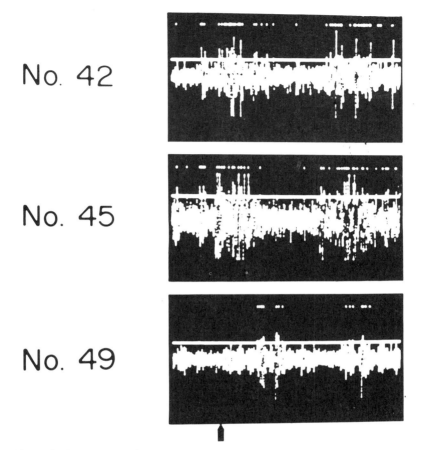

Figure 20.1. *Representative neuronal records from the anterior cingulate cortical area (AC) obtained during differential conditioning from three rabbits. The vertical calibration mark (lower right) represents 20 μV, and the horizontal (sweep) calibration represents 10 msec. The horizontal line above each record shows the Schmitt trigger setting, and the dots at the top of each record are Schmitt trigger output pulses reflecting level crossings. Onset of the CS+ is indicated by the marker at the bottom left.*

served to contribute the data of our previous studies (Gabriel *et al.*, 1977a, 1977b).

Results

BEHAVIOR

The average frequency and latency of conditioned avoidance response in pretraining (PT), three sessions of conditioning, and three sessions of overtraining are shown in Figure 20.2. The sessions of conditioning were

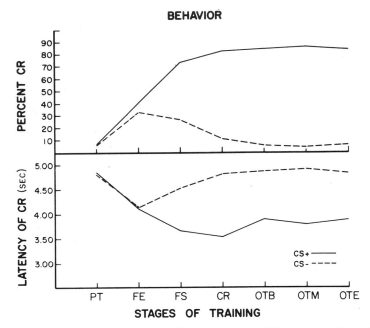

Figure 20.2. Average percentage conditioned response (CR, upper half) and latency of CR (lower half) to CS+ and CS− as a function of stage acquisition and overtraining. See text for definitions of the stages of acquisition and overtraining.

first exposure to conditioning (FE), session of first significant behavioral discrimination (FS), and session of first criterial performance (CR).[1] The sessions of overtraining presented here are the first (OTB), middle (OTM), and final (OTE) sessions.

CINGULATE CORTEX

The average magnitude of the neuronal response of cingulate cortex to CS+ and CS− is shown in Figures 20.3 and 20.4. Consistent with past findings (Gabriel *et al.*, 1977a,b), the profiles of the short-latency neuronal responses shown in Figure 20.3 indicate the triphasic nature of the response. That is, there was an initial excitation at 15–25 msec following tone-onset,

[1] The mean number of sessions prior to behavioral criterion was 3.05. It was necessary, for instances in which rabbits met criterion in the first or second sessions of conditioning, to divide the first session of conditioning into quarters or halves to obtain behavioral and neuronal values for FE and FS. This happened for 11 of the 44 rabbits. The session of first significant behavioral discrimination was defined as the session in which the percentage of conditioned responses to the CS− by at least 25 in at least one of the two 60-trial blocks of the session. This value is the minimum needed to obtain a significant χ^2 ($p = .05$) for a difference between proportions (Walker & Lev, 1953, p. 101).

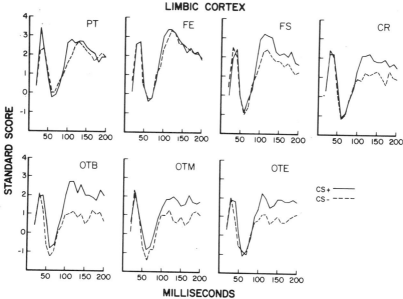

Figure 20.3. Each panel shows the average neuronal response of cingulate cortex to CS+ and CS−, in 19 consecutive 10 msec bins occurring just after tone-onset. Each panel shows data of a different stage of the behavioral task (see the text for definitions of the stages). The ordinate values are in standard score (z-score) units reflecting response magnitude relative to the pre-CS baseline (see the text for further details). Because neuronal responses did not occur in the first 10 msec interval after tone-onset, the initial bin was omitted from the analysis and from the figure. Thus, the leftmost bin in each panel of the figure represents the interval from 10–20 msec after tone-onset. However, the millisecond values on the abscissa represent elapsed time relative to tone-onset.

decreased activity (at or below baseline) from 35–75 msec, and increased activity beginning at 75 msec and continuing to 200 msec. The data shown in Figure 20.4 indicate in addition that there was a gradual decline in response magnitude from 200–400 msec. From 400–600 msec, the response remained stable, approximating the level reached at 400 msec.

The data portrayed in Figures 20.3 and 20.4 were submitted to factorial repeated-measures analysis of variance using stage of training, CS, 10 msec bin (or 100 msec period), and electrode location (posterior versus anterior cingulate cortex) as orthogonal factors. Two-tailed t-tests ($\alpha = .05$) were used to compare means associated with significant treatment effects from the overall analysis.

The analysis revealed that discriminative neuronal activity occurred in the form of a greater neuronal response to CS+ relative to CS−, within the third (75–200 msec) component of the triphasic response, during the day of first significant behavioral discrimination (FS) and in all following sessions of conditioning and overtraining. A discriminative effect was seen

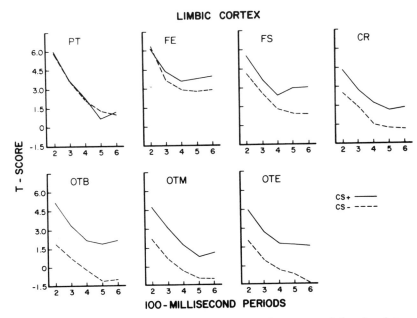

Figure 20.4. Each panel shows the average neuronal response of the cingulate cortex to CS+ and CS− over six consecutive intervals of 100 msec following tone-onset. Each panel shows the data of a different stage of the behavioral task (see the text for definition of the stages). The ordinate values are in T-score units reflecting response magnitude relative to the pre-CS baseline. The numeral 2 at the leftmost position of each abscissa represents the interval 100–200 msec. The numeral 3 represents the interval from 200–300 msec, and so on.

at longer latencies (Figure 20.4) in the same task stages. However, unlike the short-latency data, the very long latency data (300–600 msec) showed a significant discriminative effect in the first exposure to conditioning. It is interesting to note that the discriminative effect in cingulate cortex was produced by a drop relative to PT and FE in response to the CS−, rather than by an increase in response to the CS+. No discriminative effects occurred during pretraining nor were any significant effects observed in relation to stage or amount of overtraining. Finally, there were no significant effects associated with the locus factor (AC versus PC), thereby suggesting that the findings reported here hold within both loci. This suggestion was verified in separate analyses carried out on records obtained from only PC.

ANTEROVENTRAL NUCLEUS OF THE THALAMUS

A configuration of findings similar to those obtained for cingulate cortex was also obtained for AV. Thus, the profile of the neuronal response was a triphasic one, and discriminative neuronal activity occurred within the

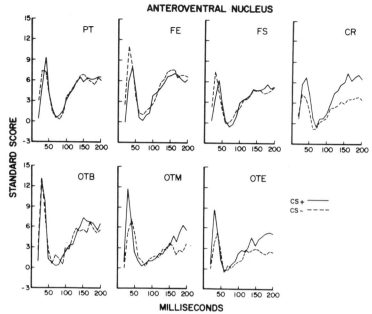

Figure 20.5. *Average neuronal response of anteroventral thalamus (AV) to CS+ and CS− for the initial 19 consecutive 10 msec bins following tone-onset. All labels are as defined in the legend of Figures 20.3 and in the text.*

third component of the triphasic response, as well as at all greater latencies (see Figures 20.5 and 20.6). In addition, AV showed no change in overall response magnitude, or in the magnitude of the discriminative effect, as a function of the stage 2 of overtraining.

The most striking difference between AV and cingulate cortex was the difference in the stage of conditioning at which significant discriminative neuronal activity first occurred. Thus, AV first manifested a significant discriminative effect during the criterial stage (CR). There was no discriminative effect in AV in the two initialmost stages (FE and FS) in which cingulate cortex showed discrimination.

The AV also differed from cingulate cortex in relation to the relative effect of CS⁺ and CS⁻. Thus whereas the discrimination in cortex was produced entirely by a drop relative to earlier stages of conditioning in response to the CS⁻, the discrimination in thalamus resulted primarily from significant increase in response to the CS⁺. This effect may be noted by comparing the results that occurred in the CR session with those of the session of first significant (FS) behavioral discrimination (Figure 20.6).

Records from cingulate cortex and from AV showed discriminative neuronal activity in the third component of the triphasic response. However, it is interesting to note that there was a significant discriminative

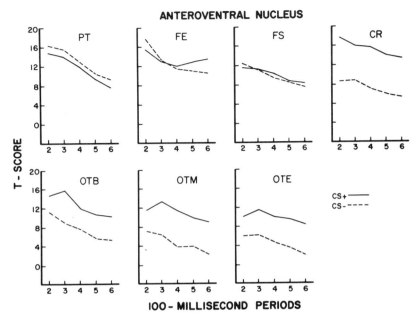

Figure 20.6. Average neuronal response of anteroventral thalamus (AV) to CS+ and CS− for six consecutive periods of 100 msec following tone-onset. All labels are as defined in the caption of Figure 20.4 and in the text.

effect in AV in the short-latency components (25 msec) during three of the four sessions (CR, OTM, and OTE) in which discrimination at longer latencies also occurred. The very short-latency effect seen in AV did not occur within cingulate cortex at any stage of the task.

Discussion

The present study indicates that neuronal activity in cingulate cortex and AV thalamus of the rabbit becomes discriminative in character during training in a discriminative avoidance task. Moreover, the discriminative effect in cingulate cortex occurs at a relatively early stage of training—a stage in which the rabbits' behavioral discrimination is incipient. The discriminative effect in AV always occurs at a later stage of training—a stage in which the rabbits display a high level of behavioral discrimination. These results provide replication of our previous findings (Gabriel et al., 1977a,b).

In addition, the present results extend our past findings in two ways. First, the occurrence of discriminative activity in PC, during the stages of acquisition associated with discriminative activity in AC, indicates that the pattern of discriminative effects observed in AC may be one that is general for all of cingulate cortex. The second way is in terms of the demonstration that discriminative neuronal activity in cingulate cortex and in AV persists without decline throughout a considerable period of overtraining (six days) in the discrimination task. Thus, our expectation that the cortical effect would drop out with overtraining was not borne out. This result suggests that the discriminative effect in AV does not serve to replace the effect in cingulate cortex. Instead, the late appearing neuronal discrimination in AV seems to represent an effect that is additional to the cortical discrimination. Nevertheless, we must reserve the possibility that further overtraining would result in a decline of neuronal discrimination in one or in both of these structures.

The early appearing discriminative neuronal activity in cingulate cortex suggests that this structure is importantly involved in mediating the acquisition of behavioral avoidance. This suggestion receives support from studies demonstrating that damage induced in cingulate cortex impairs acquisition of behavioral avoidance (e.g., Eckersdorf, 1974; Lubar & Perachio, 1965; McCleary, 1966). It is important to note additional evidence that suggests that the impairment due to induced damage in cingulate cortex may be specific to acquisition of the behavior. Deficits in retention were not found when damage was induced during the retention interval (Eckersdorf, 1974).

It is interesting to consider the possible behavioral relevance of the late-appearing discriminative neuronal activity that occurs in AV. Since this effect does not occur until well after the first occurrence of significant behavioral discrimination, it cannot be held to participate in the initialmost stages of behavioral acquisition. Given its late appearance, it seems more likely that the effect may reflect some aspect of the well-learned behavior. However, it is not clear just what that aspect may be. Thus, it is possible that discriminative neuronal activity in AV contributes to the very high level of behavioral discrimination that characterizes the criterial stages of conditioning and overtraining. On the other hand, activity in AV may signify behavioral potentiality not manifested during training, such as the retention of the behavioral discrimination over long intervals without practice. In relation to this latter suggestion, it is interesting to note that damage induced in AV in the rat has been shown to impair the retention of discriminative avoidance behavior (Rich & Thompson, 1965). Moreover, studies of the effects of human brain pathology have indicated that damage in AV and in other limbic structures of the diencephalon (e.g., mammillary bodies, mediodorsal thalamic nucleus) is associated with phenomena of human amnesia (Adams, 1969, Angelergues, 1969; Sweet, Talland, & Ervin, 1959). Thus, work on humans and animals is compatible with the second, memorial interpretation of the function of AV.

ACKNOWLEDGMENTS

The present research was supported by grants from the University Research Institute of the University of Texas at Austin. Special thanks are due Linda P. Gabriel and William Stern, who assisted in data collection. In addition, we are greatly indebted to Steven E. Saltwick, who produced the computer programs used to carry out the present studies.

REFERENCES

Adams, R. D. The anatomy of memory mechanisms in the human brain. In G. A. Talland & N. C. Waugh (Eds.), *The pathology of memory*. New York: Academic Press, 1969.

Angelergues, R. Memory disorders in neurological disease. In P. J. Vinken & G. W. Bruyn (Eds.), *Handbook of clinical neurology: Disorders of higher nervous activity* (Vol. 3). New York: American Elsevier, 1969.

Berger, T. W., Milner, T. A., Swanson, G. W., Lynch, G. S., & Thompson, R. F. Reciprocal anatomical connections between cingulate-retrosplenial cortex and anteroventral thalamus in the rabbit studied with horseradish peroxidase. *Society for Neuroscience Abstract Bulletin*, 1979, No. 888, 270. (Abstract)

Brogden, W. J., & Culler, E. A. Device for motor conditioning of small animals. *Science*, 1936, *83*, 269.

Domesick, V. B. Projections from cingulate cortex in the rat. *Brain Research*, 1969, *12*, 269-320.

Domesick, V. B. Thalamic relationships of the medial cortex in the rat. *Brain, Behavior and Evolution*, 1972, *6*, 141-169.

Eckersdorf, B. The effects of lesions in the posterior part of the gyrus cinguli on the conditioned defensive response of active avoidance in cats. *Acta Physiologica Polonica*, 1974, *3*, 105-114.

Fox, C. A., & Eichman, J. A. A rapid method for locating intracerebral electrode tracks. *Stain Technology*, 1959, *34*, 39-42.

Gabriel, M., Miller, J. D., & Saltwick, S. E. Multiple-unit activity of the rabbit medial geniculate nucleus in conditioning, extinction, and reversal. *Physiological Psychology*, 1976, *4*, 124-134.

Gabriel, M., Miller, J. D., & Saltwick, S. E. Unit activity in cingulate cortex and anteroventral thalamus of the rabbit during differential conditioning and reversal. *Journal of Comparative and Physiological Psychology*, 1977a, *91*, 423-433.

Gabriel, M., Saltwick, S. E., & Miller, J. D. Unit activity of anterior cingulate cortex in differential conditioning and reversal. *Bulletin of the Psychonomic Society*, 1977b, *9*, 207-210.

Gabriel, M., Saltwick, S. E., & Miller, J. D. Conditioning and reversal of short latency multiple-unit responses in the rabbit medial geniculate nucleus. *Science*, 1975, *189*, 1108-1109.

Gabriel, M., & Saltwick, S. E. Effects of unpaired footshock on rabbit limbic and auditory neuronal responses to tone stimuli. *Physiology and Behavior*, 1977, *19*, 29-34.

Gabriel, M., Wheeler, W., & Thompson, R. F. Multiple-unit activity of the rabbit cerebral cortex in single-session avoidance conditioning. *Physiological Psychology*, 1973a, *1*, 45-55.

Gabriel, M., Wheeler, W., & Thompson, R. F. Multiple-unit activity of the rabbit cerebral cortex during stimulus generalization of avoidance behavior. *Physiological Psychology*, 1973b, *1*, 313-320.

Lubar, J. D., & Perachio, A. A. One-way and two-way learning and transfer of an active avoidance response in normal and cingulectomized cats. *Journal of Comparative and Physiological Psychology*, 1965, *60*, 46–52.

McCleary, R. A. Response functions of the limbic system. In E. Stellar & J. M. Sprague (Eds.), *Progress in physiological psychology* (Vol. 1). New York: Academic Press, 1966.

Rich, I., & Thompson, R. Role of the hippocampo-septal system, thalamus, and hypothalamus in avoidance conditioning. *Journal of Comparative and Physiological Psychology*, 1965, *59*, 66–72.

Rose, M. Zytoarch: tektonischer Atlas der Grosshirnrinde des Kaninchens. (Cytoarchitectural atlas of the cortex of the rabbit.) *Journal of Psychology and Neurology*, 1933, *43*, 353–440.

Rose, J. E., & Woolsey, C. N. Structure and relations of limbic cortex and thalamic nuclei in rabbit and cat. *Journal of Comparative Neurology*, 1948, *89*, 279–348.

Sweet, W. H., Talland, G. A., & Ervin, F. R. Loss of recent memory following section of the fornix. *Transactions of the American Neurological Association*, 1959, *84*, 76–82.

Walker, H. M., & Lev, J. *Statistical inference.* New York: Holt, 1953.

Perception and Information Processing

PART IV

KARL H. PRIBRAM

Image, Information, and Episodic Modes of Central Processing

21

Introduction

Over the past 25 years, research on the brain mechanisms involved in learning and remembering has been rewarding beyond expectation. It was only a little over 25 years ago that Lashley uttered his famous remark that, on the basis of his lifetime of research on brain function, it was clear that "learning just could not take place." Nor was Pavlov any more successful in delineating by direct intervention in brain mechanisms the processes he and his students had so painstakingly elaborated with behavioral techniques.

All this is now changed as can be seen from the contents of this volume. In my contribution, I wish to review evidence that has accumulated around two problems. One concerns brain mechanisms in image processing and the resulting distributed semantic store. The other deals with a distinction between information and episodic processing as two different modes of learning and remembering.

Image Processing and the Distributed Memory Store

Lashley's despair was produced by his repeated findings of equivalence of function of parts of brain systems. Not only was he unable to excise any specific memory, but he was also unable to account for the facts of sensory and motor equivalence:

> These three lines of evidence indicate that certain coordinated activities, known to be dependent upon definite cortical areas, can be carried out by any part (within undefined limits) of the whole area. Such a condition might arise from the presence of many duplicate reflex pathways through the areas and such an explanation will perhaps account for all of the reported cases of survival of functions after partial destruction of their special areas, but it is inadequate for the facts of sensory and motor equivalence. These facts establish the principle that once an associated reaction has been established (e.g., a positive reaction to a visual pattern), the same reaction will be elicited by the excitation of sensory cells which were never stimulated in that way during the course of training. Similarly, motor acts (e.g., opening a latch box), once acquired, may be executed immediately with motor organs which were not associated with the act during training [Lashley, 1960, p. 240. Used with permission of McGraw-Hill Book Company].

What sort of brain mechanism could be imagined that would account for the principle that "once an associated reaction has been established, the same reaction will be elicited by the excitation of sensory cells which were never stimulated in that way during the course of training"? And what mechanism could be devised to deal with the fact that "motor acts, once acquired, may be executed immediately with motor organs which were not associated with the act during training"? What sort of mechanism of association could be taking place during learning so that its residual would, as it were, act at a distance?

The difficulties of conceptualization may be summarized as follows: During acquisition, associative processes must be operative. However, these associative processes must result in a distributed store. On the basis of Lashley's analysis, input must become dismembered before it becomes remembered. Association and distribution are in some fundamental way inexorably linked.

During the mid-1960s it became apparent that image processing through holography could provide a model for a mechanism with such "distribution by association" properties. As in the case of every novel approach, there were, of course, earlier formulations including those of Lashley that attempted to explain these aspects of brain function in terms that today we would call holographic.

Historically the ideas can be traced to problems posed during neurogenesis when the activity of relatively remote circuits of the developing nervous system must become integrated to account for such simple behaviors as swimming. Among others, the principle of chemical "resonances" that "tune" these circuits has had a long and influential life (see, e.g., Loeb, 1907; Weiss, 1939). More specifically, however, Goldscheider (1906) and Horton (1925) proposed that the establishment of tuned resonances in the form of interference patterns in the adult brain could account for a variety of perceptual phenomena. More recently, Lashley (1942) spelled out a mechanism of neural interference patterns to explain stimulus equivalence and Beurle (1956) developed a mathematically rigorous formulation of the origin of such patterns of plane wave interferences in neural tissue. But it was not until the advent of holography with its powerful damage-resistant image storage and reconstructive capabilities that the promise of an interference pattern mechanism of brain function became fully appreciated. As the properties of physical holograms became known (see Collier, Burckhardt, & Lin, 1971; Goodman, 1968; Stroke, 1966), a number of scientists saw the relevance of holography to the problems of brain function, memory, and perception (e.g., Baron, 1970; Cavanagh, 1972; Julesz & Pennington, 1965; Kabrisky, 1966; Pribram, 1966; van Heerden, 1963; Westlake, 1968).

The advent of these explanations came with the development of physical holography (e.g., Stroke, 1966) from the mathematical principles enunciated by Gabor (1948). Equally important, however, was the failure of computer science to simulate perception and learning in any adequate fashion. The problem lies in the fact that computer-based "perceptions" (e.g., Rosenblatt, 1962) were constructed on the basis of an assumed random connectivity in neural networks when the actual anatomical situation is essentially otherwise. In the visual system, for instance, the retina and cortex are connected by a system of fibers that run to a great extent in parallel. Only two modifications of this parallelity occur.

1. The optic tracts and radiations that carry signals between the retina and cortex constitute a sheaf within which the retinal events converge to some extent onto the lateral geniculate nucleus of the thalamus from where they diverge to the cortex. The final effect of this parallel network is that each fiber in the system connects ten retinal outputs to about 5000 cortical receiving cells.
2. In the process of termination of the fibers at various locations in the pathway, an effective overlap develops (to about 5 degrees of visual angle) between neighboring branches of the conducting fibers.

Equally striking and perhaps more important than these exceptions, however, is the interpolation at every cell station of a sheet of horizontally connected neurons in a plane perpendicular to the parallel fiber system. These horizontal cells are characterized by short or absent axons but spreading dendrites. It has been shown in the retina (Werblin & Dowling, 1969) and to some extent also in the cortex (Creutzfeldt, 1961) that such spreading dendritic networks may not generate nerve impulses; in fact, they usually may not even depolarize. Their activity is characterized by hyperpolarization that tends to organize the functions of the system by inhibitory rather than excitatory processes. In the retina, for instance, no nerve impulses are generated prior to the (amacrine and) ganglion cells from which the optic nerve fibers originate. Thus, practically all of the complexity manifest in the optic nerve is a reflection of the organizing properties of depolarizing and hyperpolarizing events, not of interactions among nerve impulses.

Two mechanisms are therefore available to account for the distribution of signals within the neural system. One relies on the convergence and divergence of nerve impulses onto and from a neuronal pool. The other relies on the presence of lateral (mostly inhibitory) interactions taking place in sheets of horizontal dendritic networks situated at every cell station perpendicular to the essentially parallel system of input fibers. Let us explore the possible role of both these mechanisms in explaining the results of the lesion studies.

Evidence is supplied by experiments in which conditions of anesthesia are used that suppress the functions of small nerve fibers, thus leaving intact and clearly discernible the connectivity by way of major nerve impulse pathways. These experiments have shown that localized retinal stimulation evokes a receptive field at the cortex over an area no greater than a few degrees in diameter (e.g., Talbot & Marshall, 1941). Yet, the data that must be explained indicate that some 80% or more of the visual cortex including the foveal region can be extirpated without marked impairment of the recognition of a previously learned visual pattern. Thus, whatever the mechanisms, distribution of input cannot be due to the major pathways, but must involve the fine-fibered connectivity in the visual system, either via the divergence of nerve impulses and/or via the interactions taking place in the horizontal cell dendritic networks.

Both are probably responsible to some extent. Remember that nerve impulses occurring in the fine fibers tend to decrement in amplitude and speed of conduction, thus becoming slow, graded potentials. Furthermore, these graded slow potentials or minispikes usually occur in the same anatomical location as the horizontal dendritic inhibitory hyperpolarizations and thus interact with them. In fact, the resulting microorganization of junctional neural activity (synaptic and ephaptic) could be regarded as a simple summation of graded excitatory (depolarizing) and inhibitory (hyperpolarizing) slow potential processes.

These structural arrangements of slow potentials are especially evident in sheets of neural tissue such as in the retina and the cortex. The cerebral cortex, for instance, may be thought of as consisting of columnar units that can be considered more or less independent basic computational elements, each of which is capable of performing a similar computation (Hubel & Wiesel, 1968; Mountcastle, 1957). Inputs to the basic computational elements are processed in a direction essentially perpendicular to the sheet of the cortex, and therefore cortical processing occurs in stages, each stage transforming the activation pattern of the cells in one of the cortical layers to the cells of another cortical layer. Analyses by Kabrisky (1966) and by Werner (1970) show that processing by one basic computational element remains essentially within that element, and therefore the cortex can be considered to consist of a large number of essentially similar parallel processing elements. Furthermore, the processing done by any one of the basic computational elements is itself a parallel process (see, for example, Spinelli, 1970), with each layer transforming the pattern of activity that arrived from the previous layer by the process of temporal and spatial summation, that is, the summation of slow hyperpolarization and depolarization in the dendritic microstructure of the cortex. Analyses by Ratliff (1965) and Rodieck (1965) have shown that processing (at least at the sensory level) that occurs through successive stages in such a layered neural network can be described by linear equations. Each computational element is thus capable of transforming its inputs through a succession of stages, and each stage produces a linear transformation of the pattern of activity at the previous stage.

Let us trace in detail the evidence regarding these stages in the visual system. Quantitative descriptions of the interactions that occur in the retina are inferred from the output of ganglion cells from which receptive field configurations are recorded by making extracellular microelectrode recordings from the optic nerve. The retinal interactions per se take place initially by virtue of local graded slow-wave potentials—hyperpolarizations and depolarizations that linearly sum within the networks of receptors, bipolar, and horizontal cells from which nerve impulses are never recorded. The receptive fields generated by these graded potential changes display a more or less circular center surrounded by a ring of activity of a sign opposite that of the center. This configuration has been interpreted to mean that the activity of a receptive neuron generates inhibition in neighboring neurons through lateral connectivities (e.g., Békésy, 1967; Hartline, Wagner, & Ratliff, 1956; Kuffler, 1953) perpendicular to the input channels. In view of the fact that no nerve impulses can be recorded from the cells (e.g., horizontal) that mediate the lateral inhibition, the inference can be made that the interactions among graded potentials, waveforms, are responsible (Pribram, 1971; Pribram, Nuwer, & Baron, 1974). Such waveforms need not be thought of as existing in an unstructured homogenous medium. The dendritic arborizations in which the gradual potential changes occur can act as structural wave guides. However, as Beurle (1956) has shown, such a structural medium can still give rise to a geometry of plane waves provided the structure is reasonably symmetrical. The mathematical descriptions of receptive field configurations bear out Beurle's model. Such descriptions have been given by Ratliff (1965) and Rodieck (1965). Mathematically, they involve a convolution of luminance change of the retinal input with the inferred inhibitory characteristics of the network to compose the observed ganglion cell receptive field properties.

The gist of these experimental analyses is that the retinal mosaic becomes decomposed into an opponent process by depolarizing and hyperpolarizing slow potentials and transforms into more or less concentric receptive fields in which center and surround are of opposite sign. Sets of convolutional integrals fully describe this transformation.

The next cell station in the visual pathway is the lateral geniculate nucleus of the thalamus. The receptive field characteristics of the output from neurons of this nucleus are in some respects similar to the more or less concentric organization obtained at the ganglion cell level. Now, however, the concentric organization is more symmetrical, the surround has usually more clear-cut boundaries and is somewhat more extensive (e.g., Spinelli & Pribram, 1967). Furthermore, a second penumbra of the same sign as the center can be shown to be present, although its intensity (number of nerve impulses generated) is not nearly so great as that of the center. Occasionally, a third penumbra, again of opposite sign, can be made out beyond the second (Hammond, 1972).

Again, a transformation has occurred between the output of the retina and the output of the lateral geniculate nucleus. This transformation apparatus appears to act as a rectification process. Each geniculate cell thus acts as a peephole "viewing" a part of the retinal image mosaic. This is due to the fact that each geniculate cell has converging upon it some 10,000 ganglion cell fibers. This receptive field peephole of each geniculate cell is made of concentric rings of opposing sign, whose amplitudes fall off sharply with distance from the center of the field. In these ways the transformation accomplished is like very near-field optics that describes a Fresnel hologram.

Pollen, Lee, and Taylor (1971), although supportive of the suggestion that the visual mechanism as a whole may function in a holographiclike manner, emphasize that the geniculate output is essentially topographic and punctate, is not frequency specific, and does not show translational invariance: that is, every illuminated point within the receptive field does not produce the same effect. Furthermore, the opponent properties noted at the retinal level of organization are maintained and enhanced at the cost of overall translational invariance. Yet a step toward a discrete transform domain has been taken since the output of an individual element of the retinal mosaic—a rod or cone receptor—is the origin of the signal transformed at the lateral geniculate level.

When the output of lateral geniculate cells reaches the cerebral cortex, further transformations take place. One set of cortical cells, christened "simple" by their discoverers (Hubel & Wiesel, 1968), has been suggested to be characterized by a receptive field organization composed by a literally linelike arrangement of the outputs of lateral geniculate cells. This proposal is supported by the fact that the simple-cell receptive field is accompanied by side bands of opposite sign and occasionally by a second side band of the same sign as the central field. Hubel and Wiesel proposed that these simple cells thus serve as line detectors in the first stage of a hierarchical arrangement of pattern detectors. Pollen *et al.* (1971) have countered this proposal on the basis that the output from simple cells varies with contrast luminance as well as orientation and that the receptive field is too narrow to show translational invariance. They therefore argue that an ensemble of simple cells would be needed to detect orientation. They suggest that such an ensemble would act much as the strip integrator used by astronomers (Bracewell, 1965) to cull data from a wide area with instruments of limited topographic capacity (as is found to be the case in lateral geniculate cells).

Another class of cortical cells has generated great interest. These cells were christened "complex" by their discoverers, Hubel and Wiesel, and thought by them (as well as by Pollen) to be the next step in the images processing hierarchy. Some doubt has been raised (Hoffman & Stone, 1971) because of their relatively short latency of response as to whether all complex cells receive their input from simple cells. Whether their input comes directly from the geniculate or by way of simple cell processing, however, the output from complex cells of the visual cortex displays transformations of the retinal input, characteristics of the holographic domain.

A series of elegant experiments by Fergus Campbell and his group (1974) have suggested that these complex cortical cells are spatial-frequency sensitive elements. Initially, Campbell showed that the response of the potential evoked in man and cat by repeated flashed exposure to a variety of gratings of certain spacing (spatial frequency) adapted not only to that fundamental frequency but also to any component harmonics present. He therefore concluded that the visual system must be encoding spatial frequency (perhaps in Fourier terms) rather than the intensity values of the grating. He showed further that when a square wave grating was used, adaptation was limited to the fundamental frequency and its third harmonic as would be predicted by Fourier theory. Finally, he found neural units in the cat's cortex that behaved as did the gross potential recordings.

Pollen (1973) has evidence that suggests that these spatial-frequency sensitive units are Hubel and Wiesel's complex cells, although both his work and that of Maffei and Fiorentini (1973) have found that simple cells also have the properties of spatial frequency filters, in that they are sensitive to a selective band of spatial frequencies. In addition, the latter investigators have found that the simple cells can transmit contrast and spatial phase information in terms of two different parameters of their response: Contrast is coded in terms of impulses per second and spatial phase in terms of firing pattern.

The receptive field of complex cells is characterized by the broad extent (when compared with simple cells) over which a line of relatively indeterminate length but a certain orientation will elicit a response. Pollen demonstrated that the output of complex cells was not invariant to orientation alone—number of lines and their spacing also appeared to influence response. He concluded, therefore, as had Fergus Campbell, that these cells were spatial-frequency sensitive and that the spatial-frequency domain was fully achieved at this level of visual processing. Additional corroborating evidence has recently been presented from the Pavlov Institute of Physiology in Leningrad by Glezer, Ivanoff, and Tscherbach (1973), who relate their findings on complex receptive fields as Fourier analyzers to the dendritic microstructure of the visual cortex much as we have done here.

Even more recently, series of studies from the Cambridge laboratories, from MIT, Berkeley, and Stanford University, have substantiated the earlier reports. Pribram, Lassonde, and Ptito (in preparation) have confirmed that both simple and complex cells are selective to restrictive bandwidths of spatial frequencies, but that simple cells encode spatial phase, whereas complex cells do not. Thus simple cells may be involved in the perception of spatial location, whereas complex cells are more truly "holographic" in that they are responsible for translational invariance. Schiller, Finlay, and Volman (1976a,b,c,d) have performed a comprehensive coverage of receptive field properties, including spatial frequency selectivity. Movshon, Thompson, and Tolhurst (1978a,b,c) in another set of experiments showed that receptive fields could be thought of as spatial filters (much as van Heerden, 1963, originally proposed) whose Fourier transform mapped precisely the cell's response characteristics. De Valois, Albrecht, and Thorell (1978) have taken this work even a step further by showing that whereas these cells are tuned to from .5 to 1.5 octaves of bandwidth of the spatial frequency spectrum, they are not tuned at all to changes in bar width. Finally, De Valois has tested whether the cells are selective of edges making up patterns or their Fourier transforms. The main components of the transforms of checkerboards and plaids lie at different orientations from those of the edges making up the patterns. In every case the orientation of the checkerboards or plaids had to be rotated to match the Fourier encoding and the rotation was to exact amount in degrees and minutes of arc predicted by the Fourier transform.

The results of these experiments go a long way toward validating the holographic hypothesis of brain function. However, as I have noted previously (Pribram, Nuwer, & Baron, 1974) a major problem remains even after these data are incorporated in the construction of a precise model. Each receptive field, even though it encodes in the frequency domain, does so over a relatively restricted portion of the total visual field. Robson (1975) has thus suggested that only a "patch" of the field becomes "Fourier" represented. However, this major problem has now been resolved and the solution has brought unexpected dividends. Ross (see review by Leith, 1976) has constructed holograms on the principles proposed by Bracewell (1965) and espoused by Pollen (see Pollen & Taylor, 1974). Such multiplex or strip integral holograms are now commercially available (Multiplex Co., San Francisco, California). Not only do they display all the properties of ordinary holograms, but also can be used to encode movement as well. Thus, by combining frequency encoding with a spatial "patch" or "slit" representation, a lifelike, three-dimensional moving image can be constructed.

Although detailed specification has been given for the visual system only, the foregoing analysis is in large part also relevant to the auditory system, the tactile system, and the motor system (see Pribram, 1971, for review). The recently accumulated facts concerning the visual system are the most striking because it was not suspected that spatial pattern perception would be found to be based on a stage that involves frequency analysis. The finding of the ubiquity of frequency analysis by brain tissue has made accessible explanations of hitherto inexplicable observations, such as the distributed nature of the memory trace and the projection of images away from the surface in which their representation has become encoded. The model has had considerable explanatory power.

Information and Episodic Processing

Whenever a powerful explanatory principle is discovered, there is a tendency to apply it in inappropriate situations. Image processing as used in this chapter applies *only* to what in the older neurological literature was called sensory-motor functioning. The more cognitive aspects of brain function in which the intrinsic (association) systems are implicated are served by what is now usually referred to as information processing. But even here a distinction can be drawn between the functions of the posterior and the frontal intrinsic (association) mechanisms. As will be shown, only the posterior convexity of the brain truly serves as an information processor. The frontal cortex is involved in computing familiarities from episodic variations of more or less regularly recurring organism-environment relationships. These computations were shown dependent on the operation of peripheral visceroautonomic mechanisms and the participation of the limbic forebrain.

Data will be reviewed that demonstrate that the posterior convexity of cerebral cortex is involved in the sampling of alternatives (invariant properties of a relationship between organism and environment), whereas the frontal cortex regulates behavior (e.g., habituation to repetitious episodes) that establishes a familiar context within which information can then be processed. Let me detail a representative experiment.

A modified Wisconsin General Testing Apparatus (Harlow, 1942) is used to test 12 rhesus monkeys on a complex problem. The monkeys are divided into three groups, two operated and one control, each containing four animals. The animals in one operated group had received bilateral cortical resections in the posterior intrinsic cortex and those in the other operated group, bilateral cortical resections in the frontal intrinsic cortex about 2.5 years prior to the onset of the experiment; those in the control group are unoperated. In the testing situation these animals are confronted initially with two junk objects placed over two holes (on a board containing 12 holes in all) with a peanut under one of the objects. An opaque screen is lowered between the monkey and the object as soon as the monkey has displaced one of the objects from its hole (a trial). When the screen is lowered, separating the monkey from the 12-hole board, the objects are moved (according to a random number table) to two different holes on the board. The screen is then raised and the animal is again confronted with the problem. The peanut remains under the same object until the animal finds the peanut five consecutive times (criterion). After a monkey reaches criterion performance, the peanut is shifted to the second object and testing continues (discrimination reversal). After an animal again reaches criterion performance, a third object is added. Each of the three objects in turn becomes the positive cue; testing proceeds as before —the screen separates the animal from the 12-hole board, the objects are placed randomly over 3 out of the 12 holes (with a peanut concealed under one of the objects), the screen is raised, the animal allowed to pick an object (one response per trial), the screen is lowered, and the objects moved to different holes. The testing continues in this fashion until the animal reaches criterion performance with each of the objects positive in turn. Then a fourth object is added and the entire procedure repeated. As the animal progresses, the number of objects is increased serially through a total of 12 (Figure 21.1). The testing procedure is the same for all animals throughout the experiment; however, the order of the introduction of objects is balanced—the order being the same for only one monkey in each group.

Analysis of the problem posed by this experiment indicates that solution is facilitated when a monkey attains two strategies: (a) during search, moving, on successive trials, each of the objects until the peanut is found; (b) after search, selecting, on successive trials, the object under which the peanut had been found on the preceding trial. During a portion of the experiment, searching is restricted for animals with posterior intrinsic sector ablations; and selection of the object under which the peanut had been found on the previous trial is impaired by frontal intrinsic sector ablations. The effects of the posterior intrinsic sector lesions will be dealt with first.

Figure 21.2 graphs the averages of repetitive search errors by each group. The deficit of the frontally operated group is not associated with search (a result that is discussed later). In spite of the increasing complexity of the succeeding situations, the curves appear little different from those previously reported to describe the formation of a discrimination in complex situations (Bush & Mosteller, 1951; Skinner, 1938). Although one

Figure 21.1. Diagram of the multiple object problem showing an example of the seven object situation. Food wells are indicated by dotted squares, each of which is assigned a number. The placement of each object over a food well was shifted from trial to trial according to a random number table. A record was kept of the object moved by the monkey on each trial, only one move being allowed per trial. Trials were separated by lowering an opaque screen to hide, from the monkey, the objects as they were positioned.

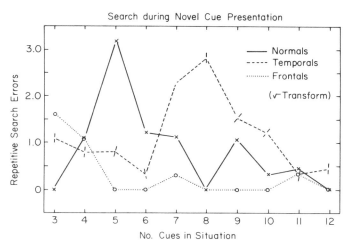

Figure 21.2. Graph of the average of the number of repetitive errors made in the multiple object experiment during those search trials in each situation when the additional, that is, the novel, cue is first added.

might, a priori, expect the number of repetitive responses to increase monotonically as a function of the number of objects in the situation, this does not happen. Rather, during one or another phase of the discrimination, the number of such responses increases to a peak and then declines to some asymptotic level (Bush & Mosteller, 1951; Skinner, 1938).

Analysis of the data of this experiment has shown that these peaks or "humps" can be attributed to the performance of the control and posteriorly operated groups during the initial trials given in any particular (e.g., 2, 3, 4 . . . cue) situation, that is, when the monkey encounters a *novel* object. The period during which the novel and familiar objects are confused is reflected in the "hump." The importance of experience as a determinant of the discriminability of objects has been emphasized by Lawrence (1949, 1950). His formulation of the "acquired distinctiveness" of cues is applicable here. In a progressively more complex situation, sufficient familiarity with *all* of the objects must be acquired before a novel object is sufficiently distinctive to be readily differentiated.

However, there is a difference between the control and the posteriorly operated groups as to when the confusion between novel and familiar objects occurs. The peak in errors for the posteriorly operated group lags behind that of the controls—a result that forced attention because of the paradoxically "better performance" of the posteriorly operated group throughout the five to six cue situations (in an experiment that was originally undertaken to demonstrate a relation between the number of objects in the situation and the discrimination "deficit" previously shown by this group).

These paradoxical results are accounted for by a formal treatment based on mathematical learning theory: On successive trials the monkeys had to "learn" which of the objects now covered the peanut and which objects did not. At the same time they had to "unlearn," that is, extinguish what they had previously learned—under which object the peanut had been and under which objects it had not been. Both neural and formal models have been invoked to explain the results obtained in such complex discrimination situations. Skinner (1938) postulated a process of neural induction to account for the peaks in errors, much as Sherrington had postulated "successive spinal induction" to account for the augmentation of a crossed extension reflex by precurrent antagonistic reflexes (such as the flexion reflex). Several of Skinner's pupils (Estes, 1950; Green, 1958) have developed formal models. These models are based on the idea that both "learning" (or "conditioning") and "unlearning" (or "extinction") involve antagonistic response classes—that in both conditioning and extinction there occurs a transfer of response probabilities between response classes. This conception is, of course, similar to Sherrington's *"this* reflex or *that* reflex but not the two together."* The resulting equations that constitute the model contain a constant that is defined as the probability of sampling a particular stimulus element (Green, 1958), that is, object, in the discrimination experiment presented here. This constant is further defined (Estes) as the ratio between the number of stimulus elements sampled and the total number of such elements that could possibly be sampled. This definition of the constant postulates that it is dependent for its determination upon both environmental and organismic factors. According to the model the rapidity of increase in errors in a discrimination series depends on this sampling ratio—the fewer objects sampled, the more delayed the peak in recorded errors. The paradox that for a portion of the experiment the posteriorly lesioned group performs better than the control group stems from the relative delay in the peak of the recorded errors of the operated group. The model predicts, therefore, that this operated group has sampled fewer objects during the early portions of the experiment. This prediction is tested as shown in Figure 21.3.

The prediction is confirmed. The posterior intrinsic sector is thus established as one of the organismic variables that determine the constant of the model. As postulated by the model, the ratio of objects sampled turns out to be more basic than the number of objects in the situation, per se.

Returning to the postsearch portion of the multiple object experiment, Figure 21.4 portrays performance following completion of search, that is, after the first response on which the peanut is found. Note that the lag

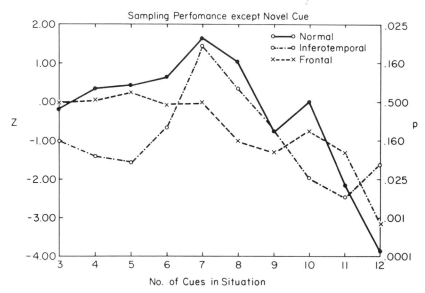

Figure 21.3. Graph of the average of the percent of the total number of objects (cues) that are sampled by each of the groups in each of the situations. To sample, a monkey had to move an object until the content or lack of content of the food well was clearly visible to the experimenter.

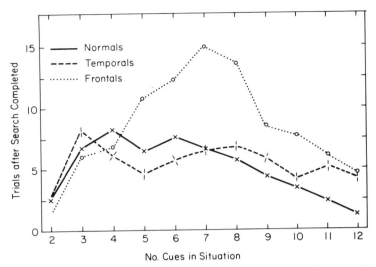

Figure 21.4. Graph of the average of the number of trials to criterion taken in the multiple object experiments by each of the groups in each of the situations after search was completed, that is, after the first correct response.

shown by the frontally operated group in reducing the number of trials taken to reach the criterion of five consecutive errorless responses (or the number of repetitive errors made) occurs *after* the peanut has been found. This group of monkeys experiences difficulty in attaining the strategy of returning on successive trials to the object under which they have, on the previous trial, found the peanut. Whatever may be the explanation of this difficulty, a precise description can be given: For the frontally operated group, "finding the peanut" does not determine subsequent behavior to the extent that "finding the peanut" determines the subsequent behavior of the normal group. In Sherrington's and in behavioristic terms, the "positive element," the response to the object, is for the frontal group inadequately reinforced by the "alliance with it" of the action, that is, finding the peanut. More generally, response probabilities of the frontal group are less affected by the outcomes of their actions (e.g., finding a peanut).

Interestingly, before the frontally operated group begins to attain the necessary strategy (after the seven cue situation), performance of this group reflects the number of alternatives in the situation. This finding suggests a parallel with analyses of the effects of outcomes developed in the theory of games and economic behavior. The effects of outcome are determined by two classes of variables: (*a*) the dispositions of the organism and (*b*) an estimate about the actions of other parts of the system. The finding that performance of the frontally operated group is related to the number of alternatives in the situation suggests that this group is deficient in evaluating the second class of variables, but this is only suggested by these results.

Support for the hypothesis that frontal lesions do not affect the dispositional variables that determine the effect of an outcome of an action comes from the results of another experiment.

In a constant (fixed) interval experiment, 10 rhesus monkeys are tested in an "operant conditioning" (Skinner, 1938) situation that consists of an enclosure in which a lever is available to the monkey. Occasionally, immediately after a depression of the lever, a pellet of food also becomes available to the monkey. The experimenter schedules the occasions on which the action of pressing the lever has the outcome that a food pellet becomes available. In this experiment, these occasions recurred regularly at a constant (fixed) interval of 2 min. The conditioning procedure, as a rule, results in performance curves (scallops) that reflect during the early portions of the interval, a slow rate of response, and during the latter portions an accelerating rate that nears maximum just prior to the end of the interval. All of the monkeys used in this experiment were trained every other day for 2 hr sessions until their performance curves remained stable (as determined by superimposition of records and visual inspection) for at least 10 consecutive hours.

Two experimental conditions were then imposed, one at a time: (a) deprivation of food for 72 and 110 hr; (b) resection of frontal and posterior intrinsic cortex. Food deprivation increases the total rate of response of all animals markedly, but does not alter the proportion of responses made during portions of the interval (Figure 21.5). Resection of the frontal intrinsic sector does not change the total number of responses, but does alter the distribution of responses through the interval—there is a marked decrease in the difference between the proportion of responses made during the various portions of the interval. Monkeys with lesions of the posterior intrinsic sectors and unoperated controls show no such changes (Figure 21.6).

The results of the constant interval experiment support the contention that the effect of an outcome of an action is influenced by variables that can be classified separately. Deprivation influences total rate of response and the frontal lesion, the distribution of that rate. Deprivation variables are akin to those that have in the past been assigned to influence the disposition of the organism. The frontal intrinsic sector lesion appears to influence the monkey's estimates about the situation. This finding is thus in accord with that obtained in the multiple object problem. Both experimental findings can be formally treated by the device of "mathematical expectation" (von Neumann & Morgenstern, 1953, ch. 1). The distribution of response probabilities in the constant interval experiment can be considered a function of the temporal "distance" from the outcome; distribution of response probabilities in the multiple object experiment is a function of the number of objects in the situation. Frontal intrinsic sector lesions interfere with those aspects of behavior that depend on an estima-

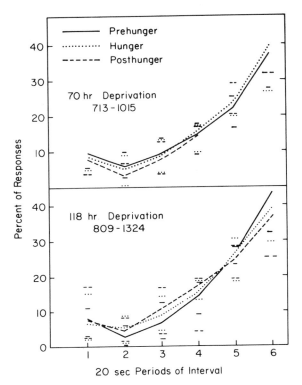

Figure 21.5. Graph showing the effect of food deprivation on monkeys' rate of lever pressing response to food (a small pellet of laboratory chow) which became available every 2 min. The change in total rate is indicated by numbers under the deprivation label. The lack of change in the distribution responses is shown by the curves. Each curve represents the average of the responses of 10 monkeys; each point represents the average rate during a period of the interval over 10 hr of testing. Variance is indicated by the short horizontal bars.

tion of the effects that an outcome of an action has in terms of the total set of available possible outcomes. The effects of frontal intrinsic sector lesions on behavior related to outcomes thus parallel the effects of posterior intrinsic sector ablations on behavior related to inputs. A general model of intrinsic sector mechanisms seems therefore to be possible. As a step toward such a model a brief review of available data follows.

The effect of frontal intrinsic sector resection on the distribution of responses in the multiple object and constant interval problems is correlated with other deficiencies in behavior that follow such resections. The most clear-cut deficiency is in the performance of delayed reaction and of alternation by subhuman primates. These problems are usually classified with those used primarily to study behavior involved in the differentiation of alternatives, although differences between the two are recognized. These differences have been conceptualized in terms of one-trial (episode specific) learning (Nissen, Riesen, & Nowles, 1938), immediate memory (Jacobsen, 1936), and retroactive inhibition (Malmo, 1942). More penetrating analyses have been accomplished for the effects of frontal intrinsic sector lesions on the performance of alternation problems (Leary, Harlow, Settlage, & Greenwood, 1952; Mishkin & Pribram, 1956). These analyses emphasize the recurrent regularities that constitute the alternation problems and suggest

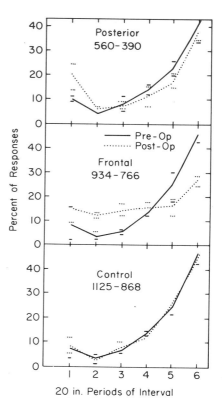

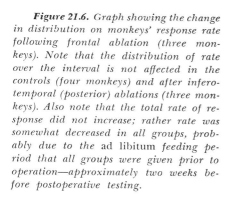

Figure 21.6. Graph showing the change in distribution on monkeys' response rate following frontal ablation (three monkeys). Note that the distribution of rate over the interval is not affected in the controls (four monkeys) and after inferotemporal (posterior) ablations (three monkeys). Also note that the total rate of response did not increase; rather rate was somewhat decreased in all groups, probably due to the ad libitum *feeding period that all groups were given prior to operation—approximately two weeks before postoperative testing.*

that such problems be considered examples of a larger class that can be distinguished from problems that require differentiation (Galanter & Gerstenhaber, 1956). Delayed reaction may also belong to the class of problems specified by episodically recurring regularities: The recurrence, at the time response is permitted, of some of the events present in the predelay situation, constitutes an essential aspect of the delay problem (Mishkin & Pribram, 1956).

These experiments have been followed up by another series that has extended the results to humans and has clearly related the deficit shown by frontally lesioned primates to their inability to compute and control the episode specific variations that occur within the context of regularly recurring variables: (Anderson, Hunt, Vander Stoep, & Pribram, 1976; Brody & Pribram, 1978; Brody, Ungerleider, & Pribram, 1977; Grueninger & Pribram, 1969; Kimble, Bagshaw, & Pribram, 1965; Konow & Pribram, 1970; Luria, Pribram, & Homskaya, 1964; Pribram, Ahumada, Hartog, & Roos, 1964; Pribram & Bagshaw, 1953; Pribram, Plotkin, Anderson, & Leong, 1977; Pribram & Tubbs, 1967; Tubbs, 1969). These studies have shown that the deficit in delayed alternation behavior produced by frontal lesions is due to proactive and retroactive interferences produced by the monotonous temporal context provided by the symmetrical intertrial interval of the classical task, since this deficit can be overcome by imposing a nonsymmetrical delay interval; that interference produced by the continuing distractions involved in manipulating the spatial context in the delayed response task is the critical variable in producing the deficit in performance of this task after frontal resection; that distractibility may be overcome by perseverative behavior under some experimental conditions; and that this continuing distractibility of frontally lesioned human and nonhuman primates is contingent on a failure to produce the viscero-autonomic concomitants of the orienting reaction that accompanies distraction in normal subjects. The results of these experiments demonstrate that the temporal and spatial ordering of behavior and of experience depends on the registration of episode specific occurrences and on composing these registrations into a context in which subsequent behavior becomes appropriate. Such controlled episodic, context-dependent processing can thus be distinguished neurally as well as conceptually from the processing of invariants—true information processing in the Shannonian (Shannon & Weaver, 1949) sense.

Conclusions

This chapter has surveyed data that distinguish three modes of central processing. Image processing was found to occur by virtue of the sensory-motor projection systems of the brain. Evidence was presented that these systems operate as frequency analyzers that operate upon periodicities in energy distributions as transduced by receptor mechanisms. It was further shown that image processing in the frequency domain, by virtue of mathematically described spread functions, distributes input and thus accounts for the distributed nature of the basic memory store.

Image processing is augmented in the primate brain by additional modes of central processing attributable to the functions of intrinsic "association" systems. Evidence was presented to show that the posterior intrinsic systems are involved in information processing, where information is defined as choice among *alternatives*. Choices were shown to depend on sampling a ratio of the number of alternatives available, the size of their ratio being a function of the operations of the posterior intrinsic systems.

By contrast, evidence was presented that showed frontal intrinsic cortex to be involved in controlling behavior dependent on variation in *recurrent* occurrences (the structure of redundancy in information theoretic terms). Control was shown dependent on computations that registered the "familiarity" among episodic variations in recurrences, in the absence of which every event became "novel" and therefore distracting. Furthermore, registration was shown contingent upon the occurrence of the visceroautonomic components of the orienting reaction.

In short, two modes in addition to image processing were identified. One of these modes depends on differentiating alternatives, the invariances operative in organism–environment relationships that constitute the information upon which knowledge is based. The other mode depends on establishing contextual familiarity from episodes in the variations that characterize recurrent regularities (the structure of redundancy). Delineating these mechanisms of central processing has gone a long way toward dispelling the despair voiced by earlier investigators of brain mechanisms involved in learning and remembering and that for a time turned psychologists away from studies of brain function.

REFERENCES

Anderson, R. M., Hunt, S. C., Vander Stoep, A., & Pribram, K. H. Object permanency and delayed response as spatial context in monkeys with frontal lesions. *Neuropsychologia*, 1976, *14*, 481–490.

Baron, R. J. A model for cortical memory. *Journal of Mathematical Psychology*, 1970, *7*, 37–59.

Békésy, G. von. *Sensory inhibition.* Princeton, N.J.: Princeton Univ. Press, 1967.

Beurle, R. L. Properties of a mass of cells capable of regenerating pulses. *Philosophical Transactions of the Royal Society of London*, Ser. B, 1956, *240*, 55–94.

Bracewell, R. *The Fourier transform and its applications.* New York: McGraw-Hill, 1965.

Brody, B. A., & Pribram, K. H. The role of frontal and parietal cortex in cognitive processing: Tests of spatial and sequence functions. *Brain*, 1978, *101*, 607–633.

Brody, B. A., Ungerleider, L. G., & Pribram, K. H. The effects of instability of the visual display on pattern discrimination learning by monkeys: Dissociation produced after resections of frontal and inferotemporal cortex. *Neuropsychologia*, 1977, *15*, 439–448.

Bush, R. R., & Mosteller, F. A model for stimulus generalization and discrimination. *Psychology Review*, 1951, *58*, 413–423.

Campbell, F. W. The transmission of spatial information through the visual system. In F. O. Schmitt and F. G. Worden (Eds.), *The neurosciences third study program.* Cambridge, Mass.: MIT Press, 1974, pp. 95–103.

Cavanagh, J. P. Holographic processes realizable in the neural realm: Prediction of short-term memory and performance. Unpublished doctoral dissertation. Carnegie-Mellon Univ. 1972.

Collier, R. J., Burckhardt, C. B., & Lin, L. H. *Optical holography.* New York: Academic Press, 1971.

Creutzfeldt, O. D. General physiology of cortical neurons and neuronal information in the visual system. In M. B. A. Brazier (Eds.), *Brain and behavior.* Washington, D.C.: American Institute of Biological Sciences, 1961.

De Valois, R. L., Albrecht, D. G., & Thorell, L. G. Spatial tuning of LGN and cortical cells in monkey visual system. In H. Spekreijse (Ed.), *Spatial contrast.* Amsterdam: Monograph Series, Royal Netherlands Academy of Sciences, 1978.

De Valois, R. L., Albrecht, D. G., & Thorell, L. G. Cortical cells: Line and edge detectors, or spatial frequency filters? In S. Cool (Ed.), *Frontiers of visual science.* New York: Springer-Verlag, 1978, pp. 544–556.

Estes, W. K. Toward a statistical theory of learning. *Psychology Review*, 1950, *57*, 94–107.

Gabor, D. A new microscopic principle. *Nature*, 1948, *161*, 777–778.

Galanter, E. H., & Gerstenhaber, M. On thought: The extrinsic theory. *Psychological Review*, 1956, *63*, 218–227.

Glezer, V. D., Ivanoff, V. A., & Tscherbach, T. A. Investigation of complex and hypercomplex receptive fields of visual cortex of the cat as spatial frequency filters. *Vision Research*, 1973, *13*, 1875–1904.

Goldscheider, A. Über die materiellen Veränderungen bei der Assoziationsbildung. *Neurol. Zentralblatt*, 1906, *25*, 146.

Goodman, J. W. *Introduction to Fourier optics*. San Francisco: McGraw-Hill, 1968.

Green, E. J. A simplified model for stimulus discrimination. *Psychology Review*, 1958, *65*, 56–63.

Grueninger, W., & Pribram, K. H. The effects of spatial and nonspatial distractors on performance latency of monkeys with frontal lesions. *Journal of Comparative and Physiological Psychology*, 1969 *68*, 203–209.

Hammond, P. Spatial organization of receptive fields of LGN neurons. *Journal of Physiology*, 1972, *222*, 53–54.

Harlow, H. F. Responses by rhesus monkeys to stimuli having multiple sign values. In *Studies in personality*. New York: McGraw-Hill, 1942, pp. 105–123.

Hartline, H. K., Wagner, H. G., & Ratliff, F. Inhibition in the eye of limulus. *Journal of General Physiology*, 1956, *39*, 651–673.

Hoffman, K. P., & Stone, J. Conduction velocity of afferents to cat visual cortex. A correlation with cortical receptive field properties. *Brain Research*, 1971, *32*, 460–466.

Horton, L. H. *Dissertation on the dream problem*. Philadelphia: Cartesian Research Society of Philadelphia, 1925.

Hubel, D. H., & Wiesel, T. N. Receptive fields and functional architecture of monkey striate cortex. *Journal of Physiology*, 1968, *195*, 215–243.

Jacobsen, C. F. Studies of cerebral function on primates. *Comparative Psychology Monograph*, 1936, *13*, 3–60.

Julesz, B., & Pennington, K. S. Equidistributed information mapping: An analogy to holograms and memory. *Journal of Optical Society of America*, 1965, *55*, 604.

Kabrisky, M. *A proposed model for visual information processing in the human brain*. Urbana: Univ. of Illinois Press, 1966.

Kimble, D. P., Bagshaw, M. H., & Pribram, K. H. The GSR of monkeys during orienting and habituation after selective partial ablations of the cingulate and frontal cortex. *Neuropsychologia*, 1965, *3*, 121–128.

Konow, A., & Pribram, K. H. Error recognition and utilization produced by injury to the frontal cortex in man. *Neuropsychologia*, 1970, *8*, 489–491.

Kuffler, S. W. Discharge patterns and functional organization of mammalian retina. *Journal of Neurophysiology*, 1953, *16*, 37–69.

Lashley, K. S. The problem of cerebral organization in vision. In *Biological symposia (Vol. 7). Visual mechanisms*. Lancaster, Pa.: Jacques Catell Press, 1942.

Lashley, K. S. Continuity theory of discriminative learning. In F. A. Beach, D. O. Hebb, C. T. Morgan, & H. W. Nissen (Eds.), *The neuropsychology of Lashley*. New York: McGraw-Hill, 1960, pp. 421–431.

Lawrence, D. H. Acquired distinctiveness of cues: I. Transfer between discriminations on the basis of familiarity with the stimulus. *Journal of Experimental Psychology*, 1949, *39*, 776–784.

Lawrence, D. H. Acquired distinctiveness of cues: II. Selective association in a constant stimulus situation. *Journal of Experimental Psychology*, 1950, *40*, 175–188.

Leary, R. W., Harlow, H. F., Settlage, P. H., & Greenwood, D. D. Performance on double alternation problems by normal and brain-injured monkeys. *Journal of Comparative and Physiological Psychology*, 1952, *45*, 576–584.

Leith, E. N. White-light holograms. *Scientific American*, 1976, *235*(4), 80–81.

Loeb, J. *Comparative physiology of the brain and comparative psychology*. Science Series. New York: Putnam, 1907.

Luria, A. R., Pribam, K. H., & Homskaya, E. D. An experimental analysis of the behavioral disturbance produced by a left frontal arachnoidal endothelloma (meningioma). *Neuropsychologia*, 1964, *2*, 257–280.

Maffei, L., & Fiorentini, A. The visual cortex as a spatial frequency analyzer. *Vision Research*, 1973, *13*, 1255–1267.

Malmo, R. B. Interference factors in delayed response in monkeys after removal of frontal lobes. *Journal of Neurophysiology*, 1942, *5*, 295–308.

Mishkin, M., & Pribram, K. H. Analysis of the effects of frontal lesions in monkeys: II. Variations of delayed response. *Journal of Comparative and Physiological Psychology*, 1956, *49*, 36–40.

Mountcastle, V. B. Modality and topographic properties of single neurons of cat's somatic sensory cortex. *Journal of Neurophysiology*, 1957, *20*, 408–434.

Movshon, J. A., Thompson, I. D., & Tolhurst, D. J. Spatial summation in the receptive field of simple cells in the cat's striate cortex. *Journal of Physiology*, 1978a, *283*, 53–77.

Movshon, J. A., Thompson, I. D., & Tolhurst, D. J. Receptive field organization of complex cells in the cat's striate cortex. *Journal of Physiology*, 1978b, *283*, 79–99.

Movshon, J. A., Thompson, I. D., & Tolhurst, D. J. Spatial and temporal contrast sensitivity of cells in the cat's areas 17 and 18. *Journal of Physiology*, 1978c, *283*, 101–120.

Nissen, H. U., Riesen, A. H., & Nowles, V. Delayed response and discrimination learning by chimpanzees. *Journal of Comparative Psychology*, 1938, *26*, 361–386.

Pollen, D. A. Striate cortex and the reconstruction of visual space. In *The neurosciences study program, III*. Cambridge, Mass.: MIT Press, 1973.

Pollen, D. A., Lee, J. R., & Taylor, J. H. How does the striate cortex begin the reconstruction of the visual world? *Science*, 1971, *173*, 74–77.

Pollen, D. A., & Taylor, J. H. The striate cortex and the spatial analysis of visual space. In F. O. Schmitt & F. G. Worden (Eds.), *The neurosciences study program, III*. Cambridge, Mass.: MIT Press, 1974, pp. 239–247.

Pribram, K. H. Some dimensions of remembering: Steps toward a neuropsychological model of memory. In J. Gaito (Ed.), *Macromolecules and behavior*. New York: Academic Press, 1966, pp. 165–187.

Pribram, K. H. *Languages of the brain*. Englewood Cliffs, N.J.: Prentice-Hall, 1971.

Pribram, K. H. Why it is that sensing so much we can do so little? In F. O. Schmitt & F. G. Worden (Eds.), *The neurosciences study program, III*. Cambridge, Mass.: MIT Press, 1974.

Pribram, K. H., Ahumada, A., Hartog, J., & Roos, L. A progress report on the neurological process disturbed by frontal lesions in primates. In I. M. Warren and K. Akert (Eds.), *The frontal granular cortex and behavior*. New York: McGraw-Hill, 1964, pp. 28–55.

Pribram, K. H., & Bagshaw, M. Further analysis of the temporal lobe syndrome utilizing fronto-temporal ablations. *Journal of Comparative Neurology*, 1953, *99*, 347–375.

Pribram, K. H., Lassonde, M. C., & Ptito, M. Intracerebral influences on the microstructure of visual cortex. In preparation.

Pribram, K. H., Nuwer, M., & Baron, R. The holographic hypothesis of memory structure in brain function and perception. In R. C. Atkinson, D. H. Krantz, R. C. Luce, & P. Suppes (Eds.), *Contemporary developments in mathematical psychology*. San Francisco: Freeman, 1974, pp. 416–467.

Pribram, K. H., Plotkin, H. C., Anderson, R. M., & Leong, D. Information sources in the delayed alteration task for normal and "frontal" monkeys. *Neuropsychologia*, 1977, *15*, 329–340.

Pribram, K. H., Spinelli, D. N., & Kamback, M. C. Electrocortical correlates of stimulus response and reinforcement. *Science*, 1967, *157*, 94–96.

Pribram, K. H., & Tubbs, W. E. Short-term memory, parsing, and the primate frontal cortex. *Science*, 1967, *156*, 1765–1767.

Ratliff, F. *Mach bands: Quantitative studies in neural networks in the retina*. San Francisco: Holden-Day, 1965.

Robson, J. G. Receptive fields: Neural representation of the spatial and intensive attributes of the visual image. In E. C. Carterette (Ed.), *Handbook of perception* (Vol. V). *Seeing*. New York: Academic Press, 1975.

Rodieck, R. W. Quantitative analysis of cat retinal ganglion cell response to visual stimuli. *Vision Research*, 1965, *5*, 583–601.

Rosenblatt, F. *Principles of neurodynamics: Perceptions and the theory of brain mechanism*. Washington, D.C.: Spartan Books, 1962.

Schiller, P. H., Finlay, B. L., & Volman, S. F. Quantitative studies of single-cell properties in monkey striate cortex. I. Spatiotemporal organization of receptive fields. *Journal of Neurophysiology*, 1976a, *39*, 1288–1319.

Schiller, P. H., Finlay, B. L., & Volman, S. F. Quantitative studies of single-cell properties in monkey striate cortex. II. Orientation specificity and ocular dominance. *Journal of Neurophysiology*, 1976b, *39*, 1320–1333.

Schiller, P. H., Finlay, B. L., & Volman, S. F. Quantitative Studies of single-cell properties in monkey striate cortex. III. Spatial frequency. *Journal of Neurophysiology*, 1976c, *39*, 1334–1351.

Schiller, P. H., Finlay, B. L., & Volman, S. F. Quantitative studies of single-cell properties in monkey striate cortex. V. Multivariate statistical analyses and models. *Journal of Neurophysiology*, 1976d, *39*, 1362–1374.

Shannon, C. E., & Weaver, W. *The mathematical theory of communication*. Urbana: Univ. of Illinois Press, 1949.

Skinner, B. F. *The behavior of organisms: An experimental analysis*. New York: Appleton-Century-Crofts, 1938.

Spinelli, D. N. Occam, a content addressable memory model for the brain. In K. H. Pribram & D. Broadbent (Eds.), *The biology of memory*. New York: Academic Press, 1970.

Spinelli, D. N., & Pribram, K. H. Changes in visual recovery function and unit activity produced by frontal and temporal cortex stimulation. *Electroencephalography and Clinical Neurophysiology*, 1967, *22*, 143–149.

Stroke, G. W. *An introduction to coherent optics and holography*. New York: Academic Press, 1966.

Talbot, S. A., & Marshall, U. H. Physiological studies on neural mechanisms of visual localization and discrimination. *American Journal of Ophthalmology*, 1941, *24*, 1255–1264.

Tubbs, W. T. Primate frontal lesions and the temporal structure of behavior. *Behavioral Science*, 1969, *14*, 347–356.

van Heerden, P. J. A new method of storing and retrieving information. *Applied Optics*, 1963, *2*, 387–392.

von Neumann, J., & Morgenstern, O. *Theory of games and economic behavior*. Princeton, N.J.: Princeton Univ. Press, 1953, pp. 19, 20; 24–28; 39–41; 60–73.

Weiss, P. *Principles of development*. New York: Holt, 1939.

Werblin, F. S., & Dowling, J. E. Organization of the retina of the mud puppy. Necturus maculosus, II. Intracellular recording. *Journal of Neurophysiology*, 1969, *32*, 339–355.

Werner, G. The topology of the body representation in the somatic afferent pathway. In *The neurosciences study program, II*. New York: Rockefeller Univ. Press, 1970.

Westlake, P. R. Toward a theory of brain functioning: A detailed investigation of the possibilities of neural holographic processes. Unpublished doctoral dissertation, Univ. of California, Los Angeles, 1968.

U. G. GASANOV
A. G. GALASHINA
A. V. BOGDANOV

A Study of Neuron Systems Activity in Learning

22

The notion that the organization of activity in the cortical neuronal systems (e.g., Asratyan, 1971; Beritashvili, 1975; Kostyuk, 1977; Livanov, 1975) is the basis of any adaptive reaction has become a solidly confirmed concept. Whereas until recently, attempts to understand systems activity had to be made on the basis of data on individual neuron functions, investigators in recent years have been persistently seeking to undertake experimental-statistical studies of integrative processes on the systems level.

The development of methodical and statistical studies of neuron systems is dictated primarily by the fact that the available information about the functional properties and plastic possibilities of an individual cell is insufficient for making judgments about multineuron integration. In the opinion of Kostyuk (1973), the recording and description of the activity of even a multiple of neurons do not yet mean that "this will enable us to understand how a new property—a system of cells—is evolved from individual nerve cells and how the cells in that system function as a unified whole" [Kostyuk, 1973, p. 115].

An experimental study of the systems activity of neurons should be possible if one employs a method that expresses quantitatively the basic property of the system—the interaction of neurons. It is in this vein that investigators in recent years have been showing interest in statistical analyses of crossover relations between two neural currents as recorded from simultaneously functioning neurons (e.g., Gasanov, 1978; Gerstein, 1970; Griffin & Horn, 1963; Livanov, 1975). It should be emphasized that crossover relations reflect a qualitatively new form of neuron activity: The dependent relationships in the microintervals do not correlate with the induced or background impulsation of any of the interacting neurons (Dickson & Gerstein, 1974; Gasanov & Galashina, 1976; Griffin & Horn, 1963; Wyman, 1966).

A basic problem in examining crossover relationships is to determine the underlying causes of the dependence of one neuron's function on another. One can assume that the common source of cortical cell excitation lies beyond the limits of the cerebral cortex (Dickson & Gerstein, 1974). However, the appearance of dependent relationships primarily between adjacent neurons (Dickson & Gerstein, 1974), as well as the data that indicate the monosynaptic character of statistical links (Asanuma & Rosen, 1973), would seem to make insufficiently convincing the arguments supporting the evidence of a strong extracortical source. Some other neighboring neuron of the pair in question might be a common source. Opposing this supposition are data which fail to demonstrate a neuron-driver of impulses when three and four neighboring impulses are simultaneously recorded, followed by a determination of their dependent relationships (Gasanov, 1978; Sil'kis, 1974).

There is experimental substantiation in the literature for the opinion that a direct link between neurons can be identified from specific parameters of crossover histograms. It is believed that the histograms with asymmetrical and short-timewise characteristics of dependent relationships indicate the existence of a functional link between the neurons in question (e.g., Dickson & Gerstein, 1974; Moore, Segundo, Perkel & Levitan, 1970).

The conditions under which we investigated the cortical cells allowed us to consider that histograms of crossover intervals are most probably statistical indicators of interneuron functional links. These conditions provided for (*a*) the recording of multineuron activity, that is, the activity of several neurons situated in direct proximity to each other; (*b*) an analysis of background activity when there is no large factor of an induced extracortical stimulation; and (*c*) an accounting of the dependent relationships in the histogram that occurred within the short time segments. Our experiments were conducted on alert cats that could move around freely in a Plexiglas box. Multineuron impulsation was led from the auditory cortex through chronically implanted Nichrome electrodes with a 50 mcr tip diameter. The crossover interval histograms, identified by impulse discriminators, were plotted on an AI-128 analyzer. Figure 22.1 shows a fragment of background multineuron activity and a schematic representation of a crossover analysis of two impulse series. Three series of impulses were usually selected from the large, medium, and small amplitude spikes. Each series was conditionally taken as an impulsation of large, medium, and small neurons with the corresponding rank numbers 1, 2, and 3.

We identified the following types of statistical interrelationships between the neurons: inhibitory, excitatory, mixed (inhibitory-excitatory or

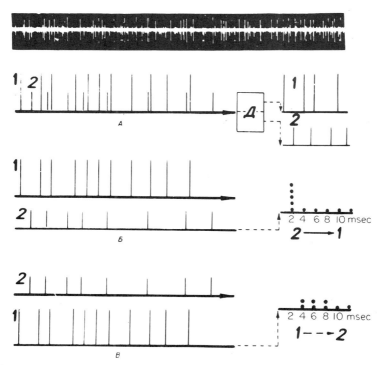

Figure 22.1. Schematic representation of impulse series selection and statistical processing of background multineuron activity in the auditory cortex of alert cats. Top: Fragment of multineuron activity. (A) Conventional representation of the activity of two neurons (1 and 2) and their subsequent selection by discriminators (D). (B) Two series of impulses, selected by amplitude, of which the first starts the analyzer's beam and the second is picked up. The crossover interval histogram shows the relationship between neuron 1 and neuron 2. (C) The same two series of impulses, but series 1 is the starter and series 2 is picked up. There are no dependent relationships in the histogram.

excitatory-inhibitory), and dependent relationships (no links). To determine the type of link in each crossover interval histogram (CIH), we computed the average number of impulses per bin and located the confidence limits in three sigmas. Spikes beyond the established limits were considered reliable deviations from the average (Figure 22.2). The bin width in all CIH was 2 msec and the analysis time was 20 msec (the smallest interval between impulses triggering the analyzer, IG in Figure 22.2).

The interneuron links differed not only in type, but also in direction. Some neuron pairs exhibited an interconnection between neurons (bilateral links), whereas others had a single (or unilateral) link, and still other neuron pairs functioned independently.

In all, five series of experiments were conducted: (*a*) automated auditory stimulation from the spikes of neuron 1 (AS-1 sound). Discrimi-

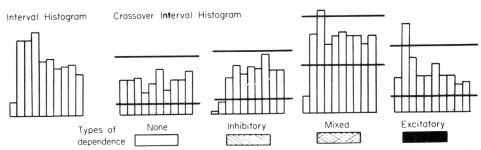

Figure 22.2. Types of crossover interval histograms. Left: *Interval histogram of one impulse series; bin width is 8 msec. The remaining histograms characterize the crossover relationships of two impulse flows; bin width is 2 ms. The horizontal lines represent the confidence limits in three sigmas.* Bottom: *Symbols denoting the types of link.*

nator-selected spikes were fed to a sonic click generator for 3–10 min. The crossover activity was analyzed before and after the automated stimulation; (b) automated stimulation with an impulse current that was transmitted to the ear muscle or to the orbicular muscle of the eye (AS-1 current); (c) rhythmic auditory stimulation with a frequency of ten per second (RS); (d) automated stimulation from the spikes of neuron 4 whose impulse differed from all three neurons being analyzed; (e) the system's activity of three neurons was studied in cats with motor-conditioned reflexes. The sonic clicks at a frequency of ten per second were combined with electrocutaneous stimulation of the anterior paw. After the conditioned reflex was secured, it was extinguished prior to the three consecutive inhibitory reactions (Figure 22.3). The segments of background neurograms that preceded the sonic stimulant by 30 sec were selected for an analysis of interneural relationships.

A total of 498 CIH was studied, of which 264 were obtained in the

Figure 22.3. Development and extinction of a motor conditional reflex. (1) Combination of sound and current; examples of myograms with the absence of a conditional response (a) and after the consolidation of the reflex (b); (2) extinction of a conditional reflex. Visible is the response developed to sound following the first sound probes without current (2) and the myogram represented during the extinguished conditioned reflex (b). The vertical lines represent the start and finish of a 5 sec sound stimulation at a frequency of 10 per second, and the horizontal line represents the stimulation of the anterior paw by an impulse current of the same frequency for .5 sec.

first four series of experiments and 234 in the experiments with conditioned reflexes. Subsequent quantitative analysis of the crossover results depended on the conditions of the experiment. In AS and RS, the total number of each type of CIH was calculated and the correlation between them was determined. Since AS and RS were studied in the course of a single experimental day, the CIH could be compared before and after stimulations with a quantitative evaluation of the degree of change in the links. A change was considered to be a restructuring of one type of interneural relationship into any other type.

In the conditioned reflex experiments, the frequency with which a given type of link appeared in individual pairs of neurons was computed, along with the total characteristics of the CIH types.

Figure 22.4 shows diagrams of total indicators of various link types in the intact cats and after isolated and combined stimulations.

A significant number of CIH did not exhibit dependent relationships in the auditory cortex of intact cats. The inhibitory type of link was predominant among the CIH with dependent relationships. There was a similar distribution of interneuron relationships in sonic and electrical stimulation from neuron 1 (AS-1 sound and AS-1 current). A marked increase in the number of excitatory interneuron relationships was noted when the cats were subjected to sonic stimulation that was not associated with the activity of the three neurons being studied (AS-4 sound and RS).

A similar picture of total indicator distribution was observed at the

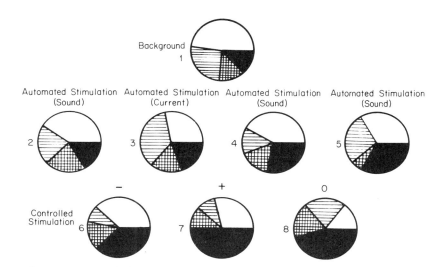

Figure 22.4. Correlation of types of crossover interval histograms by overall indicators. Designations: (1) Intact cats; (2) AS-sound; (3) AS-1 current; (4) AS-4 sound; (5) RS sound; (6) after the combinations of sound and current up to the stabilization of the conditioned reflex; (7) after the consolidation of the conditioned reflex; (8) after the extinction of the conditioned reflex. Symbol designations for the sectors of the diagram are shown in Figure 22.2.

stage of conditioned reflex development in which the combined sonic and electrocutaneous stimulations were not yet accompanied by a motor response to sound (C-1). Here, the number of combinations could go up to several dozen and even hundreds of applications. Following the consolidation of the conditioned reflex, the excitatory nature of the interneuron relationships became predominant (CR+). The excitatory influence of one neuron on another was also broadly represented in a background of the extinguished conditional reflex, but in contrast to all of the previous results there was a sharp reduction in the number of CIH with independent relationships (O).

Therefore, the total data picture indicates that the most significant changes were exhibited by the excitatory links. The most characteristic increase in those links occurred as a result of a reduction in the number of the mixed type links or because of a transition from independent relationships to dependent ones.

The excitatory relationships between the neurons became dominant after the consolidation of the conditioned reflex and when it was extinguished. In the latter case, a higher level of link activity was observed, as indicated by the sharp reduction in the number of histograms with independent relationships.

The characteristics of the interneuron relationships with respect to the total indicators leave open the question about the distribution of link types between the individual neurons participating in the microsystem. Thus, an increase or decrease in the total number of excitatory links could take place in all of the neuron pairs studied and in a certain individual pair of neurons. When the correlations of the four link types did not change following stimulations (AS-1 sound and AS-1 current), one could assume that a uniform redistribution of these links had taken place in different neuron pairs—something that is not reflected in the overall results.

Figure 22.5 shows the results of a quantitative analysis of histograms with regard to the criterion of link variability. The variability was calculated for AS and RS in which the same microsystems were examined before and after stimulation. The control data indicated that in the intact cats certain links changed in type in the course of 10–20 min. Thus, an excitatory link could change into a mixed link or dependent relationships could become independent ones. The variability for most directions (individual links) did not exceed 20–30% of the total number of interneuron relation histograms for a given direction. Those links which in 75–100% of the cases differed from the type exhibited in the initial microsystems were considered to be changed in the experiments with AS and RS.

The results from the qualitative processing of CIH with respect to variability showed that there was a different distribution of mutated links in the microsystem in different experimental series with AS and RS. At the same time, one can see from Figure 22.5 that certain links exhibited

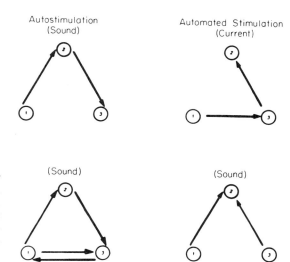

Figure 22.5. Distribution of the most reactive (70–100% mutability) links in a three-neuron microsystem after automated and rhythmic stimulations. Numbers in the circles represent the order numbers of the neurons.

an identical response to sonic stimuli. Thus, the effect of neuron 1 on neuron 2 was almost completely restructured both in automated and rhythmic sonic stimulations, and no differences were observed between AS and RS in this link.

The inclusion of neuron 3 into the system changed significantly the picture of plastic restructuring. The links in the 1–2–3 direction were changed in experiments with automated sonic stimulation from neuron 1 (AS-1 sound) spikes, and a restructuring occurred in direction 1–3–2 when the current (AS-1 current) was applied. When sonic stimulation (automated or rhythmic) did not coincide with impulses of the neurons under study, the mutability from neuron 3 to neurons 2 and 1 also increased.

The results from the experiments on AS and RS indicate that the specific systems neuron mechanisms of adaptive reactions are initially observed only at the three-neuron microsystem level. In this connection, the crossover relationships in a three-neuron microsystem evolve selectively and in a variable fashion, depending on the stimulation conditions (automated or rhythmic) and the mode of the stimulus (sound or current).

The summarized data on the systems activity of neurons after the development and extinction of a conditioned motor reflex are represented in Figure 22.6. Because the studies involving the conditioned reflexes were made over an extensive period, it was not possible to compare the same microsystems. The predominant types and directions of each of the six links were determined from histograms for different experimental days prior to the development of positive and inhibitory conditioned reflex activity.

As was mentioned previously, the interneuron links could be either unilateral or bilateral. Bilateral dependent relationships were observed in

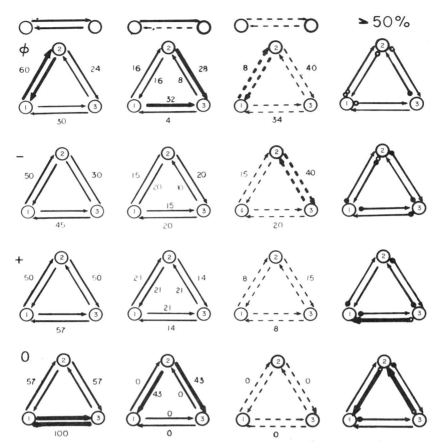

Figure 22.6. Distribution of bilateral, unilateral, and independent neuron relationships in a microsystem. Φ designates intact cats; (−) denotes the sound and current combinations up to the stabilization of the conditioned reflex; (+) denotes after the consolidation of the conditioned reflex; (0) denotes after the extinction of a conditioned reflex. The thick arrows indicate the links that differ in frequency of appearance. The numbers represent the percentage of links with similar direction.

all neuron pairs in the intact cats, but they were particularly found between neurons 1 and 2. Their frequency in this pair exceeded the frequency of similar links by two times and more in the neuron pairs 1–3 and 2–3. A gradual evening out of this factor took place in all neuron pairs in the combination process. An identical level of bilateral dependent relationships was also characteristic of combinations without an effector response and of combinations with a developed reflex. The bilateral links in one of the neuron pairs again began to predominate following the extinction of the conditioned reflex. However, in contrast to the data for intact cats, neurons 1–3 constituted this pair in these experiments.

The analysis of unilateral links disclosed the unique picture of neuron systems activity. With regard to this factor, neuron 3 in the intact cats differed significantly from the other neurons. The frequency of neuron 3 dependency on neuron 1 and neuron 2 was several times greater than the intensity of inverse effects. Following extensive combinations, the unilateral relationships also leveled out in all neuron pairs and their distribution did not depend on whether the conditioned reflex was developed or not.

After the conditioned reflex was extinguished, the number of unilateral links was significantly reduced, and the remaining links characterized the controlling role of neuron 2, both with respect to neuron 1 and neuron 3.

The distribution of independent type links differed somewhat from the previous indicators of neuron systems activity. Complete ("bilateral") independence in the intact cats was frequently observed between neurons 1–3 and 2–3. The total number of independent relationships decreased in proportion to the number of combinations, but this factor was still rather high for the neuron pair 2–3. Following the consolidation of the conditioned reflex, the number of cases where the function of two neurons was independent became minimal in all pairs. After the conditioned reflex was extinguished, not a single case of independent function in two neurons was observed.

Thus, a three-factor analysis of systems activity in interneuron relationships leads us to conclude that the development of the conditioned reflex is accompanied by activation of an ever-increasing number of potentially existing interneuron links that is expressed in the amplification and predominance of bilateral links. These links become dominant during extinction, particularly between the large (neuron 1) and the small (neuron 3) neurons.

One might ask what are characteristic dependent relationship types of bilateral and unilateral links? As was demonstrated earlier, inhibitory links were more often observed in links on hand in the intact cats, whereas excitatory links were most often observed in cats with developed responses (see Figure 22.4). In considering the physiological importance and decisive role of these link types in the systems organization of neurons, we subsequently examined their correlation in each of six link directions. Those functional relationships whose frequency exceeded relationships with the opposite sign by more than five times were considered to be dominant.

Figure 22.7 shows a graphic representation of three-neuron microsystems with dependent interneuron relationships with respect to the predominant link types under given experimental conditions. As can be seen, the interrelationships between neurons 1 and 2 in the intact cats, as well as in the direction from neuron 1 to neuron 3, were inhibitory. The remaining three links did not exhibit a significant predominance of excitatory or inhibitory influence.

Almost all the links became excitatory following combinations in a period of complete or almost complete absence of the conditioned reflex.

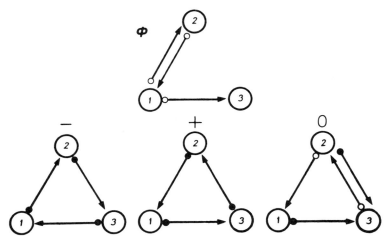

Figure 22.7. *Distribution of dominant inhibitory and excitatory types of crossover neuron relationships in a microsystem. Symbol designations are as in Figure 22.6.*

In this case, a chainwise interaction occurred in the direction of neurons 1–2–3–1.

The interaction of three neurons was also accomplished primarily by chainwise excitatory links during consolidated conditional reflexes (90–100% manifestation), but their direction was the mirror-opposite. They were now joined in a 1–3–2–1 sequence.

The distribution of dominating excitatory links was different for a developed extinctive inhibition than it was in previous experiments. They were significantly predominant in only two links: from neuron 1 to neuron 3, and from neuron 2 to neuron 3. It is apparent from the diagrams reflecting the overall data that the extinction of the conditioned reflex was also accompanied by an increase in the number of inhibitory CIH. Although the predominance of inhibitory links was not as overwhelming as the excitatory links in any of the links, they did dominate over the excitatory links in two directions. The inhibitory CIH exceeded the number of excitatory CIH by three times in the links between neuron 3 and neuron 2, and between the latter and neuron 1. A comparison of the excitatory and inhibitory interneuron relationships during extinction gives one the impression that the small neurons acquire capability of inhibiting neuron 2 through a supplementary inhibitory interneuron.

Conclusions

The data we obtained lead us to believe that the elementary system of neurons in which qualitatively new properties begin to appear should be seen as a three-neuron microsystem. Identifiable in such a microsystem are the network properties of the individual neurons, the characteristics of the functional consolidations of neurons, and the plastic interneuron relationships. Specific changes in cortical activity during adaptive responses are clearly observed on the trineuron microsystem level. Changes in the system level are manifested as a selective transformation of interneuron links following sound and current stimulation of the cats, as well as during the combination of those links with a subsequent development of the conditioned reflex.

The studies of multineuron activity confirm the data in the literature on the dependent relationships between neuron function (dimensions) and the spike amplitude generated by them (Henneman, Semjen, & Carpenter, 1965). An analysis of the interneuron relationships allows one to differentiate network properties of these neurons as well. The directional (Figure 22.6) and type (Figure 22.7) distribution of functional links clearly characterizes the different network properties in the large amplitude, medium and small amplitude spikes of the cortical cells in the intact cats.

A differentiated participation of neurons with various spike amplitudes in plastic restructuring of the microsystem was observed during AS and RS, as well as during the development of a positive and inhibitory conditioned reflex. Functional differentiation of cortical cells is particularly evident during the extinction of the conditioned reflex. This can be seen in the case of the small neurons that receive strong excitatory effects from the large and medium neurons, and subsequently inhibited the medium neurons.

One should note that all of our investigations were conducted in a background of neuron impulsation that disclosed a directional response toward learning with respect to interneuron relationships. According to the data on slow cortical reactions, Rusinov (1974) concludes that the primary most intense cerebral cortical activity during the formation of a time link takes place after the combination of stimuli. Apparently, the interneuron relationships in the background play a decisive role in the realization of a conditioned reflex on the system level. It is probable that there are two neuron mechanisms in the time link. One of them forms the memory track in the form of an organized neuron system that has coordinated and subordinated links between the neurons. The other mechanism performs a communication role and takes place in response to a conditional signal whose full functioning depends on the background interneuron relationships.

REFERENCES

Asanuma, H.a., & Rosen, J. Spread of Mono-and Polysynaptic connection within cat motor cortex. *Experimental Brain Research*, 1973, *16*, 507–520.
Asratyan, E. A. On the physiology of conditioned reflex reinforcement. *Journal of Higher Nervous Activity*, 1971, *21* (No. 1), 3–13. (In Russian)

Beritashvili, I. S. *Selected papers, neurophysiology and neuropsychology.* Moscow, 1975. (In Russian)

Dickson, J. W.a., & Gerstein, G. C. Interaction between neurons in auditory cortex of the cat. *Journal of Neurophysiology,* 1974, *37* (No. 6), 1239–1261.

Gasanov, U. G. The controlled experiment (auto-learning of nerve cells). *Progress in the Physiological Sciences,* 1978, 7 (No. 2). (In Russian)

Gasanov, U. G., & Galashina, A. G. Analysis of interneuron links in the auditory cortex of alert cats. *Journal of Higher Nervous Activity,* 1975, *25* (No. 5), 1053–1060. (In Russian)

Gasanov, U. G., & Galashina, A. G. Study of plastic changes in cortical interneuron links. *Journal of Higher Nervous Activity,* 1976, *26* (No. 4), 820–827.

Gerstein, G. L. Functional association of neurons: Detection and interpretation. In *Neurosciences communication and coding in the nervous system,* 1970, pp. 648–661.

Griffin, J. S., & Horn, G. Functional coupling between cells in the visual cortex of the unrestrained cat. *Nature,* 1963, *199,* (N4896), 876–880.

Henneman, E., Semjen, G.a., & Carpenter, D. O. Functional significance of cell size in spinal motoneurons. *Journal of Neurophysiology,* 1965, *28,* 560, 580.

Kostyuk, P. G. Basic mechanisms of neuron consolidation in a nerve center. In *Principles in the systems organizations of functions.* Moscow, 1973, pp. 115–124. (In Russian)

Kostyuk, P. G. *Physiology of the CNS.* Kiev, 1977.

Livanov, M. N. Neuron memory mechanisms. *Progress in the Physiological Sciences,* 1975, *6* (No. 3), 66–89. (In Russian)

Moore, G. P., Segundo, I. P., Perkel, D. H., & Levitan, H. *Biophysical Journal,* 1970, *10* (No. 9), 876–900.

Renaud, L.a., & Kelly, I. S. Identification of possible inhibitory neurons in the precruciate cortex of the cat. *Brain Research,* 1974, *79,* 9–28.

Rusinov, V. S. Tracking processes in the cerebral cortex and their functional importance. *Journal of Higher Nervous Activity,* 1974, *24,* (No. 3), 571–579. (In Russian)

Sil'kis, I. G. On the functional organization of monosynaptic interneuron links in the cerebral cortex. *Journal of Higher Nervous Activity,* 1974, *24,* (No. 3), 571–579.

Wyman, R. I. Multistable firing patterns among several neurons. *Journal of Neurophysiology,* 1966 (No. 5), 807–833.

M. VERZEANO

The Activity of Neuronal Networks in Cognitive Function

23

It would be encouraging to think that the accelerating pace and the magnitude of the advances made in neuroscience during the past two decades have brought this field of investigation to the dawn of an epoch of rapid expansion, analogous to that which occurred in physics in the latter part of the nineteenth and the beginning of the twentieth centuries. During a period of only a few decades, which began in the mid-1880s, physics revolutionized the concepts of matter, energy, time and space, brought to light the dual "particle-wave" nature of electromagnetic radiation, and came close to a time at which it may provide a unifying theory of electric, magnetic, gravitational, and intraatomic forces.

The findings that electron microscopy and neurochemistry are providing on the variety, the location and the ultra structure of synaptic junctions and on the nature and action of synaptic transmitters, the results obtained by the combination of behavioral and neurophysiological methods, and the study of the effects of pharmacological agents on brain function may seem to indicate that neuroscience may be standing on the threshold of a revolution of similar magnitude.

However, although the discoveries made in physics progressed *pari passu* with the development of fundamentally new concepts of the nature of the physical universe, the discoveries made in neuroscience have not yet provided any major change in the fundamental concepts of the nature of mental activity, that is, a better understanding of the neural substrate of the higher cerebral processes such as perception, integration, storage and retrieval of information, and of the relations between brain, mind, and consciousness. This leaves neuroscience in a paradoxical state of development: Its experimental discoveries are being achieved during the latter part of the twentieth century, whereas the development of its fundamental concepts remains in the middle of the nineteenth century.

There are many reasons for this discrepancy. Some of them are related to the historical development of neuroscience in the midst of constraints imposed by the philosophical outlook of past centuries, which was still prevalent in recent times; some are related to the technical difficulties inherent in investigating the neural substrate of cognitive processes; and some are related to the view that many contemporary scientists have of brain function.

Much of the available knowledge of brain function has been acquired, up to the present time, by the anatomical, physiological, or chemical analysis of the single neuron and the single synapse. This method has been enormously powerful and its use has led to an understanding of the most fundamental aspects of the structure, the electrical and chemical characteristics, and the transmission mechanisms of a large variety of neural membranes. In terms of brain function it has led to an understanding of the processes by which the information that arrives from the outside world is translated into successive codes in its passage through the different levels of the sensory pathways, and by which the information that is to be transmitted to the outside world is translated into successive codes in its passage through the different levels of the motor pathways.

It is, however, what happens between these processes of input and output that constitutes the basis of cognitive function. The achievement of perception, integration, storage, and retrieval of information requires the selection, transport, comparison, and organization of a very large quantity of sensory and memory data at very high speeds. The complexity of these operations must rely on the chemical and electrical interactions of very large numbers of axonal, synaptic, and somatic elements, interconnected in such a way as to form networks and systems of networks dedicated to particular functions. And yet, neuroscientists have tirelessly proceeded, during several decades, to investigate isolated cell after isolated cell with the hope that, any day now, the fine tip of their microelectrode will hit upon the miraculous neuron that holds the key to the understanding of cognitive function. Considering the meager results obtained so far by such methods, it should become evident that it is only by putting sufficient emphasis on the study of networks and systems of neurons that neuroscience will bring the investigation of cognitive function to the frontiers attained by the investigation of other aspects of the nervous system.

A difficulty that often precludes clear communication between investigators is the widespread misconception that the term "neuronal network" refers to a formless mass of cells that operates as an undifferentiated unit, all the neurons in it doing the same thing at the same time, in relation to a particular cerebral process. The evidence available so far, however, indicates that the different types of neurons in a network may belong to different systems, assemblies, and subassemblies, each one contributing in a highly organized way to the work of the whole. For this reason the study of the isolated neuron and of the isolated synapse will undoubtedly have to continue, in conjunction with the study of networks and systems, in order to determine the type, the location, and the interconnections of the neurons that belong to a particular system, and the kind of synaptic transmitters involved in the interactions of particular networks.

Another difficulty in the development of more advanced concepts of the neural basis of cognitive function has been the view, which has been prevalent for a long time, that the brain is a static system. In a sensory receiving area, for example, some neurons are silent and some neurons are spontaneously active. As information that arrives from the outside world reaches the area, some of the silent neurons become active and some of the active neurons become silent or change their activity in a way related consistently to the characteristics of the sensory stimulus. In this view, the activity of these neurons, continually flickering in a static firmament, forms the basis of brain function.

The concept of the brain as a dynamic system is based on the fact that, in addition to the neurons whose activity goes on and off with the arrival of the incoming information, other neuronal systems maintain a continual and rhythmic barrage of impulses that "sweep" or "scan" through the networks of the sensory receiving neurons, interacting with them, influencing them, and being influenced by them (Dill, Vallecalle & Verzeano, 1968; Negishi & Verzeano, 1961; Verzeano, 1970; Verzeano & Negishi, 1960). Evidence from several sources, which has accumulated during the last 25 years, indicates that, in the cortex and in the thalamus, this periodically circulating activity is related to several basic cerebral processes such as the generation of the electroencephalogram (Andersen & Andersson, 1968; Mescherskii, 1961; Verzeano, 1956, 1963; Verzeano & Calma, 1954) and the control of wakefulness and sleep (Verzeano & Negishi, 1959, 1961; Verzeano, 1963), and that it may be implicated in cognitive function (Verzeano, 1965, 1977; Verzeano, Dill, Navarro & Vallecalle, 1970).

The circulating activity can be best observed by recording the activity of several neurons, simultaneously, with arrays of microelectrodes. This is illustrated in Figure 23.1, which shows several examples of recordings obtained from various regions of the brain of the cat, by means of arrays of microelectrodes, under different conditions of wakefulness or sleep, in experiments in which each microelectrode detected the activity of a single neuron. By looking simultaneously at the activity of a few, isolated neurons, by means of microelectrodes whose tips are separated by small distances,

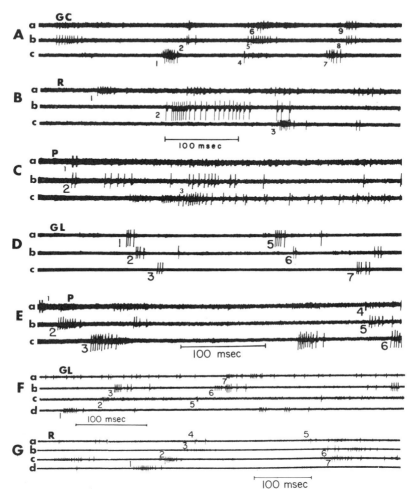

Figure 23.1. Movements of activity in the cerebral cortex and the thalamus of the cat: awake under gallamine tetraiodide (flaxedil) in **A, B, C**; in slow wave sleep in **D**; and under light pentobarbital anesthesia in **E, F, G**. Recordings obtained from single neurons by means of arrays of microelectrodes (a, b, c or a, b, c, d), from the cingulate gyrus (A), the n. reticularis thalami (B and G), the pulvinar (C and E), and the n. geniculatus lateralis (D and F). Movements of activity in the direction c-b-a occur repeatedly in A and F (1, 2, 3–4, 5, 6, etc.); in the direction a–b-c in B to E; in the direction d–c-b-a with a reversal to the direction a-b-c in G. Distances between microelectrode tips such as a to b, b to c, and so on: 100 to 200 μ. The time calibration under B refers to A and B; the one under E refers to C, D, and E. (From Verzeano & Negishi, 1960, Negishi & Verzeano, 1961.)

only a small network can be observed. But, even a small network, when it is activated by a volley of circulating activity passing through, shows a great deal of temporal and spatial coherence in the discharge of its neurons.

The extension of the observed network can be increased by having each microelectrode in the recording array detect the activity of several neurons, rather than isolating one of them. It is much easier in this way to determine the relations between the characteristics of the circulating neuronal activity and those of the gross waves that develop within the same network, as well as the changes occurring in the activity of the network in the transition from wakefulness to sleep, as illustrated in Figure 23.2. This figure shows recordings obtained by means of an array of three microelectrodes (tracings 2, 3, and 4) implanted in the association cortex of an unanesthetized, unrestrained cat. The tracings were obtained in the state of relaxed wakefulness. An example of the arrival of the circulating activity, in regular sequence, at the tip of each microelectrode in the array is shown at a, b, and c.

Once the existence of periodic movements of activity through the cortical and thalamic networks is established, questions arise with regard to its consistency: Is it a fleeting phenomenon that occurs sporadically or is it a consistent process that develops in an organized way? Systematic

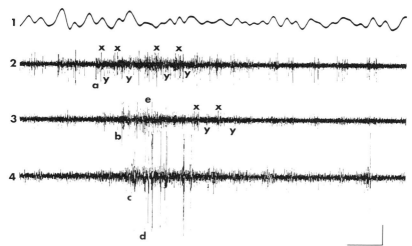

Figure 23.2. Gross waves (1) and neuronal activity (2, 3, 4) in the association cortex of the waking, unrestrained, unanesthetized cat, recorded by means of an array of microelectrodes displayed anteroposteriorly along a straight line. Distances between microelectrodes: 308 μ between 2 and 3; 168 μ between 3 and 4. The anterior microelectrode (2) was 208 μ deeper than the other two. Note: Activity appearing in sequence at the tips of the microelectrodes (at a, b, c); the intervals (x-y) between alternating clusters of action potentials of higher and lower amplitudes. The gross wave (1) was recorded by the anterior microelectrode. Calibration is 100 msec, 200 μV. (From Rinaldi et al., 1977.)

studies of recordings, obtained over long periods, from the cat as well as from the monkey (Verzeano, 1963; Verzeano, Laufer, Spear & McDonald, 1970), have determined that the process is highly organized and that it remains highly consistent as long as the animal remains in a steady state of wakefulness or sleep; as the state changes, some of the characteristics of the circulating activity change and, again, remain consistent as long as the new state continues.

More recent quantitative studies based on the methods of autocorrelation, cross-correlation, and spectral analysis have confirmed in all respects the consistency and the high degree of organization of the movement of activity through the cortex and the thalamus (Rinaldi, Juhasz & Verzeano, 1976, 1977). This is illustrated in Figure 23.3, which shows the

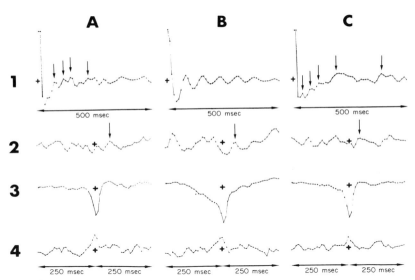

Figure 23.3. Consistency of rhythmicity and of time and phase relations in the activity of the neuronal networks in the visual cortex of the unrestrained, unanesthetized cat, in the states of wakefulness (A), slow wave sleep (B), and paradoxical sleep (C). (1): Autocorrelations of gross waves; (2): crosscorrelations between gross waves and clusters of action potentials of amplitudes higher than 80 µV; (3): crosscorrelations between clusters of action potentials of amplitudes higher than 80 µV and clusters of action potentials of amplitudes of 20–30 µV, recorded by a single microelectrode; (4): crosscorrelations between clusters of action potentials of amplitudes higher than 80 µV recorded by the same microelectrode and clusters of action potentials in the same amplitude range recorded by another microelectrode located 330 µ away and posterior to the first. The gross waves correlated on lines 1 and 2 were recorded by a gross electrode located between the two microelectrodes and 0.5 mm above them. The arrows in the autocorrelograms of line 1 indicated the presence of two super-imposed frequencies; 25 and 12.5 Hz in A; 12.5 Hz in B; and 25 Hz and 5.2 Hz in C (theta range). The arrows in the crosscorrelograms of line 2 indicate the time by which the waves lead the clusters of action potentials. All correlograms were computed from recordings of 90 sec duration. Each cross represents zero time and zero correlation. (From Rinaldi et al., 1977.)

analysis of a series of samples of activity, each of a duration of 90 sec, obtained by an array of microelectrodes, from the visual cortex of the unanesthetized, unrestrained cat, in the states of wakefulness (A), slow wave sleep (B), and paradoxical sleep (C). The autocorrelograms of line 1 show the rhythmicities of the gross waves: 25 Hz and 12.5 Hz in wakefulness; 12.5 Hz in slow wave sleep; 25 Hz and 5.2 Hz in paradoxical sleep. The cross-correlograms of line 2 show the relations between the rhythmicity of the gross waves and that of the sequence of clusters of action potentials recorded by one of the microelectrodes in the array, which represent the activity of one group of neurons in the network. It can be seen that in all three states there is some correlation between the development of the two activities even though the phase relation between them may change from one state to another. The cross-correlograms of line 3 show the relations between sequences of clusters of action potentials of two different amplitudes, representing the activity of two different groups of neurons, surveyed in the same network and at the same time by one of the microelectrodes in the array. In this case the discharges of the two groups of neurons remain 180 degrees out of phase in all three states. They correspond to clusters of neuronal discharges of high and low amplitude, alternating at equal intervals and recorded by one microelectrode in the array, such as those which are indicated as x and y in Figure 23.2.

The cross-correlograms of line 4 show the relations between the activity of *one* of the two groups of neurons whose activities were correlated on line 3 and a *third* group of neurons whose activity was being simultaneously recorded by the tip of another microelectrode in the array, located at a distance of 330 μ from the tip of the first. In this case the discharges of these two groups remained nearly in phase in all three states. Thus, by compiling the information provided by the cross-correlograms of lines 2, 3, and 4, one can see that the three groups of neurons surveyed by the two microelectrodes showed activities consistently related among themselves as well as with the gross waves during periods extending as long as 90 sec. Other studies conducted in the same series of experiments (Rinaldi *et al.*, 1976) have shown that such correlations may remain consistent for periods of several hours in the cortex as well as in the thalamus. The degree of correlation and the amount of lead or lag between one process and another may change in the transition from wakefulness to slow wave sleep or to paradoxical sleep, but the consistency of the movements of activity through the networks and its relations with the gross waves is maintained.

A mechanism by which the circulation of activity could be generated has been postulated by Verzeano (1977). According to this hypothesis (Figure 23.4) impulses circulate within a series of nested feedback loops located within each thalamic nucleus and within each cortical region to which the nucleus projects. Impulses circulating through longer feedback loops, located between the thalamic nuclei and between the thalamus and

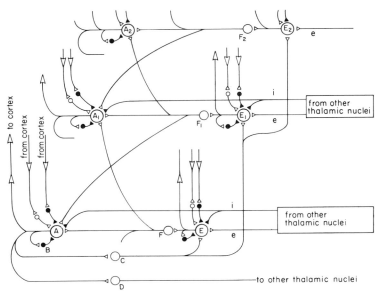

Figure 23.4. Proposed mechanism for the generation of circulating activity in thalamic nuclei. Large circles: *Principal neurons;* medium and small circles: *interneurons.* Light circles and light triangles: *Excitatory neurons and excitatory synapses;* black circles and black triangles: *inhibitory neurons and inhibitory synapses.* Neuron A discharges and activates interneurons B, C and D, and sends impulses to the cortex; interneuron B, activated by a recurrent collateral, causes the early inhibition of neuron A; interneuron C activates neurons E, E_1, E_2 each of which, via interneurons such as F, F_1, and F_2, activates several neurons such as A_1, A_2, and so on, thereby accelerating the circulation process along the line A, A_1, A_2. After discharging and activating neurons A_1, A_2, and so on, neurons E, E_1, E_2 are inhibited via their recurrent collaterals; interneuron D transmits the activity to another thalamic nucleus where similar processes take place. From other thalamic nuclei as well as from the cortex, excitatory and inhibitory fibers (e, i) carry inhibitory and excitatory impulses responsible for the late and prolonged inhibitory phases and for the reactivation of the process. (From Verzeano, 1977.)

cortex, would coordinate the rhythmicity of this thalamo-cortical system. Such a mechanism would explain not only the circulation of activity, but, by the interplay of summations of excitatory and inhibitory postsynaptic potentials resulting from the circulation of impulses over such loops, it would also explain the very close time and phase relations between circulating neuronal activity and the periodic waves of the electroencephalogram.

Thus, the evidence provided so far indicates that highly organized movements of activity through thalamic and cortical networks occur in a consistent way, that they may last through periods extending over several hours, that they are closely associated with the processes that control thalamo-cortical rhythmicities, and that a physiological mechanism to explain their origin, maintenance, and consistency can be postulated on the basis of very solid anatomical and physiological data.

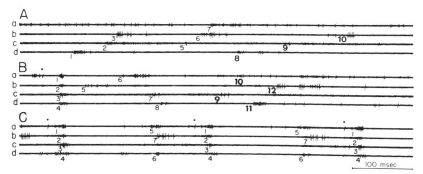

Figure 23.5 *Interaction between circulating activity and activity engendered by incoming sensory information. Neuronal activity recorded by means of an array of four microelectrodes (a, b, c, d) displayed along a straight line, from the lateral geniculate body of the cat, under light barbiturate anesthesia. (A) Spontaneous activity; (B) activity evoked by stimulation with a single flash of light (dot); (C) activity evoked by repetitive flashes. 1, 2, 3, and so on, indicate circulation of activity through the network. The order in which neurons are activated (i.e., the pathway of circulating activity) changes from spontaneous to evoked activity (A to B) and with the frequency of stimulation (B to C). Distance between microelectrode tips: a to b, 156 μ; b to c, 160 μ; c to d, 156 μ. (From Negishi and Verzeano, 1961.)*

Additional evidence, which indicates that these movements of activity may be implicated in cognitive function, is provided by the following findings:

1. The circulation of neuronal activity through cortical and thalamic networks interacts with the activity engendered by incoming sensory information, and its characteristics are modified by it. This is illustrated in Figure 23.5, which shows this interaction as it occurs in the lateral geniculate body of the cat. In A, the circulation of activity occurs spontaneously and its movement through the network can be seen in the sequences of bursts of discharge recorded from the single neurons at 1, 2, 3–5, 6, 7–and 8, 9, 10. In B, stimulation with a brief flash of light generates an incoming volley of impulses at 1, 2, 3, 4, followed by an early sequence of bursts of discharge at 5, 6, 7, 8 and by a later one at 9, 10, 11, 12. The order in which the neuronal discharge appears in different parts of the network has changed radically under stimulation. The circulating activity has been modified by the incoming sensory information. In C, as the frequency of the visual stimulation increases, a new sequence appears in the discharge of the neurons at 5, 6, 7, which remains constant as long as the characteristics of the stimulus remain the same. The circulating activity has, again, been modified to correspond to a new change in the characteristics of the incoming information, and the modification continues to be carried through the network with each stimulus. Similar interactions between circulation of activity and activity engendered by incoming information have been found to occur in the cortex (Negishi & Verzeano, 1961).

2. Theta wave activity, which is known to be related to processes such as consolidation and learning (Berry & Thompson, 1978; Landfield, 1976a,b; Winson, 1978) is always accompanied by a greatly increased regularity in the circulation of activity that develops within the very same network from which the theta wave is recorded, wherever the network may be located: in the cortex, the thalamus or the hippocampus (Berry, Rinaldi, Thompson & Verzeano, 1978; Rinaldi et al., 1977). This is illustrated in Figures 23.6, 23.7, and 23.8.

Figure 23.6 shows the analysis of a series of samples of activity, each of a duration of 90 sec, obtained by means of an array of microelectrodes, from the associational cortex of the unanesthetized, unrestrained cat, in the stages of wakefulness (A), slow wave sleep (B), and paradoxical sleep (C). The autocorrelograms of line 1 show the degree of rhythmicity of the gross waves. The spectral distributions of line 2 show the most prevalent frequencies present in the gross waves. The cross-correlograms of lines 3, 4, and 5 are analogous to those of lines 2, 3, and 4 of Figure 23.3, and show correlations between waves and neuronal activity (line 3) and correlation between the activity of several groups of neurons in the network (lines 4 and 5). The most remarkable finding shown in this figure is that when a dominant theta rhythm appears in the neocortex, as it occasionally does in paradoxical sleep, the cross-correlation between the theta waves and the neuronal activity, as well as the cross-correlations between the activities of the different groups of neurons in the network, which indicate the consistency of the circulation of activity, are extremely high (column C).

Figure 23.7 shows recordings of the gross waves (A) and of the activity of three single neurons (B,C,D) obtained by means of an array of three microelectrodes from the granule cell layer of the dentate gyrus of the rabbit. The circulation of activity through the network is shown by the regular sequence of discharge of the action potentials of the three neurons at 1, 2, 3, 4, 5, 6, and so on. The association of each passage of activity through the network and the theta waves can be seen at 1–10, 4–11, 7–12, and so on.

Figure 23.8 shows the analysis of a series of samples of activity, each of a duration of 90 sec, obtained by an array of three microelectrodes from the granule cell layer of the hippocampus of the rabbit. (A) shows a series of autocorrelograms that indicate a very high degree of rhythmicity of the theta wave (1), recorded by one of the microelectrodes in the array, and of the discharge of two different groups of neurons, recorded by another microelectrode in the array (2 and 3). (B) shows a series of cross-correlograms that indicate a very high level of correlation between the theta wave and the activity of three different groups of neurons, each one recorded by a different microelectrode in the array. (C) and (D) show a series of cross-correlograms that indicate high levels of correlation between the activities of several groups of neurons (crosscorrelated two at a time), recorded by the

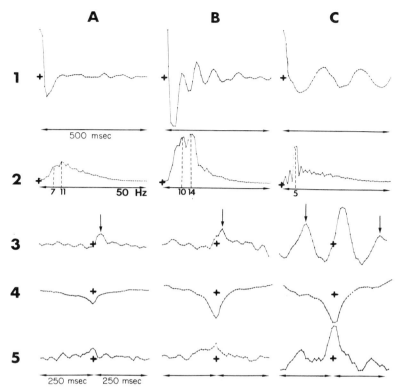

Figure 23.6. Appearance of dominating theta activity in the neocortex of the unrestrained, unanesthetized cat. (A) *Wakefulness;* (B) *slow wave sleep;* (C) *paradoxical sleep.* Line 1: *autocorrelation of gross waves;* line 2: *spectral distribution (power spectrum) of gross waves;* line 3: *crosscorrelations of gross waves and clusters of action potentials of amplitudes higher than 70 µV recorded by one microelectrode;* line 4: *crosscorrelations between clusters of action potentials of amplitudes higher than 70 µV and clusters of action potentials of amplitudes of 20 to 30 µV, recorded by the same microelectrode;* line 5: *crosscorrelations between clusters of action potentials of amplitudes higher than 70 µV recorded by the same microelectrode and clusters of action potentials of the same amplitudes recorded by another microelectrode located 300 µ away and posterior to the first. The gross waves analyzed on lines 1 and 2 were recorded by the anterior microelectrode in the array. Note the broad spectral distribution of the waves in wakefulness, showing peaks at 7 Hz and 11 Hz and the narrow spectral distribution in paradoxical sleep at 5 Hz. Note the high degree of correlation between gross waves and clusters of action potentials occurring at times that correspond to the theta range (arrows, at C 3). Each cross represents zero time and zero correlation. (From Rinaldi et al., 1977.)*

three microelectrodes in the array, and separated by spike amplitude discrimination.

Figure 23.8E shows a *multiple* correlogram relating the simultaneous activities of six groups of neurons and the theta wave, obtained by using an averaging process, similar to that which is used in computing poststimulus

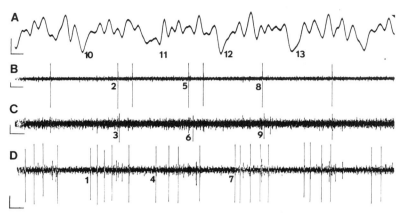

Figure 23.7. Circulation of neuronal activity related to the theta rhythm in the hippocampus. Recordings obtained, by means of an array of three microelectrodes, oriented along a mediolateral line (B, C, D) from the granule cell layer of the dentate gyrus of the rabbit under halothane anesthesia. Each microelectrode records the activity of a single neuron. The sequence in which the three neurons discharge (1, 2, 3; 4, 5, 6, etc.) as well as its correlation with the theta rhythm (1, 2, 3–10; 4, 5, 6–11, etc.) is very consistent. Distances between microelectrode tips: B to C, 170 μ; C to D, 200 μ. Wave shown (A) was recorded by electrode C. Calibration is 100 μV, 50 msec. (From Berry et al., 1978.)

histograms; however, instead of a stimulus pulse, the sweep of the computer is triggered by the action potentials of one group of neurons, whereas the action potentials of all six groups, including that which provides the trigger, are simultaneously averaged with respect to time on six different computer channels. Thus, should a correlation exist between the activity of the group of neurons that provides the trigger (Group 1) and that of a second group, one or several peaks showing maximum probabilities of discharge, should appear at the times at which the discharge of the second group follows that of the first (such as x on line 3). Similarly, should a correlation exist between the activity of the group of neurons that provides the trigger and that of a third group, one or several peaks showing maximum probabilities of discharge should appear at the times at which the discharge of the third group follows that of the first (such as y on line 2). The interval between the maximum probability of discharge in one channel and the maximum probability of discharge in another channel indicates the time relation between the discharge of the two groups of neurons averaged on those two channels (such as x and y on lines 3 and 2). The correlation between the activity of any group of neurons and the theta rhythm is evaluated by averaging the gross waves on a seventh computer channel (W). The interval between the maximum probability of discharge in any of the six neuron channels and the maximum amplitude of the average wave on the seventh channel indicates the time and phase relation between neuronal activity in that channel and the gross wave.

The appearance of regular temporal and spatial sequences in the dis-

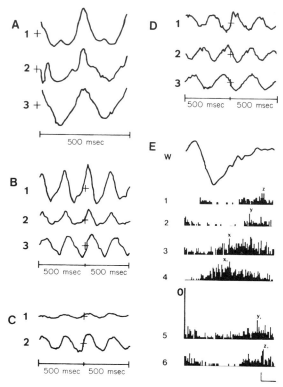

Figure 23.8. Correlation analysis of recordings obtained from the same region and with the same electrode arrangement as those shown in Figure 23.7. (A) Autocorrelograms showing the degree of rhythmicity of the waves (1) recorded by one microelectrode, and of the activities of two groups of neurons (2 and 3) recorded by another microelectrode in the array. (B) Cross-correlograms showing the time and phase relations between the wave and the activities of three different groups of neurons (1, 2, 3), with each one recorded by a different microelectrode in the array. (C) and (D) Cross-correlograms showing the relation between activities of various goups of neurons recorded by the three microelectrodes and separated by amplitude discrimination. (E) Multiple correlograms showing the relations between the activities of six groups of neurons (1–6) and the wave (W); the computer is triggered by the action potentials of Group 1 at time zero; the activity of each group and that of the wave are averaged on separate channels; sequential distributions of maximum probabilities of discharge such as O–X–Y–Z and $O-X_1-Y_1-Z_1$ indicate sequential activation of the different groups of neurons, that is, circulation of activity through the network (see text). Calibration: 6 active potentials, 50 msec. The results presented in this figure were obtained from samples of activity of 90 sec duration. (From Berry et al., 1978.)

tribution of the maximum probabilities of discharge (such as O–x_1–x–y–z or O–x_1–y_1–z_1) indicates the development of highly organized movements of activity through the network; their time and phase relations with the average wave (W) indicate the degree of correlation between these movements and the theta activity.

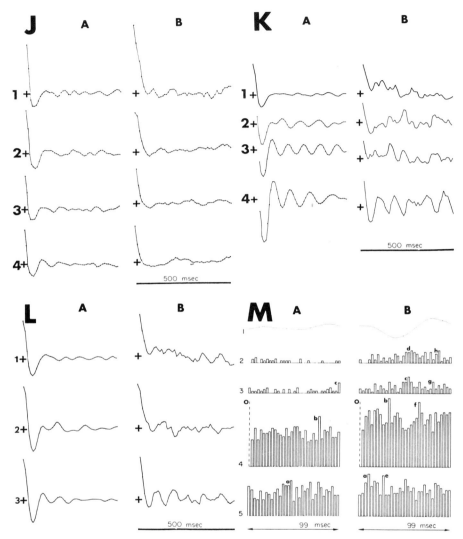

Figure 23.9. Changes in the rhythmicities of the gross waves and of the associated neuronal discharge, and enhancement of circulation of activity induced by conditioning of the cat, with a brief flash of light followed 500 msec later by a brief electrical pulse to the sciatic nerve. (J) Autocorrelograms showing the rhythmicity of the waves (A) and of the neuronal discharge (B) in a control animal in which the two stimuli are specifically unpaired and no conditioning develops; 1, spontaneous activity before any stimulation; 2, 3, and 4, activity recorded progressively later in the process. (K) Autocorrelograms showing the gradual increase in rhythmicity of the waves (A) and of the neuronal discharge (B) in another animal; 1, spontaneous activity before any stimulation; 2, 3, 4, activity recorded progressively later in the process, following the trials in which both stimuli were present and appropriately paired. (L) Autocorrelograms showing the gradual increase in the rhythmicity of the waves (A) and of the neuronal discharge (B), in the same animal and

3. The rhythmicity of the gross waves and that of the associated neuronal discharge, as well as the circulation of activity through the networks, are greatly enhanced during the process of conditioning. This is illustrated in Figure 23.9.

Figure 23.9K shows a series of autocorrelograms that indicate the changes in the degree of rhythmicity of the waves (A) and of the neuronal discharge (B) that appear in the cortex of the cat (visual area II) as the conditioning of the animal, by means of paired sensory stimuli, progresses from the early to the late stages. The stimuli are a brief flash of light followed, 500 msec later, by a brief electric pulse applied to the sciatic nerve. In both columns, A and B, line 1 shows the autocorrelograms that indicate the degree of rhythmicity of the spontaneous waves (A) and of the spontaneous neuronal discharge (B) at the very beginning of the experiment, before any stimulation is initiated. As the conditioning proceeds, lines 2, 3, and 4 show the progressive increase in the degree of rhythmicity of the waves and of the neuronal discharge during those trials in which the stimuli are paired and an electric pulse to the sciatic nerve actually follows the flash of light.

In Figure 23.9L, both columns show, on lines 1, 2, and 3, the progressive increase in the level of rhythmicity of the waves (A) and of the neuronal discharge (B) during those trials in which the flash of light occurs, but the pulse to the sciatic nerve is omitted.

Figure 23.9M shows the evolution of the circulation of neuronal activity, in the same animal, at the same cortical location and during the same conditioning procedure as that shown in the previous two figures. Column A shows a multiple correlogram relating the simultaneous activities of four groups of neurons and of the gross wave, computed early in the conditioning process, from trials in which the sciatic stimulus is omitted. The appearance of a regular temporal and spatial sequence in the distribution of maximum probabilities of discharge (at O–a–b–c), indicates some movement of activity through the neuronal networks, correlating to a small degree with the gross wave (on line 1). Column B shows a similar multiple

Figure 23.9 (continued)

the same conditioning procedure as shown in K; 1, 2, 3, activity recorded progressively later in the process, following the trials in which the light stimulus was present but the sciatic stimulus was omitted. (M) Multiple correlograms computed from recordings obtained from the same animal, during the same conditioning procedure as that illustrated in K and L, showing the relations among the activities of four groups of neurons (2, 3, 4, and 5) and the gross wave (1), at an early stage (A) and at late stage (B) of the conditioning process; movements of activity through the network can be seen at O–a–b–c in the early stage and at O–a–b–c–d and e–f–g–h in the later stage. Autocorrelations as well as multiple correlations were computed over periods of 5.5 sec for each trial. Five such periods (27.5 sec) were averaged to obtain the total correlogram at each stage of conditioning. The method for obtaining the multiple correlograms is the same as that used in obtaining the results shown in Figure 23.8E (see text). (From Rinaldi and Verzeano, in preparation) correlogram, computed late in the conditioning process. It can be seen that, at this stage, the intensity of neuronal discharge (indicated by the height of the vertical bars) and the periodicity of the gross wave have increased, the movements of activity through the network occur more frequently and more regularly (at O–a–b–c–d and at e–f–g–h), and the degree of correlation between them and the wave is considerably higher.

Figure 23.9J shows a series of autocorrelograms computed from data obtained in another animal, in a control experiment, in which every procedure was identical to that illustrated in Figure 23.9K, L, and M, except that the light stimulus and the electric pulse to the sciatic nerve were specifically unpaired, so that no conditioning would develop. It can be seen that in this case, the rhythmicity of the waves (A) and that of the neuronal discharge (B) shows little change as the experiment progresses from the recording of the spontaneous activity before stimulation (1) to its later stages (2, 3, 4).

4. Pharmacological agents such as pentylenetetrazol which, at subconvulsive doses, are known to facilitate memory consolidation (Krivanek & McGaugh, 1968), have a significant action on the movements of activity through the neuronal networks of the cortex and of the thalamus (Negishi, Bravo, & Verzeano, 1963; Verzeano, Laufer, Spear & McDonald, 1970): They enhance the regularity and the coherence of the circulating activity and they increase the number of neurons encompassed within its pathway. This is illustrated in Figure 23.10, which shows the action of subconvulsive doses of pentylenetetrazol on the activity of the nucleus centralis lateralis of the thalamus. After the injection, the clustering of neuronal action potentials is enhanced, and the rate at which neurons are activated (neurons activated/unit time) within the pathway of circulating neuronal activity becomes more marked. When the preconvulsive state is reached (D), the passages of circulating activity through the network are very rhythmic, very rapid, and involve a large number of neurons discharging at a very high rate. The progressive increase in the rate of neuronal discharge under the influence of pentylenetetrazol can be seen with particular clarity by following the evolution of the pattern of discharge of the single neuron shown on tracing b from A to D.

A question arises with respect to the specific role which organized movements of activity through neuronal networks may play in cognitive function. The concept that "reverberating circuits" may be implicated in memory consolidation and learning was advanced many years ago by Young (1938), Hilgard and Marquis (1940), and Hebb (1949). According to this concept, the neural activity that represents the events to be memorized would be maintained by reverberation along chains of neurons until permanent storage is achieved.

This concept, fruitful as it was in focusing the investigative effort on chains of neurons, remained unsatisfactory because the simplicity of the

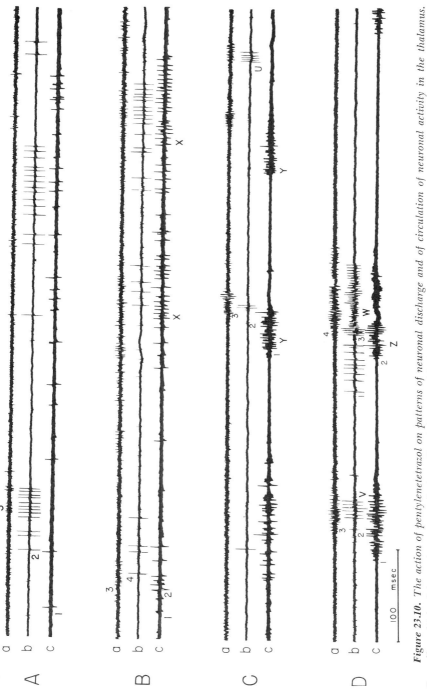

Figure 23.10. The action of pentylenetetrazol on patterns of neuronal discharge and of circulation of neuronal activity in the thalamus. Recording obtained from the n. centralis lateralis of the cat, with three microelectrodes displayed along a straight line. Distances between tips: a to b = 120 μ; b to c = 130 μ. (A) Control record before the administration of pentylenetetrazol; (B) after pentylenetetrazol, 4 mg/kg; (C) and (D) after additional doses of 8 mg/kg. Increased rates of neuronal discharge can be seen at u,z;w,x,y, and z; increased rate at which neurons are activated can be seen at v, w, y, and z; increased regularity in circulation of neuronal activity can be seen (at 1-2-3) in C and D. (From Negishi, Bravo, & Verzeano, 1963.)

mechanism that was proposed did not match the enormous complexity of the processes to be explained.

With the discovery of the movements of activity through the networks of the mammalian brain it was possible to advance a hypothesis that would account for a number of complex events related to cognitive function.

According to this hypothesis (Verzeano, 1977), illustrated in the greatly simplified diagram of Figure 23.11, the movements of activity proceed through very complex networks composed of nested loops located in the thalamus and in the cortex, such as those outlined in Figure 23.4. These networks produce a pattern of excitation and inhibition distributed over a large number of neurons that, as it circulates through the thalamic, cortical, and cortico-thalamic loops, forms a *moving matrix* that continually scans the neurons of the sensory systems, as these are being activated by incoming sensory information. The interaction between the two systems "modulates" the moving matrix and impresses upon it a new pattern of excitation and inhibition that corresponds to the characteristics of the peripheral stimuli. As the matrix continues to move, it carries this new pattern to other systems of neurons.

Such a process could serve at least two functions. It could transport the incoming sensory information from the receiving networks to other networks, where it would be compared with information previously stored, an operation which may be needed for perception; and it could transport it, iteratively, over the networks in which it would ultimately be retained, an operation which may be needed for memory consolidation and learning.

Figure 23.11. Proposed mechanism for the formation of a modulated circulating matrix. The upper part of the diagram (above the horizontal dotted line) represents cortical sensory receiving and other networks. The lower part represents thalamic networks. Large circles represent principal neurons, small circles represent interneurons. Impulses arriving through the sensory pathways over specific afferents (SA) activate specific interneurons (S), which can be excitatory or inhibitory (as shown by the light/dark circles and light/dark synaptic terminals). These interneurons impose a pattern of activity on the principal neurons, which corresponds to the characteristics of the sensory stimulus. In this highly simplified diagram, this pattern is represented by the levels of activity in neurons C_1, C_2, C_3.

Neurons T_1, T_2, and T_3 belong to a thalamic network that generates circulating impulses (arrows) as indicated in Figure 23.4. These impulses, transmitted to the cortex, activate nonspecific interneurons I_1, I_2, I_3 whose output sweeps over neurons C_1, C_2, C_3, and "scans" their state (the pattern of activity that corresponds to the peripheral stimulus), causing them to discharge at a rate related to their level of excitation, and to activate, via interneurons t_1, t_2, t_3, thalamic neurons T_7, T_8, T_9. These thalamic neurons, in turn, activate, via interneurons I_7, I_8, I_9, another group of cortical neurons, C_4, C_5, C_6, and reproduce, in this cortical network, their own pattern of activity. Thus the original pattern of activity C_1, C_2, C_3, engendered by the sensory stimulus, has been transported, under thalamic control and by means of circulating matrices, from one group of cortical neurons to another. By a similar mechanism, involving thalamic neurons T_{10}, T_{11}, T_{12}, the pattern of activity is further transported to cortical network C_7, C_8, C_9 and to other cortical networks over which the activity may have to circulate during the process of consolidation.

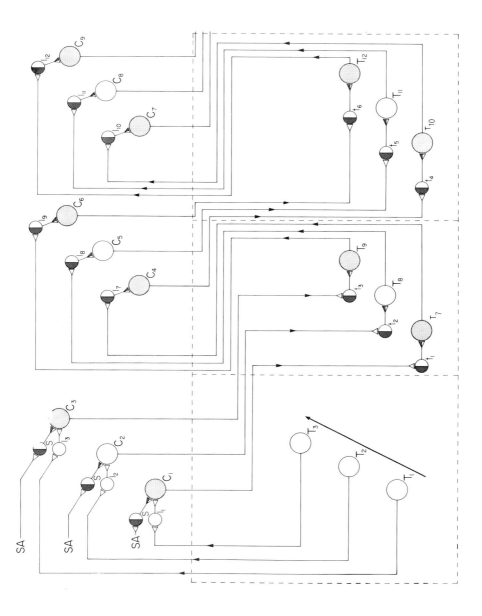

The circulation of the moving matrix involves systems of networks, each one containing large numbers of neurons. Every time it passes through such a system of networks a volley of activity produces excitatory and inhibitory interactions determined by the type of neurons, the type of synapses, and the connectivity that form the system.

Because of the very large number of connections involved and the variability of levels of excitation and inhibition, at any given passage over a network, new neurons may be incorporated into the matrix and others may be left out, in a random fashion. For this reason the moving matrix should be looked upon as a stochastic process: While there may be a considerable amount of variability in the behavior of one of its individual neurons, the matrix system as a whole, statistically viewed, should perform its appropriate function.

Since movements of neuronal activity related to the theta rhythm occur in the hippocampus as well as in the cortex and in the thalamus, the moving matrix hypothesis can be extended to all three structures, in conformity with the view that cognitive function, including memory, is based on the coordinated action of several regions of the brain.

The study of mechanisms of such complexity will require the observation of the activity of hundreds of neurons, in one or several structures simultaneously. The techniques available so far, which permit the simultaneous observation of the activities of a few neurons or groups of neurons, are limited and imperfect. If the neuronal basis of cognitive function is going to be effectively studied, the development of more powerful methods for the analysis of the activity of neuronal networks is imperative.

REFERENCES

Andersen, P., & Andersson, S. A. *Physiological basis of the alpha rhythm.* New York: Appleton-Century-Crofts, 1968.

Berry, S. D., Rinaldi, P., Thompson, R. F., & Verzeano, M. Analysis of temporal relations among units and slow waves in rabbit hippocampus. *Brain Research Bulletin,* 1978, *3,* 509–518.

Berry, S. D., & Thompson, R. F. Prediction of learning rate from the hippocampal electroencephalogram. *Science,* 1978, *200,* 1298–1300.

Dill, R. C., Vallecalle, E., & Verzeano, M. Evoked potentials, neuronal activity and stimulus intensity in the visual system. *Physiology and Behavior,* 1968, *3,* 797–802.

Hebb, D. O. *The organization of behavior.* New York: Wiley, 1949.

Hilgard, E. R., & Marquis, D. G. *Conditioning and learning.* New York: Appleton-Century-Crofts, 1940.

Krivanek, J., & McGaugh, J. L. Effects of pentylenetetrazol on memory storage in mice. *Psychopharmacologia,* 1968, *12,* 303–321.

Landfield, P. W. Synchronous EEG rhythms: Their nature and their possible functions in memory, information transmission and behavior. In W. H. Gispen (Ed.), *Molecular and functional neurobiology.* Amsterdam: Elsevier, 1976a.

Landfield, P. W. Computer determined EEG patterns associated with memory facilitating drugs and with ECS. *Brain Research Bulletin,* 1976b, *1,* 9–17.

Mescherskii, R. M. The vectorgraphical characteristics of spontaneous rabbit brain cortex activity. *Sechenov Physiological Journal of the USSR*, 1961, *47*, 419–426. (In Russian)

Negishi, K., Bravo, M. C., & Verzeano, M. The action of convulsants on neuronal and gross wave activity in the thalamus and in the cortex. *Electroencephalography and Clinical Neurophysiology*, (Suppl.) 1963, *24*, 90–96.

Negishi, K., & Verzeano, M. Recordings with multiple micro-electrodes from the lateral geniculate body and the visual cortex of the cat. In R. Jung & H. Kornhuber (Eds.), *The visual system: Neurophysiology and psychophysics*. Berlin: Springer-Verlag, 1961.

Rinaldi, P., Juhasz, G., & Verzeano, M. Analysis of circulation of neuronal activity in the waking cortex. *Brain Research Bulletin*, 1976, *1*, 429–435.

Rinaldi, P., Juhasz, G., & Verzeano, M. Circulation of cortical and thalamic neuronal activity in wakefulness and in sleep. *Electroencephalography and Clinical Neurophysiology*, 1977, *43*, 248–259.

Rinaldi, P., & Verzeano, M. Manuscript in preparation, 1980.

Verzeano, M. Activity of cerebral neurons in the transition from wakefulness to sleep. *Science*, 1956, *124*, 366–367.

Verzeano, M. The synchronization of brain waves. *Acta Neurologica Latinoamericana*, 1963, *9*, 297–307.

Verzeano, M. Proceedings of the third conference on learning, remembering and forgetting. In D. P. Kimble (Ed.), *Readiness to remember*. New York: Gordon and Breach, 1965, 477–507.

Verzeano, M. Evoked responses and network dynamics. In R. E. Whalen, R. F. Thompson, M. Verzeano, & N. M. Weinberger (Eds.), *The neural control of behavior*. New York: Academic Press, 1970, pp. 27–54.

Verzeano, M. The activity of neuronal networks in memory consolidation. In R. Drucker-Colin & J. L. McGaugh (Eds.), *Neurobiology of sleep and memory*. New York: Academic Press, 1977, pp. 75–97.

Verzeano, M., & Calma, I. Unit activity in spindle bursts. *Journal of Neurophysiology*, 1954, *17*, 417–428.

Verzeano, M., Dill, R., Navarro, G., & Vallecalle, E. The action of metrazol on spontaneous and evoked activity. *Physiology and Behavior*, 1970, *5*, 1099–1102.

Verzeano, M., Laufer, M., Spear, P., & McDonald, S. The activity of neuronal networks in the thalamus of the monkey. In K. H. Pribram & D. E. Broadbent (Eds.), *Biology of memory*. New York: Academic Press, 1970.

Verzeano, M., & Negishi, K. Neuronal activity and states of consciousness. *Proceedings of international congress of physiology*, Buenos Aires, 1959.

Verzeano, M., & Negishi, K. Neuronal activity in cortical and thalamic networks. A study with multiple microelectrodes. *Journal of General Physiology* (Suppl.), 1960, *43*, 177–195.

Verzeano, M., & Negishi, K. Neuronal activity in wakefulness and in sleep. In G. E. W. Wolstenholme & M. O'Connor (Eds.), *The nature of sleep* (Ciba Symposium). London: Churchill, 1961, pp. 108–130.

Winson, J. Loss of hippocampal theta rhythm results in spatial memory deficit in the rat. *Science*, 1978, *201*, 160–162.

Young, J. Z. The evolution of the nervous system and of the relationship of organism and environment. In G. R. de Beer (Ed.), *Evolution*. New York: Oxford Univ. Press, 1938, pp. 179–204.

D. G. SHEVCHENKO

Activity of Visual Cortex Neurons in Systems Processes of Behavioral Act Interchange

24

The selective organization of physiological processes into an integral functional system to achieve a useful adaptive result is the basis of goal-directed behavior, according to the functional system theory of P. K. Anokhin. In the continuum of integral behavior, individual behavior acts replace each other at the times an animal achieves stepwise results. The result of one behavioral act is the simultaneous trigger stimulus for another. The interchange in the organization of the physiological processes that correspond to various behavioral acts is accomplished during the systems processes of afferent synthesis and decision making that take place between the time the results of the preceding act are achieved and the beginning of the action undertaken to achieve the next result in the continuum of goal-directed behavior (Anokhin, 1973; Shvyrkov, 1978).

A comparison of overall electrical activity in different brain structures and individual time intervals of a behavioral act has shown that a primary response and negative evoked potential (EP) are developed between the trigger stimulus and the onset of movement as recorded by EMG activity (Ikeda, 1973; Peimer, 1971; Shvyrkov, Bezdenezhnykh, 1973; and others). The EP components and corresponding phases of neuron activations turn out to be widely generalized and synchronized in different brain regions (Aleksandrov, 1975; John & Morgades, 1969; Livanov, 1975; Olds, Disterhoft, Segal & Kornblith, 1972; Shevchenko, 1975). Individual activation phases that have been observed even in the same neuron have exhibited differences, both with respect to their dependence on current afferentation parameters (Shvyrkov, 1973) and with respect to their association with future events (Aleksandrov, 1975; Shevchenko, 1976; Shvyrkova & Shvyrkov, 1975), and even with respect to their sensitivity to iontophoretically applied biologically active substances (Bezdenezhnykh & Shvyrkov, 1976).

All of these data that have been obtained both in our laboratory and in others allow us to conclude that, in the first place, the time organization of neurophysiological processes in the behavioral act has a uniform structure in different regions of the brain, and, in the second place, the various EP components and their corresponding phases of individual neuron activations are of variable significance in the behavioral act. The comparison we made in our laboratory between the time structure and properties of individual EP components and neuron activation phases and the structure and properties of systems processes in the functional system of an elementary behavioral act enabled us to conceive EP components and phases of individual neuron activation in a behavioral act as correlates of specific systems processes (Anokhin, 1973; Shvyrkov, 1973). As seen from this viewpoint, on the system level, a primary response is a correlate of the process of collating the real result with the action results acceptor or with the goal of the preceding behavioral act. A negative evoked potential and corresponding neuron firing constitute a correlate of simultaneously occurring processes of afferent synthesis; that is, a synthesis of different kinds of information, such as motivation, the situation and memory material, and decision making. The activations corresponding to a late positive component, and later ones, correspond to the systems processes of mobilizing and realizing the actuating mechanisms that have been organized within the systems processes of the results acceptor and the action program (Figure 24.1).

The association between the activations that develop during the transition processes of afferent synthesis and decision making and the nature of neuron activity in preceding and ensuing behavior is a question of great significance in studying the transition of one specific aggregate of active neurons into another during the interchange of behavioral acts. In the present work, we analyzed the relationship between the involvement of visual cortex neurons in transitional systems processes and their participation in the actuating mechanisms of preceding and ensuing behavioral acts.

Rabbits were trained to obtain food by pressing a lever that was placed in a corner opposite the food box in the experimental chamber. A flash of light that appeared either to the right or left of the lever was made to synchronize with the pressing of the lever. Depending on the side from which the light came, the rabbit was supposed to approach the food box from either the right or left, accordingly. When the rabbit made an incorrect approach, the food box would be closed and the rabbit would have to press the pedal again. The rabbit's behavior was very cyclic. It went from the food box to the pedal, and after pressing it and producing the light, the animal returned to the food box and so on. We identified several

24. Visual Cortex Neurons in Systems Processes of Behavioral Act

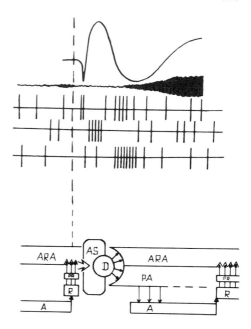

Figure 24.1. Correlation of evoked potential components and neuron phase activations with ganglionic mechanisms of a behavioral act's functional system (Shvyrkov, 1977). From top to bottom: Evoked potential circuit, EMG-activity, phase activations of neurons corresponding to primary positive, negative, and late positive EP. Diagram of sympathetic mechanisms of a functional system. AS, Afferent synthesis; D, decision making; ARA, acceptor of action results; PA, action program; A, section; R, result; PR, parameters of results. Vertical dotted lines indicate flash of light. (From Shvyrkov, 1977.)

behavioral acts in this cycle: The approach to the lever from the food box, pressing the lever and producing the light, and the approach to the food box from the lever and taking the food. In this experiment, we recorded the times at which the lever was pressed, the light was produced, the rabbit's snout was lowered into the food box, and the times at which the rabbit moved. To accomplish this, we attached a miniature incandescent lamp to the rabbit's head, and placed photoelectric plates in the corners of the cage that recorded the beginning and direction of the animal's turns. The impulse activity of the visual cortex neurons was recorded by glass microelectrodes, filled with a 3M solution of KCl, and whose tip diameter was about 1 mc. A micromanipulator was attached to the skull (Grinchenko & Shvyrkov, 1974). The impulse and overall electrical activity of the visual cortex, along with markings, were recorded on a four-channel recorder, subsequently reproduced on paper at a tenfold reduced velocity and finally processed on a minicomputer. A raster analysis and histograms were prepared for the times of all stepwise results, and the evoked potential from the light flashes was averaged. The interval from the flash of light to the animal's turning toward the right or left of the food box was considered to be the time of afferent synthesis and decision making. The interval from the beginning of the animal's turn to the lowering of its snout into the food box was considered to be the time required for the realization of the actuating mechanisms of this behavioral act.

The primary response and negative EP, complicated by various subcomponents, corresponded to the interval between the flash of light and the rabbit's turning toward the food box in all of the experiments conducted on seven rabbits. Early phases coincident with a particular EP component in response to the flash of light were detected in 18 out of the 68 analyzed neurons. Primary activations were recorded in 10 neurons, firings corresponding to negative EP were recorded in 12, and firings corresponding to late positive EP were recorded in 5 neurons. The same neuron could exhibit several activation phases at once.

Twelve neurons were activated in the act of pressing the lever where the flash of light was the result, and 22 neurons were activated in the next act—the approach to the food box where the flash of light served as the trigger stimulus. A comparison of the aggregate of neurons activated in the actuating mechanisms of these acts and the early activation phases indicated that they do not coincide. Only two neurons that were active during the pressing of the lever, and six neurons that were active during the rabbit's approach to the food box, had early phases. The remaining ten neurons that had early phases neither participated in the preceding nor ensuing act relative to the flash of light. Figure 24.2 shows a neuron that fired during descending negative and positive EP, and that was neither activated during the pressing of the lever nor during the rabbit's approach to the food box. Conversely, the neurons that were active both in the preceding and ensuing acts could have been inactive during the transition processes (Figure 24.3). This figure shows a neuron that was activated during the pressing of the lever and the rabbit's turn toward the food box. The described neuron exhibited activity inhibition following the flash of light. Thus, the aggregate of neurons that is active during the transition processes is different from the one that is active in the actuating mechanisms of the preceding and ensuing behavioral acts. This aggregate is consequently not merely the sum of neurons "ending" their activity in the preceding behavior and "beginning" their activity in the ensuing behavior.

The specificity of the activations occurring during the transition processes is once again underscored by the fact that these activations are associated timewise with the trigger stimulus. The activations that are developed during the implementation of the actuating mechanisms regularly precede the appearance of the results, as is demonstrated in Figure 24.3. A "preactivation" before the flash of light is identified in the upper histogram, averaged from the flash of light. Identified in the lower histogram, averaged from the lowering of the snout into the food box, is a "preactivation" before the snout is lowered into the food box.

Only 3 of the 18 neurons with phase activations did not participate at all in the actuating mechanisms of any behavioral acts. Activations were detected in the remaining 15 neurons, except the phase ones, either in adjoining acts or in other stages of behavior (approach to the lever or the taking of food). Thus, the processes occurring during the development of

24. Visual Cortex Neurons in Systems Processes of Behavioral Act

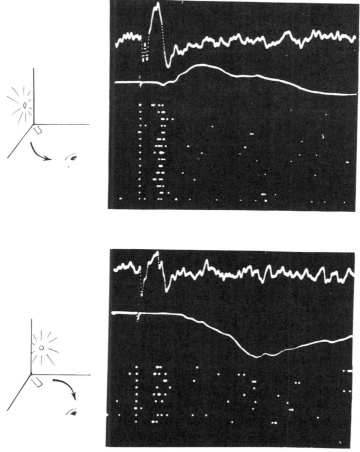

Figure 24.2. Postlight phase activation in a neuron neither participating in a preceding nor ensuing behavioral act. In each block from top to bottom: *Averaged EP in response to the light, markers for the light, and rabbit's turns to the food box, raster of neuron activity.* Upper block: *Rabbit turns to the left (diagram top left);* lower block: *rabbit turns to the right (diagram bottom left). Calibration is 200 msec.*

EP in which those neurons are involved are specific, but not the neurons themselves.

The phase activations changed regardless of the activations corresponding to the actuating mechanisms when the side from which the flash of light came was changed and, correspondingly, the side from which the rabbit approached the food box. Thus, in three out of the six neurons that had both types of activations, only those that corresponded to the realization of the actuating mechanisms were extinguished when the rabbit

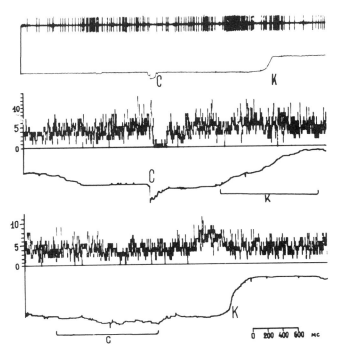

Figure 24.3. Activity inhibition after light flashes in a neuron active before the light and before lowering of snout into food box. From top to bottom: Neurogram; light markers (c) and markers for lowering the snout into the food box (k); histogram plotted from the light markers; mixed markers and dispersed markers for snout lowering into food box; histogram plotted from markers for lowering snout into food box; dispersed markers for light flashes and mixed markers for lowering snout into food box. Channel band in histograms is 5 msec.

approached the food box from different sides. All, or only one of the early activations, were extinguished in the other three neurons, as shown in Figure 24.4. This figure shows a neuron in which a primary activation was recorded and in which the firings corresponded to negative EP when the rabbit approached the food box from the right. This neuron exhibited only a primary activation when the rabbit approached the food box from the left.

In comparing the presence of phase activations in relation to the side from which the light came and the side from which the rabbit turned toward the food box, we found that in seven neurons primary activations were recorded in both types of acts; that is, during the rabbit's approach to the food box from any side, and that in three neurons such activations were recorded in only one type. The firings that corresponded to negative EP were recorded in only two neurons during both types of acts, and in 10 neurons during only one type. Late activations were recorded in only one

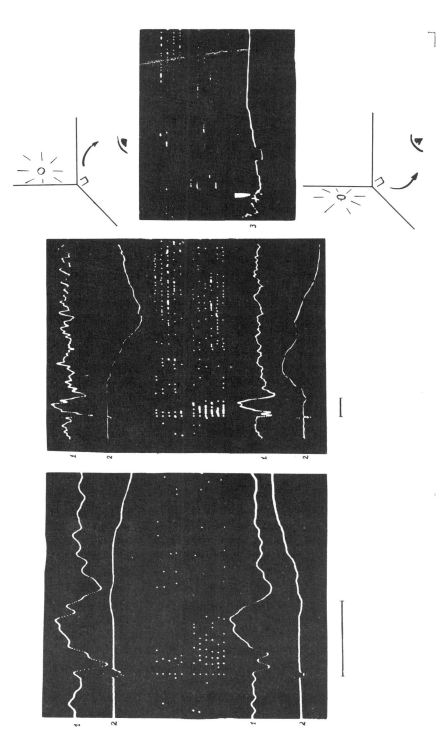

Figure 24.4. Change in activation phases of a visual cortex neuron during turns to the food box from various sides. In the center along the horizontal—Rasters for the activity of the same neuron; upper section of the rasters—turns to the left. Designations: (1) Averaged EP in response to the light; (2) markers for the light and rabbit turns toward the food box (bottom—to the right, top—to the left); (3) markers for the flash of light and lowerings of the rabbit's snout into the food box (combined lowerings designated by arrow. Here the preactivations are also identified). Left-hand and middle columns: Coincidence for light flashes; right-hand column: coincidence for lowerings of snout into food box. Calibrations are 200 msec.

neuron when the rabbit turned to either side, and in four neurons when the rabbit turned toward only one side; that is, in one type of act. Thus, different activation phases are associated in various ways with the nature of memory-derived behavior. At the time of the primary activations, the kind of act that will evolve is not yet known, but whether the rabbit will turn toward the food box from the right or left side can be predicted from the activity of the neurons that fire in accordance with the negative sign of EP and during the late positive sign.

A particularly clear association between individual activation phases and ensuing behavior was demonstrated when several phases were recorded at once in a single neuron, as, for example, is shown in Figures 24.4 and 24.5. Figure 24.5 shows the phase activity of a neuron in one of the rare instances in which the rabbit was observed to have made incorrect turns toward the food box (a turn to the left after a flash of light from the right). These instances demonstrated particularly well the association between the firings corresponding to negative and late positive EP and the nature of a future behavior, but not with the side from which the flash of light was

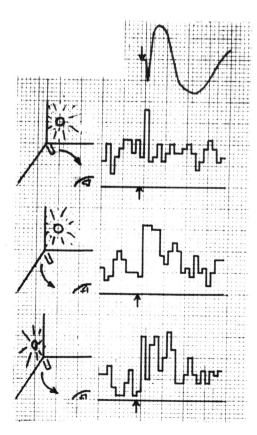

Figure 24.5. Phase activations of the same visual cortex neuron in various types of behavior. Top: Evoked potential circuit. Light designated by arrow. The activation identified in the upper histogram corresponds to the primary positive EP when the rabbit turned toward the food box from the right following a right-handed flash (diagram in top left). The middle and lower histograms show activations corresponding to primary positive, negative, and late positive EP when the rabbit turned left (middle histogram denotes following right-handed flashes; lower histogram denotes after left-handed flashes; diagram on the left). Channel band is 20 msec. Along the vertical is number of spikes in the channel (first square—1 spike), N = 12.

presented. When a primary activation was recorded in any type of behavior, firings corresponding to negative and late positive EP were recorded only when the rabbit turned to the left (middle and lower histograms), although in one type of behavior, the light came from the right (middle histogram) and in another, it came from the right (lower histogram). Consequently, the specificity of transition processes is particularly well defined in individual phases of activation.

Conclusions

Neuron activity in various stages of animal goal-directed behavior has been recorded by a whole series of investigators who have demonstrated that individual neurons may change their activity in either all or some stages of behavior (Andrianov & Fadeev, 1976; Ranck, 1976). In order to study the mechanisms for the transfer of one aggregate of active neurons into another during the interchange of behavioral acts, we compared the activity of neurons in the transition processes of afferent synthesis and decision making to their activity in preceding and ensuing acts. We based our evaluation of neuron involvement in the transition processes on B. V. Shvyrkov's hypothesis on the correlation between the time structure of components and the development of systemic mechanisms of a behavioral act's functional system, and on the presence of phase activations in neurons that correspond to a particular EP component.

The specificity of phase activations that we observed and the differences between the aggregate of neurons yielding such activations and the aggregate of neurons accomplishing goal-directed behavior probably reflects the fact that the involvement of neurons into a system in various developmental stages of a behavioral act's functional system is accomplished in accord with different informational factors, such as goal, medium, and movement (Shvyrkov, 1978). The goal and medium remained constant under the conditions of our experiments when the rabbit made different turns toward the food box. Only the trigger stimulus and the means of achieving the result were changed.

Inasmuch as the aggregate of neurons that yielded primary activations had little to do with the selected means of achieving the result, one would think that this aggregate is selected in accordance with both possible means and, consequently, in accord with both types of light flashes. The firings corresponding to negative and late positive EP are already significantly associated with one of the means of achieving the result, but the aggregate of firing neurons does not yet coincide with that that leads to the achievement of the goal. This stage corresponds to decision making; that is, to fewer variables of organizations of the actuating mechanisms for the anticipated action. The aggregate of neurons that are active at this time is probably created in accord with a turn only toward one side by various means. Finally, the aggregate of neurons that are active during the realization of the actuating mechanisms has already been created in accordance with only one method of accomplishing the turn.

Thus, during the interchange from one behavioral act to another, the various activation phases of neurons probably reflect sequential stages in the organization of elements into a system in accordance with one specific mode by which a specific goal in a specific medium is achieved.

REFERENCES

Aleksandrov, Yu. I. Changes in the configuration of conditional reflexes of visual cortex neurons of a rabbit's brain changes in reinforcement parameters. *Journal of Higher Nervous Activity*, 1975, *25*, 760. (In Russian)

Andrianov, V. V., & Fedeev, Yu. A. Impulse activity of visual cortex neurons in sequential stages of instrument behavior. *Journal of Higher Nervous Activity*, 1976, *26*, 916. (In Russian)

Anokhin, P. K. Fundamental questions in the general theory of functional systems. In *Principles of the systems organization of functions*. Moscow, 1973. (In Russian)

Anokhin, P. K. Systems analysis of the conditional reflex. *Journal of Higher Nervous Activity*, 1973, *23*, 229 (In Russian)

Bezdenezhnykh, B. N., & Shvyrkov, V. B. Changes in phase-induced activity of cortical neurons during the iontophoretic application of glutamate, gamma-aminobutyric acid, and atropine. *Neurophysiology*, 1976, *8*, 192.

Grinchenko, Yu. V., & Shvyrkov, V. B. A simple micromanipulator for examining neuron activity in rabbits in free movements. *Journal of Higher Nervous Activity*, 1974, *24*, 870. (In Russian)

Ikeda, T. *Folia Psychiatrica et Neurol.*, Japan, 1973, *27*, 3.

John, E. R. Switchboard versus statistical theory of learning and memory. *Science*, 1972, *177*, 850.

John, E. R., & Morgades, P. P. Neural correlates of conditional responses studied with multiple chronically implanted moving microelectrodes. *Experiments in Neurology*, 1969, *23*, 412.

Livanov, M. N. Neuronal mechanisms of memory. *Progress in Physiology*, 1975, *6*, 66. (In Russian)

Olds, J., Disterhoft, J. K., Segal, M., & Kornblith, C. L. Learning centers of rat brain mapped by measuring latencies of conditioned unit responses. *Journal of Neurophysiology*, 1972, *35*, 202.

Peimer, S. I. Time and space field distribution of evoked potentials and the reprocessing of information in the human brain. In *Mechanisms of induced brain potentials*. Leningrad, 1971, p. 21.

Ranck, J. B. Behavioral correlates and firing repertoires of neurons in the dorsal hippocampal formation of septum of unrestrained rats. In *The hippocampus* (Vol. 2). New York: Plenum, 1976.

Shevchenko, D. G. Study of reticular formation neurons in the rabbit's midbrain in the defensive conditional reflex. *Journal of Higher Nervous Activity*, 1975, *25*, p. 727. (In Russian)

Shevchenko, D. G. Reticular formation neurons in the mechanisms of decision-making. In *Problems of decision-making*. Moscow, 1976, p. 210.

Shvyrkov, V. B. Neuronal mechanisms of learning as a component of a behavioral act's functional system. In *Principles of the systems organization of functions,* 1973, p. 156. (In Russian)

Shvyrkov, V. B. *Neurophysiological study of systems mechanisms of behavior.* Moscow, 1978.

Shvyrkov, V. B. On the relationship between physiological and psychological processes in the functional system of the behavioral act. *Studia Psychologica,* 1977, *19*(2), 89–95.

Shvyrkov, V. B., & Bezdenezhnykh, B. N. The role of conditional and unconditional stimulant analyzers in the functional system of a behavioral act. *Journal of Higher Nervous Activity,* 1973, *23,* 15. (In Russian)

Shvyrkova, N. A., & Shvyrkov, V. B. Activity of visual cortex neurons in eating and defensive behavior. *Neurophysiology,* 1975, *7,* 100. (In Russian)

DAVID F. LINDSLEY
KENT M. PERRYMAN
DONALD B. LINDSLEY

Brain Mechanisms of Attention and Perception[1]

25

Under the direction of Professor Sokolov, Dr. Olga Vinogradova and I studied the orienting reflex properties of single visual neurons in the striate cortex of restrained, unanesthetized rabbits (Vinogradova & Lindsley, 1964). About half of 147 cells in the primary visual cortex showed a response to light flash that was unchanged during prolonged repetition of the stimulus. The other half showed a change in response to flash and/or click with stimulus repetition. Two types of extinction were observed. One group of cells, including a neuron that responded to movement of a light spot in one direction but not in the opposite direction, responded to a new stimulus (either visual or auditory) and ultimately ceased responding with repetition of the stimulus, presumably as the novelty of the stimulus was lost. Another group responded to the stimulus by exhibiting a decrease in spontaneous activity and showed a suppression of response that increased in duration with stimulus repetition. These labile or plastic changes in the responses of primary visual cortical neurons to stimulus repetition presumably reflect the extinction of the orienting reflex at the single cell level.

Professor K. L. Chow, Dr. Morton Gollender, and I extended this work on the modifiability of neuronal responses in the lateral geniculate nucleus, one synaptic station below the visual cortex (Chow, Lindsley, & Gollender, 1968). Of 145 geniculate cells that were studied in 23 unanesthetized cats, nine neurons showed one response pattern to a light

[1] This research was supported by USPHS Grants NS 8552 and MH 25938 to D. B. Lindsley and supplemented by funds from The Grant Foundation.

spot presented to the contralateral eye before pairing with a diffuse light flash to the ipsilateral eye, and a different pattern after pairing. Only two of the nine cells showed a response to light flash alone. The modified responses after pairing took the form of either latency shifts or pattern alterations. Having gained some understanding of the primary visual system (geniculo-striate), I have become interested in the second visual system (tecto-pulvinar-extrastriate). The second visual system appears to play a role in visual attention and, in particular, in foveation of visual stimuli of interest to the animal, that is, moving the eyes so as to place the image of the object on the fovea. It is now well known that superior colliculus cells respond to eye movements (EMs), and that the pulvinar nucleus of the thalamus receives projections from the superior colliculus. Dr. Kent Perryman, my father, Dr. Donald Lindsley, and I have been studying the responses of pulvinar neurons to EMs and flashes of light. Eye movements play a significant role in visual perception and may be considered also an index of visual attention.

Extracellular microelectrode recordings were made in the inferior, lateral, and medial divisions of the pulvinar nucleus of chronic, restrained squirrel monkeys during spontaneous EMs. Figure 25.1 shows a sketch of the animal preparation. During recording sessions the monkey sat in an upright position with head and eyes at the center of a 6-in, translucent plastic hemisphere illustrated by the stippled circle. The body was supported in a padded contour chair, and the head was rigidly attached to a stereotaxic frame by means of bolts embedded in an acrylic pedestal mounted on top of the monkey's head. The animal's eyes were free to move in any direction within the externally illuminated, translucent hemisphere that constituted a homogenous visual field or Ganzfeld. Eye movement electrooculograms (EOGs) resulting from horizontal saccades were recorded from AgAgCl electrodes held at the external canthi of the eyes by plastic adjustable arms shown in Figure 25.1. Poststimulus time histograms (PSTHs) were triggered either by EOGs or by light flashes. Experimental single unit recording and data storage and analysis procedures have been described previously by Perryman and Lindsley (1977).

PSTHs to spontaneous eye movements were compared under three conditions: light-adapted Ganzfeld, dark-adapted state, and with brief flashes in the dark. Figure 25.2 shows the PSTHs of two cells in the inferior pulvinar to EMs in the light, in the dark, and to a brief flash. The initiation and duration of the EM are indicated by the EOG tracings above each of the upper four PSTHs. The PSTHs of the neuron on the left (D5–4B) showed responses under all three conditions, whereas the neuron on the right (C14–4) did not respond to the flash. About 10% (56) of cells in a restricted area of the lateral and inferior pulvinar showed EM responses in the light. Figure 25.3 illustrates the distribution of responses under the

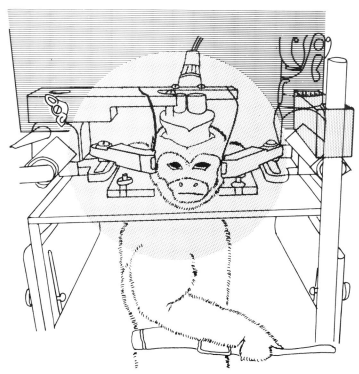

Figure 25.1. Illustrations of chronic, restrained, squirrel monkey preparation. The head of the monkey was located at the center of a 6 in., translucent plastic hemisphere shown by the stippled circle. The body was supported in a padded contoured chair, and the head was rigidly attached to a stereotaxic frame by three bolts embedded in an acrylic pedestal mounted on top of the head. Penetrating this pedestal and the skull were two lucite cylinders through which microelectrodes could be lowered repeatedly for recording in different regions of the pulvinar. The animal's eyes were free to move in any direction within the externally illuminated, translucent hemisphere which constituted a homogenous visual field or Ganzfeld. Horizontal saccades were recorded by means of AgAgCl disc electrodes held by adjustable plastic arms to the sides of the head at the external canthi as shown. (From data of Perryman, Lindsley, & Lindsley.)

three conditions. Figure 25.3A shows that 35 cells exhibited excitatory responses to spontaneous EMs in the light, and 21 cells showed inhibitory responses. Figure 25.3B shows that 18 cells (about one-third of the 56 EM light-responsive cells) responded to EMs in the dark, and all of them were in the inferior pulvinar; 6 exhibited excitatory responses and 12 showed inhibitory responses. About two-thirds (37) of the 56 cells also responded to light flashes delivered in the dark-adapted state; 21 of them were excitatory and 16 inhibitory (Figure 25.3C). Nine of the 56 cells did not respond to EMs in the dark or to light flashes and showed only EM re-

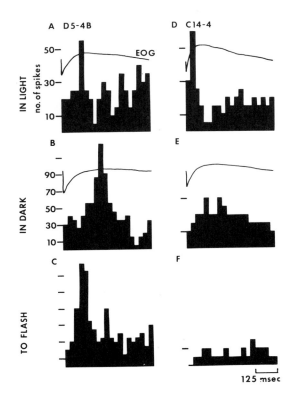

Figure 25.2. Poststimulus-time histograms (PSTHs) for two inferior pulvinar cells to spontaneous eye movements (EMs) under three conditions: light-adapted Ganzfeld, dark-adapted state, light-flash in the dark. PSTHs were triggered at the onset of horizontal saccades by electroculograms (EOGs at the top of the upper four histograms), or to light flashes. Note that the neuron on the left (D5–4B) shows responses under all three conditions, whereas the neuron on the right (C14–4) did not respond to the flash. The PSTHs represent the summed responses to 25–50 EMs and to 50 light flashes. The ordinate shows number of spikes; time calibration is 125 msec. (From data of Perryman, Lindsley, & Lindsley.)

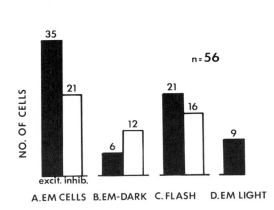

Figure 25.3. Proportion of pulvinar eye movement (EM) responsive cells which give excitatory or inhibitory responses under three conditions: (A) light-adapted Ganzfeld (all 56 EM cells); (B) dark-adapted state (18 EM-dark cells; (C) light flash (37 flash cells); (D) EM responses in lighted Ganzfeld only with no response to EMs in the dark or light flash (9 EM-light cells). Black bars indicate excitatory-type responses and white bars inhibitory-type responses. Total cells studied, $N = 56$. Note that of the 56 neurons with postsaccadic EM responses in the light, about one-third showed EM responses in the dark (B), about two-thirds responded to light flash (C), and 9 cells responded to EMs in the light only (D). (From data of Perryman, Lindsley, & Lindsley.)

sponses in the light; all of these were of the excitatory type (Figure 25.3D). Unlike the EM cells of the inferior and lateral pulvinar, there were no EM responsive cells in the medial pulvinar.

None of the 56 EM cells in the pulvinar showed presaccadic responses; all showed postsaccadic responses. Figure 25.4 shows peristimulus-time histograms from neurons at different levels of the visual system. The EM is indicated by the small vertical line under each histogram. Figure 25.4A is an example of a postsaccadic response of a pulvinar cell (C9–17D); this neuron had an inhibitory-excitatory response. Peristimulus-time histograms from superior colliculus cells showed responses both preceding and following the onset of the EMs (Figure 25.4B, cell A4–1D). The histogram of Figure 25.4C shows the activity from a lateral geniculate cell (DTr1–1D) during EMs; no EM-responsive cells were found in the lateral geniculate nucleus.

Figure 25.5 shows the distribution of EM-responsive cells in the pulvinar. Neurons responding to EMs were localized to a columnar region about 1 mm in diameter in the lateral (diagonal lines) and the inferior (black bar) pulvinar. Cells responding to EMs in the dark were found only in the inferior pulvinar (black bar). Neurons outside of this column in the inferior, lateral, and medial pulvinar did not respond to EMs (stippled areas). EM-responsive cells were randomly distributed within the column.

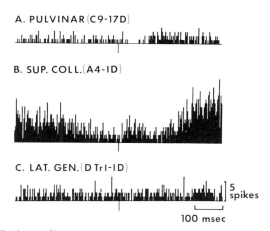

Figure 25.4. Peri-saccadic activity in the pulvinar nucleus, superior colliculus, and the lateral geniculate nucleus of the thalamus recorded during spontaneous horizontal eye movements (EMs). EMs are indicated by vertical lines below each peri-saccadic time histogram. (A) The response of pulvinar cell C9–17D is typical of pulvinar neurons which showed only post-saccadic responses to EMs; this neuron shows an inhibitory-excitatory pattern of response to EMs. (B) The response of superior colliculus cell A4–1D is typical of superior colliculus cells which showed both pre- and post-saccadic responses as indicated. (C) The activity of lateral geniculate cell DTr1–1D is typical of lateral geniculate cells in showing no responses to EMs. Ordinate calibration represents five spikes; time calibration is 100 msec. (From Perryman, Lindsley, & Lindsley.)

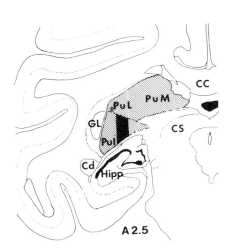

Figure 25.5. Regions of pulvinar where eye movement (EM) responsive cells were found. The cells responding to EMs were restricted to a cylindrical region 1 mm in diameter extending from the medial border of the lateral pulvinar (PuL) downward through the inferior pulvinar (PuI). Cells responding to EMs in the dark were found only in the inferior pulvinar (solid black region). Cells responding to EMs in the light were found both in lateral pulvinar (diagonal lines) and in the inferior pulvinar (solid black). Stippled regions of pulvinar were unresponsive to EMs. PuI-inferior pulvinar; PuL-lateral pulvinar; PuM-medial pulvinar; GL-lateral geniculate; CS-superior colliculus; Cd-caudate nucleus; Hipp-hippocampus; CC-corpus callosum. This cross-section of the squirrel monkey brain is at A 2.5. (Data from Perryman, Lindsley, & Lindsley.)

Except for the localization of dark EM cells to the inferior pulvinar, there was no further anatomical segregation of EM cells within this region in terms of dark and/or light EM cells or excitatory or inhibitory responses.

Of the more than 500 cells studied within the column extending vertically from the inferior pulvinar upward into the lateral pulvinar, only 56 or about 10% were responsive to eye movements in the light. Of these 56 neurons one-third responded to eye movements in the dark, and all of them were located within the region of the inferior pulvinar. About two-thirds of the 56 cells also responded to flashes of light. All of the cells which responded to EMs in the dark also responded to light flashes, but not all cells that responded to light flashes responded to EMs in the dark. Because 18 of the 56 light-responsive cells also responded to EMs in the dark (in the absence of light or light flashes), it seems clear that these neurons can be activated by oculomotor-related input. All EM cells in the pulvinar showed only postsaccadic responses. Such oculomotor input could come from proprioceptive feedback from the extraocular muscles or could emanate from oculomotor nuclei, superior colliculus, intralaminar nuclei, frontal eye fields, or other brain areas associated with eye movement-related activity. These cells that responded to EMs in the dark and were activated by light flashes in the absence of EMs may be capable of integrating oculomotor and visual information.

In further support of the position that the inferior pulvinar neurons that responded to EMs in the dark may be involved in integrating oculomotor and visual information, there is anatomical evidence in the rhesus monkey for projections from the deeper layers of the superior colliculus to the inferior pulvinar and medial region of the lateral pulvinar (Benevento & Fallon, 1975). These regions of the pulvinar correspond in general to the columnar region in which all of the EM-responsive cells have been found. The activity of the cells in the deep layers of the superior colliculus has been associated with EMs, manifesting both pre- and postsaccadic discharges (Goldberg & Wurtz, 1972; Robinson & Jarvis, 1974; Schiller & Koerner, 1971; Wurtz & Goldberg, 1972; Wurtz & Mohler, 1976).

There is additional evidence for oculomotor input to pulvinar neurons. Nine cells responded only to EMs in the light and not to light flashes or EMs in the dark (see bar graph, Figure 25.3D). These cells were not responding to variations in contrast, because the retinal field was uniformly illuminated (Ganzfeld). There was no difference in spontaneous activity during the dark-adapted and light-adapted conditions that might account for the EM responses in the light on the basis of tonic visual (retinal or corticofugal) influences. These nine EM light-responsive cells may have reached threshold for discharge through spatial-temporal convergence of retinal and oculomotor inputs, so that EM-related activity only occurred in the light-adapted state.

The majority of EM cells (37 of 56, Figure 25.3C) could be activated by a visual flash. Generally, the excitatory and/or inhibitory responses associated with flashes were more prominent and had a shorter latency than those related to EMs. Thus, visual stimulation seemed to exert a stronger influence on neural activity in the pulvinar than did oculomotor input. The variations in response to visual and oculomotor input may be related to different types of pulvinar neurons, since two distinct sizes of cells have been observed in the pulvinar nucleus of the squirrel monkey (Mathers, 1971, 1972, 1973) and the rhesus monkey (Campos-Ortega & Hayhow, 1972). The differences in cell size or type may be associated with differential processing of oculomotor and visual information. For example, the inferior pulvinar neurons that responded to EMs in the dark may differ from other inferior and lateral pulvinar cells that responded only to EMs in the light. There may be stages of visual-oculomotor processing in this region of the pulvinar, with EM cells that respond in the dark representing an earlier state of processing of oculomotor information and converging with visual information on pulvinar neurons that respond only to EMs in the light. Such visual input could come from striate and extrastriate projections to the pulvinar nucleus. On the basis of the foregoing evidence and the possibilities proposed, it is suggested that the inferior and lateral pulvinar are involved in the integration of visual and oculomotor information (Perryman, Lindsley, & Lindsley, 1976, 1980).

Anatomical and physiological data indicate that the pulvinar nucleus is part of a second visual system, interposed between superior colliculus and extrastriate cortex, as contrasted with the primary or geniculo-striate system. Two studies in rhesus monkeys have failed to demonstrate visual discrimination deficits following pulvinar lesions. Chow (1954) and Mishkin (1972) reported that retention of visual discriminations learned preoperatively was not impaired when tested following lesions of the pulvinar.

Thompson and Myers (1968), however, found that medial and lateral pulvinar lesions in monkeys caused an impairment in retention of visual discriminations learned preoperatively. This result may be questioned due to the possibility of extrapulvinar invasion of visual pathways (Chalupa, Coyle, & Lindsley, 1976). These lesion studies employed a testing procedure that permitted the monkey a relatively long time to scan the discriminative stimuli before making a response. In contrast, Chalupa et al. (1976) found visual discrimination deficits following pulvinar lesions in Macaca nemestrina monkeys when very brief stimuli were used. Four monkeys with only inferior pulvinar lesions and one monkey with inferior plus medial and lateral pulvinar lesions were markedly impaired in the postoperative learning of a visual pattern discrimination, using an "N" versus a "Z" discrimination task presented for 10 msec. Monkeys with medial or lateral pulvinar lesions showed no deficit in learning ability compared to unoperated control monkeys. Chalupa et al. (1976) believe that this task is highly demanding of visual attention and have suggested that the impairment in visual pattern discrimination may be due in part to deficits in visual attention. Additional support for this derives from the fact that two of the monkeys with inferior pulvinar lesions, which learned to discriminate only after many trials, subsequently showed marked impairment in performance when extraneous and distracting stimuli (annular surrounds and superimposed interference grids) were introduced. The performance of monkeys with medial and lateral pulvinar lesions, as well as the control monkeys, was only temporarily disrupted by this procedure. Similar interpretations of impairment in visual discrimination, caused by irrelevant stimuli, have been advanced by Diamond and associates (1971, 1973) in the tree shrew and by Butter (1969) in the monkey. Thus, the Chalupa et al. (1976) study adds further support to the concept of a second visual system in which the inferior pulvinar plays a significant role in visual pattern discrimination highly demanding of alertness and attention.

The work we have reported here on pulvinar neuronal responses associated with eye movements and visual flashes indicates that the pulvinar of primates may be an important region for the integration of visual and oculomotor information. In addition, since eye movements can be considered in part an index of visual attention, it is also suggested that the pulvinar may be involved in visual attention. To investigate further the role of the pulvinar in visual attention, experiments are being designed to study the responses of pulvinar cells in primates trained to make pattern discriminations to brief duration visual stimuli.

ACKNOWLEDGMENTS

The pulvinar data reported here were obtained in collaboration with Drs. Kent M. Perryman and Donald B. Lindsley. More detailed results will be available shortly (Perryman, Lindsley, & Lindsley, 1980).

I would like to take this opportunity to indicate my appreciation for the USA–USSR Scientific Exchange Agreement that has permitted us to participate in a symposium with our Russian colleagues. Special thanks are due to Professors R. F. Thompson and B. F. Lomov who organized the program. A similar exchange agreement permitted me to spend a year with Professor E. N. Sokolov of the Department of Physiology of Higher Nervous Activity of Moscow State University in 1961–1962. From the time I first began working in the field of neuroscience, my basic interest has been in brain mechanisms of attention and perception.

REFERENCES

Benevento, L. A., & Fallon, J. H. The ascending projections of the superior colliculus in the rhesus monkey (Macaca mulatta). *Journal of Comparative Neurology*, 1975, *160*, 339–362.

Butter, C. M. Impairments in selective attention to visual stimuli in monkeys with inferotemporal and lateral striate lesions. *Brain Research*, 1969, *12*, 374–383.

Campos-Ortega, J. A., & Hayhow, W. R. On the organization of the visual cortical projections to the pulvinar in Macaca mulatta. *Brain, Behavior, Evolution*, 1972, *6*, 394–423.

Chalupa, L. M., Coyle, R. S., & Lindsley, D. B. Effect of pulvinar lesions on visual pattern discrimination in monkeys. *Journal of Neurophysiology*, 1976, *39*, 354–369.

Chow, K. L. Lack of behavioral effects following destruction of some thalamic association nuclei in monkey. *A.M.A. Archives of Neurology and Psychiatry*, 1954, *71*, 762–771.

Chow, K. L., Lindsley, D. F., & Gollender, M. Modification of response patterns of lateral geniculate neurons after paired stimulation of contralateral and ipsilateral eyes. *Journal of Neurophysiology*, 1968, *31*, 729–739.

Goldberg, M. E., & Wurtz, R. H. Activity of superior colliculus in behaving monkeys. II. Effect of attention on neuronal responses. *Journal of Neurophysiology*, 1972, *35*, 560–574.

Harting, J. K., Glendenning, K. K., Diamond, I. T., & Hall, W. C. Evolution of the primate visual system: anterograde degeneration studies of the tecto-pulvinar system. *American Journal of Physical Anthropology*, 1973, *38*, 383–392.

Killackey, H., Snyder, M., & Diamond, I. T. Function of striate and temporal cortex in the tree shrew. *Journal of Comparative and Physiological Psychology*, 1971, *74* (No. 1, Pt. 2, Monograph), 1–29.

Mathers, L. H. Tectal projection to the posterior thalamus of the squirrel monkey. *Brain Research*, 1971, *35*, 295–298.

Mathers, L. H. Ultrastructure of the pulvinar of the squirrel monkey. *Journal of Comparative Neurology*, 1972, *146*, 15–42.

Mathers, L. H., & Rapisardi, S. C. Visual and somatosensory receptive fields of neurons in the squirrel monkey pulvinar. *Brain Research*, 1973, *64*, 65–83.

Mishkin, M. Cortical visual areas and their interactions. In A. G. Karczmar & J. C. Eccles (Eds.), *Brain and Human Behavior*. New York: Springer-Verlag, 1972, pp. 187–208.

Perryman, K. M., Lindsley, D. F., & Lindsley, D. B. Pulvinar unit responses associated with eye movements in squirrel monkeys. *Neuroscience Abstract*, 1976, *II*, 279.

Perryman, K. M., & Lindsley, D. B. Visual responses in geniculo-striate and pulvino-extrastriate systems to patterned and unpatterned stimuli in squirrel monkeys. *Electroencephalography and Clinical Neurophysiology*, 1977, *42*, 157–177.

Perryman, K. M., Lindsley, D. F., & Lindsley, D. B. Pulvinar neuron reponses to spontaneous and voluntary eye movements and light flashes in squirrel monkeys. *Electroencephalography and Clinical Neurophysiology*, 1980, in press.

Robinson, D. L., & Jarvis, C. D. Superior colliculus neurons studied during head and eye movements of the behaving monkey. *Journal of Neurophysiology*, 1974, *37*, 533–540.

Schiller, P. H., & Koerner, F. Discharge characteristics of single units in the superior colliculus of the alert monkey. *Journal of Neurophysiology*, 1971, *34*, 920–936.

Thompson, R., & Myers, R. E. Brainstem mechanisms underlying visually guided responses in the rhesus monkey. *Journal of Comparative and Physiological Psychology*, 1968, *74*, 487–498.

Vinogradova, O. S., & Lindsley, D. F. Extinction of reactions to sensory stimuli in single neurons of visual cortex in unanesthetized rabbits. *Federation Proceedings*, 1964, *23*, T241–T246. (English translation of article in Pavlov, *Journal of Higher Nervous Activity*, 1963, *13*, 207.)

Wurtz, R. H., & Goldberg, M. E. Activity of superior colliculus in behaving monkey. IV. Effects of lesions on eye movements. *Journal of Neurophysiology*, 1972, *35*, 587–596.

Wurtz, R. H., & Mohler, C. W. Organization of monkey superior colliculus: enhanced visual response of superficial layer cells. *Journal of Neurophysiology*, 1976, *39*, 745–765.

MICHAEL E. GOLDBERG
DAVID LEE ROBINSON

Behavioral Modulation of Visual Responses of Neurons in Monkey Superior Colliculus and Cerebral Cortex

26

Study of the sensory properties of single neurons has led to a significant understanding of the way in which the brain handles visual information. Hubel and Wiesel (1968) have demonstrated that single neurons have well-defined requirements for stimulus features and spatial localization, and they have shown that in the monkey striate cortex there is a functional architecture that provides an anatomical foundation for the physiological properties of these neurons, at least in terms of binocular interaction and stimulus orientation (Hubel, Wiesel, & Stryker, 1978). Wurtz (1969) showed that the properties so elegantly demonstrated by Hubel and Wiesel (1968) in the anesthetized preparation are present also in the awake, behaving monkey.

The brain, however, does not just absorb visual information. It actively filters and selects significant stimuli from the field for further analysis, and the question that the visual system must answer is not only what is the stimulus, but how relevant is the stimulus for behavior. In order to answer this latter question, the brain could utilize neurons whose visual excitability is modifiable by behavioral factors. The nature of this excitability change is of great importance: It may be a qualitative change, such as a change in receptive field requirements or even an abolition of the limits of a receptive field; or it may be a quantitative change, such as an increase in the discharge rate evoked by a stimulus within the preexisting field. We have studied two areas of the brain of the rhesus monkey (*Macaca mulatta*), the superior colliculus and the posterior parietal cortex, where the quantitative effectiveness of a visual stimulus is related directly to the importance of that stimulus to the monkey's behavior. The receptive field properties of the neurons in these areas are not significantly modified by the behavioral context, but the behavioral factor acts as a gain control on the effect of the visual stimulus. In the superior colliculus, the response of neurons to visual stimuli is enhanced when the stimulus becomes a target for an eye movement, but not under circumstances when the animal responds to the stimulus in a manner that does not require an eye movement. In the posterior parietal cortex, visual neurons have an enhanced discharge to the visual stimulus whenever the animal uses the stimulus as data for the generation of behavior, regardless of whether or not the stimulus is a target for eye movement or merely a signal for some other response.

Methods

We trained rhesus monkeys to fixate a spot of light using the methods developed by Wurtz (1969). While the monkeys gaze at the spot, they do not break fixation to look at a second spot of light flashed on a screen. The second spot can then serve as a stimulus for analyzing visual properties of a neuron while the animal voluntarily holds his eyes still. For other experiments the monkeys learned to make saccadic eye movements from the first spot to a second spot when the fixation point went off as the target appeared. In a third task the monkey learned to respond to the dimming of the large peripheral stimulus without breaking fixation from the central stimulus (Wurtz & Mohler, 1976). We then prepared the trained monkeys for chronic neurophysiological recording using the methods of Evarts (1966). We used platinum-iridium microelectrodes (Wolbarsht, MacNichol, & Wagner, 1961) to record extracellular single neuron activity, and permanently implanted silver-silver chloride electrodes to record horizontal and vertical electrooculograms (Bond & Ho, 1970). We then recorded the activity of single neurons in the superior colliculus and posterior parietal cortex of the rhesus monkeys while the monkeys were either fixating a single spot or making an eye movement from one spot to another. A PDP 11/10 computer controlled the animal's behavior, sampled and analyzed eye movements, and digitized single-unit pulses. The computer calculated rasters, histograms, and cumulative histograms on-line, and stored them on a disk. We made electrolytic lesions at the site of certain cells, and located the lesions histologically.

Results

In both the superior colliculus and the posterior parietal cortex there are neurons with visual receptive fields (Cynader & Berman, 1972; Goldberg & Robinson, 1977; Goldberg & Wurtz, 1972a; Robinson, Goldberg,

26. Visual Responses of Neurons in Colliculus and Cerebral Cortex

& Stanton, 1979; Yin & Mountcastle, 1977). Each superior colliculus has a representation of the contralateral visual hemifield, with a foveal enlargement anterior in the colliculus. Unlike the neurons in the striate cortex, collicular neurons for the most part are not fastidious about stimulus orientation or color, and in the monkey only a small percentage of neurons are directionally selective. The receptive fields of the neurons are much larger than those in striate cortex and in fact increase with increasing depth in the colliculus (Goldberg & Wurtz, 1972b; Humphrey, 1968). Figure 26.1 shows a typical receptive field of the pandirectional neuron in the monkey superior colliculus. These cells respond to the onset of a

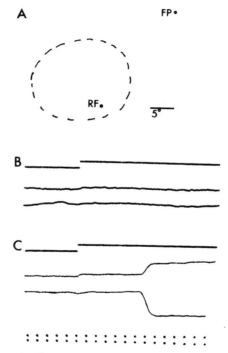

Figure 26.1. Schematic diagram of experimental conditions illustrating eye movements in no-saccade and saccade cases. Dashed circle in (A) shows the receptive field of a neuron in the monkey's left lower visual field, as defined by an excitatory response to the onset of a 1 × 1 degree spot of light. FP describes the position in the field of the fixation point. RF indicates the location of the spot that was used as the visual stimulus and saccade target in (B) and (C). (B) shows the horizontal (upper) and vertical (lower) electrooculograms in the no-saccade case. The animal had begun gazing at the fixation point before the figure began. The stimulus came on when the horizontal artifact above the electrooculogram traces began, and stayed on. There was no significant eye movement to break fixation and look at the receptive field stimulus. (C) shows the eye movements during the saccade case. At the beginning of the horizontal artifact the receptive field stimulus came on and the fixation point disappeared. At some time later the monkey made a saccade to fixate the stimulus. Time dots are 50 msec apart. (Reproduced with permission from Goldberg & Wurtz, 1972b.)

small spot of light within a large excitatory area, and they also have an inhibitory surround, but no other requirements for stimulus parameter. Ninety percent of the cells in the superficial layers of the superior colliculus are pandirectional, and 10% are directionally selective.

Fifty percent of the cells in the superficial layers of the colliculus have discharge properties that are modifiable by the behavioral significance of the stimulus. We analyzed the passive visual properties by flashing a test stimulus on the tangent screen while the animal was looking at a fixation point. The test stimulus was irrelevant to the animal's behavior, and the animal learned to make no response to the stimulus. If the fixation point disappeared when the test stimulus came on, the animal made a saccade

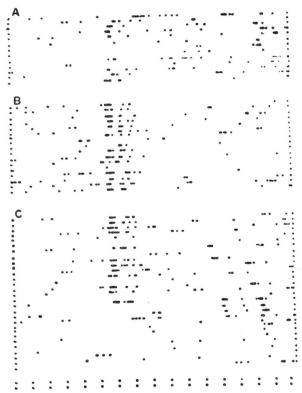

Figure 26.2 Saccade-related enhancement of neuron in monkey superior colliculus. (A) shows the response of the neuron in the no-saccade case. (B) shows the enhanced response of the same nueron in the saccade case. (C) shows the waning of the response when the animal returns to the no-saccade case. Each dot of the raster display represents a single unit discharge, or the beginning or end of each line. Each line represents an 800 msec epoch. Successive lines are synchronized on the onset of the receptive field or saccade stimuli. Time dots are 50 msec apart. (Reproduced with permission from Goldberg & Wurtz, 1972b.)

to fixate the test stimulus. In this paradigm the identical stimulus became a target for an eye movement instead of an irrelevant stimulus. The diagram in Figure 26.1 shows eye movements of the monkey in first, the fixation ("no-saccade") case and in second, the saccade case. Note that in the reaction time following the onset of the stimulus and preceding the eye movement the retinal situation is the same, but the behavioral situation is quite different: In the first case the animal is presumably ignoring the stimulus, and in the second case the animal is preparing to make an eye movement to fixate it.

In the superficial layers of the colliculus 50% of the cells show an enhanced response to the stimulus when the stimulus is the target for a saccadic eye movement, as opposed to the case when the animal fixates a fixation point and ignores the stimulus. Figure 26.2 shows the response of a neuron in the superior colliculus to a stimulus that is not the target for an eye movement (Figure 26.2A), and the response of the same neuron to the same stimulus when the stimulus is the target for an eye movement (Figure 26.2B). The response of the neuron is greatly enhanced when the stimulus in the receptive field is the target for a behavioral act. When the task reverts to the no-saccade task, the response returns to the no-saccade level after a number of trials. As long as the saccade task continues, the enhanced response does not habituate. If one examines the discharge pattern of this type of neuron in total darkness in relation to a spontaneous eye movement made in the absence of a visual stimulus, the neuron not only shows no discharge, but its background activity may be inhibited (Goldberg & Wurtz, 1972a; Robinson & Wurtz, 1976). Thus the enhancement effect is not a summation of a sensory response and a motor command as can be found in the intermediate layers of the superior colliculus (Wurtz & Goldberg, 1972), but is rather a behavioral effect that operates to modify the visual sensory input. The enhancement effect is selectively related to the case when the monkey makes an eye movement to the receptive field of the cell. If one studies the response of a neuron to the stimulus in the receptive field in the case when the animal makes a saccade to a simultaneously presented stimulus that is far from the receptive field of the cell, there is no enhancement of the response to the stimulus. The excitatory receptive fields of the cell stay roughly the same size in the saccade case and the no-saccade case, although the margins of the field may be slightly extended (Goldberg & Wurtz, 1972a; Wurtz & Mohler, 1976), but the basic topography of the retinotopic map in the monkey superior colliculus is maintained during this kind of goal-directed behavior. Figure 26.3 shows the responses of a neuron in the superior colliculus to stimuli inside and outside of the receptive field during saccade and non-saccade trials. It is of interest that the margins of the receptive field are not affected by the behavior, but the intensity of the discharge is.

Recent experiments by Hyvärinen and Poranen (1974) and Mount-

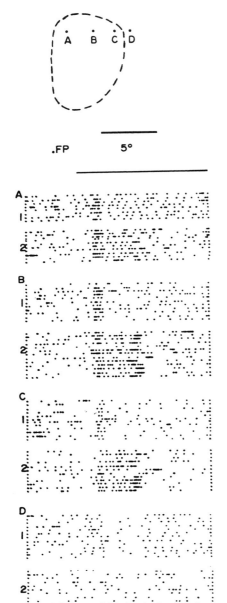

Figure 26.3. Maintenance of receptive field topography in the presence of response enhancement. The diagram shows the receptive field of a superior colliculus neuron as described using a 1×1 degree spot stimulus. Saccade and no-saccade cases were studied using stimuli at points A, B, and C inside the receptive field, and point D outside it. Each pair of responses shows the no-saccade (1) and the saccade (2) results at each point in the field. Thus, A1 shows the no-saccade and A2 the saccade cases for the stimulus at point A. Successive lines of each raster are synchronized on the onset of the receptive field or saccade stimulus. (Reproduced with permission from Goldberg & Wurtz, 1972b.)

castle (1976) and his colleagues (Lynch, Mountcastle, Talbot, & Yin, 1977; Mountcastle, Lynch, Georgopoulous, Sakata, & Acuna, 1975) have shown that neurons in the posterior parietal cortex (area 7) of the rhesus monkey discharge in relation to saccadic and pursuit eye movement and visual fixation. We found that every neuron associated with visual behavior had a visual receptive field and that, like the superior colliculus, half of these neurons have a behavioral enhancement (Goldberg & Robinson, 1977; Robinson, Goldberg, & Stanton, 1978). Parietal cells have receptive fields that are large, predominantly but not exclusively contralateral, and even less fastidious about stimulus parameters than those of cells in the colliculus. The neurons respond better to large, bright stimuli and have remarkably long visual latencies, most in the range of 80–140 msec from the onset of the stimulus. Figure 26.4 shows the response of a parietal neuron to a stimulus in its receptive field in the no-saccade (Figure 26.4A) and saccade (Figure 26.4B) conditions. The neuron shows a markedly enhanced response when the animal uses the stimulus in its receptive field as the target for an eye movement. Enhancement in the parietal cortex, like that in the superior colliculus, is spatially specific.

Eye movements, however, are not the only form of visual behavior. We frequently perform acts based on visual cues that are not dependent upon making a targeting movement toward that cue. In the superior colliculus, stimuli that are important for behavior but do not serve as the target for eye movement do not evoke enhanced responses (Wurtz &

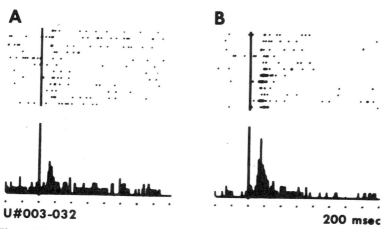

Figure 26.4. Saccade-related response enhancement in area 7 of the monkey. (A) shows the response of a posterior parietal neuron when the animal does not make an eye movement to fixate the stimulus. (B) shows the enhanced response of the same neuron to the same stimulus when the animal makes a saccade to the stimulus. Successive trials are synchronized on the onset of the stimulus, and the eye movements occur (in B) on the average 250 msec after the stimulus onset. The histograms sum the responses in the rasters above. Bin width is 24 msec.

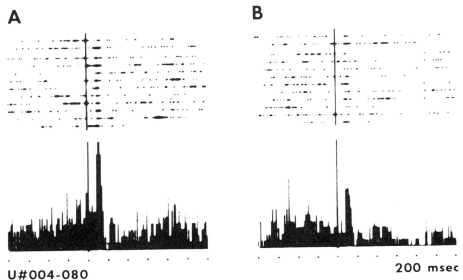

Figure 26.5. Saccade-independent enhancement in area 7 of the monkey. (A) shows the response of the neuron to the onset of a stimulus to which the monkey may have to respond by releasing a lever when the stimulus dims. The animal does not make an eye movement to fixate the stimulus. Stimulus onset occurs at the vertical line in each trial. (B) shows the response of the same neuron to the same stimulus in a series of trials in which the animal did not have to respond to the stimulus.

Mohler, 1976). In area 7, however, behaviorally relevant stimuli evoke enhanced responses regardless of whether or not they are eye movement targets. We trained monkeys to respond to a change in intensity of one of two simultaneously presented stimuli by releasing a lever. One was a large stimulus placed in the periphery, and the second was a .1 degree of arc spot. The latter was so small that the animal could only perform the task by fixating the spot. The former was large enough so it could perceive the intensity change with its peripheral vision. Both stimuli came on before either dimmed, and at the onset of the peripheral stimulus the animal could not predict which of the stimuli would dim, so we can assume that the animal was attending to both stimuli. The animal also performed series of trials in which he did not have to respond to the peripheral stimulus, and this series provided a control in which the previously significant stimulus was no longer important to the monkey. Every parietal neuron that showed eye movement-related enhancement had attentional enhancement in the task when the animal responded to but did not make a movement toward the stimulus. Figure 26.5 shows an example of such a response enhancement.

Thus cells in the parietal lobe and the superior colliculus have their basic visual properties quantitatively modified when the stimulus is important to some aspect of behavior. In the superior colliculus, the behavioral modification occurs only when the relevant stimulus is going to be the target for an eye movement. In the parietal cortex, the enhancement occurs with eye movements and with visual tasks that do not require a targeting movement. This may indicate that the superior colliculus participates in a functional subsystem of the visual system involved in eye movements (Shvyrkov, this volume), whereas the parietal cortex is involved more generally in the process of visual attention. In each case, however, the effect of the behavioral modulation is to change the quantitative aspects of the visual excitability, but not to interfere with the basic receptive field properties of the neuron. Behavioral modifiability is not a universal property of visual neurons. Attempts to demonstrate spatially specific behavioral responsiveness in striate (Wurtz & Mohler, 1977) and prestriate cortices (Baizer & Robinson, 1974) have been unsuccessful.

REFERENCES

Baizer, J. S., & Robinson, D. L. Activity of prestriate neurons in behaving monkey. *Abstract 35, Society for Neuroscience Abstracts*, 1974.
Bond, H. W., & Ho, P. Solid miniature silver-silver chloride electrodes for chronic implantation. *Electroencephalography and Clinical Neurophysiology*, 1970, 28, 206–208.
Cynader, M., & Berman, N. Receptive-field organization of monkey superior colliculus. *Journal of Neurophysiology*, 1972, 35, 187–201.
Evarts, E. V. Methods for recording activity of subcortical neurons in moving animals. *Methods of Medical Research*, 1966, 11, 241–250.
Goldberg, M. S., & Robinson, D. L. Visual responses of neurons in monkey inferior parietal lobule: The physiologic substrate of attention and neglect. *Neurology*, 1977, 27, 350.
Goldberg, M. E., & Wurtz, R. H. Activity of superior colliculus in behaving monkey: I. Visual receptive fields of single neurons. *Journal of Neurophysiology*, 1972a, 35, 542–559.
Goldberg, M. E., & Wurtz, R. H. Activity of superior colliculus in behaving monkey: II. Effect of attention on neuronal responses. *Journal of Neurophysiology*, 1972b, 35, 560–574.
Hubel, D. H., & Wiesel, T. N. Receptive fields and functional architecture of monkey striate cortex. *Journal of Physiology* (London), 1968, 195, 215–243.
Hubel, D. H., Wiesel, T. N., and Stryker, M. P. Anatomical demonstration of orientation columns in macaque monkey. *Journal of Comparative Neurology*, 1978, 177, 361–380.
Humphrey, N. K., Responses to visual stimuli of units in the superior colliculus of rats and monkeys. *Experimental Neurology*, 1968, 20, 312–340.
Hyvärinen, J., & Poranen, A. Function of the parietal associative area 7 as revealed from cellular discharges in alert monkeys. *Brain*, 1974, 97, 673–692.
Lynch, J. C., Mountcastle, V. B., Talbot, W. H., & Yin, T. C. T. Parietal lobe mechanisms for directed visual attention. *Journal of Neurophysiology*, 1977, 40, 362–389.
Mountcastle, V. B. The world around us: Neural command functions for selective attention. The F. O. Schmitt Lecture in Neuroscience for 1975. *Neuroscience Research Progress Bulletin*, 1976, 14 (Suppl.), 1–47.
Mountcastle, V. B., Lynch, J. C., Georgopoulous, A., Sakata, H., & Acuna, C. Posterior parietal association cortex of the monkey: Command functions for operations within extrapersonal space. *Journal of Neurophysiology*, 1975, 38, 871–908.

Robinson, D. L., & Wurtz, R. H. Use of an extraretinal signal by monkey superior colliculus neurons to distinguish real from self-induced stimulus movement. *Journal of Neurophysiology*, 1976, *39*, 852–870.

Robinson, D. L., Goldberg, M. E., & Stanton, G. B. Parietal association cortex in the primate: Sensory mechanisms and behavioral modulations. *Journal of Neurophysiology*, 1978, *41*, 910–932.

Wolbarsht, M. L., MacNichol, E. F., & Wagner, H. G. Glass insulated platinum microelectrode. *Science*, 1960, *132*, 1309–1310.

Wurtz, R. H. Visual receptive fields of striate cortex neurons in awake monkeys. *Journal of Neurophysiology*, 1969, *32*, 727–742.

Wurtz, R. H., & Goldberg, M. E. Activity of superior colliculus in behaving monkey. III. Cells discharging before eye movements. *Journal of Neurophysiology*, 1972, *35*, 575–586.

Wurtz, R. H., & Mohler, C. W. Enhancement of visual responses in monkey striate cortex and frontal eye fields. *Journal of Neurophysiology*, 1976, *39*, 766–772.

Yin, T. C. T., & Mountcastle, V. B. Visual input to the visuomotor mechanisms of the monkey's parietal lobe. *Science*, 1977, *197*, 1381–1383.

Hormonal, Pharmacological, and Developmental Factors

PART V

ROBERT L. ISAACSON

Limbic System Contributions to Goal-Directed Behavior

27

It sometimes seems that I have been studying the effects of brain-damaged animals for a lifetime, although, on reflection, the "lifetime" amounts only to about 20 years. Much of what my students, colleagues, and I have learned relates to changes in behavior that are related to motivation, learning, and memory. At one time I thought that by studying these changes carefully we could learn something about the functions of a particular brain area, in my case the hippocampus, a major component of the limbic system. I am less convinced of this than I was some years ago, and would endorse the statement of Mountcastle (1978) that it may be possible to localize a lesion but not a function. However, the data we have generated do tell us a great deal about how behavior is organized and do carry important lessons for all who are interested in behavioral analysis.

One of the puzzling aspects of studying the behavior of brain lesions in animals is the behavioral variability found after what are by all usual histological evaluations identical lesions. This variability is found after all types of lesions, but was formally noted by Thomas (1971) in the study of rats with lesions of the fornix. We have repeatedly found variability in behavior after bilateral hippocampal lesions that we have not been able to relate to either the amount or the nature of the lesions.

The explanation of this variability must begin with the individual genetic invitations given every individual and their interactions with both the pre- and postnatal environments.

Figures 27.1 and 27.2 show in a schematic fashion these interactions.

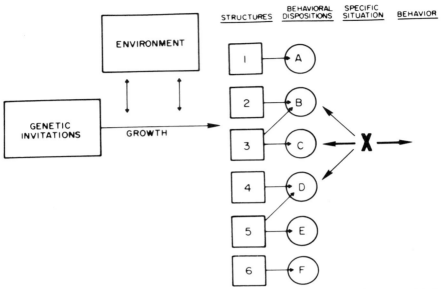

Figure 27.1. Schematic drawing of the way in which genetic and environmental influences interact to produce a set of neural structures and "behavioral dispositions" (microstructures?) in a particular individual (A). (From Isaacson, 1975.)

Furthermore, the squares in each figure relate structures, meaning neural structures, with specific behavioral dispositions. The arrangement of the arrows in these figures is meant to indicate that behavioral dispositions reflect activity in several anatomic (or biochemical) systems and that a one-to-one correspondence between any structure or system and a manner of behavior should not be expected.

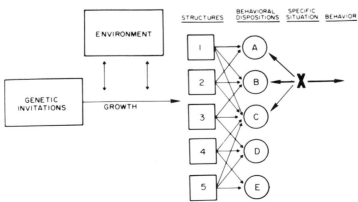

Figure 27.2. Schematic drawing of the development of neural structures in a different individual (B). (From Isaacson, 1975.)

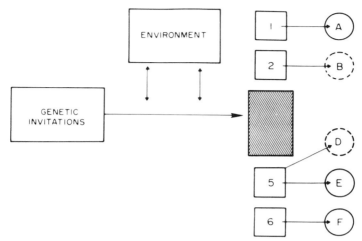

Figure 27.3. Hypothetical effects of brain damage of areas 3 and 4 in individual (A). This damage eliminates behavioral capacity (C) and changes (B) and (D). (From Isaacson, 1975.)

Figures 27.3 and 27.4 show the hypothetical effects of removing areas or systems in two individuals. The areas removed would be the same (blocks 3 and 4), but the results would be quite different behaviorally. In individual (A), some behavioral capacities are lost and others altered; in (B) no capacity is lost, but all but one affected. However, this type of analysis is helpful in understanding the variability of behavior after brain damage, but ignores the time course of anatomical, physiological, and behavioral changes found after brain damage.

Brain damage whether it be through experimental lesions, accidental

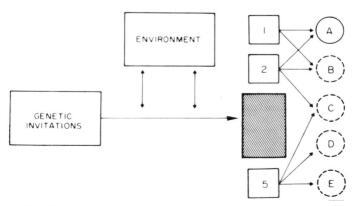

Figure 27.4. Hypothetical effects of brain damage of areas 3 and 4 in individual (B). In this case no behavioral capacity is lost but a number are altered. (From Isaacson, 1975.)

trauma, or disease initiates a series of events in the remaining structures. These events change over the course of time and although we can make a lengthy list of such changes (e.g., Isaacson, 1975), we do not really know all that occur and understand only dimly the known ones. Although fundamentally an act of faith, it seems likely that behavioral effects of brain damage must reflect the alterations in the anatomy, physiology, and chemistry in the remaining areas that have been disturbed by the damage.

There is no way in which comparable lesions can be made to two animals, despite the greatest of precision of surgery. Because of the individuality of the animals, the progressive series of anatomical and physiological changes initiated will be quite different. As always, we must resist the temptation to separate behavioral and physiological variables. Alterations in behavior imply alterations in physiological functions.

Scientists working in the relationship between the biology of the organism and behavior all too frequently forget one of the major sources of this variability, namely, the genetic makeup of the individual. One example of this potent variable is shown in Figure 27.5, where in this study (Isaacson & McClearn, 1978) two selectively bred strains of mice were tested for open-field activity. These are the high and low activity strains of DeFries and Hegmann (1970). As can be seen, the selectivity for the high and low activity lines is quite distinct. In this study one group of

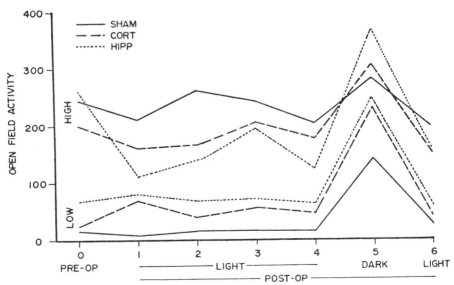

Figure 27.5. Preoperative and postoperative locomotor activity in sham operated, hippocampally lesioned (HIPP) and neocortically lesioned (CORT) mice of the high and low activity lines developed by DeFries. Conditions of testing are described in the text. (From Isaacson & McClearn, 1978.)

animals from each line were given hippocampal lesions and one group from each line were given lesions of the overlying cortex. In these mice, either type of lesion decreased the activity of the high activity mice but increased that of the low activity line. These tests were conducted in a reasonably well-lit environment, but on the fifth testing session the animals were evaluated under subdued illumination. As can be seen, the nearly dark testing conditions greatly changed the behavior of animals with hippocampal lesions. They evidenced a great deal more activity, regardless of their selection line than they did in the light. Two important things can be learned from this experiment. The first is that the effect of hippocampal destruction, and probably any other type of brain damage, depends upon the genetic endowment of the animals. Second, these effects of genetic inheritance and the lesion interact with testing conditions to produce a particular behavior. This indicates that it is very difficult, if indeed possible, to talk about *the effects* of any particular form of brain damage. Indeed, the work of Donovick and his associates with animals with septal lesions (e.g., Donovick, Burright, & Swidler, 1973) indicates that environmental conditions given the animals before and/or after septal area damage produce tremendous influences on the behavior exhibited by the animals after such damage.

Another neglected factor in the field of brain behavior studies is the sex of the animals being tested. In Figure 27.6, some data from a study by Isaacson, Yongue, and McClearn (1978) are presented. In this example, the effect of apomorphine on male and female Long Evans rats is shown. The animals were tested in a 16-hole open field situation similar to that used by File and Wardill (1975) as is shown in Figure 27.7. In Figure 27.6 the number of exploratory responses made by the animals at different doses of apomorphine are shown. Exploratory responses were defined in terms of the animals poking their heads into one of the holes during the testing. As can be seen, the female animals responded in an exaggerated fashion to low doses of the drug. A close examination of the literature of the effects of brain damage reveals numerous instances of differences between male and female animals in response to brain lesions as well as pharmacological interventions. Yet, we tend to ignore these differences and to base our conclusions on data obtained only from males.

If the hypothesis that brain damage initiates a series of progressive changes in both physiology and behavior of animals is correct, this should be readily found when animals are tested at different times during the postoperative or posttraumatic period. In Figure 27.8, one such example is shown. This is from a study by Dr. Linda Lanier Spear and myself (Lanier & Isaacson, 1975) in which animals were repeatedly tested after bilateral hippocampal lesions. It shows the changes in movements of the animals in a large open field at different times after surgery. Animals with nearly complete bilateral hippocampal lesions and with ventral hippocampal

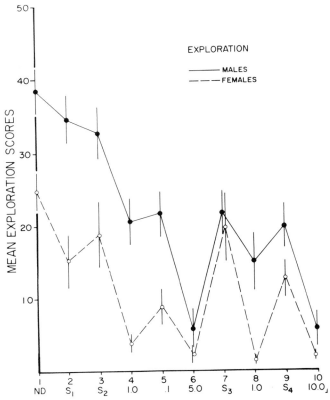

Figure 27.6. The exploration (head-dipping) scores of male and female rats tested in the 16-hole box given different doses of apomorphine. ND: No injection given; S_1, S_2, S_3, and S_4: days when saline injections were given. Other numbers on x-axis indicate doses of apomorphine given in mg/kg. (From Isaacson, Yongue, & McClearn, 1978.)

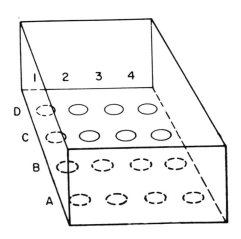

Figure 27.7. The 16-hole box used to test for locomotion, exploration, and grooming. (From Isaacson & Green, 1978.)

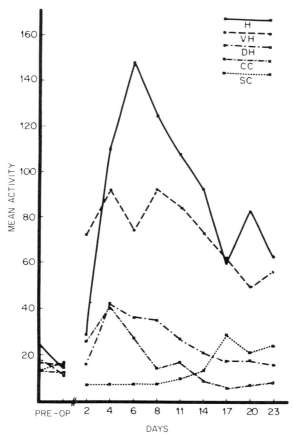

Figure 27.8. *Locomotor activity at various postoperative days of animals with nearly complete (H) dorsal (DH) and ventral (VH) hippocampal lesions; those with lesions restricted to the neocortex (CC) and sham-operated controls (SC). (From Lanier & Isaacson, 1975.)*

lesions show a peak in activity about six days after the lesion that becomes reduced over time. Very little effect is found in animals with dorsal hippocampal lesions or control animals with lesions involving only the cortex overlying the hippocampus. This progressive change in locomotion after the lesion does not depend upon the repeated testing of the animals. When groups of animals are tested only once at one of these postoperative times, almost the same results are observed. The only difference is in the pronounced decline in locomotor activity found approximately two weeks after hippocampal damage that does not occur to nearly the same extent. This supports the observations of Donovick and his co-workers in regard to the importance of the postoperative experiences of the animal in determining the final outcome of brain damage. These data not only illustrate a progressive change after bilateral hippocampal damage, but also point out the behavioral differences between destruction of the dorsal and ventral aspects of the structure.

If, as hypothesized, the effect of brain damage is to induce changes in remaining neural systems, many of the usual consequences of brain damage may not be absolutely necessary results of the damage. Of course, we already know this, since pre-and postoperative environments and treatments do influence the ultimate consequences of brain damage as shown previously. A related question is to what degree pharmacological interventions can also alter the progressive changes after brain damage and the final or long-term nature of the changes. It may even be possible to intervene with the remaining brain systems to produce a state approaching the damage conditions for these secondary systems with the proper drugs and amounts. We have tried to do this by using animals with bilateral hippocampal lesions. We consider this a useful preparation because the behavioral consequences of such damage have been so well studied.

One of the most common characteristics of bilateral hippocampal damage is the unusually high rate of response found in intermittent schedules of reinforcement in the operant situation (see Isaacson, 1974, for a review). One of these schedules is the delayed response schedule in which an animal must wait 20 sec between bar presses in order to produce a reinforcement (DRL-20 schedule). A deficit in this task was first reported by Clark and Isaacson (1965) and has been found by many investigators since then. In 1973 Van Hartesveldt (unpublished observations) found that the abnormally high rate of response found in such animals after this schedule could be reduced by the administration of chlorpromazine. In an attempt to determine if chlorpromazine's effect was due to actions on dopaminergic systems, these original observations were confirmed and extended by Schneiderman and Isaacson (1975). Fish (1976) went on to study the effect of haloperidol on animals with such lesions. Her data are shown in Figure 27.9. The upper portion of this figure shows the extremely high rate of response generated by the animals on this schedule, and the lower portions show the reduced number of reinforcements obtained by the animals. As also shown in Figure 27.9, the administration of several doses of haloperidol produces a dramatic decrease in the rate of response but only small changes in the number of rewards obtained by the animals. Consideration of interresponse times indicates that the drug does not greatly alter them. The main effect of the drug is to produce an overall suppression of response rate, bringing it in the range produced by normal animals without the drug. However, the effect of the drug on the rate of response of animals with hippocampal lesions is by no means a complete panacea. The conclusion to be drawn is that the effect of haloperidol or chlorpromazine is primarily on the rate of response. These drugs do not greatly improve the

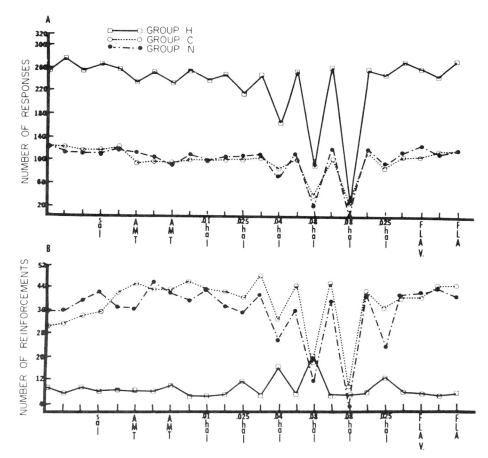

Figure 27.9. Number of responses made (upper) and number of reinforcements obtained (lower) by animals with hippocampal lesions (H), neocortical lesions (C), and normal animals (N) on the DRL-20 task. Drugs given are sal: saline; AMT: alpha-methyl-paratyrosine; hal: haloperidol in amounts indicated (mg/kg); and FLA: FLA-63. (From Fish, 1976.)

animals' ability to time their responses in order to obtain rewards on the DRL-20 schedule.

For a great many reasons, it seems likely that the effect of hippocampal damage is related to an exaggerated responsiveness of at least one component of the ascending forebrain dopaminergic systems. This model emphasizes the role of fibers from the ventral hippocampus and adjacent subiculum that reach the nucleus accumbens (Swanson & Cowan, 1975). The general hypothesis under which I have been working for the past several years is that destruction of the ventral hippocampus produces an enhanced sensitivity in the forebrain basal ganglia areas, probably the nucleus accumbens. It is known to receive direct afferents from the ventral hippocampal region, although secondary changes through multisynaptic pathways could occur in the caudate as well.

Reinstein, Isaacson, and Dunn (1979) undertook a more direct evaluation of this hypothesis by giving animals ^{3}H-2-deoxyglucose (2-DG) two weeks after large bilateral hippocampal lesions by using the method of Delanoy and Dunn (1978). The animals were sacrificed 45 min after injection of the tritiated 2-GD at a postoperative interval of two weeks. The results indicate that the destruction of the hippocampus produced a selective decrease in activities of the hypothalamus and the olfactory bulb. With this method we did not study the nucleus accumbens. This, of course, is something that we will be undertaking in the near future.

Certain areas of the brain, especially those of the limbic system, are sensitive to circulating hormones. Receptor and binding sites have been demonstrated in such areas, which suggest that pituitary, adrenal, sexual, and other hormones may influence the brain's activities by modulation of neural activities in these areas. The hippocampus is one of the limbic system regions that has been shown to be associated with hormonal actions. The study of hormonal effectiveness after hippocampal destruction represents another approach to the study of changes that occur after brain damage.

We have recently tried to clarify that the relationship between the hippocampus and the behavioral effects of the ACTH fragments are without corticosteroid effects, for example, ACTH 1–10, ACTH 4–9, ACTH 1–16, 7-d-phe (ACTH 4–10) (de Wied, 1974). Also stimulating our interest is the fact that many of the effects produced by ACTH fragments without adrenal effects are similar to some of the behavioral consequences of bilateral hippocampal lesions. The enhanced resistance to extinction found under several training conditions by ACTH 4–10 are a classic example. Furthermore, the peripheral administration of this fragment is without effect in animals with dorsal hippocampal lesions (Van Wimersma Greidanus, Bohus, & de Wied, 1974).

Recently, another effect of ACTH fragments has come under study. This is the excessive grooming obtained when certain ACTH fragments are injected into the ventricular system. This excessive grooming occupies the animals for approximately an hour or more after interventricular[1] injection (in the foramen of Monro) when placed in a small observation chamber. The ACTH fragments 7-d-phe 4–10, ACTH 1–16, and ACTH 1–24 are all very effective in producing this behavior, although ACTH 4–10, itself, is not. It is of interest that grooming can also be induced by mor-

---

[1] The word interventricular is used rather than intraventricular since the injection is made at the intersection of the lateral ventricles.

phine, Beta-endorphin, and other morphinelike agents. Recently, we have shown that this effect can be blocked by near total lesions of the hippocampus (Colbern, Isaacson, Bohus, & Gispen, 1977) but not by lesions restricted to the dorsal hippocampus. Lesions of that area, as well as the septal area and other brain sites investigated, did not inhibit the excessive grooming in response to interventricular injection of ACTH 1–24.

When we first discovered that large bilateral lesions of the hippocampus affected ACTH-fragment-induced grooming, we became concerned that perhaps the only effect was one of intensifying the predominant response of the animal in the test situation. Since the tests are usually made in a highly restricted environment, a small Plexiglas observation cage, we were concerned that the hormone fragment only potentiated the most likely response to occur in the situation. Accordingly, we undertook to determine if excessive grooming would be produced in other situations in which other responses occurred with a higher probability than did grooming. We tested animals with implanted cannulae into the interventricular foramen in the open field with 16 holes in the floor shown in Figure 27.7. After several days of preliminary testing during which we recorded locomotion, exploration, rearing, and grooming, we tested them after interventricular injection of ACTH 1–24 (Isaacson & Green, 1978). As can be seen from Figure 27.10 the greatest change was a very large increase in grooming associated with a reduction of the other behaviors. Since these other behaviors had been much more prominent than grooming before the administration of the hormone, we conclude that the effect of the hormone is not simply to enhance the most likely response to be made in the situation but, rather, is specific to the grooming behavior. Grooming can be induced by the transport and handling of the animal to the testing chamber. An example of this is shown in Figure 27.11 (Colbern, Isaacson, Green & Gispen, 1978). In this study, the difference between being tested in the novel environment and the home cage is illustrated in the bottom two lines of the graph. The heightened grooming score found in the novel environment can be thought of as an after-effect to novelty. Although we have not studied it systematically in this laboratory, it occurs as an after-effect to many different types of stressful conditions, both natural and experimentally induced (e.g., Fentress, 1968; MacFarland, 1966; van Ierssel & Bol, 1958). This after-effect to novelty seems to be mediated by ACTH released from the pituitary since it is greatly reduced in hypophysectomized animals and in animals that have been treated with antiserum to human ACTH (Dunn, Green, & Isaacson, 1979). These data suggest that the after-effect to novelty or stress as reflected in grooming is mediated by ACTH released from the pituitary and is transported to the brain. These grooming effects are also subject to interference by treatment with naloxone at doses that do not produce a general debilitation of the animal. Grooming, as an after-effect to novelty,

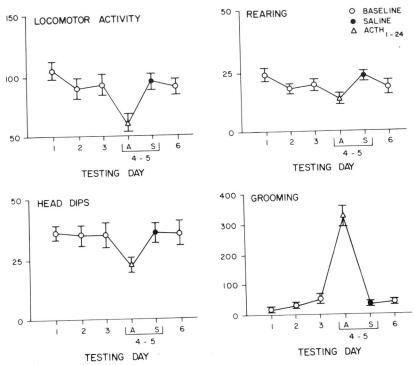

Figure 27.10. *Locomotor activity, rearing, head dipping, and grooming measured in the 16-hole box before and after the interventricular administration of ACTH 1-24. (A) Day of ACTH 1-24 administration; (S) day of saline administration. (From Isaacson & Green, 1978.)*

is reduced by doses of this drug that fail to influence locomotion, exploration, or rearing in the 16-hole open-field test chamber.

These data suggest that it is possible to develop hypotheses concerning the interrelationship of the limbic system, the basal ganglia, and at least certain aspects of the hormonal systems. It appears that the hormone fragment ACTH 4–10 has its effects on nuclei of the septal area, the dorsal and posterior regions of the dorsal thalamus. The ACTH fragment ACTH 1–24 also seems to influence these structures since it does affect avoidance learning and extinction, but in addition it has effects on the ventral aspects of the structure. This is the area from which fibers arise that project to the nucleus accumbens. It is this portion of the structure, that is, the ventral regions of the CA1 and subiculum, that are thought to be essential for the elicitation of the grooming response by effective ACTH fragments. The activation of the grooming response depends on basal ganglia dopaminergic activities, but not on the presence of ACTH fragments themselves, since direct application of ACTH 1–24 to the basal

27. Limbic System Contributions to Goal-Directed Behavior

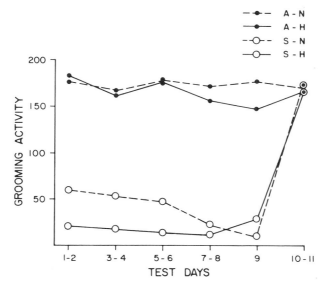

Figure 27.11. Grooming activity found in novel testing situations (N) or in home cages (H) after interventricular injection of ACTH 1-24 (A) or saline (S). (From Colbern et al., 1978.)

ganglia structures does not induce grooming (Wiegant, Cools, & Gispen, 1977). However, disrupting the balance of output of the caudate and the nucleus accumbens disturbs the conditions required for the elicitation of peptide-induced grooming. This disruption of the balance of output between these basal ganglia structures may be one consequence of bilateral hippocampal destruction.

What is not yet known is whether there is a progressive change in the effect of hippocampal destruction on ACTH-induced grooming, as might be expected on the basis of progressive changes in other behaviors that have been observed after such damage. Furthermore, if it is the case that destruction of the hippocampus produces some of its alterations in behavior by eliminating normal input to nucleus accumbens, and thus initiating secondary changes in the region, we must learn the range of behavioral alterations affected in this fashion.

This approach offers new suggestions as to consequences of the lesions in different aspects of the nervous system represented and allows systematic data to be gathered concerning the interactions among the basal ganglia and limbic systems, an interaction largely ignored in the past. What is important to point out is that lesions in any brain structure will undoubtedly induce secondary effects in all remaining systems directly interconnected with it and likely in some systems with less than direct connections. Many of these changes will affect behavior and the physiological responsiveness of the animal to different types of testing situations. Because of this, as well as because of the individual variability found among individual subjects, it is not possible to say that a lesion in any of the particular areas will categorically disturb any particular motivational or goal-directed act, but rather begin alterations of function within many systems. We must understand changes in goal-directed behaviors in terms of these aberrant functions of multiple systems. The precise nature of these changes will depend upon the animal, its experiences, its altered internal environment, and the specific conditions of testing.

REFERENCES

Clark, C. V., & Isaacson, R. L. Effect of bilateral hippocampal ablation on DRL performance. *Journal of Comparative and Physiological Psychology*, 1965, *59*, 137–140.

Colbern, D., Isaacson, R. L., Bohus, B., & Gispen, W. H. Limbic-midbrain lesions and ACTH-induced excessive grooming. *Life Sciences*, 1977, *21*, 393–402.

Colbern, D. L., Isaacson, R. L., Green, E. J., & Gispen, W. H. Repeated intraventricular injections of ACTH 1–24: The effects of home or novel environments on excessive grooming. *Behavioral Biology*, 1978, *23*, 381–387.

DeFries, J. C., & Hegmann, J. P. Genetic analysis of open field behavior. In G. Lindzey & D. D. Thiessen (Eds.), *Contributions to behavior-genetic analysis*. New York: Appleton-Century-Crofts, 1970. Pp. 23–56.

Delanoy, R. L., & Dunn, A. J. Brain deoxyglucose uptake after footshock. ACTH analogs, α-MSH, corticosterone, or LVP. *Pharmacology, Biochemistry, and Behavior*, 1978, *9*, 21–26.

de Wied, D. Pituitary-adrenal system hormones and behavior. In F. O. Schmidt & F. G. Worden (Eds.), *The Neurosciences. Third Study Program*. Cambridge, Mass.: MIT Press, 1974. Pp. 653–666.

Donovick, P. J., Burright, R. G., & Swidler, M. A. Presurgical rearing environment alters exploration, fluid consumption, and learning of septal lesioned and control animals. *Physiology and Behavior*, 1973, *11*, 543–553.

Dunn, A. J., Green, E. J., & Isaacson, R. L. Intracerebral adrenocorticotropic hormone mediates novelty-induced grooming in the rat. *Science*, 1979, *203*, 281–283.

Fentress, J. C. Interrupted ongoing behavior in two species of vole (*Microtis agrestis* and *Clethrionomys brittannicus* II): Extended analysis of motivational variables underlying fleeing and grooming. *Journal of Animal Behaviour*, 1968, *16*, 154–167.

File, S. E., & Wardill, A. G. Validity of head-dipping as a measure of exploration in a modified hole board. *Psychopharmacologia*, 1975, *44*, 53–59.

Fish, B. S. *Catecholamine modulation of behavior following bilateral hippocampal damage*. Unpublished doctoral dissertation, University of Florida, 1976.

Isaacson, R. L. *The limbic system*. New York: Plenum, 1974.

Isaacson, R. L. The myth of recovery from early brain damage. In N. R. Ellis (Ed.), *Aberrant development in infancy*. Hillsdale, N.J.: Lawrence Erlbaum Associates, 1975. Pp. 1–26.

Isaacson, R. L., & Green, E. J. The effect of ACTH 1–24 on locomotion, exploration, rearing, and grooming. *Behavioral Biology*, 1978, *24*, 118–122.

Isaacson, R. L., & McClearn, G. E. The influence of hippocampal damage on locomotor activity of the inbred mouse. *Brain Research*, 1978, *150*, 559–567.

Isaacson, R. L., Yongue, B., & McClearn, D. Dopamine agonists: Their effect on locomotion and exploration. *Behavioral Biology*, in press.

Lanier, L. P., & Isaacson, R. L. Activity changes related to the location of lesions in the hippocampus. *Behavioral Biology*, 1975, *13*, 59–69.

MacFarland, P. J. The role of attention in the disinhibition of displacement activity. *Quarterly Journal of Experimental Psychology*, 1966, *18*, 19–30.

Mountcastle, V. B. Brain mechanisms for directed attention. *Journal of the Royal Society of Medicine*, 1978, *71*, 14–28.

Reinstein, D. K., Isaacson, R. L., & Dunn, A. J. Regional changes in Z-deoxyglucose uptake after neocortical and hippocampal destruction. *Brain Research*, 1979, *175*, 392–397.

Schneiderman, B., & Isaacson, R. L. Pharmacologic changes in performance of normal and brain damaged rats. *Behavioral Biology*, 1976, *17*, 197–211.

Swanson, L. W., & Cowan, W. M. A note on the connections and development of nucleus. accumbens. *Brain Research*, 1975, *92*, 324–330.

Thomas, G. J. Maze retention by rats with hippocampal lesions and with fornicotomies. *Journal of Comparative and Physiological Psychology*, 1971, *75*, 41–49.

van Ierssel, J. J. A., & Bol, A. C. Preening of two tern species. *Behaviour*, 1958, *13*, 1–88.

Van Wimersma Greidanus, Tj. B., Bohus, B., & de Wied, D. Differential localization of the influence of lysine vasopressin and ACTH 4–10 on avoidance behavior; a study in rats bearing lesions in the parafascicular nuclei. *Neuroendocrinology*, 1974, *14*, 280–288.

Wiegant, V. M., Cools, A. R., & Gispen, W. H. ACTH-induced excessive grooming involves brain dopamine. *European Journal of Pharmacology*, 1977, *41*, 343–345.

LINDA PATIA SPEAR

A Psychopharmacological Approach to Memory Processing[1]

28

The search for the physiological basis of memory has had a long history. Presumably, an organism is able to remember learned information over a relatively long retention interval due to some relatively stable and long-lasting change in nervous tissue that occurs during and/or soon after the learning situation. In theories of the physiological basis of memory, changes occurring at the specialized synaptic junctures between neurons have traditionally been a major suspect: "While changes in the characteristic of the neuron are an entirely possible basis for learning, there are certainly many more possibilities for memory if the changes are in the synapse, because there are many more synapses than there are cells" (Neal Miller, in Kimble, 1965, pp. 148–149). Synaptic changes occurring during learning have been postulated to occur presynaptically and/or postsynaptically by increases in membrane adhesiveness between juncture membranes, growth of new synaptic connections, alterations in the threshold for response, and so on (Kimble, 1965). In common among these synaptic theories of the memory trace is the suggestion that synaptic connections are strengthened in pathways that are repetitively activated during learning. Until recently, however, although functional changes in synaptic efficacy have been shown to occur with alterations in synaptic activity (Eccles, 1961), there was little empirical evidence to specify the physiological basis for such plasticity. In this chapter, the problem of the physio-

---

[1] The research reported in this chapter was supported in part by Grant 1 R03 MH29834-01 MSM from the National Institute of Mental Healths Grant 0005-02-040-76 0 from the Research Foundation of the State University of New York, and National Science Foundation Grant BNS 78-02360.

logical basis of synaptic plasticity will be approached from a psychopharmacological perspective. Recent advances in psychopharmacology have suggested a means by which repetitive activation might strengthen synaptic efficacy.

Modulations in Receptor Sensitivity and Alterations in Synaptic Efficacy

POSTSYNAPTIC RECEPTORS

In the psychopharmacological field, plasticity at the synaptic level has been studied extensively. Since the early work of Anderson (1904), who examined alterations in epinephrine sensitivity after denervation, a vast amount of psychopharmacological, behavioral, electrophysiological, and neurochemical evidence has accumulated to suggest that alterations in postsynaptic receptor sensitivity occur with alterations in neurotransmitter availability. In general, this work has shown that manipulations that chronically decrease the amount of neurotransmitter reaching the postsynaptic receptors increase the sensitivity of those receptors (e.g., Arbuthnoff, 1976; Burkard & Bartholini, 1974; Gianutsos & Moore, 1977; Iversen & Creese, 1975; Tarsy & Baldessarini, 1974; Ungerstedt, Ljungberg, Hoffer, & Siggins, 1975). In addition, although it has not been so extensively investigated, some evidence also suggests the inverse—that chronically increasing the amount of neurotransmitter available to the postsynaptic receptors decreases their sensitivity (Baudry, Martres, & Schwartz, 1976). However, these modulatory changes in postsynaptic sensitivity occurring after alterations in neurotransmitter availability have not been considered to be candidates for the possible physiological mechanism of the synaptic "trace." For in their plasticity lies a great paradox: Such changes in sensitivity induced in postsynaptic receptors are in a homeostatically directed manner, not in a synaptic facilitatory fashion. With repetitive activation of a synaptic link, postsynaptic sensitivity is decreased, presumably leading to a decrease in efficacy of the activated pathways. This is opposite to what would be expected in synaptic theories of the memory trace wherein repetitive activation should increase synaptic efficacy of affected pathways.

AUTORECEPTORS

Carlsson and associates (1972), on the basis of psychopharmacological and neurochemical evidence, postulated the existence of a new type of receptor that could resolve the preceding paradox. It now appears that all neurotransmitter receptors are not located postsynaptically. "Autoreceptors" or "presynaptic receptors" are regulatory receptors believed to be present on the prejunctional or dendritic membranes of neurons. These receptors appear to be responsive to the neurotransmitter that is released by the neuron on whose membranes they are located. Electrophysiological (Aghajanian & Bunney, 1977; Bunney & Aghajanian, 1975; Walters, Bunney, & Roth, 1975) and neurochemical (Carlsson, 1975; Iversen, Rogawski, & Miller, 1976; Roth, Walters, Murrin, & Morgenroth, 1975) evidence suggests that these autoreceptors may form part of a feedback system regulating neurotransmitter synthesis and neuronal activity. Large amounts of neurotransmitter in the synaptic region may activate these receptors and lead ultimately to a decrease in neurotransmitter synthesis and utilization. In times of low neurotransmitter availability in the synaptic region, these receptors would be activated to a lesser degree; consequently, neurotransmitter synthesis would be derepressed and neuronal activity increased. These autoreceptors, like postsynaptic receptors, also modulate their sensitivity in response to alterations in neurotransmitter availability (Baudry, Costentin, Marçais, Martres, Portais, & Schwartz, 1977, Costentin, Marçais, Protais, & Schwartz, 1977; Langer & Dubocovich, 1977; Martres, Costentin, Baudry, Marçais, Protais, & Schwartz, 1977; Nowycky & Roth, 1977).

Alterations in autoreceptor sensitivity, however, differ from alterations in postsynaptic receptor sensitivity in at least two important respects. First, it appears that autoreceptors are much more sensitive to modulation than postsynaptic receptors (Martres et al., 1977). Thus, alterations in autoreceptor sensitivity may be much more likely to occur under physiological conditions than alterations in postsynaptic sensitivity. Second, although the net effect of alteration in postsynaptic receptor sensitivity is homeostatically directed, the net effect of modulating autoreceptor sensitivity is in a synaptic facilitatory fashion. When much neurotransmitter is chronically available in the synaptic region (as would be the case with repetitive synaptic activation), the autoreceptors would become less sensitive. Stimulation of these hyposensitive autoreceptors would result in less of a decrease in neurotransmitter synthesis and utilization than stimulation of normally sensitive autoreceptors. Thus, the net effect of modulation in autoreceptor sensitivity would be to facilitate synaptic efficacy of activated synaptic pathways. Perhaps autoreceptor hyposensitivity develops normally as a consequence of repeated synaptic activation. Such a process may provide a simple mechanism for increasing synaptic efficacy in pathways repetitively activated during learning. (This point was also made by Martres et al., [1978] in a very elegant study on autoreceptor function.)

The Developmental Psychobiology Approach

One other indirect bit of evidence that modulation of autoreceptors helps strengthen neuronal circuits repetitively activated during learning is found in the literature of developmental psychobiology. Although young animals may be able to learn a number of tasks fairly quickly, they seem to have a relatively difficult time retaining this information over long intervals of time—a phenomenon termed "infantile amnesia" by Freud (Campbell & Coulter, 1976; Campbell & Spear, 1972; Spear & Campbell, 1979). The developing animal may therefore be a tool with which to study memory processing: By assessing what ontogenetic physiological events are correlated with gradual maturation of retention capabilities, one may derive useful hypotheses as to the biological basis of learning and retention.

Although varying from task to task, adult memory capabilities do not typically develop in rats until 25–35 days postnatally. Surprisingly, few developmental milestones are reached in the nervous system during this time interval. A notable exception is the modulatory components of presynaptic and postsynaptic neurons. For example, work in my laboratory (Shalaby, Spear, & Brick, 1978) suggests that modifications in postsynaptic receptor sensitivity found typically in adults after chronic alteration in neurotransmitter availability are not seen in young (preweanling) animals. Although adult animals develop dopamine receptor supersensitivity after chronic treatment with the dopamine receptor-blocking agent haloperidol, young animals given haloperidol chronically before weaning develop an apparent long-lasting dopamine receptor hyposensitivity when measured pharmacologically (see Table 28.1). In addition, during this 25–35 postnatal daytime interval, feedback control of neurotransmitter synthesis reaches adult levels (Kellogg, Lundborg, & Ross, 1972; Kellogg & Wennerström, 1974). As discussed previously, autoreceptors appear to play an important role in such feedback regulatory activity. Moreover, we have found no psychopharmacological evidence for dopaminergic autoreceptor functioning until 28–35 days postnatally (Shalaby & Spear, 1979).

Thus, at least two modulatory devices in normal neurotransmitter systems (feedback control of neurotransmitter synthesis and alterations in receptor sensitivity after alterations in neurotransmitter availability to the

Table 28.1
Psychopharmacological Responsiveness of Adult and Infant Animals after Chronic Haloperidol Treatment

	After adult chronic haloperidol	After infant chronic haloperidol		
		Tested P25	Tested P37	Tested P49
Baseline activity	Hyperactive	Hyperactive	No effect	Hyperactive
R. to amphetamine	Accentuated	Attenuated	No effect	Attenuated? (Dose–R. curve shifted to right)
R. to haloperidol	Attenuated	Accentuated	No effect	Accentuated

receptors) appear to develop more slowly than other aspects of the neurotransmitter systems (Lanier, Dunn, & Van Hartesveldt, 1976). Ontogenetically, neurotransmitter systems may become functional prior to the time that they are able, presynaptically and postsynaptically, to respond in a modulatory manner to alterations in synaptic neurotransmitter availability. Maturation of these modulatory systems correlates temporally with the ontogeny of adult memory capabilities. Perhaps the lack of such synaptic forms of plasticity is a contributory factor to infantile amnesia.

Psychopharmacological Studies

How does one begin to test the hypothesis that modulation in the sensitivity of regulatory autoreceptors may play a role in strengthening pathways that are repetitively activated during learning? Many approaches are possible, and necessary. Although we are utilizing both neurochemical and psychopharmacological approaches to this question, the psychopharmacological approach will be the focus of this chapter. Classically, psychopharmacological experiments have examined the influence of drugs administered before or shortly after a learning task on later retention of that task (for reviews of this literature, see Alpern & Jackson, 1978; Zornetzer, 1978). When using this procedure, certain drugs can be found to enhance, and other drugs to impair, retention of the learned task. Typically, no alteration in retention is reported if drug administration is delayed until several hours posttraining (see Alpern & Jackson, 1978). Our approach differs from the classic psychopharmacological approach in that we are interested in whether the alterations that occur at the synaptic level after chronic administration of drugs could be used as a model with which to consider the biological substrates of learning and memory. There are interesting similarities between learning and retention processes and alterations in the nervous system induced by chronic administration of psychoactive drugs. For example, both the retention of a learned task and the production of drug tolerance presumably require some long-lasting change in the nervous system that can be blocked by administration of protein synthesis inhibitors at certain critical stages (e.g., Barondes, 1970; Cohen, Keats, Krivoy & Ungar, 1965; Feinberg & Cochin, 1977). Moreover, alterations in receptor sensitivity after drug administration can also be blocked by protein synthesis inhibition (Costentin, Marçais, Protais, Baudry, De La Baume, Martres, & Schwartz, 1977).

Perhaps during learning of any particular task, certain pathways in the nervous system are repetitively activated. This repetitive activation may result in facilitation of the synaptic efficacy of the involved pathways by modulation of autoreceptor sensitivity. By facilitating the synaptic efficacy of these pathways, learning occurs. A "memory" could thus be strengthened by restimulation of critical pathways involved (by administering new learning trials) or by restimulating some critical portion of these pathways. Behaviorally, this could be accomplished by a "reactivation" or "reminder" treatment wherein part of the learning situation is reintroduced. (It is especially notable for this conference that such a mechanism was suggested in 1863 by Sechenov [1965]). Another approach for the latter method would be to stimulate part of the critical pathways involved by administration of a psychoactive drug that acts upon certain of these critical synaptic links. Pathways involving many neurotransmitter systems (both known and unknown) may be involved in the learning of any particular task, and the critical pathways as well may be different, and utilize different neurotransmitter systems, for different learning tasks. But the hypothesis is that strengthening certain of these critical pathways during the retention interval (by a reactivation treatment or by chronic administration of a psychoactive agent) may enhance retention of the learned task if enough of these critical synaptic links are restimulated by the experimental treatment, and thus "strengthened" by reuse. The mechanism we chose for "strengthening" these synapses is psychopharmacological and it mimics procedures used to establish physical dependence to a drug—chronic drug administration during the retention interval. The use of this type of chronic procedure is not unprecedented and is beginning to be utilized elsewhere with a variety of drugs and training tasks (Alpern & Crabbe, 1972; Stripling & Alpern, 1974).

In our initial study, adult Sprague-Dawley albino rats were trained on an active avoidance task to a criteria of five consecutive correct avoidances and were given a retention test 7 days later. Injections of saline, 2 or 5 mg/kg dl-amphetamine, were given either once or twice a day on Days 1–5 of the retention interval. Half of the animals receiving one injection a day were injected in the morning; the rest were injected in the evening. Thus, the first injection was given approximately 24 hr after initial training and the last approximately 48 hr prior to the retention test. The retention test consisted of 10 nonshocked trials.

With both doses of amphetamine, twice daily injection or one daily injection in the evening markedly enhanced retention. Groups given one daily injection of amphetamine in the morning, however, showed poorer retention than the controls. As seen in Figure 28.1, this effect was seen on the first test trial and (Figure 28.2 and 28.3) generally continued over the subsequent testing trials.

The observed diurnal effects were very interesting. A number of investigators have reported that there is a diurnal rhythm in sensitivity to amphetamine (Evans, Ghinsell, & Patton, 1973; Lew, 1977; Scheving, Vedra, & Pauly, 1968). Indeed, it has been reported that amphetamine introduced at the time of our morning injections does not affect catechol-

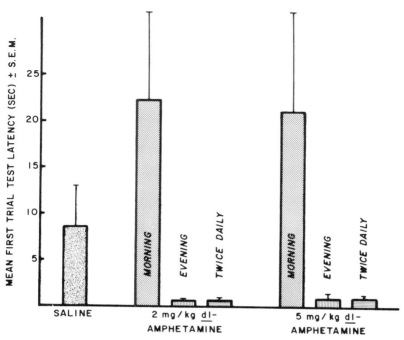

Figure 28.1. Mean first trial test latency on the retention test for animals given saline (twice daily), 2 or 5 mg/kg dl-amphetamine on Days 1–5 of the retention interval.

amine levels, although it does decrease catecholamine levels when administered at the time of our evening injections (Lew, 1977). Some of these changes in sensitivity to drugs at different phases of the diurnal cycle may be due to actual alterations in receptor availability (Kebabian, Zata, Romero, & Axelrod, 1975; Romero, 1976; Romero & Axelrod, 1975). Perhaps amphetamine enhances retention when given in the evening because the receptors of the dopamine and norepinephrine systems are most sensitive at that time, whereas not enhancing retention when given at the onset of the light cycle due to receptor insensitivity at that time.

This study is currently being replicated with controls for (a) order of morning and evening drug injections within the retention interval and (b) activity levels on the retention test (using a combined active avoidance/passive avoidance retention test). The data gathered so far with these modified procedures are consistent with those of the previous work. This approach is also being extended to developing animals. Preliminary data have indicated that retention may not be enhanced in weanling rats given amphetamine chronically during the retention interval. Indeed, this is what might be expected if the maturation of autoreceptors is related to the ontogeny of memory capabilities.

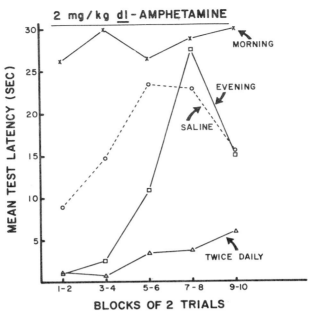

Figure 28.2. Mean test latency on the retention test in blocks of two trials for animals given saline (twice daily) or 2 mg/kg dl-amphetamine on Days 1–5 of the retention interval.

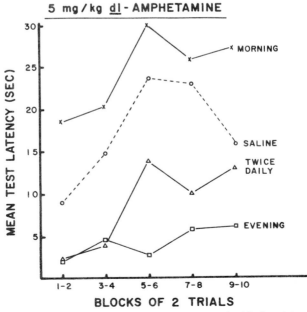

Figure 28.3. Mean test latency on the retention test in blocks of two trials for animals given saline (twice daily) or 5 mg/kg dl-amphetamine on Days 1–5 of the retention interval.

Summary

Autoreceptor hyposensitization may develop normally as a consequence of repeated utilization of synapses. Such a process may constitute a simple mechanism for stabilization of neuronal circuits repetitively activated during a learning situation. Alterations that occur at the synaptic level after chronic drug administration may be a model with which to investigate biological substrates of learning and memory processes. Chronic amphetamine administration during the retention interval seems to enhance the retention of adult rats, but not perhaps that of weanling rats whose synaptic modulatory devices are not yet mature. Use of such a psychopharmacological model in combination with neurochemical approaches may be helpful in investigations of the biological bases of learning and retention.

ACKNOWLEDGMENTS

Some of this work was conducted in collaboration with Norman E. Spear, Greg J. Smith, Ismail A. Shalaby, and Norman G. Richter.

REFERENCES

Aghajanian, G. K., & Bunney, B. S. Dopamine "autoreceptors": Pharmacological characterization by microiontophoretic single cell recording studies. *Naunyn-Schmiedeberg's Archives of Pharmacology*, 1977, *297*, 1–7.

Alpern, H. P., & Crabbe, J. C. Facilitation of the long-term store of memory with strychnine. *Science*, 1972, *177*, 722–724.

Alpern, H. P., & Jackson, S. J. Stimulants and depressants: Drug effects on memory. In M. A. Lipton, A. DiMascio, & K. F. Killam (Eds.), *Psychopharmacology: A generation of progress*. New York: Raven Press, 1978.

Anderson, H. K. The paralysis of involuntary muscle, with special reference to the occurrence of paradoxical contraction. Part I. Paradoxical pupil dilation and other ocular phenomena caused by lesions of the cervical sympathetic tract. *Journal of Physiology*, 1904, *30*, 290–310.

Arbuthnoff, G. W. Supersensitivity of dopamine receptors. In H. F. Bradford & C. D. Marsden (Eds.), *Biochemisty and neurology*. New York: Academic Press, 1976.

Barondes, S. H. Some critical variables in studies of the effect of inhibition of protein synthesis on memory. In W. L. Byrne (Ed.), *Molecular approaches to learning and memory*. New York: Academic Press, 1970.

Baudry, M., Costentin, J., Marçais, H., Martres, M. P., Protais, P., & Schwartz, J. C. Decreased responsiveness to low doses of apomorphine after dopamine agonists and the possible involvement of hyposensitivity of dopamine "autoreceptors." *Neuroscience Letters*, 1977, *4*, 203–207.

Baudry, M., Martres, M. P., & Schwartz, J. C. Modulation in the sensitivity of nonadrenergic receptors in the CNS studied by the responsiveness of the cyclic AMP system. *Brain Research*, 1976, *116*, 111–124.

Bunney, B. S., & Aghajanian, G. K. Evidence for drug actions on both pre- and postsynaptic catecholamine receptors in the CNS. In E. Usdin & W. E. Bunney (Eds.), *Pre- and postsynaptic receptors.* New York: Dekker, 1975.

Burkard, W. P., & Bartholini, G. Changes in activation of adenylate cyclase and of dopamine turnover in rat striatum during prolonged haloperidol treatment. *Experientia,* 1974, *30,* 685–686.

Campbell, B. A., & Coulter, X. The ontogenesis of learning and memory. In M. R. Rosenzweig & E. L. Bennett (Eds.), *Neural mechanisms of learning and memory.* Cambridge, Mass.: MIT Press, 1976.

Campbell, B. A., & Spear, N. E. Ontogeny of memory. *Psychological Review,* 1972, *79,* 215–236.

Carlsson, A. Receptor-mediated control of dopamine metabolism. In E. Usdin & W. E. Bunney (Eds.), *Pre- and postsynaptic receptors.* New York: Dekker, 1975.

Carlsson, A., Hehr, W., Lindquist, M., Magnusson, T., & Atack, C. V. Regulation of monoamine metabolism in the central nervous system. *Pharmacological Review,* 1972, *24,* 371–384.

Cohen, M., Keats, A., Krivoy, W., & Ungar, G. Effects of actinomycin D on morphine tolerance. *Proceedings of the Society for Experimental Biology and Medicine,* 1965, *119,* 381–384.

Costentin, J., Marçais, H., Protais, P., Baudry, M., De La Baume, S., Martres, M. P., & Schwartz, J. C. Rapid development of hypersensitivity of striatal dopamine receptors induced by alpha-methylparatyrosine and its prevention by protein synthesis inhibitors. *Life Science,* 1977, *21,* 307–314.

Costentin, J., Marçais, H., Protais, P., & Schwartz, J. C. Tolerance to hypokinesis elicited by dopamine agonists in mice: Hyposensitization of autoreceptors? *Life Science,* 1977, *20,* 883–886.

Eccles, J. C. The effects of use and disuse on synaptic function. In J. F. Delafresnaye (Ed.), *Brain mechanisms and learning.* Springfield, Ill.: Thomas, 1961.

Evans, H. L., Ghinsell, W. B., & Patton, R. A. Diurnal rhythm of methamphetamine, p-chloromethamphetamine and scopolamine. *Journal of Pharmacology and Experimental Therapeutics,* 1973, *186,* 10–17.

Feinberg, M. P., & Cochin, J. Studies on tolerance. II. The effect of timing on inhibition of tolerance to morphine by cycloheximide. *Journal of Pharmacology and Experimental Therapeutics,* 1977, *203,* 332–339.

Gianutsos, G., & Moore, K. E. Dopaminergic supersensitivity in striatum and olfactory tubercle following chronic administration of haloperidol or clozapine. *Life Science,* 1977, *20,* 1585–1592.

Iversen, S. D., & Creese, I. Behavioral correlates of dopaminergic supersensitivity. In D. B. Calne, T. N. Chase, & A. Barbeau (Eds.), *Dopaminergic mechanisms—advances in neurology* (Vol. 9). New York: Raven Press, 1975.

Iversen, L. L., Rogawski, M. A., & Miller, R. J. Comparison of the effects of neuroleptic drugs on pre- and post-synaptic dopaminergic mechanisms in the rat striatum. *Molecular Pharmacology,* 1976, *12,* 251–262.

Kebabian, J. W., Zatz, M., Romero, J. A., & Axelrod, J. A. Rapid changes in rat pineal B-adrenergic receptor: Alterations in L-[^{3}H] alprenolol binding and adenylate cyclase. *Proceedings of the National Academy of Sciences,* 1975, *72,* 3735.

Kellogg, C., Lundborg, P., & Ross, B. E. Ontogenic changes in cerebral homovanillic acid concentrations in response to haloperidol treatment. *Brain Research,* 1972, *40,* 469–475.

Kellogg, C., & Wennerström, G. G. An ontogenetic study of the effect of catecholamine receptor-stimulating drugs on the turnover of noradrenaline and dopamine in the brain. *Brain Research,* 1974, *79,* 451–464.

Kimble, D. P. (Ed.). *Learning, remembering and forgetting: The anatomy of memory* (Vol. 1). Palo Alto, Calif.: Science and Behavior Books, 1965.

Langer, S. Z., & Dubocovich, M. L. Subsensitivity of presynaptic α-adrenoceptors after exposure to noradrenaline. *European Journal of Pharmacology*, 1977, *41*, 87–88.

Lanier, L. P., Dunn, A. J., & Van Hartesveldt, C. Development of neurotransmitters and their function in brain. *Reviews of Neuroscience*, 1976, *2*, 195–256.

Lew, G. M. Temporal differences in the effects of amphetamine on norepinephrine. *General Pharmacology*, 1977, *8*, 109–111.

Martres, M. P., Costentin, J., Baudry, M., Marçais, H., Protais, P., & Schwartz, J. C. Long-term changes in the sensitivity of pre- and postsynaptic dopamine receptors in mouse striatum evidenced by behavioral and biochemical studies. *Brain Research*, 1977, *136*, 319–337.

Nowycky, M. C., & Roth, R. H. Presynaptic dopamine receptors. Development of supersensitivity following treatment with Fluphenazine Decanoate. *Naunyn-Schmiedeberg's Archives of Pharmacology*, 1977, *300*, 247–254.

Romero, J. A. Influence of diurnal rhythms on biochemical parameters of drug sensitivity: The pineal gland as a model. *Federal Proceedings*, 1976, *35*, 1157.

Romero, J. A., & Axelrod, J. A. Regulation of sensitivity to beta-adrenergic stimulation in induction of pineal-N-acetyltransferase. *Proceedings of National Academy of Sciences*, 1975, *72*, 1661.

Roth, P. H., Walters, J. R., Murrin, L. C., & Morgenroth, V. H. Dopamine neurons: Role of impulse flow and presynaptic receptors in the regulation of tyrosine hydroxylase. In E. Usdin & W. E. Bunney (Eds), *Pre- and postsynaptic receptors*. New York: Dekker, 1975.

Scheving, L. E., Vedra, D. F., & Pauly, J. E. Daily circadian rhythm in rats to *d*-amphetamine: Effects of blinding and continuous illumination on the rhythm. *Nature*, 1968, *219*, 621–622.

Sechenov, I. M. *Reflexes of the brain*. Cambridge, Mass.: MIT Press, 1965. (Originally published, 1863.)

Shalaby, I. A., Spear, L. P. Psychopharmacological effects of apomorphine during ontogeny. *Society for Neuroscience*, Atlanta, November, 1979.

Shalaby, I. A., Spear, L. P., & Brick, J. Long-term effects of haloperidol administration during development. *Society for Neuroscience*, St. Louis, November, 1978.

Spear, N. E., & Campbell, B. A. (Eds.). *Ontogeny of learning and memory*. Hillsdale, N.J.: Lawrence Erlbaum Associates, 1979.

Stripling, J. S., & Alpern, H. P. Nicotine and caffeine: Disruption of the long-term store of memory and proactive facilitation of learning in mice. *Psychopharmacologia*, 1974, *38*, 187–200.

Tarsy, D., & Baldessarini, R. J. Behavioral supersensitivity to apomorphine following chronic treatment with drugs which interfere with the synaptic function of catecholamines. *Neuropharmacology*, 1974, *13*, 927–940.

Ungerstedt, U., Ljungberg, T., Hoffer, B., & Siggins, G. Dopaminergic supersensitivity in the striatum. In D. B. Calne, T. N. Chase, & A. Barbeau (Eds.), *Dopaminergic mechanisms—advances in neurology* (Vol. 9). New York: Raven Press, 1975.

Walters, J., Bunney, B., & Roth, R. Piribedil and apomorphine: Pre- and postsynaptic effects on dopamine synthesis and neuronal activity. In D. B. Calne, T. N. Chase, & A. Barbeau (Eds.), *Dopaminergic mechanisms—advances in neurology* (Vol. 9). New York: Raven Press, 1975.

Zornetzer, S. F. Neurotransmitter modulation and memory: A new neuropharmacological phrenology? In M. A. Lipton, A. DiMascio, & K. F. Killam (Eds.), *Psychopharmacology: A generation of progress*. New York: Raven Press, 1978.

NANCY J. KENNEY

A Case Study in the Neuroendocrine Control of Goal-Directed Behavior: The Interaction between Angiotensin II and Prostaglandin E_1 in the Control of Water Intake[1]

29

The study of the role of hormones in the initiation and modification of neural activity and, consequently, behavior is a relatively new field. Early work in the area was concerned mainly with the interactions between cortical structures and gonadal hormones in the control of sexual behavior (Beach, 1942). In recent years this field has been burgeoning and the intricacies of the interactions between the endocrine system and the brain in the control of a wide variety of behaviors are being uncovered. In this chapter I am going to concentrate on the role of two hormones, angiotensin II (A II) and prostaglandin E (PGE), in the central control of a specific behavior, the ingestion of water, as an example of a situation in which the nervous and endocrine systems work together to regulate goal-directed activity.

A II is an octapeptide derived from the action of renal renin on the plasma protein angiotensinogen (forming A I) and the further reduction of this peptide to its active form (A II) by converting enzyme. A II acts within the brain to elicit a variety of effects associated with hydro-mineral homeostasis. Included in these actions are the release of antidiuretic hormone, glucocorticoids and aldosterone, a centrally mediated increase in blood pressure that is independent of the peripheral arteriolar constric-

[1] Much of the work reviewed here was undertaken while the author was the recipient of a National Institutes of Child Health and Human Development National Research Service Award (HD02273). More recent work and the preparation of this manuscript were supported by a Scholarly Development Award from the Graduate School of Arts and Sciences at the University of Washington.

tion also resulting from A II stimulation, and the centrally mediated increase in water intake that I will be discussing in this chapter.

Studies by a variety of investigators strongly suggest that A II may be involved in a central neurological system that initiates drinking. A II is the most potent dipsogen known (Epstein, Fitzsimons, & Rolls, 1970; Fitzsimons & Simons, 1969). With direct application to the subfornical organ, a circumventricular organ located on the dorsal surface of the third ventricle that is reportedly a receptor site for A II-induced drinking, A II is effective in eliciting drinking by water-replete rats at doses as low as 10^{-16}–10^{-15} moles (.1–1.0 pg, Simpson, 1977; Simpson, Epstein, & Camardo, 1978). And infusion of the specific A II receptor antagonist, saralasin, into the lateral cerebral ventricle suppresses water intake of water-deprived rats (Malvin, Mouw, & Vander, 1977).

It was recently suggested that the prostaglandins, particularly those of the E series, might serve as a natural satiety signal in situations in which drinking was initiated by A II (Epstein & Kenney, 1977; Kenney & Epstein, 1975, 1978). The prostaglandins are a group of 20-carbon carboxylic acids. E-series prostaglandins are distinguished from the other subgroups by the presence of an 11-hydroxy and a 9-keto group. In a variety of organs including the brain, prostaglandins of the E series are produced and released in response to A II infusion (Aiken & Vane, 1973; McGiff, Crowshaw, Terragno, & Lonigro, 1970; Phillips & Hoffman, 1977). Although PGE is rapidly degraded and most likely cannot pass from the peripheral circulation to the central nervous system, it is known to be present in brain tissue of many species including the rat and the rabbit (Ambache, Reynolds, & Whiting, 1963; Kataoka, Ramwell, & Jessup, 1967). The enzyme system necessary for the production of the prostaglandins is also found in brain tissue (Pappius, Rostworowski, & Wolfe, 1974; Wolfe, Pappius, & Marion, 1976). In many cases the effects of PGE counter those of A II. For example, as previously noted, A II infusion results in peripheral arteriolar constriction and a resulting increase in blood pressure. PGE_1, on the other hand, causes peripheral vasodilatation and a resultant decrease in blood pressure. Finally, PGE has been determined to be involved in the central regulation of body temperature (Feldberg, 1975), a process that is closely allied with thirst both behaviorally and physiologically (Hainsworth, Stricker, & Epstein, 1968).

General Methodology

The standard preparation used for the study of the role of PGE in the central control of body water involved male Sprague-Dawley or Long-Evans rats fitted with unilateral in-dwelling, stainless-steel cannulae. These cannulae permitted the delivery of the hormones directly into the lateral cerebral ventricle while the animals were awake and moving freely. For the complete implant methodology see Miselis and Epstein (1975) and Kenney and Epstein (1978). In most cases, the animals were tested in a three-day experimental series with a test day on which the prostaglandin was paired with the A II, preceded and followed by a control day on which the carrier solution for the prostaglandin was paired with the dipsogen. This 3-day testing procedure assured that any decrease in intake observed on the test day was a direct result of the prostaglandin and was not due to reduced patency of the cannulae. The single exception to the 3-day rule was in the case of animals tested with hypovolemia induced by the subcutaneous injection of polyethylene glycol. Because of the toxicity of this substance, animals were tested only twice, once with PGE_1 and once with the prostaglandin vehicle, with one week elapsing between repeated tests. The PGE_1 and the vehicle were administered in counterbalanced order such that one-half of the animals were tested first with the prostaglandin and one-half with the carrier solution. Animals were tested for drinking in response to intraventricular (IVT) A II following the final test to assure the patency of their cannulae.

A II was dissolved in isotonic saline and delivered at a dose of 5 ng/μl per rat. The prostaglandins were dissolved in a sodium carbonate-ethanol solution (nine parts .2% sodium carbonate to one part 90% ethanol). The dose of the prostaglandins varied from 1 ng to 5000 ng (5 μg). The prostaglandin was always delivered in 1 μl solvent. All injections into the lateral ventricle were administered at the rate of .1 μl/sec, and normally the prostaglandin injection immediately preceded the injection of the dipsogenic agent. Animals were tested in their home cages with food present. Water was presented by means of chemical burets calibrated in .1 ml units. Test sessions lasted 20 min. Latency to the onset of drinking was noted and cumulative water intake was measured at 5-min intervals.

In a preliminary experiment, one group of rats was fitted with both the ventricular cannulae and jugular-vein catheters of silastic tubing. These animals received their A II intravenously at a dose of 64 ng/min for 20 min. Infusion rate was approximately 10 μl/min. PGE_1 was administered into the lateral cerebral ventricle in doses of 1, 10, 100, or 1000 ng at the onset of the A II infusion.

Effect of PGE_1 on A II-Induced Water Intake

Treatment with 1 μg PGE_1 reduced significantly the amount of water ingested in response to IVT A II (Figure 29.1). It did so by shortening the length of the drinking bout and not by increasing the animals' latency to drink. On the PGE_1-treatment day, most of the rats had completed their drinking by the end of the first 5-min segment of the test session. On control days drinking continued well into the second test-session segment.

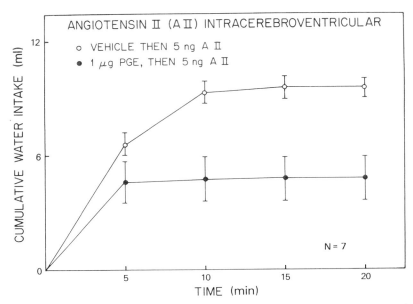

Figure 29.1. *Effect of intraventricular prostaglandin E_1 on cumulative water intake following the intraventricular injection of angiotensin II. Data are presented as means $\pm$ standard errors. (From Kenney & Epstein, 1978.)*

Average time from the end of the A II injection to the onset of drinking was nearly identical on the control and test days. The average latency was 59.3 ± 6.1 sec on the days on which the dipsogen was paired with the carrier solution and 59.3 ± 6.8 sec on days on which the A II was paired with the prostaglandin itself.

When both the A II and the PGE_1 were injected into the lateral ventricle, the prostaglandin effectively suppressed water intake at a dose as low as 10 ng (Figure 29.2). The intake-suppressant effect of ventricular PGE_1 was not limited to those cases in which the A II was administered into the brain, however. Although these are only preliminary data, it was apparent that IVT PGE_1 also reduced the amount of water ingested in response to the intravenous infusion of A II (Figure 29.3). The dose–response relationship in this case was identical to that seen with IVT A II. Treatment with 10, 100, or 1000 ng PGE_1 resulted in a significant and substantial decrease in intake. As with IVT A II, the 1 ng dose of PGE_1 was ineffective in reducing intake to intravenous A II.

Specificity to the E-Prostaglandins

The effectiveness of the prostaglandins in reducing A II-induced water intake is solely a property of the E-series prostaglandins (Epstein & Kenney, 1977; Kenney & Epstein, 1978). $PGF_{2\alpha}$, a prostaglandin that like

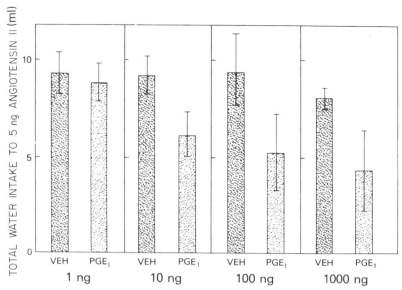

Figure 29.2. *Suppression of intraventricular angiotensin-induced drinking by various doses of prostaglandin E_1 (From Kenney & Epstein.)*

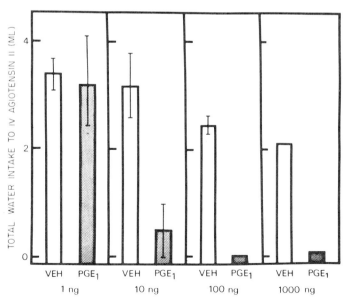

Figure 29.3. *Suppression of drinking induced by the intravenous infusion of angiotensin II by intraventricular prostaglandin E_1.*

PGE is found in brain tissue (Ambache et al., 1963; Kataoka et al., 1967), had no effect on IVT A II-induced drinking. Neither did PGB_1, a prostaglandin that is nearly chemically identical to PGE_1. PGA_1, which may in actuality be an impure form of PGE_1, was slightly effective in reducing A II-induced water intake when it was administered at an extremely high dose (5 μg). A reduction of the PGA_1 does to 1 μg eliminated its effectiveness, however. PGE_2 appeared to equal PGE_1 in its effectiveness in reducing water intake resulting from A II stimulation.

Bradykinin, a naturally occurring polypeptide, stimulates the synthesis and release of endogenous PGE (McGiff, Terragno, Malik, & Lonigro, 1972). When injected into the lateral ventricle, bradykinin suppresses water intake induced by IVT A II injection (Figure 29.4). Rats treated with 5 μg bradykinin drank significantly less in response to 5 ng A II than they did when the A II was paired with the bradykinin carrier solution, isotonic saline. The reduction in intake was evident at the end of the first 5-min segment of the test session and remained throughout the session (average total water intake 10.0 ± 1.5 ml on the control day versus 5.6 ± 1.8 ml on the test day ($t = 10.5$, $p < .01$). Given the similarity between the effect of this peptide and that of exogenous PGE_1, it is possible that the effect of bradykinin may be mediated through the release of endogenous PGE, although this possibility clearly warrants further analysis.

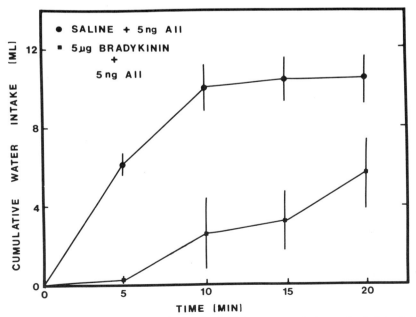

Figure 29.4. Effect of intraventricular bradykinin on cumulative water intake following the intraventricular injection of angiotensin II.

Effect of PGE_1 on Food Intake

The suppression of water intake resulting from PGE_1 treatment is not due to general malaise. Rats treated with this substance did not reduce the amount of food eaten following a deprivation period if the animals were tested under the appropriate conditions. In rats, the ingestion of food is normally tied to the ingestion of water. A rat offered dry laboratory chow will reduce its food intake markedly if it is denied access to water. Food-deprived rats treated with centrally administered PGE_1 and offered food under standard laboratory conditions did reduce the amount of food ingested in a test session (Baile, Simpson, Bean, McLaughlin, & Jacobs, 1973; Kenney & Epstein, 1978). This effect, however, was apparently secondary to the PGE-suppression of water intake under these conditions. When the rats were hydrated by gavage prior to food presentation, the need for water during the ingestion of food was reduced and PGE had no effect on the amount of food eaten (Kenney & Epstein, 1978).

Specificity of the PGE Effect to A II-Initiated Drinking

Not only is the intake-suppressant effect of PGE_1 specific to water intake, but it also appears to be related solely to those conditions in which the thirst is elicited by the action of A II. A number of commonly used dipsogenic agents are known to cause the release of endogenous A II (Abdelaal, Mercer, & Mogenson, 1976). These include hypovolemia induced by the subcutaneous injection of the hyperoncotic colloid, polyethylene glycol, and water deprivation. The first of these thirst stimuli produces a marked increase in plasma A II levels, whereas the latter results in a much smaller but still significant rise in plasma A II. Other thirst stimuli, most notably cellular dehydration induced by the injection of hypertonic saline, are totally independent of the release of A II. We found that the effectiveness of IVT PGE_1 in suppressing water intake to each of these agents is directly related to their effectiveness in releasing endogenous A II (Table 29.1; Kenney & Epstein, 1976, 1978). When paired with hypovolemia, PGE_1 was an effective suppressor of intake, thus causing a significant reduction in the total amount of water ingested during the test session at doses of 100 ng and 1 μg. The 10 ng dose may also have been effective against this dipsogen, but the analysis of those data was difficult due to apparent order and practice effects under those conditions (Kenney & Epstein, 1978). Water deprivation-induced thirst was also affected by IVT PGE_1, but, in this case, the lowest significantly effective dose of the prostaglandin was 1 μg. Water intake due to cellular dehydration was unaffected by IVT PGE_1 even at the highest dose tested (1 μg). Thus, it appears that when a dipsogenic stimulus causes the release of endogenous

Table 29.1
Percentage Change from Control Levels in Average Total Water Intake (20 min into the Test Session) and Number of Animals Reducing Intake in Response to Various Doses of PGE_1 Paired with Different Dipsogenic Agents.

	PGE_1 dose					
	10 ng		100 ng		1000 ng	
Dipsogen [a]	Percentage change	No. decreasing/ No. tested	Percentage change	No. decreasing/ No. tested	Percentage change	No. decreasing/ No. tested
Hypovolemia (polyethylene glycol)	+7 or −35 [b]	4/10	−72	9/11	−84	10/10
Water deprivation	−6	5/7	−13	4/6	−52	7/7
Cellular dehydration (hypertonic saline)	+2	2/6	0	2/5	+3	3/6

[a] Dipsogens are listed in order of their reported effectiveness in releasing endogenous A II (Abdelaal et al., 1976).
[b] Dependent on order of testing with PGE_1 or its carrier solution.

A II, PGE_1 reduces the amount of water ingested in response to that agent. And the degree of suppression is apparently directly related to the amount of A II released. When a dipsogenic stimulus is totally independent of A II release, the resultant thirst is unaffected by PGE_1.

Summary and Conclusions

PGE_1 is antidipsogenic. When administered into the lateral cerebral ventricle of rats, it reduces the amount of water ingested in response to A II. Its effects are not malaise dependent; the ingestion of food is not impaired by even high doses of the prostaglandin if the animal is adequately hydrated prior to food presentation. PGE_1 is not a universal antidipsogen. Its effectiveness in reducing water intake appears to be related to the degree to which the dipsogenic agent releases endogenous A II. When the thirst is independent of A II, as is the case with intracellular dehydration, water intake remains unaffected by even high doses of PGE_1.

This analysis of the interactions between angiotensin II and prostaglandin E suggests that the central systems involved in the regulation of water intake are quite complex. Instead of satiety being represented by the simple fading of a salient thirst cue, it may be reflected in a complex neuroendocrine interaction in which rises in plasma A II that give rise to water intake also cause the synthesis and release of PGE within the brain. This newly synthesized PGE may then act either on the same neural site(s) as A II or on separate but interacting loci to signal satiety and inhibit this goal-directed behavior.

ACKNOWLEDGMENTS

The author wishes to thank Elizabeth Barrett, Steven J. Fluharty, Eve Perara, and Priscilla W. Wright for their technical assistance and Dr. John Pike of the Upjohn Company for supplying the prostaglandins.

REFERENCES

Abdelaal, A. E., Mercer, P. F., & Mogenson, G. J. Plasma angiotensin II levels and water intake following β-adrenergic stimulation, hypovolemia, cellular dehydration, and water deprivation. *Pharmacology, Biochemistry and Behavior*, 1976, *4*, 317–321.

Aiken, J. W., & Vane, J. R. Intrarenal prostaglandin release attenuates the renal vasoconstrictor activity of angiotensin. *Journal of Pharmacology and Experimental Therapeutics*, 1973, *184*, 678–687.

Ambache, N., Reynolds, M., & Whiting, J. M. C. Investigation of an active lipid in aqueous extracts of rabbit brain, and some further hydroxyacids. *Journal of Physiology*, 1963, *166*, 251–283.

Baile, C. A., Simpson, C. W., Bean, S. M., McLaughlin, C. L., & Jacobs, H. L. Prostaglandins and food intake of rats: A component of energy balance regulation. *Physiology and Behavior*, 1973, *10*, 1077–1085.

Beach, F. A. Central nervous mechanisms involved in the reproductive behavior of vertebrates. *Psychological Bulletin*, 1942, *39*, 200–226.

Epstein, A. N., Fitzsimons, J. T., & Rolls, B. J. Drinking induced by injection of angiotensin into the brain of the rat. *Journal of Physiology*, 1970, *210*, 457–474.

Epstein, A. N., & Kenney, N. J. Suppression of angiotensin-induced thirst by the E-prostaglandins. In J. P. Buckley & C. Ferrario (Eds.), *International symposium on the central actions of angiotensin and related hormones*. New York: Pergamon Press, 1977.

Feldberg, W. Body temperature and fever: Changes in our views during the last decade. *Proceedings of the Royal Society of London, Series B*, 1975, *191*, 199–229.

Fitzsimons, J. T., & Simons, B. J. The effects on drinking in the rat of intravenous infusions of angiotensin given alone or in combination with other thirst stimuli. *Journal of Physiology*, 1969, *203*, 45–57.

Hainsworth, F. R., Stricker, E. M., & Epstein, A. N. The water metabolism of the rat in the heat: Dehydration and drinking. *American Journal of Physiology*, 1968, *214*, 983–989.

Kataoka, K., Ramwell, P. W., & Jessup, S. Prostaglandins: Localization in subcellular particles of the rat cerebral cortex. *Science*, 1967, *157*, 1187–1189.

Kenney, N. J., & Epstein, A. N. The antidipsogenic action of prostaglandin E_1. *Neuroscience Abstracts*, 1975, *1*, 469.

Kenney, N. J., & Epstein, A. N. An elaboration on the antidipsogenic action of prostaglandin E_1. *Neuroscience Abstracts*, 1976, *2*, 430.

Kenney, N. J., & Epstein, A. N. Antidipsogenic role of the E-prostaglandins. *Journal of Comparative and Physiological Psychology*, 1978, *92*, 204–219.

Malvin, R. L., Mouw, D., & Vander, A. J. Angiotensin: Physiological role in water-deprivation-induced thirst of rats. *Science*, 1977, *197*, 171–173.

McGiff, J. C., Crowshaw, K., Terragno, N. A., & Lonigro, A. J. Release of a prostaglandin-like substance into renal venous blood in response to angiotensin II. *Circulation Research*, 1970, (Suppl. 1), 26–27; I 121–I 130.

McGiff, J. C., Terragno, N. A., Malik, K. U., & Lonigro, A. J. Release of a prostaglandin E-like substance from canine kidney by bradykinin: Comparison with eledoisin. *Circulation Research*, 1972, *31*, 36–43.

Miselis, R. R., & Epstein, A. N. Feeding induced by intracerebroventricular 2-deoxy-D-glucose in the rat. *American Journal of Physiology*, 1975, *229*, 1438–1447.

Pappius, H. M., Rostworowski, K., & Wolfe, L. S. Biosynthesis of prostaglandin $F_{2\alpha}$ and E_2 by brain tissue *in vitro*. *Transactions of the American Society for Neurochemistry*, 1974, *5*, 119.

Phillips, M. I., & Hoffman, W. E. Sensitive sites in the brain for the blood pressure and drinking responses to angiotensin II. In J. P. Buckley & C. Ferrario (Eds.), *International symposium on the central actions of angiotensin and related hormones*. New York: Pergamon Press, 1977.

Simpson, J. B. Localization of dipsogenic receptors for angiotensin II. In J. P. Buckley & C. Ferrario (Eds.), *International symposium on the central actions of angiotensin and related hormones*. New York: Pergamon Press, 1977.

Simpson, J. B., Epstein, A. N., & Camardo, J. S., Jr. The localization of dipsogenic receptors for angiotensin II in the subfornical organ. *Journal of Comparative and Physiological Psychology*, 1978, *92*, 581–608.

Wolfe, L. S., Pappius, H. M., & Marion, J. The biosynthesis of prostaglandins by brain tissue *in vitro*. In B. Sammuelsson & R. Paoletti (Eds.), *Advances in prostaglandin and thromboxane research* (Vol. 1). New York: Raven Press, 1976.

K. V. SHULEIKINA-TURPAEVA

Goal-Directed Behavior in Ontogenesis

30

The functional system of food acquisition is the only active form of goal-directed behavior present in most newborn birds and mammals. According to the theory of systemogenesis (Anokhin, 1964), the ontogenetic processes of functional systems are accomplished in strict accord with a given species' ecology through the selective and heterochronic maturation of the individual elements. This chapter is concerned with a study of the basic stages of food acquisition, the composition of the sensory stimuli, and its guiding and neurophysiological mechanisms. *Ficudula hypoleuca* nestlings and 30-day-old kittens were used as the study subjects.

The alimentary behavior of the nestlings was studied in their natural habitat, a national forest. The experimental chamber comprised an ordinary hollowed-out tree with its back side removed and replaced by a light-insulated chamber built onto the tree that housed the experimenter (Khayutin & Dmitrieva, 1976).

Immediately after hatching, and in the ensuing days, the alimentary reaction of the nestlings consisted of a rapid lifting of their heads with upwardly outstretched necks and simultaneously wide-opened beaks (Figure 30.1). This was accompanied by vocalization beginning with the 4th to 5th day. In the course of the first eight days, the food-providing birds flew through the trough and fed the nestlings from places that were strictly fixed for each food-providing bird. We established that the probability of obtaining food for nestlings located in different sections of the nest was not identical. If we conditionally relate the perimeter of the nest to the perimeter of the face of a clock (Figure 30.2) and let the various sections

Figure 30.1. Periods of nestling alimentary behavior determined by changes in trigger stimuli. The alimentary reaction was induced by (I) sound; (II) change in luminosity; (III) change in luminosity and moving silhouette of bird; (IV) moving silhouette of bird only. Along the vertical is shown days of fledglings' nestling life.

of the nest correspond to the positions of the hour hand with the entrance to the nest placed over the 12, and if we designate the section of the nest in which the bird sits as the number 12, we then can say that the zone of maximum reinforcement corresponds to the position of 6 o'clock, which is exactly opposite the bird's entry point, with two submaxima to the right and the left of this section (3 and 9). Our study has shown that the nest-

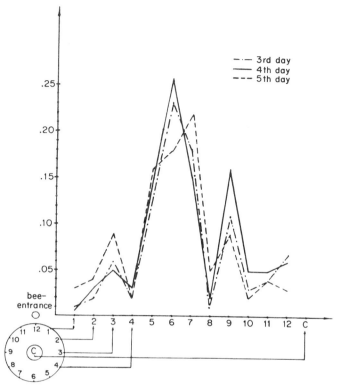

Figure 30.2. Probability of nestlings' obtaining food in different zones of the nest. Ordinate is probability of obtaining food in each zone of the nest; abscissa is zone of the nest. Age of nestlings is 3–5 days.

lings were able to obtain food regularly by constantly shifting about the perimeter of the nest. The nestling located in the zones of maximum reinforcement (6, 3, 9) exhibited the maximum level of alimentary motivation. Thus, the alimentary reaction has two phases: (a) raising the head with wide open beak; (b) crawling away from the feeding zone. This can be seen clearly in the EMG recordings that were made from electrodes chronically implanted into the nestling's cervical muscle region (Figure 30.3).

The question arises as to what are the trigger stimuli that direct the nestling's behavior, and does the composition of those stimuli change in the course of the 15-day nesting period?

From the 1st to the 5th day, the nestlings are blind, and the alimentary trigger stimulus is a complex of natural sounds that accompany the arrival of the food-providing bird. A spectral analysis of these sounds on a sonograph demonstrated the following (Figure 30.4).

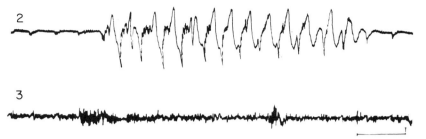

Figure 30.3. *Electrophysiological correlates of a nestling's alimentary reaction. (1) Stimulation marker (exclusion of light); (2) vocalization accompanying alimentary reaction; (3) electromyogram of cervical muscles. Age of nestling is 5 days. Calibration is 50 mcV; 1 sec.*

1. Low frequency component (.2–1 kHz) is created by the flapping of the approaching bird's wings, the scratching of claws as the bird lands at the entrance, and the rustling noises made by the bird as it jumps and moves about the nest (A).
2. High frequency component (1–4.8 kHz) is the low "clattering" sound that is the direct alimentary signal of the adult bird (B).

The nestlings open their eyes after the 5th to 6th day. From this time on, the trigger stimulus is no longer a sound but a diffuse change in illumination. As the adult bird flies into the birdhouse, the entry hole is covered up by its body. During this time (100–150 msec), the illumination in the birdhouse falls from 150 to 2–5 lux. This induces the alimentary reaction of the nestlings.

By the 8th day, the clattering alimentary signal of the adult bird (either the natural one or reproduced on tape) ceases to cause an alimentary reaction in the nestlings and the decrease in luminosity in the nest becomes the natural trigger stimulus.

Beginning with the 9th day, the nestling alimentary reaction changes somewhat. The upward extension of the head and the passive opening of

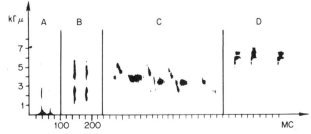

Figure 30.4. *Dynamic spectrogram of type-specific signals of adult birds significant to nestling behavior. (A) Sounds accompanying arrival of bird with food; (B) "alimentary" signals; (C) species song; (D) alarm signal. (From Kay Electric, 7029A.)*

the mouth are replaced by an active taking of food. The nestlings turn their heads to meet the mother bird, "assault" her beak, and pull out the food. A frame-by-frame analysis indicates that this form of behavior is maintained by two stimuli: the trigger (drop in luminosity) and the directing stimulus (the moving silhouette of the bird).

On the 12th to 15th day (time of flight) the adult bird's silhouette becomes a trigger and a directing signal as well. The diffuse change in luminosity loses its signal importance. The sequential change in the significance of all of the aforementioned stimuli is shown in Figure 30.1.

The developmental characteristics of alimentary behavior were also studied in the kitten. Beginning at birth and throughout the entire nursing period, the food-acquisition process is an exemplary model of goal-directed behavior with a high level of alimentary motivation. It represents a stagewise operating functional system with an interchange of sensory transmissions that carry information about the results that change from one stage to the next.

The first stage is distance search. The action result here is the identification of the mammary gland region. Information about results comes primarily from the olfactory receptors. The second stage is contact search. The result is the identification of the nipple. Information about results is transmitted by the olfactory and tactile (from the lip region) afferents. The third stage is the grasping of the nipple, where information about the results comes from the tactile receptors of the lip and tongue. Finally, the 4th stage is sucking. The final result of sucking is the passage of milk into the oral cavity, the esophagus, and the stomach. Information about the achievement of that result is formed by the sum of signals transmitted from the tactile and gustatory receptors of the lips and tongue and from the proprioceptors of the masticatory musculature.

Chronic experiments were performed on kittens that included the recording of EEGs in all stages of food acquisition in order to clarify the neurophysiological mechanisms of alimentary behavior.

An overall EEG of newborn animals in the wakeful resting or sleeping state is represented by low-voltage activity with a narrow frequency range (Figure 30.5A). During distant and contact search, the amplitude of potentials increased sharply from 30–50 to 100–150 μv, and the high-amplitude slow waves at 3–6 and 8–10 per second became the dominating EEG type (b). These fluctuations became particularly well defined during the grasping of the nipple, after which the amplitude fell sharply and remained at a low level during the entire period of sucking. A tendency toward synchronization at 10–20 per second was observed in the structures transmitting impulses from the oral afferents during sucking, and a sinusoidal rhythm of 2–4 per second (B, D) was observed in the formations of the lower cord.

Changes that reflect a state of physiological hunger could be recorded

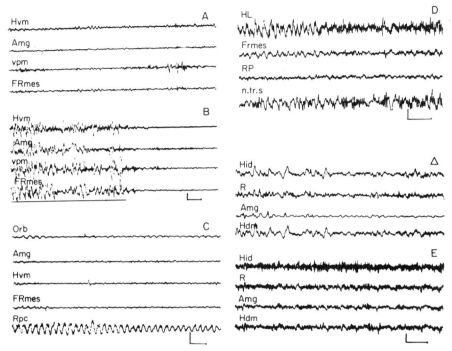

Figure 30.5. *EEG correlates of a nestling's alimentary behavior. (A) EEG background outside alimentary reactions; (B) during search of mother's nipple and sucking (search underscored); (B, D) selective synchronization of EEG during sucking; ($\triangle$, E) EEG background in the satiated ($\triangle$) and hungry 1-day-old kitten (E). Age of animals is 3 days (A–B), 40 days (D–E). Calibration is 50 mcV; 1 sec. Symbol designations: (see original).*

in the first days of the kitten's life. Such changes included faster oscillations ($\triangle$,E) in the frequency spectrum with a simultaneous amplification of the slow high-voltage "search" activity that was now recorded spontaneously, not only in connection with the direct search for the nipple.

Specially conducted experiments allowed us to hypothesize that high-frequency activity reflects a general hunger, nonspecific activation, whereas slow spindles constitute the EEG correlate of the motivational component of alimentary stimulation (Anokhin & Shuleikhina, 1977).

By utilizing the EEG data, we were able to study the more analytical processes and characteristics of the properties of the individual neurons included in the actuating and sensory apparatus of food acquisition.

In the course of studying the actuating apparatus of food acquisition, we recorded motor neurons at the C_{1-3} level that innervate the cervical muscles and accomplish the nipple-seeking movements. The motor neurons of the nuclei of the facial nerve (VII) and the sublingual (XII) nerve produce the sucking movements. The responses of those nerves to stimulation of the oral afferents can be identified very early, and are even recorded in the fetus in the form of single or group firings or as tonic-type reactions in the form of a retardation or extinction of the initial rhythm.

We identified a selectivity of response with respect to the localization of an applied stimulus. The greatest percentage of responses (52% in the fetus) was elicited by stimulation of the tongue. The smallest percentage was elicited by stimulation of the supercilium. For example, the responses of the VII nerve neurons to stimulation of the tongue acquired a more mature form of compact bundles earlier than responses to the stimulation of the supercilium (Figure 30.6).

Assuming that the receptive zone of the tongue is one of the basic sensory approaches of the alimentary functional system, we studied the electrophysiology of neurons in the nucleus of a single bundle (n. tr. s.) and a complex of trifacial nuclei (n.V) that transmit information from

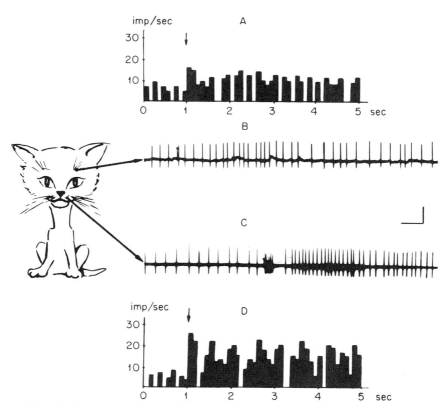

Figure 30.6. Responses of the N. facialis *neuron during variable localization of the stimuli. (A, B) Tonic type of response during electrical stimulation of the supercilium; (C, D) more mature physical reaction during stimulation of whiskers. Age of kitten is 10 days. Calibration is 500 mV; 1 sec.*

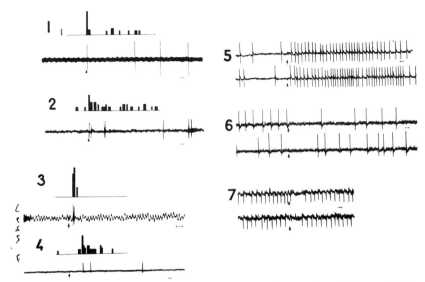

Figure 30.7. Neuron reaction of N. caudalis trigemini to stimulation of the lingual nerve. (1, 2) Short latent physical responses; (3, 4) long latent reactions; (5-7) tonic activational and inhibitory responses. Age of animals is 1-4 days. Ordinate in histograms is 10 imp. Calibration is 50 msec.

the gustatory and tactile receptors of the tongue. From 33 to 35% of the neurons were shown to respond to stimulation of the *chorda tympani* and *n. lingualis* during the first days of the kitten's life. We recorded various types of responses that included both inhibitory and activational responses. Of greatest interest was the physical type of reaction: a single spike or a bundle of impulses followed by a secondary activation (Figure 30.7, 1, 2). This type of response has been described as specific for corresponding neurons in the adult animal. These reactions were represented to the greatest extent in the caudal nucleus of the V nerve complex and in the n. tr. s. (33%). Similar reactions were observed in a few cases in the oral section of the trigeminal complex in the newborn kitten. They were less well defined and had a long latent period.

The aforementioned data allow us to conclude that the nuclei we examined do not develop simultaneously in ontogenesis. The more rapid maturation of the caudal section of nerve V and the lag of the oral section of nerve V attest to the heterochronism of facial sensitivity. This is especially indicative of the earlier maturation of crude (in response to touch) proprioceptive and gustatory sensitivity, and of a later, fine, tactile discriminatory sensitivity. This corresponded to our earlier findings (Shuleikina, 1972) about the change in the signal significance of these modalities during the maturation of alimentary behavior.

These data indicate that the information from the oral afferent com-

30. Goal-Directed Behavior in Ontogenesis

prise an important link in the sensory processes of alimentary behavior. It would be interesting to know how the alimentary reactions might be affected by a cessation of that sensory transmission. With this in mind, we performed a series of experiments on a kitten that provided for the exclusion of the perioral zone and tongue by anesthesia (application of a tampon soaked in a 2% cocaine solution to the tongue or lips for 1–2 min).

The local disconnection of the lingual receptors led to intense activation of food-acquisition behavior (Figure 30.8). The kitten started to burrow frantically into the mother's fur while continuously searching over an extensive period of time. The capability of orientational search was observed. The kitten searched only in the region of the nipples, although in a number of instances its striving for the preferred nipple was lost. When the kitten seized the nipple, it did not calm down, became fussy, sucked interruptedly, often broke away, and again started to search.

A local exclusion of the perioral zone (lips) also led to an intense increase in food-acquisition activity. However, the goal-directiveness of the search was disrupted. The kitten started to seek out the nipples along the entire body of the parent, including the back, near the tail, or head. The capability of grasping the nipples was extinguished, and despite a long search and close localization of the nipple, the kitten did not grasp it and was not in a position to start sucking.

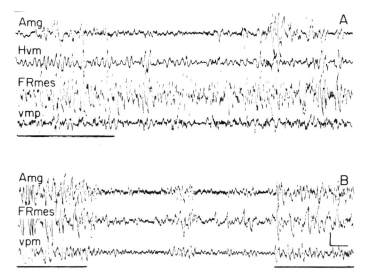

Figure 30.8. High-voltage slow EEG and absence of potential amplitude fall during transition from search to sucking with anaesthetized tongue. (A) Search and sucking in the anesthetized kitten; (B) search and sucking in the intact kitten (search underscored). Age of kitten is 20 days. Symbol designations are the same as in Figure 30.5. Calibration is 50 mcV; 1 sec.

We were further interested in tracking the EEG changes that occur following anesthesia. What kind of EEG would now correspond to search and sucking and, most importantly, would the drop in the amplitude of potentials be retained when the nipple is grasped and the sucking movements begin? This series of experiments was conducted with anesthesia of the tongue since, as has been demonstrated, sucking does not occur during perioral anesthesia.

Search activity proceeded with ever-increasing intensity in the initial seconds following the application of the cocaine, but epileptoid activity was observed in the structures conducting impulses from the tongue (orbital cortex, thalamus, and ventral reticular nucleus of the medulla oblongata adjoining the nucleus of a single bundle). After several minutes of continuous search, a high-voltage slow "search" activity started to register in all leads.

The generalized search spindles were observed to the greatest degree when the maximum effect of the anesthesia was reached. If the nipple had not been grasped at that stage, we did not during subsequent sucking observe any change in the amplitude of potentials that is characteristic of the transitions from search to sucking, and the EEG remained high voltage. Sucking at this stage was of short duration. Without obtaining feedback about grasping or the course of sucking, the kitten frequently broke away, sought, grasped again, fretted, lost the nipple, and again started to search for it. During this time, the EEG recorded a continuous flow of high-voltage "search" spindles.

Irreversible exclusion of the oral afferents, accomplished by bilateral sectioning of the tympani chorda and the lingual nerve, produced the most dramatic consequences of deprivation. This led to a sharp fall in alimentary motivation in the 1- to 5-day-old kittens. The kittens moved about actively two days after the operation, and the orienting response was conserved. Food-acquisition behavior in those kittens, however, was almost extinguished. The animals did not seek the nipples or attempt to suck at all, or if they did, this was done very slowly and sluggishly. The average search time in the operated kittens was twice as great as the search time in the intact kittens. The former sucked for brief periods not exceeding .5–1.0 min. During the first 2 to 3 days after the operation, the animals lost 40% of their weight, and if the surgery was performed before the 5th day of life, the animals died of emaciation 2 to 3 days after the operation. It is important to note that under these conditions the operated animals avidly gulped milk fed from a pipette. Dissection disclosed no postsurgical complications. A control sectioning of the *n. infraorbitalis* did not cause animal death.

All the kittens operated on at the age of 20 days or more survived the surgery. Reduced food-acquisition behavior was weakly exhibited, and completely disappeared 3 to 4 days after the operation.

Consequently, sensory transmissions from the receptors of the oral zone and tongue play a decisive role in the realization of food-acquisition behavior. This is of the greatest significance in the first days of life. Inasmuch as this was demonstrated by sensory deprivation, we formulated an inverse problem that called for studying the effect that heightened sensory activation would have on the ontogenetic process. These experiments were conducted on nestlings by combining variously aged nestlings into one family.

A portion of 4- to 5-day-old fledglings in a nest was replaced by nestlings from other nests that were 4 days younger. Under these conditions, the younger nestlings were placed in an enriched medium since they were surrounded by older birds with more active motility, vocalization, and different trigger stimuli. Thus, the alimentary signal in the 1- to 4-day-old nestlings was a sound stimulus, whereas that signal was diffuse change in luminosity in the 5- to 6-day-old nestlings. We found that although the younger nestlings lagged behind the development of the older ones in the beginning, as indicated by their weight curve, their development had become equal by the last days of nesting and they reached the weight of the older nestlings. The younger birds opened their eyes earlier and they left the nest at the same time as the older nestlings, that is, 4 days earlier than they would have otherwise. The acceleration processes were well reflected in the alimentary activity of the younger nestlings (Figure 30.9). In the initial days of the experiment, the average frequency of feedings per nestling was significantly greater in the older birds, but later that correlation was evened out, and beginning with the 11th day, the correlation was reversed (b). The same characteristic was observed in the amount of food consumed. In the first 3 days the older nestlings received more food per bird than did the younger ones, but later that relationship was evened out, and in the last five days before flight the younger nestlings were receiving more food than the older ones (A). One of the mechanisms that sustained this process was the behavioral specificity of the younger nestlings who, after their routine alimentary reaction, did not crawl away, as they usually did, but remained in the zone of maximum alimentary reinforcement.

We also observed an acceleration of the maturation processes when sensory transmission was amplified in acute experiments on the neuron level as well. Thus, when we applied a series of stimulants we were able to record not only the direct response to a stimulus but a general amplification of background rhythm in the interstimulant interval. The episodic nature of the background rhythm changed to a continuous rhythm during stimulations of the fetal tongue in neurons of the reticular formation of the medulla oblongata (Figure 30.10). The same characteristic was demonstrated in the cortical neurons of the newborn kitten. When the *locus coeruleus* was stimulated, the immature episodic form of rhythm became

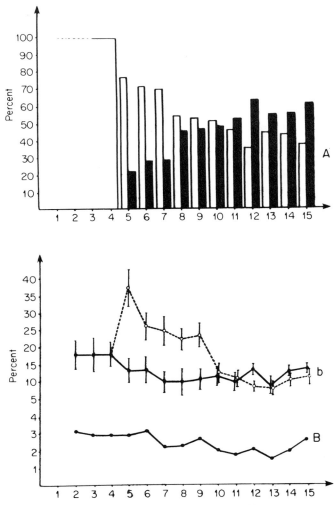

Figure 30.9. Dynamics of food distributions among nestlings of two age groups in an extension of joint habitation in one nest. (A) Relative amount of food obtained by the nestlings of the older (clear columns) and younger (black columns) age groups. Ordinate is amount of food (1%); abscissa is days of nesting life. (b) is average intervals between the receipt of food by the same nestling of the older (black circles) and younger (clear circles) age groups. Ordinate is time in minutes; abscissa is days of nesting life. (B) Average intervals between the arrivals of birds with food.

more mature—a continuous or bundle rhythmic form. The latter type of impulse usually appears in month-old kittens. It is important to note that this kind of activation of cellular firing activity is observed for a long period—up to 10 min, and is maintained following the cessation of the stimulation. Inasmuch as the *locus coeruleus* is an important adren-

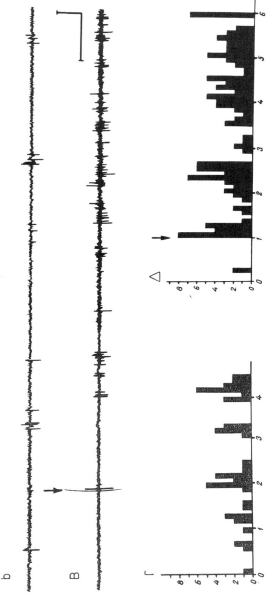

Figure 30.10. Phenomenon of background rhythm activation in the interstimulant interval. (b, Γ) Firing activity prior to stimulation of the tongue; (B, △) in a background of lingual stimulation. Neuron from the N. reticularis magnocellularis medulla; feline fetus is 54 days old. Ordinate is number of impulses; abscissa is time in seconds. Summation for five presentations. Calibration is 500 mV; 1 sec.

ergic structure of the brain, one can assume that the activation of a neuron's firing activity is accomplished with the participation of an adrenergic chemical mechanism, as we have demonstrated in earlier works (Shuleikina & Raevskii, 1974).

The acceleration of electrophysiological phenomena can also be tracked by the evoked potential method. Two monophasic negative oscillations are recorded in the normal 4-day-old kittens in the motor sensory region of the cortex when the *n. ischiadicus* is stimulated. However, if the lingual nerve, projecting into the orbital cortex, is stimulated before that time, or if strychnine is applied to the orbital cortex, a more mature form of response that includes a positive phase will appear in the motor sensory cortex in response to stimulation of the *n. ischiadicus*.

Thus, sensory impulse transmissions play a decisive role in the organization of behavioral ontogenesis up to the neuron level processes that are the basis of that behavior. A sensory overflow transmission is capable of causing acceleration, but sensory deprivation causes behavioral disintegration. This is observed primarily in the early stages of ontogenesis.

It may be assumed that the influence of a sensory transmission is reflected primarily in the maturation rate of a particular activity, while the total sequence of behavioral stages is formed independently and is determined by a genetic program (Khayutin & Dmitrieva, 1976).

REFERENCES

Anokhin, P. K. Systemogenesis as a general regulator of brain development. In W. Himwich & H. Himwich (Eds.), *Progress in Brain Research*, 1964, 9, 54–75.

Anokhin, P. K., & Shuleikina, K. V. System organization of alimentary behavior in the newborn and the developing cat, *Developmental Psychology*, 1977, *10*(5), 385–419.

Khayutin, S. N., & Dmitrieva, L. P. The role of natural sensory factors in the organization of alimentary behavior as illustrated in the development of nestlings, *Journal of Higher Nervous Activity*, 1976, *26* (No. 5), 931–1051. (In Russian)

Shuleikina, K. V. *Systems organization of alimentary behavior*. Moscow, 1972.

Shuleikina, K. V., & Raevskii, V. V. Microelectrophoretic study of brain stem neurons in fetal and newborn cat. In L. Jilek & S. Trojan (Eds.), *Ontogenesis of the brain* (Vol. 2). Universitas Carolina Pragensis, 1974, pp. 67–76.

ELLIOTT M. BLASS

The Ontogenesis of Suckling, A Goal-Directed Behavior[1]

31

The description and analyses of ontogenetic processes have been acknowledged by Eastern and Western psychobiologists alike as vital for understanding goal-directed behavior. The attraction to developmental events is natural. The ontogenetic landscape is dynamic; the morphological, physiological, and behavioral characteristics of developing organisms are continuously changing. This dynamism, which approaches stability in adult behavior, is often thought to contribute significantly to the appearance and expression of normal adult-motivated behavior (Schneirla, 1965); indeed, many studies have been designed to reveal the effects of early experiences on adult behavior. This may be called the longitudinal approach to the study of developmental processes. Harlow's (Harlow, Harlow, & Hansen, 1963) famous studies demonstrating fundamental aberrations in social, reproductive, and material behaviors by rhesus monkeys reared in isolation exemplifies this paradigm: Specifically the effects of a manipulation made during infancy are evaluated when the animal reaches adulthood. The longitudinal approach has been venerated in the psychiatric literature and has held a central position in theories about human cognitive (Piaget, 1953) and animal social development (Schneirla & Rosenblatt, 1963). At its best, it identifies and focuses on behavioral events that either *maintain* an ongoing behavior, *facilitate* the appearance and asymptotic performance, but not the quality, of a behavior, or *induce* some of

[1] Research in the author's laboratory was supported by National Science Foundation Grant BMS 75–01460 and National Institute of Arthritis Metabolism and Digestive Diseases Grant AM 18560.

the qualitative characteristics of a behavioral sequence (see Gottlieb's [1976] excellent monograph on this issue). The longitudinal approach takes into consideration the animal's sensory capabilities and ecological stresses and does not fall into the trap of assuming that the altricial neonate is an incompletely formed adult whose adult processes are primitive and incomplete.

A second way to study developmental processes is to treat ontogenesis as a problem in comparative behavior. Here emphasis is placed on the systematic differences in the animal's environment and behavior that occur from developmental epoch to epoch. According to the comparative approach, changes in behavior at a particular time continue to remain important as potential determinants of future behavior, but the changes must be viewed as important in and of themselves for the animal's survival at that time. These changes reflect the developing organism's adaptations to alterations in its mother, siblings, nest, its own expanding sensorimotor capacities, and, by definition, its increasingly complex environment. It follows that each developmental epoch [2] has its own characteristics and that survival to adulthood reflects the efficacy of physiological-behavioral adjustments to the stresses heralded by each epoch.

An analogy with the embryological, morphological, and functional development of the kidney illustrates how both the longitudinal and cross-sectional approaches are necessary to understand developmental processes. Briefly, renal development in higher vertebrates occurs through three different, more or less clearly defined, tubular systems, with each successive system occupying a different morphological, cephalocaudal location (see Fraser [1950] for review). The structure of each succeeding tubular system becomes increasingly complex and differentiated, culminating in the third system, the metanephrotic renal system. This becomes functional postnatally as the kidney of higher vertebrates. There are two especially relevant points concerning the present discussion. First, the tubular systems of the pronephros and mesenephros (the first two kidneys) disappear by birth. All that remains morphologically is the collecting duct that started in the pronephritic stage. The pronephritic duct elongates and eventually becomes connected to the developing epithelial vesicles of the newly developed mesenephrotic kidney. This duct then becomes available to the metanephrotic kidney as the conduit for urine excretion. Thus the longitudinal analysis of renal embryology provides a basis for understanding the development of collecting duct formation and the induction of the characteristics of each kidney.

[2] The idea of a developmental time frame or epoch must be used with considerable caution. I do not mean to imply that transitions are abrupt and occur at specifiable intervals. The terms are used tentatively as first approximations to emphasize major differences in morphology, physiology, and behavior that are obvious across an interval of several days (in rats at least).

The second, and complementary, approach studies renal function at each embryological stage. The pronephros becomes functional only in the most archaic living vertebrates (e.g., the elasmobranch fishes) and never in the embryos of amniotes. The mesenephros, however, does become functional in certain mammalian embryos, the pig, for example. This functional development is revealing because it occurs only in those fetuses in which the connection between maternal and fetal tissue is *not* well established, that is, cannot adequately remove metabolic wastes. In mammals with close exchanges, that is, those with hemochorial placentas, the mesenephros apparently serves to pass its duct along to the metanephrotic stage and does not function renally. In short, abiding exclusively by the longitudinal analysis may cause us to miss certain events that are related to functional changes at each stage of development. The cross-sectional approach is complementary, because differences from stage to stage may reveal the bases of changes in either system or organism behavior.

The two conceptually distinguishable approaches merge, of course, when the individual "cross-sections" of behavioral development are analyzed successively. The characteristics of each epoch can be studied as can changes as the animal matures. The biases that developed at one point in time may be passed along to the succeeding point, and so on, so that an event early in the animal's lifetime may exert a profound effect on its adult behavior (cf. Daly, 1973). This is clearly seen in the ontogeny of ingestive behavior. The elegant study of Hall (1975) revealed that infant rats do not have to suckle beyond the second day of life in order to eat and drink normally at the time of weaning. Animals raised via intragastric intubation adjusted their food and water intake normally to a variety of nutritional and hydrational demands. In short, the expression of normal feeding and drinking does not depend upon suckling experience (beyond Day 2 at least). Yet the studies of Galef and his colleagues (see Galef [1976] for review) have clearly demonstrated that the suckling and feeding events that occur before weaning help determine the foods that the rat will *choose* to eat.

Accordingly, we have allowed both of these paradigms to direct our analyses of development. We have chosen to study the ontogeny of ingestive behavior, which includes suckling, infantile feeding and drinking, and the more mature forms of feeding and drinking seen in the juvenile. We have sought to identify the characteristics of developmental transitions in behavior to determine experiential events that might naturally and meaningfully contribute to these transitions. We have chosen to study suckling behavior in rats because more is known about the ingestive behavior of rats than about that of any other species. To the extent that suckling is the infant's only source of obtaining nutrients, this consideration was not minor.

In order to identify the characteristics of suckling in rats and their

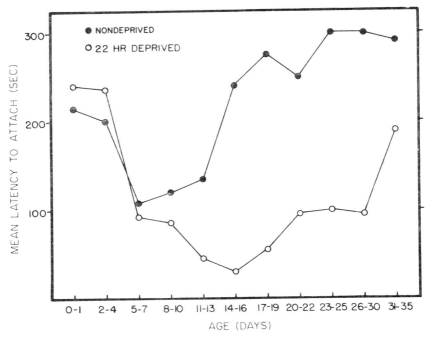

Figure 31.1. Mean latency to attach to the nonlactating teat of anesthetized rat dams by their deprived (O) and nondeprived (●) young of various ages. (From Hall et al., 1977)

transitions to adult ingestion, Hall, Cramer, and Blass (1975, 1977) allowed nondeprived rats and siblings deprived of their mother for 4, 8, or 24 hr to attach to the nipples of their anesthetized mothers. (Barbiturate anesthetization also blocks milk letdown in our test conditions.) Figure 31.1 shows a very clear ontogenetic time course of nipple attachment. Until about 10–12 days of age nipple attachment behavior is not related to privational status. By about 14–15 days postpartum and beyond, the animal's immediate privational history severely affects its relationship vis-à-vis the mother. These findings suggested that internal events contributed to suckling control starting at about two weeks postpartum. This idea was supported by Hall and Rosenblatt (1977, 1978), who demonstrated that neither gastric fill (1977) nor the nutrient quality of the gastric preloads (1978) affect nipple attachment or the amount of milk derived through an oral cannula during the act of suckling. Adult levels of control were reached at about 20 days postpartum. Blass, Beardsley, and Hall (1979) have shown that the gut hormone cholecystokinin, a demonstrated inhibitor of feeding in adult rats, does not influence milk intake until about two weeks postpartum, at which time it starts to reduce milk obtained via the tongue cannula while suckling. Like gastric distension, cholecystokinin effectiveness reaches adult levels by about 20 days of age. This conclusion is supported by Bruno (1977), who found that neither cellular nor extracellular dehydration affects nipple attachment until at least 2 weeks postpartum, at which time nipple attachment latencies are exaggerated by cellular dehydration. Extracellular fluid depletion exaggerated nipple attachment latencies starting at about 20 days of age.

In short, a number of independent lines of evidence support the idea that nipple attachment and the volume of milk consumed via suckling are not directly affected by the internal stimuli that signify either hydromineral or nutritional deficits or surfeits until about two weeks postpartum. More complete discussions of this ontogenetic pattern and its implications have been presented at length by Blass, Hall, and Teicher (1979) and Hall and Rosenblatt (1978).

Given that the consummatory aspects of nipple attachment are not affected by internal status until about 2 weeks of age, it becomes of interest to determine whether and when the appetitive or goal-directed facet of suckling came under internal control. Kenny and Blass (1977) devised a Y-maze that allowed neonatal rats to suckle a non-milk-producing nipple on one side of the maze or root into the rat fur, without attachment, on the other side of the maze. We found that rats from 7 to 23 days of age chose to suckle the nonlactating nipple. Thus, the nipple, which elicits the integrated suckling act, is a sufficient incentive for instrumental learning in rats with very differing physical and social attributes (e.g., furless versus furred; socially passive versus hyperactive). Blass, Stoloff, and Kenny (1977) provided infant rats a choice between a nonlactating nipple on one side of the maze and a lactating one on the other side. Their results are shown in Figure 31.2. Rats 10 to 12 days of age chose indiscriminately between milk-producing and nonproducing nipples. By 15 days of age a majority of the rats chose the lactating side and by 20 days of age all rats chose the nipple that yielded milk. The parallel between the ontogenies of the consummatory and appetitive components of nipple-seeking behavior is striking. It implies that not only is intake volume coming under the control of internal events but that the stimulus array that reinforces approach behavior is changing and becoming increasingly more complex. The mother, in addition to being defined as a source of contact, warmth, and so on, is now also seemingly responded to as a nutritional source. This has important implications for the weaning process because the mother is becoming increasingly less accessible and her milk supply is dwindling.

We then determined if this preference of nutritive over nonnutritive suckling reflected the infant rat's suckling experiences during the time, beginning at Days 14–15, when internal stimuli gained control over suckling or whether this change in behavior would occur independently of normal suckling and milk ingestion experiences. Accordingly, rats were removed from their dams on the 12th day postpartum and kept alive and in excellent health by chronic intragastric intubation through Day 16. Figure 31.3 demonstrates that neither the suckling act per se nor the pair-

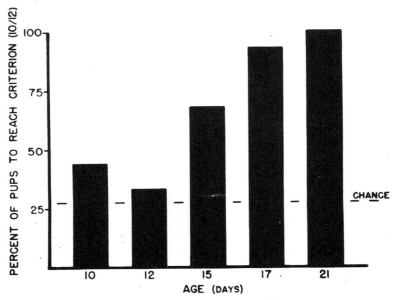

Figure 31.2. *Percentage of rats of various ages that preferred a lactating to a non-lactating nipple. Each column represents 12 rats.*

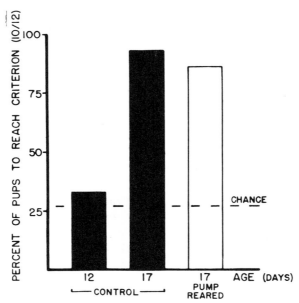

Figure 31.3. *Preference for a lactating nipple of 12- and 17-day-old rats reared normally (shaded columns) or isolation-reared for Days 11–17 postpartum (open column).*

ing of suckling with its postingestive consequences is necessary for the expression of nutrient control over nipple preference.

Stoloff, Blass, and Kenny (personal communication, 1978) determined how the increasing unavailability of the dam and the increasing utilization of free food and water by the neonate becomes expressed in the infant's relationship with the mother. We allowed rats to choose between suckling (either a lactating or nonlactating nipple) and eating either a liquid or solid diet. Figure 31.4 demonstrates the subtlety of the choices made by the neonates. Virtually all 24 hr-deprived 17-day-old rats chose the nipple, regardless of its nutritional status, over eating a palatable liquid diet (they had at least 24 hr of exposure to this diet prior to deprivation). Rats 21 days of age were both similar to and different from 17-day-old rats. They were similar in that milk availability did not differentially affect behavior as 72% and 60% learned to the nipple side in the

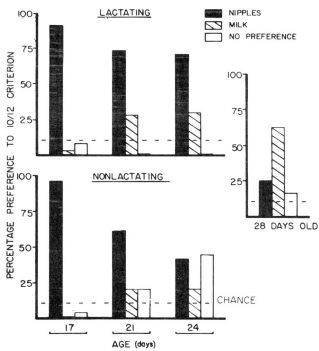

Figure 31.4. Percentages of rats 17–28 days of age that went to the side of a "Y-maze" that permitted suckling (black columns); that went to the side containing milk (lined columns); or that did not demonstrate a preference for either side (open columns). Performance when the nipple was lactating is seen in the upper panel; when nipple was not lactating is seen in the lower panel. Data from 28-day-old rats are presented individually because even though 25% of these rats chose the side of the maze containing the lactating nipple, none of these rats suckled. All younger rats suckled after choosing the appropriate side.

lactating and nonlactating conditions, respectively. They were different in that significantly more rats chose either to eat or to divide their choices between suckling and feeding relative to 17-day-old rats. Twenty-four-day-old rats were different yet. For the first time the milk-producing status of the nipple became important in this task. When the nipple yielded milk, it remained highly preferred to eating a liquid diet. When the nipple did not produce milk, however, it was no longer preferred, for a majority of the rats either ate or divided their choices between suckling and feeding.

Twenty-eight-day-old rats differed from the others, too. Fifty percent ate, but about 24% continued to go to the side of the maze that accommodated the mother. But these animals did not suckle, even though the nipple was lactating. They merely sat in the goal compartment. To analyze this behavior further we removed the gauze screen separating the milk from the dam. Under these circumstances the pups unanimously chose the side that permitted them to eat in the presence of the mother and did so.

Thus the cross-sectional approach to the ontogeny of suckling behavior has yielded a portrait of successive liberation from maternal sensory dominance. Prior to 12 days of age, rats, regardless of their hydrational and nutritional status, rapidly attach to the nonlactating nipples of their anesthetized mother. By Day 15 their choice of nipple reflects its lactational status, even though the nonlactating nipple remains preferred to feeding. By 24 days of age internal controls determine whether the rat suckles or eats: A lactating nipple is reliably chosen over free food, but a nonlactating nipple is not. Finally, 25% of the Day 28 rats deprived of mother and ingestion for 24 hr prefer to sit by the mother, neither suckling nor feeding.

The circle is almost complete. From Hall (1975) we have learned that extensive suckling is not necessary for the animal to know *how* to eat; Galef (1976) has demonstrated the influence of experience on *what* to eat. Using the cross-sectional approach, we have started to show the changing relationship with the mother that culminates in weaning. Because important behavioral differences occur in a matter of days, we may now seek differences, if any, in the mother's behavior or milk letdown patterns that might induce these changes in the young. The longer lasting effects on ingestive and other social behaviors of these putative environmental changes beckon exploration.

REFERENCES

Blass, E. M., Hall, W. G., & Teicher, M. H. The ontogeny of suckling and ingestive behaviors. In J. M. Sprague & A. N. Epstein (Eds.), *Progress in psychobiology and physiological psychology* (Vol. 8). New York: Academic Press, 1979.

Blass, E. M., Beardsley, W., & Hall, W. G. Age-dependent inhibition of suckling by cholecystokinin. *American Journal of Physiology*, 1979, *5*, E567–E570.

Blass, E. M., Stoloff, M. L., & Kenny, J. T. Nutritive vs. nonnutritive suckling: The ontogeny of preferences in the albino rat. Paper read at Annual Meeting of the Psychonomic Society. Washington, 1977.

Bruno, J. P. Body fluid challenges inhibit nipple attachment in preweanling rats. Paper presented at the Eastern Psychological Association meeting, April 1977.

Daly, M. Early stimulation of rodents: A critical review of present interpretations. *British Journal of Psychology*, 1973, *64*, 435–460.

Fraser, E. A. The development of the vertebrate excretory system. *Biological Reviews*, 1950, *25*, 159–187.

Galef, B. G. The social transmission of acquired behavior: A discussion of tradition and social learning in vertebrates. In J. S. Rosenblatt, R. A. Hinde, E. Shaw, & C. Beer (Eds.), *Advances in the study of behavior* (Vol. 6). New York: Academic Press, 1976.

Gottlieb, G. The role of experience in the development of behavior and the nervous sysstem. In G. Gottlieb (Ed.), *Neural and behavioral specificity: Studies on the development of behavior and the nervous system*. New York: Academic Press, 1976, pp. 1–25.

Hall, W. G. Weaning and growth of artificially reared rats. *Science*, 1975, *190*, 1313–1315.

Hall, W. G., Cramer, C. P., & Blass, E. M. Developmental changes in suckling of rat pups. *Nature*, 1975, *258*, 318–320.

Hall, W. G., Cramer, C. P., & Blass, E. M. The ontogeny of suckling in rats: Transitions toward adult ingestion. *Journal of Comparative and Physiological Psychology*, 1977, *91*, 1141–1155.

Hall, W. G., & Rosenblatt, J. S. Suckling behavior and intake control in the developing rat pup. *Journal of Comparative and Physiological Psychology*, 1977, *91*, 1232–1247.

Hall, W. G., & Rosenblatt, J. S. Development of nutritional control of food intake in suckling rat pups. *Behavioral Biology*, 1978, *24*, 413.

Harlow, H. F., Harlow, M. K., & Hansen, E. The maternal affectional system of rhesus monkeys. In H. L. Rheingold (Ed.), *Maternal behavior in mammals*. New York: Wiley, 1978, pp. 254–281.

Kenny, J. T., & Blass, E. M. Suckling as an incentive to instrumental learning in preweanling rats. *Science*, 1977, *196*, 898–899.

Piaget, J. *The origin of intelligence*. London: Routledge & Kegan Paul, 1953.

Schneirla, T. C. Aspects of stimulation and organization in approach/withdrawal processes underlying vertebrate behavioral development. In D. S. Lehrman, R. A. Hinde, & E. Shaw (Eds.), *Advances in the study of behavior* (Vol. 1). New York: Academic Press, 1965.

Schneirla, T. C., & Rosenblatt, J. S. "Critical" periods in the development of behavior. *Science*, 1963, *139*, 1110–1115.

RALPH R. MILLER

Infantile Forgetting of Acquired Information[1]

32

The ability of animals to retain acquired information is an essential component of the learning process. One potentially useful approach to understanding long-term memory is to study its ontogeny in developing organisms Across diverse species and tasks, young animals are frequently reported to be deficient relative to adults in retention after initial acquisition has been equated (e.g., Campbell & Coulter, 1976; Campbell & Spear, 1972). This deficiency, often called "infantile amnesia," was first observed in humans, and was thought to arise from uniquely human factors such as language development, sexually induced repression of early memories, or the emergence of strong hemispheric dominance; however, such mechanisms are not likely to be critical determinants, for infantile amnesia is also observed in infrahuman species.

Current efforts to explain infantile amnesia can be divided into two basic categories: maturational and experiential. In the former category are those hypotheses that assume that the deficit is due to the occurrence of rapid maturational processes such as myelination, synaptogenesis, and changes in perception due to increasing body size that are relatively independent of specific experiential events. In the latter category are hypotheses that focus on the consequences of specific experiences such as retroactive stimulus interference that may result from stimuli encountered during retention periods. Maturational and experiential explanations rest, repec-

---

[1] This research was supported by National Science Foundation Grant BMS75-03383 and a grant from the CUNY Faculty Research Award Program. The author was supported by Research Scientist Development Award K2-MH00061.

tively, on the parallelism during infancy between decreasing neural growth rate and decreasing average frequency of novel interactions with the environment capable of producing retroactive interference on the one hand and decreasing forgetting with age at acquisition on the other hand. As is usually the case when behavioral differences are analyzed in terms of nature and nurture factors, it is likely that both kinds of factors as well as their interaction contribute to infantile amnesia, but the relative degree of each contribution is of considerable interest for both theoretical and practical (educational) reasons. Of course, evaluating the contributions of each of these two gross categories of sources is only a first step toward understanding infantile amnesia; we must then determine what are the specific maturational and experiential factors underlying the phenomenon. Although infantile amnesia is obviously undesirable in places such as the classroom, it is worth noting that it can also be highly adaptive for the developing organism. The behavioral requirements of an adult animal are quite different from those of a young organism, and indiscriminate transfer of information with its concomitant behavior from infancy to adult would likely prove detrimental.

Few efforts have been made to date to identify the causes of infantile amnesia, and those that are published are subject to problems of interpretation. The most frequently cited study is one by Campbell, Misanin, White, and Lytle (1974) in which greater infantile amnesia was observed in the altricial rat than in the precocial guinea pig. Noting the greater postnatal growth of the nervous system in the rat, those authors interpreted their data as supporting the importance of maturational factors in explaining infantile amnesia. Unfortunately, they used only two tasks, both involving avoidance of footshock, so that their conclusion is subject to the problem of task inequivalence that plagues most cross-species comparisons.

In our laboratory, we have recently completed two sets of studies, the first correlative and the second interventive, investigating the contribution of maturational and experiential factors in producing infantile amnesia. The first series (Miller & Berk, 1977) took advantage of the peculiar features of metamorphosis, which is a time of rapid morphological and neurological development, but, due to decreased motor activity, is also a time of relatively few interactions with the environment (Figure 32.1). Poor retention of information over metamorphosis would suggest the importance of maturational factors, whereas good retention over metamorphosis would underscore the role of experiential factors in producing infantile amnesia.

Alloway (1972) reported "good" retention in the mealworm, *Tenebrio molitor*, but included no baseline data concerning retention in nonmetamorphosing *Tenebrio* to use as a basis for comparison. Our research used an amphibian rather than an insect (although neither the change in neural growth rate nor the change in activity level over vertebrate metamorphosis is as extreme as that in most insect species) and included the necessary

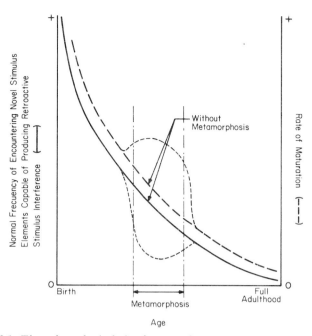

Figure 32.1. These hypothetical developmental curves represent rate of maturation and frequency of encountering novel stimulus elements in a typical laboratory or field environment. Relative starting and terminating points and the exact slopes of the curves have been chosen for clarity; only the approximate parallelism of the two curves is necessary to illustrate that both predict infantile amnesia. Because of this, forgetting due to intrinsic maturational changes is difficult to distinguish from forgetting due to retroactive stimulus interference in developmental studes using nonmetamorphosing species (heavy lines). Metamorphosis, which elevates the rate of neural maturation and reduces interaction with potential sources of retroactive interference (dashed lines), or, alternatively, the manipulation of the retention environment (not illustrated), offers potential solutions to this problem.

measures of baseline retention. The African claw-toed frog, *Xenopus laevis*, was selected because it is fully aquatic throughout its life, thereby minimizing changes in the animal's perception of stimuli and response repertoire over metamorphosis. As metamorphosis in *Xenopus* occurs primarily between developmental Stages 54 and 63 (Nieuwkoop & Faber, 1967), we trained animals at Stage 54 and tested them at Stage 63. In our laboratory, this transition normally took 35 days. Extrapolating 35 days backward from Stage 54, our animals were in Stages 48 or 49 (hereafter Stage 48.5), and extrapolating 35 days forward from Stage 63, our subjects were in Stage 66. Therefore, retention baselines were provided by training *Xenopus* in Stages 48.5 and 63 and testing them 35 days later in Stages 54 and 66, respectively.

Considerable pilot work aimed at obtaining equal acquisition and

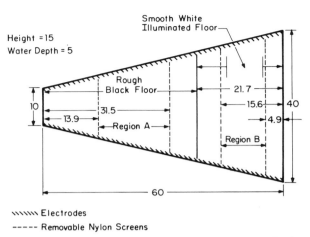

Figure 32.2. Top view of the apparatus used in the metamorphosis studies. Dimensions are in centimeters. The trapezoidal shape of the trough yielded a shock current density that decreased linearly by a factor of four moving from the narrow end to the wide end of the apparatus. Other than the stainless steel electrodes and the emery cloth rough black floor, all surfaces were made of Plexiglas. The removable nylon screens were the only obstacles to an animal's having complete access to the water-filled part of the apparatus.

"short-term" retention (measured one day after training) in larval and adult *Xenopus* finally resulted in the apparatus illustrated in Figure 32.2. In the absence of shock, no differences in position preferences appeared across age levels. Training consisted of placing a squad of eight animals in the center of the black region with the shock being pulsed on for .5 sec once every second. By swimming to the wider section of the apparatus, the subjects were able to reduce the current density experienced. The number of *Xenopus* on the black side was recorded every 2 hr starting 2 hr into the session that ran for 24 hr. These long sessions permitted the use of weak shock that never exceeded the tetanization threshold. Retention was evaluated by savings during a 24 hr retraining session. Parametric studies found that larvae were less sensitive to shock than adults, and, to obtain equal performance during training and short-term retention testing, larvae required 48.8 V across the electrodes, whereas adults received only 31.3 V, both delivered through 12 kΩ.

The design and performance data of the central study in this series are presented in Table 32.1. The completely ambulatory (CA) animals were free to position themselves anywhere in the apparatus; hence, they could form associations both between their behavior and shock intensity and between spatial cues and shock intensity. Groups with nylon screens (NS) in place were each yoked to a CA group; they were restrained in Region A until a majority of their CA paired group was observed on the white side, and then were restrained in Region B for the remainder of the 24 hr training

Table 32.1

Xenopus Retention: Design and Mean Percentage Subjects Observed on Black Side [a]

Group [b]	Initial treatment [c]	Test score [d]
	Stage 48.5 (48.8 V)	Stage 54 (48.8 V)
L-L(CA)	Completely ambulatory (26.1)	19.6
L-L(NS)	Nylon screens in place	20.8
L-L(NS + FR)	Nylon screens in and black floor removed	23.8
L-L(NT)	None	23.7
	Stage 54 (48.8 V)	Stage 63 (31.3 V)
L-A(CA)	Completely ambulatory (24.2)	13.3 *
L-A(NS)	Nylon screens in place	17.8
L-A(NS + FR)	Nylon screens in and black floor removed	19.2
L-A(CA + FS)	Completely ambulatory with floor switched (12.8)	31.7 *
L-A(NT)	None	22.7
	Stage 63 (31.3 V)	Stage 66 (31.3 V)
A-A(CA)	Completely ambulatory (23.1)	12.2 *
A-A(NS)	Nylon screens in place	16.4
A-A(NS + FR)	Nylon screens in and black floor removed	18.1
A-A(NT)	None	20.4

[a] Initial treatment and testing were separated by 35 days. Metamorphosis occurs between Stage 54 and Stage 63.

[b] Abbreviations: L = larva, A = adult, CA = completely ambulatory, NS = nylon screen, FR = black floor removed, FS = floor switched, NT = no training.

[c] During initial treatment, each squad consisted of 8 *Xenopus*.

[d] During testing, all subjects were completely ambulatory; each squad consisted of 6 *Xenopus*.

* $p < .05$, two-tailed, relative to appropriate NS + FR control group.

session. As the NS groups were severely restricted in their freedom to explore the current density gradient, the associations primarily available to them were between shock intensity and spatial cues; that is, the Pavlovian component of the CA training experience. A third set of groups was treated the same as the NS groups, but the black emery cloth flooring was removed (NS + FR). As floor color and texture provided the principal spatial cues in our apparatus, these animals served essentially as controls for changes in sensitivity due to training shock. A further set of groups that received initially no training (NT) was also included. Finally, to test for the importance of cues provided by the emery cloth *per se*, a single group was trained at Stage 54 like the NS + FR groups, except that the emery cloth on the floor was switched (NS + FS) to the low shock intensity side during training. All groups were tested for savings (retrained) 35 days later in the CA condition.

The data suggested a trend within each age level, with the CA animals showing the greatest savings, the NS animals next best, followed by the

NS + FR and NT animals. Apparently both instrumental and classically conditioned associations contributed to the test performance of the CA groups. However, the only significant savings relative to the NS + FR controls of the same age were seen in the CA *Xenopus* that metamorphosed during the retention interval and in the CA *Xenopus* that were adults throughout the retention interval. The savings in these two CA groups were significantly greater than the nonsignificant savings seen in the CA animals that remained larvae throughout the retention period. The inferior retention of this latter group relative to the other CA groups is evidence of infantile amnesia in *Xenopus*. That the retention by the metamorphosing animals was more similar to that of their adult counterpart than to their larval counterparts suggests that experiential factors played a more important role in producing the observed infantile amnesia than did maturational factors.

There are two major difficulties with our metamorphosis studies. First, the results may not be generalizable to mammals. Second, the conclusions must be qualified owing to the correlative nature of the metamorphosis variable. To address these issues, we performed a second set of experiments this time using mammalian subjects and manipulating experiental factors during the retention interval.

In this series of studies, rats were used in a lick suppression task with a 2400-Hz tone as the CS and footshock as the US. This measure was selected as pups and adults have a similar behavioral baseline of 6 to 7 licks per second. An auditory CS was chosen in order to minimize any potential stimulus generalization decrement effects due to growth (Perkins, 1965) and the 1.5 mA (2.0 sec) footshock has been reported to be equally aversive to rat pups and adults (Campbell & Teghtsoonian, 1958). Individual wire-mesh cages provided physically similar retention environments. To reduce differences in novelty of this housing condition, the rat pups were weaned and placed in their individual wire-mesh cages at 17 days of age, 2 days before training. Ambient room temperature for the pups was 26°C (rather than the 20°C provided for the adults) to counteract their imperfect thermoregulation. The early weaning of the pups produced small initial losses in weight relative to unweaned control animals, but these losses were always recovered by the day of testing. To minimize differences in motivation, all subjects were maintained on *ad libitum* rat food and water throughout the study; licking in the test situation was of a condensed milk solution that was highly appetitive to rats of all ages. Experience in drinking this solution in the test enclosure, a wire-mesh cage, began 4 days before training, with condensed milk intake restricted to prevent it from constituting a significant part of any animal's diet. To assure maximal associative learning to the CS as opposed to apparatus cues of the training environment (a small Plexiglas enclosure with a grid floor), all subjects were exposed to the training apparatus for three days prior to the conditioning trials.

As in any study of ontogenetic differences in retention, the major problem proved to be equating initial acquisition. Ideally, equal training experience *and* equal initial performance are desired. As these two conditions can rarely be obtained, equal performance appears to be the more important criterion. Studies in our laboratory parametrically varying the number of CS-US pairings and testing two days after training found that adults learned the association somewhat faster than pups; with three pairings for the pups and two for the adults, both groups displayed appreciable and equal suppression to the tone during testing. Ceiling effects were avoided by permitting each animal up to 1 hr to resume licking in the presence of the tone. In practice almost no animals reached this ceiling.

We first wanted to determine if our preparation was sensitive to infantile amnesia. Therefore, separate groups of pups and adults were tested 2, 8, 16, 32, and 64 days after conditioning. As before, no difference was seen on the 2nd day. However, pup retention declined significantly by the 8th day and continued to decline thereafter reaching baseline by the 32nd day. Although the adults displayed a slight decrement in retention over time, the difference between the 2nd day and any subsequent day failed to achieve significance.

Having found our situation sensitive to infantile amnesia, the question was whether we could induce a similar infantile retention deficit relative to adults on a retention test two days after training by subjecting *all* animals to some sort of interfering stimulus during the retention interval. As the CS itself seemed likely to be the strongest source of interference, separate groups of pups and adults received 0, 2, 4, 8, or 12 exposures of 15 sec duration to the CS on the day between training and testing. This procedure constituted a form of latent extinction, for the lick tube was not present during the presentations. As before, with no CS presentations during the retention interval, no differences in retention over two days were evident. However, as can be seen in Figure 32.3, with as few as two CS extinction presentations, the pups yielded a significant retention deficit, whereas the adults failed to show significant extinction even with 12 CS presentations (although there was a nonsignificant trend towards extinction). Most important, pup retention was inferior to adult retention at 2, 4, 8, and 12 CS exposures. Clearly pups proved to be more susceptible to extinction in this situation.

As it is unlikely that animals undergo explicit extinction trials during the "typical" retention interval preceding manifestation of infantile amnesia, the greater propensity of pups toward extinction can bear on infantile amnesia only if experienced latent extinction generalizes to other stimuli. To probe the extent of such generalization, we conditioned pups as before,

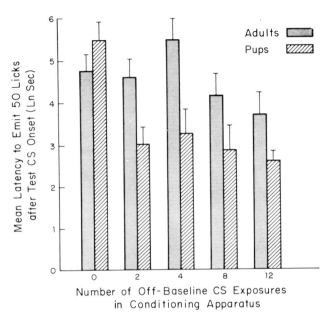

Figure 32.3. Latent extinction in rats as a function of age and number of CS exposures. Standard errors are represented by brackets at the top of the bars. All groups of pups receiving nonreinforced CS exposures differed significantly from the zero exposure pups. None of the adult groups differed from one another.

using a 2400 Hz CS, and then, on the day between conditioning and testing, subjected the pups to two 15 sec exposures to either 2400, 4058, 6861, or 11,600 Hz tones. When tested with the 2400 Hz tone, the first three groups displayed significant extinction and all differed from the 11,600 Hz group, which in turn did not differ from a control group receiving no treatment during the retention interval. No adults were included in this study as they had failed to display extinction even with 12 exposures to the original training stimulus in the previous study.

One possible explanation of the tendency of pups to extinguish faster than adults is that they may not differentiate training from extinction contexts as readily as adults, that is, the pups may be poorer than adults at discriminating between the training and the extinction stimulus complexes. To examine this likelihood, we compared adults to pups on generalization of excitation. If pups are poorer at discriminating between stimuli than adults, they ought to yield a broader generalization gradient. Explicitly, eight separate groups of adults and pups were trained with a 2400 Hz CS as before and then, two days later, were tested for lick suppression to either 2400, 4058, 6861, or 11,600 Hz tones. As can be seen in Figure 32.4, the two age levels yielded similar generalization gradients with only the 11,600 Hz groups of each age level differing significantly from the ap-

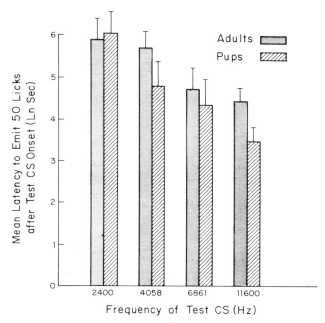

Figure 32.4. Generalization of conditioned lick suppression in rats as a function of age. The training tone was 2400 Hz. Both adults and pups tested with a 11,600 Hz tone differed from those tested with the 2400 Hz tone. No other groups differed significantly from their 2400 Hz age-mates.

propriate 2400 Hz control groups. Clearly the pups did not display a broader generalization gradient than the adults.

At this time we are unable to explain why pups have a greater propensity toward latent extinction than adults. However, it appears likely that this susceptibility is responsible at least in part for poor infantile retention. The presence of this source of infantile amnesia is consistent with the metamorphosis studies in suggesting that experiential factors play an important role in producing the infantile amnesia syndrome.

In closing, it is important to our understanding of infantile amnesia to note that several investigators (e.g., Haroutunian & Riccio, 1977; Spear & Parsons, 1976) have identified various treatments capable of restoring memory in young rats suffering from deficient retention. These pharmacological and environmental manipulations during the retention interval often exclude the possibility of new acquisition. This suggests that infantile amnesia, at least in part, is due to a failure to retrieve information still present in the young organism, rather than an irreversible loss of information. In this respect, infantile amnesia appears to share common properties with both "spontaneous" forgetting in mature animals and subjects suffering from experimental amnesia induced by electroconvulsive shock, antimetabolites, and hypothermia (Miller & Springer, 1973).

ACKNOWLEDGMENTS

Alvin Berk and Michael Vigorito assisted both in designing the studies and with the data collection. Nancy Marlin and John Sullivan were kind enough to critically read a preliminary version of the manuscript, and Joan Wessely assisted in its preparation.

REFERENCES

Alloway, T. M. Retention of learning through metamorphosis in the grain beetle (*Tenebrio molitor*). *American Zoologist*, 1972, *12*, 471–477.

Campbell, B. A., & Coulter, X. The ontogeny of learning and memory. In M. R. Rosenzweig & E. L. Bennett (Eds.), *Neural mechanisms of learning and memory*. Cambridge, Mass.: MIT Press, 1976, 209–235.

Campbell, B. A., Misanin, J. R., White, B. C., & Lytle, L. D. Species differences in ontogeny of memory: Support for neural maturation as a determinant of forgetting. *Journal of Comparative and Physiological Psychology*, 1974, *87*, 193–202.

Campbell, B. A., & Spear, N. E. Ontogeny of memory. *Psychological Review*, 1972, *79*, 215–236.

Campbell, B. A., & Teghtsoonian, R. Electrical and behavioral effects of different types of shock stimuli on the rat. *Journal of Comparative and Physiological Psychology*, 1958, *51*, 185–192.

Haroutunian, V., & Riccio, D. C. Effect of arousal conditions during reinstatement treatment upon learned fear in young rats. *Developmental Psychobiology*, 1977, *10*, 25–32.

Miller, R. R., & Berk, A. M. Retention over metamorphosis in the African claw-toed frog. *Journal of Experimental Psychology: Animal Behavior Processes*, 1977, *3*, 343–356.

Miller, R. R., & Springer, A. D. Amnesia, consolidation, and retrieval. *Psychological Review*, 1973, *80*, 69–79.

Nieuwkoop, P. D., & Faber, J. (Eds.), *Normal table of Xenopus laevis (Daudin)* (2nd ed.). Amsterdam: North-Holland, 1967.

Perkins, Jr., C. C. A conceptual scheme for studies of stimulus generalization. In D. I. Mostofsky (Ed.), *Stimulus generalization*. Stanford, Calif.: Stanford Univ. Press, 1965.

Spear, N. E., & Parsons, P. J. Analysis of a reactivation treatment: Ontogenetic determinants of alleviated forgetting. In D. L. Medin, W. A. Roberts, & R. T. Davis (Eds.), *Processes of animal memory*. Hillsdale, N.J.: Lawrence Erlbaum Assoc., 1976, 135–165.

BARRY D. BERGER
DANIEL MESCH
RICHARD SCHUSTER

An Animal Model of "Cooperation" Learning[1]

33

Perhaps the most singular contribution that psychology as a discipline has made to our understanding of brain function has been the development and analysis of objective tests based on the behavior of lower animals to be used as tools and measuring devices for the experimental study of goal-directed behavior. Here we can only acknowledge the pioneering classic contributions of Pavlov, Thorndike, Miller, Skinner, and Olds, to name but a few, who have provided animal models for studies designed to reveal ultimately the higher nervous activity associated with normal and pathological human behavior.

It is noteworthy that these classical tools in the repertoire of the neuroscientist and indeed the animal models reported in this symposium are based on the behavior of *single* animals in relatively well-defined and easily measurable paradigms. There are clear advantages to these models in that they make it possible to accurately manipulate, measure, and control the various features of the experimental environment. Indeed, perhaps it is because of the need for objective definition of contingencies, of isolating relevant stimuli, or identifying and measuring particular responses, and of automating these various procedures that animal models of complex behaviors have derived primarily from ethology and the emerging field of sociobiology (Lorenz, 1974; Tinbergen, 1974; Wilson, 1975) rather than from experimental psychology.

However, this neglect of animal models of complex forms of social

[1] This research was sponsored by United States–Israel Binational Foundation, Grant 331, and by the Gulton Foundation.

behavior and of other forms of multimotivated behavior has its disadvantages. Much of human and animal behavior—normal and pathological—takes place within a social context, and may conceivably be based on fundamentally different processes from those involved in simple behaviors. Not having adequate tools to measure these phenomena virtually excludes a major area of investigation from our experimental analysis. Furthermore, simple behaviors may be insensitive or refractory to certain subtle effects of drugs, brain lesions, and brain stimulation and may account for, at least in part, the so-called negative results or lack of effects that are often obtained with some of these manipulations in standard operant tasks.

Thus, it would appear that the development of an animal model of social behavior based on the interaction between organisms and its objective and automated measurement could be an important new contribution to the study of behavior and its physiological substrates. We are particularly pleased to take the occasion of this cooperative symposium between the United States and the Soviet Union to present the highlights of our first efforts in this direction.

In the first experiment we introduce a new learning paradigm that is based on the behavior of pairs of animals (rats) working together for food reward. In corollary studies we attempt to determine that successful performance in this task is due to coordinated activity of the animals and not to some extraneous or incidental stimulus. Finally, we present preliminary findings that animals "recovered" from bilateral lesions of the lateral hypothalamus are unable to learn (or perform) the task when it is based on coordination between pairs of rats, but learn normally when the same task is converted into a "conventional" single animal test.

Methods and Procedures

The animals were adult, naive, male albino rats, Sprague-Dawley derived, obtained from a local supplier. They were maintained on approximately 20 gm Purina Lab Chow per day with water available *ad libitum*. The rats were housed one to a cage except in specified cases when they were housed in pairs. The apparatus (Yissum Company, University of Haifa) consisted of a rectangular Plexiglas chamber, separated into two compartments (A and B) of unequal size (Figure 33.1). A feeder, with either one or two dippers for the delivery of .1 cc sweetened condensed milk (two parts water to one part milk), was located at one end of the chamber together with a magazine lamp that was lit during the 6 sec dipper cycle. A signal light was located on the opposite end of the apparatus. The stainless steel grid floor was divided into three separate hinged sections, each connected to microswitches that served as sensors for determining on which of the sections of the chamber the animals were located.

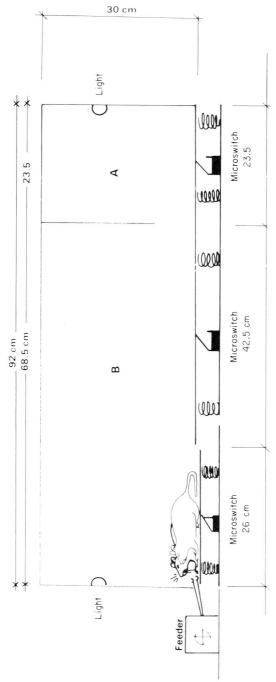

Figure 33.1. "Cooperation" learning apparatus for the rat. In order to obtain milk reward both animals must coordinate entry first into the A compartment and then into the feeder section of the B compartment.

The animals were placed in the apparatus in pairs. Milk reward was delivered if both rats shuttled together from the dipper end of the apparatus into the smaller (A) compartment, and then back again into the dipper area of the larger (B) compartment in that sequence. The signal light in the A compartment was turned off as both rats entered the A compartment, and the magazine light and feeder mechanisms were turned on when both rats shuttled back into the feeder end of the B compartment. At the end of the feeder cycle, the magazine light was turned off and the signal light was turned on to start a new trial. There were 50 such trials per session for ten sessions.

The location of each of the two animals was monitored by the microswitches connected to the three hinged floor sections. Thus, when both animals entered the A compartment, *only* the A compartment microswitch was activated. When both animals then moved into the feeder area at the other end of the box, *only* that floor section was activated. These two unique states of the floor positions occurring in sequence represented "coordinated behavior" and constituted the criteria for the delivery of reinforcement. The latency in seconds for the completion of each of these two components was recorded.

When the two animals were not together or when they did not run in sequence, this indicated lack of coordination. In terms of the apparatus, the conditions of maximal lack of coordination was defined as simultaneous activation of the two floor sections at opposite ends of the chamber. Such occurrences were designated "errors."

Results and Discussion

In the early sessions, the pairs of rats were inefficient at obtaining reinforcement that required coordinated activity. This was characterized by long latencies and large numbers of errors. Moreover, when the animals were observed, it appeared that the location of a given animal was unrelated to the position of its partner; each moved about independently of the other.

As training progressed, latencies and errors decreased (Figure 33.2) and the pairs of rats shuttled together, apparently in a coordinated fashion. By the end of the tenth session, the animals were completing 50 trials within about 250 sec and were making only about one error per trial compared with about 2700 sec and four to five errors per trial in the first two sessions.

It is tempting to speculate that the improvement in performance in this paradigm is due to the learning of a coordination strategy. However, it is also possible that other factors are involved. For example, the animals may have learned to pace themselves individually so as to optimize rein-

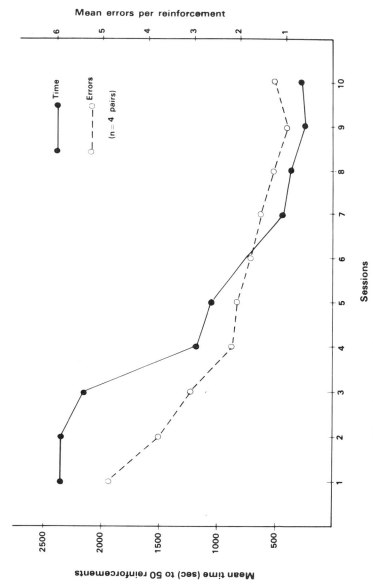

Figure 33.2. Acquisition of "cooperation" learning in the rat.

forcement in some adventitious fashion. Another possibility is that individual rats may have learned to enter the A compartment, wait until the signal lamp was turned off (which occurred when the second animal happened to join him in the compartment), and use that signal as a cue to shuttle into the feeder area.

In order to evaluate these and other alternative explanations to "coordination learning," several corollary experiments were conducted. In one such study, an opaque barrier was placed lengthwise in the apparatus, separating the two rats. If acquisition were based on pacing, or on incidental and adventitious cues, the animals should be able to learn despite the barrier between them. If, on the other hand, learning were based on coordination with the behavior of the partner, learning should be impaired. The results supported an interpretation that the behavior of the partner was the relevant stimulus (i.e., coordination learning) since the opaque barrier greatly impaired acquisition under the conditions tested (Figure 33.3). An intermediate deficit was obtained when the barrier was transparent rather than opaque, suggesting that visual cues from the behavior of the partner rat played an important role in acquisition of this task.

In a second study along similar lines, the signal lamp in compartment A was removed after the pairs of rats had acquired the task. If the onset of the signal lamps had served as the cue to enter compartment A and its offset as a signal to shuttle into the feeder area, removal of this critical stimulus should greatly disrupt performance. This, however, was not the case. Performance of the partners was unaffected by the removal of the signal lamps, suggesting that the behavior of the partner rat, and not some other feature of the situation, mediates the task.

Other experiments in our laboratories reinforced the notion of a social component involved in this learning paradigm. In several replications we have observed consistently that rats housed together learn the coordination task more quickly than do animals that are housed singly. Furthermore, if the same partners are kept constant throughout the experiment, learning is facilitated relative to conditions in which each rat is tested with a new partner each session. Finally, we have evidence that animals previously trained in the coordination task facilitate the acquisition of naive animals, suggesting that imitation or even active "instruction" may take place.

From these behavioral experiments we therefore tentatively conclude that the task presented here is based on "coordination" learning. Moreover, according to the classification of Hake and Vukelich (1972), the paradigm may fit the category of "cooperation" learning since their criteria for a cooperative task—that reward is at least in part dependent on the responses of both partners, and that there is an equitable division of responses and reinforcers—are met in our procedure. Aside from early experimental studies on cooperation behavior among primates (i.e., Boren, 1966; Crawford, 1941) and field observations (Brown, 1975; Wilson, 1975), little if any

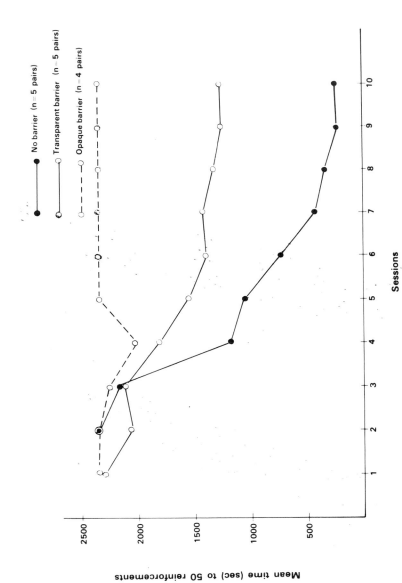

Figure 33.3. Deficit in acquisition of "cooperation" learning by restricting contact between partners.

work has satisfied criteria for demonstrating cooperation behavior experimentally in the lower animals. The attempts by Miller and Dollard (1941), Daniel (1942, 1943), Baron (1957), Taylor and Erspamer (1971), and Epley and Rosenbaum (1975) to demonstrate cooperative behavior in the rat were inconclusive either because the animals failed to learn the task or because some process other than "cooperation" was found to mediate the behavior (Marcucella & Owens, 1975; Hake & Vukelich, 1972; Cohen, 1978). Our procedure may thus be more suitable than these others as an animal model for the experimental study of social behavior because the animals learn our task readily and because the paradigm appears to fulfill the formal criteria for cooperative behavior.

We have begun a program in our laboratories, using the rat cooperation-learning paradigm, to assess the effects of brain damage on social behavior. Our first experiments have evaluated rats with lesions of the lateral hypothalamus, an area of the brain involved in many regulatory and motivated behaviors (Anand & Brobeck, 1951; Teitelbaum & Epstein, 1962; Davenport & Balagura, 1971; Marshall, Turner, & Teitelbaum, 1971; Marshall & Teitelbaum, 1974).

In these preliminary studies, bilateral lesions of the lateral hypothalamus were made in adult male rats and recovery from the resulting aphagia, adipsia, and anorexia was monitored according to the method of Teitelbaum and Epstein (1962). Various groups of control animals received sham lesions, unilateral lateral hypothalamic lesions, or bilateral lesions about 1 mm dorsal to the lateral hypothalamic placement. Behavioral testing was begun on the average about two months following surgery; by this time the lesioned animals had recovered feeding and drinking sufficiently so as to maintain themselves on normal laboratory chow and tap water. All animals were maintained on approximately 20 gm of chow with water *ad libitum* for the duration of the study.

In some experiments, lesioned animals were tested for acquisition of cooperation learning; in other studies, animals that had already been trained in the procedure before surgery then lesioned and tested for retention or performance of the already learned behavior. In both cases, the performance of other lesioned and controlled animals was evaluated in the shuttle box modified for the behavior of single animals rather than for pairs. In this instance the social component was absent, but the same shuttle response was required, the same visual stimuli were employed, and the same milk reward was used as in the two animal situation.

The results indicate that lesions of the lateral hypothalamus cause a specific deficit in the acquisition and the performance of the cooperative task, but that lesioned animals are normal or even improved in the same task where cooperation behavior is not required. In the acquisition problem (Figure 33.4), the four pairs of bilaterally lesioned rats showed virtually no improvement over the ten sessions. In the retention problem (Figure

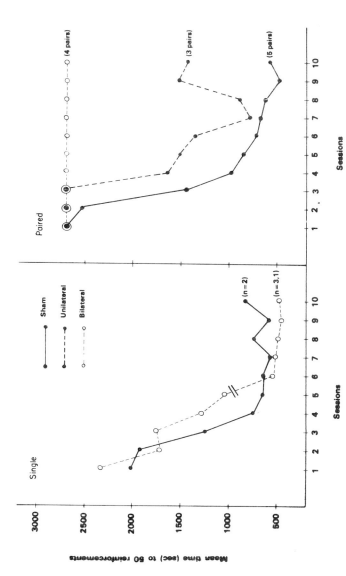

Figure 33.4. Right: *Deficit in "cooperation" learning in rats "recovered" from lesions of the lateral hypothalamus.* Left: *No deficit is observed when the cooperation component is absent in the same task.*

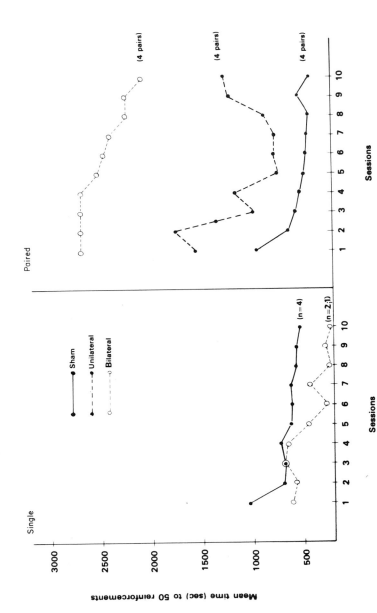

Figure 33.5. Right: Deficit in retention or performance of "cooperation" learning in rats trained in the task prior to lesions of the lateral hypothalamus. Left: No deficit is observed when the cooperation component is absent in the same task.

33.5), four different pairs of bilaterally lesioned animals began to show some improvement only in the later sessions despite the fact that these pairs had achieved a high degree of coordinated behavior prior to lesion. Sham operates showed considerable savings, achieving their preoperative baseline values after only one or two retention sessions. Unilaterally lesioned rats exhibited an intermediate impairment.

These data suggest a specific social behavior deficit associated with lesions of the lateral hypothalamus. The unimpaired performance of lesioned animals in the single animal shuttle box paradigm indicates that there is nothing obvious related to stimulus, motor, or motivational components of the apparatus or procedure that could account for the lesioned-induced deficit in cooperative behavior in the paired-animal task. Moreover, observation of the lesioned animals in the apparatus suggested a specific deficit in coordination with the behavior of the partner rat. Often these animals would shuttle through the apparatus, each independent of the other, and ultimately undergo extinction when reinforcement was not delivered.

In addition to, or instead of, a social behavior deficit, it is possible that other lesion-induced processes may be involved in the impairments in the two animal situation. For example, animals with lesions of the lateral hypothalamus and animals with 6-hydroxdopamine lesions of these and other catecholamine-containing systems of the brain exhibit stimulus neglect (Marshall et al., 1971; Marshall & Teitelbaum, 1974), and even possible stimulus rejection (Shallert & Wishaw, 1979). These deficits could possibly affect coordination specifically in the two-animal task. Another possibility—that simple differences in task difficulty can explain the deficit in the two animal task and not in the single animal situation—is unlikely for several reasons. First, we have no conclusive evidence that the cooperative task is much more difficult for the rat than the single animal task. Second, lesions of the lateral hypothalamus cause deficits in certain simple behaviors such as the acquisition of taste aversions (Roth, Schwartz, & Teitelbaum, 1973; Schwartz & Teitelbam, 1974) and do not always cause deficits in other, even complex, behaviors such as food-reinforced visual discrimination tasks (Davenport & Balagura, 1971).

These data from animals with lesions of the lateral hypothalamus, and recent experiments using 6-hydroxdopamine lesions would thus suggest that some component or components of coordination behavior in the rat may be mediated by a lateral hypothalamic catecholamine system. Precisely what these components are—social behaviors, stimulus neglect, or stimulus aversion—have not yet been conclusively identified. In any event, our preliminary work suggests that our animal model of coordination behavior may provide an interesting tool for the analysis of complex forms of normal and pathologic goal-directed behavior and its psychobiological correlates and, in addition, could provide a link between the disciplines of experimental psychology and ethology.

REFERENCES

Anand, B. K., & Brobeck, N. J. Hypothalamic control of food intake. *Yale Journal of Biology and Medicine,* 1951, *24,* 123–140.
Baron, A. An analysis of a parasitic social relation in pairs of rats. *Dissertation Abstracts,* 1957, *17,* 3094–3095.
Boren, J. J. An experimental social relation between two monkeys. *Journal of the Experimental Analysis of Behavior,* 1966, *9,* 691–700.
Brown, J. L. *The evolution of behavior.* New York: Norton Press, 1975.
Cohen, D. *Rat's learning of cooperation under unequitable allocation of rewards.* Unpublished master's thesis, Univ. of Haifa, 1978.
Crawford, M. P. The cooperative solving of chimpanzees of problems requiring serial responses to colour cues. *Journal of Social Psychology,* 1941, *13,* 259–280.
Daniel, W. J. Cooperative problem solving in rats. *Journal of Comparative Psychology,* 1942, *34,* 361–368.
Daniel, W. J. Higher order cooperative problem solving in rats. *Journal of Comparative Psychology,* 1943, *35,* 297–305.
Davenport, L. D., & Balagura, S. Lateral hypothalamus: Reevaluation of function in motivated feeding behavior. *Science,* 1971, *172,* 744–746.
Epley, S. W., & Rosenbaum, M. E. Cooperative behavior in rats: Effect on extinction of shock onset condition during acquisition. *Journal of Personality and Social Psychology,* 1975, *31,* 453–458.
Hake, D. F., & Vukelich, R. A classification and review of cooperation procedures. *Journal of Experimental Analysis of Behavior,* 1972, *18,* 333–343.
Lorenz, K. Z. Analogy as a source of knowledge. *Science,* 1974, *185,* 229–234.
Marcucella, H., & Owens, K. Cooperative problem solving by albino rats: A reevaluation. *Psychological Reports,* 1975, *37,* 591–598.
Marshall, J. F., & Teitelbaum, P. Further analysis of sensory inattention following lateral hypothalamic damage in rats. *Journal of Comparative and Physiological Psychology,* 1974, *86,* 375–395.
Marshall, J. F., Turner, B. H., & Teitelbaum, P. Sensory neglect produced by lateral hypothalamic damage. *Science,* 1971, *174,* 523–525.
Miller, N. E., & Dollard, J. *Social Learning and Imitation.* New Haven, Conn.: Yale Univ. Press, 1941.
Roth, S. R., Schwartz, M., & Teitelbaum, P. Failure of recovered lateral hypothalamic rats to learn specific food aversions. *Journal of Comparative and Physiological Psychology,* 1973, *83,* 184–197.
Schallert, T., & Wishaw, I. Q. Two types of aphagia and two types of sensorimotor impairment after lateral hypothalamic lesions: Observations in normal-weight, dieted, and lateral rats. *Journal of Comparative and Physiological Psychology,* in press.
Schwartz, M., & Teitelbaum, P. Dissociation between learning and remembering in rats with lesions in the lateral hypothalamus. *Journal of Comparative and Physiological Psychology,* 1974, *87,* 384–398.
Taylor, J., & Erspamer, R. A method for the measurement of cooperative behavior in albino rats. *The Psychological Record,* 1971, *21,* 121–124.
Teitelbaum, P., & Epstein, A. N. The lateral hypothalamic syndrome: Recovery of feeding and drinking after lateral hypothalamic lesions. *Psychological Review,* 1962, *69,* 74–90.
Tinbergen, N. Ethology and stress diseases. *Science,* 1974, *185,* 20–27.
Wilson, E. O. *Sociobiology, The new synthesis.* Cambridge, Mass.: Harvard Univ. Press, 1975.

Human Psychophysiology, Information Processing, and Language

PART VI

JOHN I. LACEY
BEATRICE C. LACEY

The Specific Role of Heart Rate in Sensorimotor Integration[1]

34

Introduction: Heart Rate in Somatic "Arousal" Patterns

In our early studies of physiological changes that accompany mental activity and painful stimulation (Lacey, Bateman, & Van Lehn, 1953; Lacey & Lacey, 1958, 1962), patterns of response were found in which individuals were overreactive in some variables, of average reactivity in other variables, and underreactive in still others. Individuals differed, one from the other, in these patterns. The patterns were reproducible over different tasks and over a 4-year period.

When we used tasks that called upon a wider variety of behavioral activities, we found another source of patterning. Different kinds of tasks seemed to produce different response patterns (Lacey, 1972; Lacey, Kagan, Lacey, & Moss, 1963; Lacey & Lacey, 1970). Tasks that involved primarily attention to external environmental events produced a pattern of response in which heart rate decelerated and blood pressure either decreased or showed insignificant changes. In comparing these results with those found with effortful "mental" activity and painful stimulation, a psychophysiological continuum was suggested: At one end there is decreased heart rate and blood pressure that accompany goal-directed attention to external events; at the other end there is increased heart rate and blood pressure when careful attention to external stimuli might interfere with the goal-

---

[1] This research was supported by National Institute of Mental Health Grant MH623.

directed behavior. Other measures, such as respiration, pupillary diameter, and electrodermal activity do not respond differently in these two classes of behavior.

How generalizable is the phenomenon? Why is it that only these cardiovascular subsystems respond in this way? What is their specific role in behavior? What do these responses tell us about brain mechanisms in goal-directed behavior?

Anticipatory Heart-Rate Changes in Preparation For Goal-Directed Behavior

The first significant lead to answering such questions emerged from simple visual reaction time experiments (Lacey & Lacey, 1970), using fixed intervals of 4 sec between a visual warning signal and a different visual stimulus requiring the rapid release of pressure on a lever. A highly reliable cardiac deceleration was found during the 4 sec interval. The heart progressively slowed from one beat to the next over the course of several cardiac cycles. The deceleration amounted to as much as 40 beats per minute in selected subjects and trials. The average is about 6 to 10 beats per minute. These decelerations could not be attributed to respiratory or blood pressure changes. We found also that reaction times were faster the greater the anticipatory decrease in heart rate and the lower the heart rate level achieved. The correlations, although consistent, were typically small. Therefore, we sought other means of exploring the functional interrelationships of cardiac activity and behavior.

ATTENTION AND RESPONSE-INTENTION

In reaction time experiments the behavioral goal is to respond rapidly to stimuli administered at times chosen by the experimenter. What happens when we reverse the order of events by having significant environmental inputs follow, instead of precede, the motor act, and when the response-intention is invoked not at times chosen by the experimenter, but at times more freely chosen by the subject himself?

We have studied this problem in three experiments. In the first two experiments, the subjects pressed telegraph keys for monetary rewards, on operant schedules that required responses to occur only at specified intervals. In one experiment (Lacey & Lacey, 1970), the interval was from 15–19 sec. In another experiment (Lacey & Lacey, 1973), it was 30 sec or greater. In both experiments we found that the heart showed a beat-by-beat progressive slowing in advance of the "voluntary" motor act.

In the third experiment (Lacey & Lacey, 1978), we made the time of execution of the motor act even less under the control of the experimenter.

34. Heart Rate in Sensorimotor Integration

Each of 20 male subjects self-initiated brief tachistoscopic exposures of a clock face outlined by small lamps placed in labeled clock positions 1 through 12. Each subject worked at his own pace, lifting his finger gently from a touch plate whenever he felt ready to receive another brief, instantaneously administered stimulus. His task was to report which light was off, a task made difficult by keeping the exposure times so short that success was achieved only 50–70% of the time.

Figure 34.1 shows results for a typical subject. The top trace shows the systematic evolution of a seven beat per minute deceleration that be-

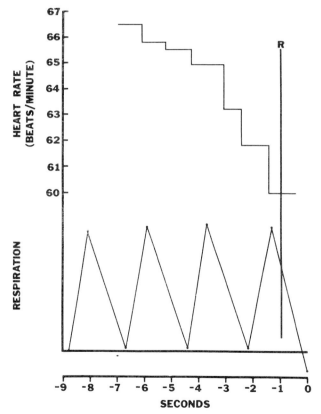

Figure 34.1. Computer-averaged curves for one typical subject showing deceleration in advance of self-initiated motor responses, which resulted in a brief tachistoscopic stimulus and lack of concomitant respiratory changes. "R" shows the time of the subject's response. Respiratory curve shows times of onset and duration of inspiration (upward deflection) and expiration (downward deflection) and relative amplitude of chest movement. (From "Two-way communication between the heart and the brain: Significance of time within the cardiac cycle," by B. C. Lacey & J. I. Lacey, American Psychologist, 1978, 33, 99–113. Copyright 1978 by The American Psychological Association. Reprinted by permission.)

gins some 7 sec before the subject "decides" to unleash a stimulus. The time of the subject's response is shown by the long vertical line headed by the letter "R." At the bottom is a computer-averaged respiratory curve that faithfully reproduces inspiratory onsets and durations, expiratory onsets and durations, and relative amplitude of chest movement in arbitrary units. It is obvious that the changes in heart rate are not artifacts of respiration. The anticipatory decelerations again were highly reliable, occurring in 18 out of 19 subjects for whom complete data were available.

THE SIGNIFICANCE OF TIME WITHIN THE CARDIAC CYCLE

The final experiment to be reported in this series is again a reaction time experiment, this time using audition and omitting the warned foreperiods, since it was now clear that the subject needed no warning signal in order to initiate anticipatory cardiac decelerations. It introduces the first of two new and significant phenomena: variations of heart period that are dependent on time within the cardiac cycle at which stimulus inputs or motor outputs occur.

In this experiment, intertrial intervals averaged about 7.0 sec. Each of the over 380 reaction times was reported to the subject together with approval for short reaction times and disapproval for long ones. The subjects responded typically with vigor, rapidity, and involvement. The auditory stimulus was presented either at the R-wave or 350 msec after the R-wave in counterbalanced order.

Figure 34.2 shows, for one representative subject, the average heart rate for seven successive cardiac cycles, from three cycles before to three cycles after the imperative stimulus. Time, on the X axis, is given in successive cardiac cycles, from left to right, with 0 indicating the cycle of the imperative stimulus. As can be seen, there is a smooth increase in duration of successive cardiac cycles before the auditory signal to respond. This cardiac deceleration is continued during the 0 cycle for 350 msec delays, but is aborted for 0 msec delays! An acceleration is seen after the motor response. The greater and more delayed prolongation of heart period for the 350 msec delays than for 0 msec delays is highly reliable. In 104 individual cases, 91 show the effect; $p < 10^{-9}$ by the binomial expansion. We turn now to a detailed consideration of this phenomenon.

Variations in Heart Period: A Function of Event-Timing Within the Cardiac Cycle

The immediate prolongation of the cardiac cycle when stimuli occurred 350 msec after the R-wave is only an example—a selected point—of a more general functional relationship: *Stimulus inputs or motor outputs occurring early in the cardiac cycle prolong the heart period in which*

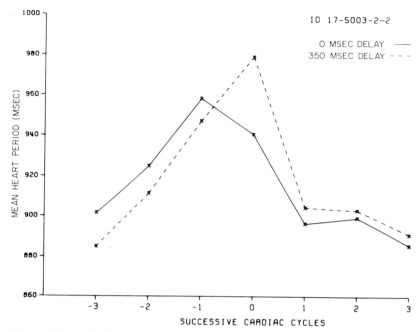

Figure 34.2. This shows for a single subject in an auditory reaction time experiment, the mean heart period from three cardiac cycles before to three cycles after the stimulus to respond. The cycle in which the stimulus fell is labeled "0."

they occur more than do stimuli falling late in the cycle; inputs or outputs that fall late in the cycle prolong the subsequent cycle. This functional relationship was found both in a simple visual reaction time experiment and in the experiment with self-initiated tachistoscopic exposures. It has been described in detail by Lacey and Lacey (1977). The function is not precisely monotonic and, particularly in subjects with slow heart rates, the function first increases and then decreases. The just-described differential effects of stimulus-response episodes at the R-wave, or at 350 msec after the R-wave, reflect the early increase in the function.

Without decisive proof as yet, we tentatively ascribe the function to the operation of cerebro-vagal pathways for the following reasons. First, not all stimuli produce such effects. For example, in the reaction time experiment, the warning signal did not do so, although the subsequent signal-to-respond as quickly as possible did do so. Hence it can be argued that some evaluation is made by the subject of the significance of the stimulus-event and that therefore the initiating mechanism is central. Furthermore, as we shall describe, new data show that both task difficulty and expectancy modify the magnitude of the function, thus clearly invoking cortical functions.

That the final pathway is vagal is suggested first of all by the promptness with which the heart is affected—within one cardiac cycle. Sympathetic cardiac efferents, as it is well established, cannot do this, whereas vagal stimulation does have such a prompt effect. Moreover, our results in the human being with behaviorally significant events reproduce in almost all details the temporal dependencies found in acute animal experiments upon stimulation of the vagus (Brown & Eccles, 1934; Dong & Reitz, 1970; Levy, Iano, & Zieske, 1972; Levy, Martin, Iano, & Zieske, 1969, 1970; Reid, 1969). These investigations show conclusively that electrical stimulation of the vagus early in the cardiac cycle prolongs that cycle, but has little effect on the subsequent cycle. Stimulation later in the cycle fails to affect that cycle but prolongs the next one, the more so the later the stimulation. The mechanism, briefly described, starts with activation of the vagal efferents that results in the release of acetylcholine (ACh), the neurotransmitter substance that directly affects the sinoatrial node. Since ACh decreases the slope of diastolic depolarization, its availability at the pacemaker early in diastole will prolong the cycle more than if ACh becomes available late in the cycle. However, since ACh is rapidly hydrolyzed, it will be more effective in prolonging the subsequent cycle if it arrives late in a cycle than if it arrives earlier, and thus has more time to be dissipated. Sympathetic terminals, on the other hand, utilize norepinephrine, which is removed much more slowly. This accounts for the lack of temporal dependency for sympathetically produced cardiac responses (Levy, 1977).

CEREBRAL MEDIATION OF TIME-DEPENDENT CARDIAC SLOWING

We currently are studying some aspects of cortical mediation of this effect. We were struck by the fact that the warning signal in the reaction time experiment failed to produce time-dependent slowing of the heart rate, and we sought a technique that would enable comparison of responses to stimuli that, although physically similar, differ in significance or relevance for the assigned task. For this purpose, we chose an experimental procedure often used in studying the late positive component ("P300") that appears in averaged scalp-recorded evoked potentials. P300 is known to be related to some aspects of event significance.

In our experiment, tones of 100 msec duration at 72 db (A) re 20 $\mu N/M^2$ were presented through earphones at 3 sec intervals. Eighty percent of the tones were of one frequency. They are called "standard" tones. Twenty percent were of another frequency. These are called "rare" tones. Two rare tones never appeared consecutively. The number of standard tones intervening between rare tones varied from two to six. At the end of each of six blocks of trials, subjects were given 25 cents if they reported correctly the number of rare tones in that block.

We first conducted two pilot experiments (Lacey & Lacey, 1978). In one experiment, tones of 1000 Hz and 1200 Hz were used. The difference of 200 Hz was easy for subjects to discriminate. In the second experiment, the two frequencies were 1000 Hz and 1050 Hz, a difference of only 50 Hz. Figure 34.3 shows the results of both experiments. For each, the average heart period is shown as a function of the quintile or fifth of the cardiac cycle in which the stimulus occurred. The panel on the right is for the easy discrimination experiment. The three lines show the results for three sets of tones: standard tones that just preceded the rare tone, rare tones, and standard tones immediately subsequent to rare tones. From results reported in experiments with P300, we would have expected to find a good cardiac cycle effect only for rare tones. This, however, we did not find. Instead, as the figure shows, preceding standard tones yielded a linearly decreasing function, $p < .01$. Heart periods were longer for stimuli appearing early in the cardiac period. The effect for rare tones, although clear, is not quite as smooth and was significant only in the first half of the experiment. The result for subsequent standard tones suggests no significant variation.

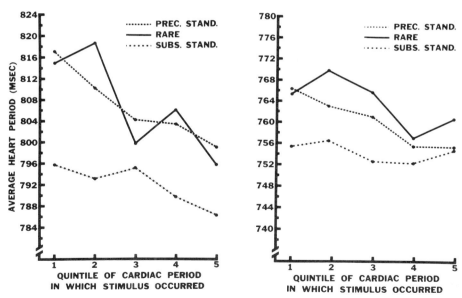

Figure 34.3. Relationship of concurrent heart period to relative time of occurrence of (a) standard tones immediately preceding rare tones, (b) rare tones, and (c) standard tones immediately subsequent to rare tones. Results of an "easy" discrimination experiment in which standard and rare tones differed by 200 Hz are shown in the right panel; results of a more difficult discrimination experiment in which standard and rare tones differed by 50 Hz are shown in the left panel. $N = 10$ in both experiments. (From "Two-way communication between the heart and the brain: Significance of time within the cardiac cycle," by B. C. Lacey & J. I. Lacey, American Psychologist, 1978, 33, 99–113. Copyright 1978 by The American Psychological Association. Reprinted by permission.)

The panel on the left presents the results for the experiment with increased difficulty of discrimination. The increase in difficulty resulted in an enhanced effect for both rare and preceding standard stimuli; both trends were significant: $p < .01$ and $p < .02$, respectively. Although there seems to be a small effect for standard tones subsequent to rare tones, this was not statistically significant.

The fact that a standard tone evokes the effect when it precedes a rare tone but not when it follows a rare tone suggests that expectancy may be a critical variable. We reasoned that subjects come to appreciate the sequencing and probabilities of the tone series; as intervening standard tones increased in number, the subject's expectancy of a rare tone would grow. Therefore, preceding standard tones, as well as rare tones, evoked the cardiac cycle effect. Since, however, two rare tones never succeeded one another, the standard tones subsequent to rare tones would be of no importance to the subject and thus would produce no cycle effect. In the major experiment now under way (Lacey & Lacey, 1978), we are studying specifically the role of expectancy.

The results for the first 38 subjects are shown in Figure 34.4. The data continue to show remarkable consistency. The cardiac cycle effect for rare tones was significant at the .01 level and for preceding standard tones at the 10^{-5} level. As the graph shows, the function for preceding

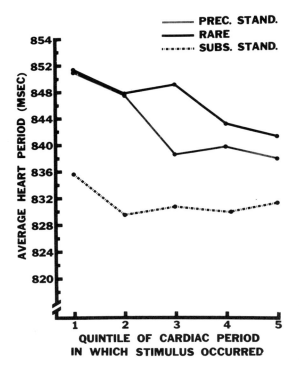

Figure 34.4. Relationship of concurrent heart period to relative time of occurrence of three tones in a group of 38 subjects. Conditions are as in the more difficult discrimination experiment shown on the left in Figure 34.3 (From "Two-way communication between the heart and the brain: Significance of time within the cardiac cycle," by B. C. Lacey & J. I. Lacey, American Psychologist, *1978*, 33, 99–113. Copyright 1978 by The American Psychological Association. Reprinted by permission.)

standard tones again was somewhat more regular than for rare tones. Subsequent standard tones again did not yield statistical significance.

The "growth of expectancy" can be studied by grouping trials according to their position in the series of standard tones. Figure 34.5 shows the results. We have grouped trial positions of standard tones in pairs for simplicity of presentation. Thus, standard tones in positions five and six would be the last in the series of intervening standard tones and expectancy of a rare tone would be high by this time. Note first that heart rate level is slower with "increasing expectation." Second, note that time-dependent cardiac slowing also is a function of expectation. It is clearly larger for trials three and four than for trials one and two. There is also an additional but small increase in effect for trials five and six.

One of our more persistent aims is to demonstrate differential activation of simultaneously recorded physiological response systems. For there always is the high likelihood that although different changes may be evoked by the same event, those changes reflect different behaviorally related functions of the nervous system. The current experiment enables us to look at the differences between averaged evoked potentials recorded from vertex to linked ears and the cardiac slowing effect.

As would be expected from our design, we found clear-cut P300s to rare tones in almost all subjects. But, again as we would expect from

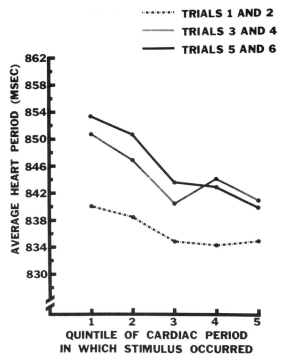

Figure 34.5. Effect of expectancy on the relationship between concurrent heart period and relative times of occurrence of standard tones. Standard-tone trials have been grouped according to their position between rare-tone occurrences (N = 38). (From "Two-way communication between the heart and the brain: Significance of time within the cardiac cycle," by B. C. Lacey & J. I. Lacey, American Psychologist, 1978, 33, 99–113. Copyright 1978 by The American Psychological Association. Reprinted by permission.)

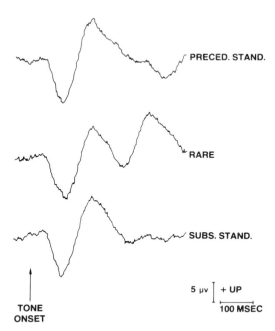

Figure 34.6. Averaged evoked potentials (vertex to linked ears) for a single typical subject for preceding standard, rare, and subsequent standard stimuli. Each curve is based on approximately 135 trials; the few trials with eye movements or blinks were not included. Only rare trials show a large clear P300.

previous reports of the behavior of P300, we found that P300s to preceding standard stimuli were either absent or were markedly smaller than those evoked by rare stimuli. Figure 34.6 shows one such typical response: a large P300 to rare tones, no P300 to preceding standard tones, and either no P300 or a very small one to subsequent standard tones. In Figure 34.7, also typical, we see again a relatively large P300 to rare tones and markedly reduced or questionable responses to preceding and subsequent tones.

Cardiac cycle effects and the endogenous potential known as P300, then, clearly reflect somewhat different aspects of the processing and evaluating in which subjects are engaged as they judge whether a tone is rare or standard. The two responses can be said to be associated, in that rare stimuli evoke both effects in most subjects. They are dissociated in that rarity is sufficient but not necessary to evoke the cardiac cycle effects: The cardiac responses of subjects to preceding standard stimuli are at least as good and probably better than to rare stimuli. On the other hand, P300 is clearly most evident on rare tone trials.

Why Should Significant Sensorimotor Events Slow the Heart?

Why, to put the question more generally, do we so consistently observe prompt cardiac slowing in states of expectant attention or preceding

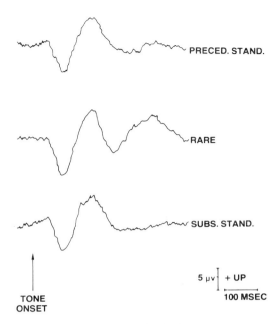

Figure 34.7. Averaged evoked potentials for another typical subject as in Figure 34.6. Again P300 is prominent on rare trials only.

simple goal-directed motor outputs? Why do we not find the more commonly observed increases in heart rate and in blood pressure? Does a slowed heart rate have any particular significance for sensorimotor behavior? Our supposition is that changes in heart rate and blood pressure have a regulatory effect on the brain and hence on behavior. A brief explanation of the basis for this speculation is in order.

At the expulsion of blood into the arterial tree with each heartbeat, sensitive strain receptors scattered along the arterial tree are stimulated. These specialized interoreceptors—the baroreceptors—have their major concentration in the aortic arch and the carotid sinus. They are extremely sensitive to changes in blood pressure. At each pressure wave produced at each heartbeat, the thin wall of the carotid sinus, for example, is distended, stretched, and distorted, and a burst of impulses is produced in the baroreceptor visceral afferent pathways. It is no surprise to find that activity of these negative feedback pathways has important regulatory effects on the cardiovascular system itself. But activity of these baroreceptors also has widespread effects on noncardiovascular functions of the nervous system. Since the early review by Heymans and Neil in 1958, a steadily increasing number of neurophysiological studies has shown that stimulation of the baroreceptor afferents results in widespread inhibition of a variety of other functions, including skeletal muscular activity and electroencephalographic activity. For example, a study (Coleridge, Coleridge, & Rosenthal, 1976) showed that inflation of the carotid sinus resulted in prolonged inhibition of pyramidal cell activity in the motor cortex. Similar effects have been shown in sensory pathways by studying single cell activity in the cuneate nucleus after cutaneous stimulation, conditioned by prior or simultaneous stimulation of vago-aortic afferents (Gahery & Vigier, 1974). The mechanism underlying this effect is presynaptic inhibition or primary afferent depolarization.

The inference from these and many other studies is that the central nervous system is so connected to the baroreceptor afferents that sensory and motor functions can be inhibited by increases in baroreceptor activity, and that, by inference, sensorimotor activity can be facilitated by decreases in baroreceptor activity.

It has been extremely difficult to secure decisive data that clearly suggest the operation of such a mechanism in the intact human. Recent data from our laboratory, however, reveal a phenomenon that seems to us to demand such an explanation, although many neurophysiological details need to be uncovered.

NONRANDOM VARIATIONS IN THE PLACEMENT OF SENSORIMOTOR RESPONSES IN THE CARDIAC CYCLE

In the first of our experiments using operant schedules, not only did we find the cardiac cycle effect already described, we also found a complementary effect. Not only did the placement of the "voluntary" act within the cardiac cycle affect the duration of that cardiac cycle, but the overall prevailing level of heart rate determined *where in the cycle the subject would "choose" to emit his motor response*. The effect is shown in Figure

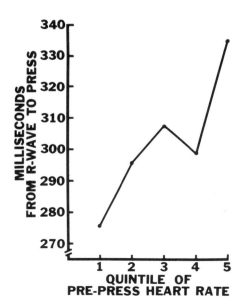

Figure 34.8. With a DRL 15″, LH4″ operant-conditioning schedule (52 subjects), the median absolute times of response, timed from the R-wave, increased as the heart rate increased. The linear trend is significant—$p < .001$. For the group, the time of response changes by approximately 5 msec per unit change in heart rate.

34. Heart Rate in Sensorimotor Integration

34.8. This is the average curve for the 52 subjects in that study who were maintained on the operant schedule until they had executed 60 successfully timed responses. Since the cardiac cycle within which the response was emitted was disturbed by that response, we used as an estimate of the prevailing heart rate the cardiac cycle just preceding the response. These cardiac cycles, translated into heart rate, are called prepress heart rates. The distribution of 60 prepress heart rates for each subject was divided into five equal parts (quintiles). For each quintile of the heart rate distributions, there was a distribution of elapsed times from R-wave to motor response, measured to the nearest millisecond. The medians of these distributions were then plotted as a function of the heart rate quintiles. What the figure shows is that for the group as a whole, as prepress heart rate increased, the key-presses occurred later and later in the cardiac cycle, with some slight irregularity. In the 52 subjects in this experiment, the average postponement is about 5 or 6 msec per unit change in heart rate. It ranged between subjects from about 40 msec per unit change in heart rate to no change. The group results were statistically significant, that is, $p < .001$. Thus, even very small changes in heart rate—say six beats per minute—could result in behaviorally significant changes in the time of response-emission—say 30 to 240 msec.

Precisely the same results were secured in the experiment in which subjects self-initiated brief tachistoscopic stimuli. Figure 34.9 shows the results for this group of subjects. Once more we see a rather large increase in absolute time within the cardiac cycle at which a "voluntary" response

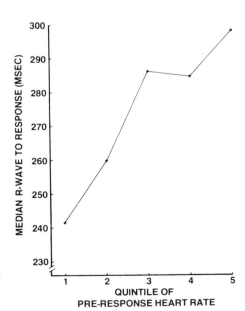

Figure 34.9. In a self-initiated tachistoscopic exposure experiment (60 subjects), responses occurred later and later in the cardiac cycle as heart rate increased.

is emitted as the prevailing level of heart rate increases, $p < .01$. Thus, with slow heart rates the subject emits responses early in the cycle and, as heart rate increases, the subject emits responses progressively later in the cardiac cycle.

Why does this happen? Either there is some totally unsuspected interaction between central structures mediating motor behavior and central structures mediating heart rate, or as we suspect, there is a direct interference by heart rate with freedom to emit motor responses.

These results suggest that variations in heart rate alone have a pronounced effect upon the pattern of discharge of baroreceptor afferents and upon central nervous system structures. We are pursuing intensively investigations into the distribution of interspike intervals in the whole carotid sinus nerve as a function of variations in frequency alone of the pressure wave, thus imitating the effects of the heart in a controlled fashion. The results are only preliminary, but they suggest strongly that accelerative and decelerative changes in heart rate have different effects on baroreceptor activity and that the effects depend upon which subpopulation of fibers in the carotid sinus nerve one looks at. As these investigations progress and as we are able to establish the relationships of such variations to the functions of central nervous structures, we hope to get some insight into the operation of this visceral afferent feedback pathway and to establish clearly the reasons for, and the results of, the characteristic and specific behavior of heart rate in simple sensorimotor activity.

REFERENCES

Brown, G. L., & Eccles, J. C. The action of a single vagal volley on the rhythm of the heart beat. *Journal of Physiology (London)*, 1934, *82*, 211–240.

Coleridge, H. M., Coleridge, J. C. G., & Rosenthal, F. Prolonged inactivation of cortical pyramidal tract neurones in cats by distension of the carotid sinus. *Journal of Physiology (London)*, 1976, *256*, 635–649.

Dong, E., Jr., & Reitz, B. A. Effect of timing of vagal stimulation on heart rate in the dog. *Circulation Research*, 1970, *27*, 635–646.

Gahery, Y., & Vigier, D. Inhibitory effects in the cuneate nucleus produced by vago-aortic afferent fibers. *Brain Research*, 1974, *75*, 241–246.

Heymans, C., & Neil, E. *Reflexogenic areas of the cardiovascular system*. Boston, Mass.: Little, Brown, 1958.

Lacey, B. C., & Lacey, J. I. Change in heart period: A function of sensorimotor event timing within the cardiac cycle. *Physiological Psychology*, 1977, *5*, 383–393.

Lacey, B. C., & Lacey, J. I. Two-way communication between the heart and the brain: Significance of time within the cardiac cycle. *American Psychologist*, 1978, *33*, 99–113.

Lacey, J. I. Some cardiovascular correlates of sensorimotor behavior: Examples of visceral afferent feedback. In C. H. Hockman (Ed.), *Limbic system mechanisms and autonomic function*. Springfield, Ill.: Thomas, 1972.

Lacey, J. I., Bateman, D. E., & Van Lehn, R. Autonomic response specificity: An experimental study. *Psychosomatic Medicine*, 1953, *15*, 8–21.

Lacey, J. I., Kagan, J., Lacey, B. C., & Moss, H. A. The visceral level: Situational determinants and behavioral correlates of autonomic response patterns. In P. H. Knapp (Ed.), *Expression of the emotions in man.* New York: International Univ. Press, 1963.

Lacey, J. I., & Lacey, B. C. Verification and extension of the principle of autonomic response stereotypy. *American Journal of Psychology,* 1958, *71,* 50–73.

Lacey, J. I., & Lacey, B. C. The law of initial value in the longitudinal study of autonomic constitution: Reproducibility of autonomic responses and response patterns over a four-year interval. *Annals of the New York Academy of Sciences,* 1962, *98,* 1257–1290; 1322–1326.

Lacey, J. I., & Lacey, B. C. Some autonomic-central nervous system interrelationhips. In P. Black (Ed.), *Physiological correlates of emotion.* New York: Academic Press, 1970.

Lacey, J. I., & Lacey, B. C. Experimental association and dissociation of phasic bradycardia and vertex-negative waves: A psychophysiological study of attention and response-intention. *Electroencephalography and Clinical Neurophysiology,* 1973 (Suppl. 33), 281–285.

Levy, M. N. Parasympathetic control of the heart. In W. C. Randall (Ed.), *Neural regulation of the heart.* New York: Oxford Univ. Press, 1977.

Levy, M. N., Iano, T., & Zieske, H. Effects of repetitive bursts of vagal activity on heart rate. *Circulation Research,* 1972, *30,* 186–195.

Levy, M. N., Martin, P. J., Iano, T., & Zieske, H. Paradoxical effect of vagus nerve stimulation on heart rate in dogs. *Circulation Research,* 1969, *25,* 303–314.

Levy, M. N., Martin, P. J., Iano, T., & Zieske, H. Effects of single vagal stimuli on heart rate and atrioventricular conduction. *American Journal of Physiology,* 1970, *218,* 1256–1262.

Reid, J. V. O. The cardiac pacemaker: Effects of regularly spaced nervous input. *American Heart Journal,* 1969, *78,* 58–64.

FRANCES K. GRAHAM

Control of Reflex Blink Excitability[1]

35

Blinking is part of the generalized startle reflex—a reflex that can be distinguished both from defensive or nociceptive reflexes and from the orienting reflex (Sokolov, 1963). It includes a distinctive pattern of widespread flexor contraction as well as a short latency, rapidly habituating acceleration of heart rate (Landis & Hunt, 1939). A longer latency, slowly habituating heart rate acceleration accompanies the defensive reflex, whereas a rapidly habituating heart rate deceleration accompanies the orienting reflex (Graham & Clifton, 1966; Graham & Slaby, 1973). The stimulus for startle is also generalized, in the sense that startle can be evoked by stimuli in any modality. But the stimulus has very specific characteristics. There must be a sufficiently large change in stimulation occurring within approximately 10 msec (Berg, 1973; Fleshler, 1965).

The startle reflex can be predictably modulated by the unlearned effects of prior stimulation. The phenomenon of reflex modulation was known in the nineteenth century (e.g., Sechenov, 1965) and work in the early part of the twentieth century demonstrated its potential for understanding sensory and complex processes. However, interest in behavior of conditioned reflexes replaced interest in unconditioned reflexes and the latter have been largely neglected until the past decade when work has resumed in several laboratories (e.g., Davis, 1974; Hoffman & Wible, 1970; Ison & Hammond, 1971). This chapter will describe effects that suggest

[1] This work was supported by research funds from The Grant Foundation, Public Health Service Grant HD01490, National Science Foundation Grant BMS75-17075, and by Research Scientist Award K3-MH21762.

that the blink reflex in humans can be controlled by at least three different systems for processing sensory inputs: a system responsive to transients, a system responsive to steady-state intensity, and an orienting system that, as Sokolov (1963) has described, is responsive to the information in, or signaled by, a stimulus.

We measure blinking by tracking lid position. A plastic piece taped to the lid is connected by thread with the arm of a potentiometer whose output is computer-sampled once a millisecond. The computer is programmed to recognize a movement as small as 33 μm, to read the latency from stimulus onset to the onset of movement, and to compute the maximum size and velocity of movement (Graham, Putnam, & Leavitt, 1975). We are also measuring activity of the fast and slow motor units of orbicularis oculi, the muscle controlling blink. We are interested in these two kinds of motor unit because their pattern of activity and order of recruitment vary, depending on central commands. Slow activity is very prominent when a subject blinks on instruction. It is almost absent in a spontaneous blink. It varies in a reflex blink, not only with the eliciting conditions, but also with how the motoneuron pool is preconditioned (Silverstein & Graham, unpublished data).

Because of time limitations, the preconditioning studies to be discussed will consider only changes in lid movement and, in some cases, heart rate; also the discussion will deal only with studies of the awake human adult, although the method has been used with sleeping adults (Silverstein & Graham, unpublished data) and with young infants (Strock, Graham, & Zeigler, 1977). Our technique is to precede an adequate startle stimulus with a weak prestimulus that cannot, itself, elicit a blink. Each subject always receives a control condition, in which the startle-eliciting stimulus occurs without any prestimulation and one or more prestimulus conditions. Figure 35.1 shows the first four trials of a subject who received, on Trial 3, the control condition in which a 104 db 50 msec burst of white noise was presented alone. The blink measured approximately 6 mm and its latency was 68 msec. The other three trials were prestimulus trials in which onset of a 70 db, 1000 Hz tone preceded the startle pulse by 30 to 60 msec. The prestimulus lasted 20 msec in Trial 1 and was sustained throughout the interval in Trials 2 and 4. With prestimulation, the size of the response was much reduced, whereas its latency was shortened. The effects were present on the first trial when there had been no prior pairing of stimuli.

These findings are typical of the effects of acoustic prestimuli at short lead intervals, as can be seen in Figure 35.2. Figure 35.2 shows effects of a 70 db tone prestimulus, and visual and tactual prestimuli that were matched psychophysically to be equally intense. All had rise times less than 1 msec and the startle stimulus was the 104 db noise burst in each case. The curves are the averages of two experiments using nine subjects

35. Control of Reflex Blink Excitability

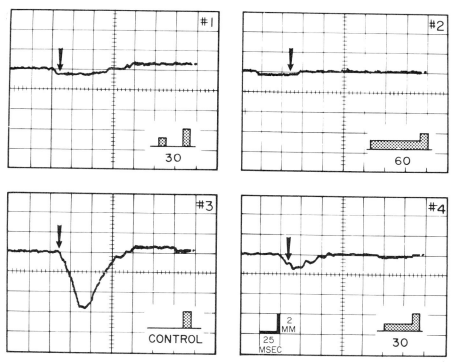

Figure 35.1. Oscilloscope tracings of the blink reaction on the first four trials of a single subject. Stimulation delivered on each trial is diagrammed in the lower right of each frame. Traces begin with onset of the startle stimulus and the arrow marks the latency to the onset of the control response in Trial 3.

each in the case of acoustical (from Graham & Murray, 1977) and visual prestimuli, and the average of a single nine-subject experiment in the case of tactual prestimuli. Each subject received 81 trials, 9 trials in each condition, with the presentation order determined by a Latin-square design.

It is apparent that there is an early period of facilitation, more pronounced with visual and tactual prestimuli, which is followed by a period of inhibition that affects the size but not the latency of the response. The inhibition is just as marked with the 20 msec prestimulus as when the prestimulus is sustained throughout the interval. Thus, all that is needed to produce inhibition is a transient change in stimulation that occurs shortly before the startle stimulus. In fact, the change need not be a stimulus onset. Inhibition can also be produced by the offset of a stimulus or by a change in its quality (Stitt, Hoffman, Marsh, & Boskoff, 1974; Zeigler & Graham, unpublished data). This is true despite the fact that inhibition is greater with the onset of a more intense stimulus (Graham & Murray, 1977; Hoffman & Wible, 1970). Not only does prestimulation lasting longer than 20 msec not increase inhibition, but it

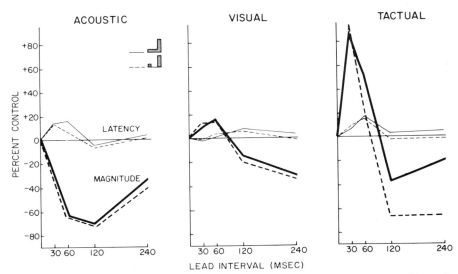

Figure 35.2. Blink magnitude and latency in response to 104 db, 50 msec white noise following 70 db, 1000 Hz tone, and equally intense visual or tactual prestimuli at lead intervals from 30 to 240 msec. Prestimuli were either 20 msec in duration or were sustained throughout the interval.

has the opposite effect. By 60 msec, sustained prestimuli produced relatively more facilitation of latency and magnitude even when the overall effect was inhibitory. The differences are small but reliable and become more pronounced at longer intervals.

Figure 35.3 shows the effects on blink size and latency when prestimuli lead the startle-eliciting stimuli by 2000 msec. The sustained prestimulus elicits a larger and faster blink, whereas the effect of the transient stimulus varies depending on whether or not orienting, indexed by cardiac deceleration, occurs during the lead interval. The first row shows that when there is no orienting, the brief prestimulus has no reliable effect. These results are obtained when subjects receive only two conditions—a control condition and one prestimulus condition (Brown, 1975). For one group the prestimulus was brief; for the other it was sustained. In both groups, the startle stimulus always occurred 2 sec after prestimulus onset. In such simple, highly predictable situations, prestimuli do not provide much information and thus they do not elicit an orienting response.

What happens when the conditions do elicit orienting? Figure 35.4 shows results of an experiment using the nine-condition design with lead intervals ranging from 200 to 2000 msec. Under these conditions, the prestimuli do not predict very well when the startle stimulus will follow, and we found that heart rate began to decelerate sharply as soon as a prestimulus began (Graham, Putnam, & Leavitt, 1975). Blinking was also

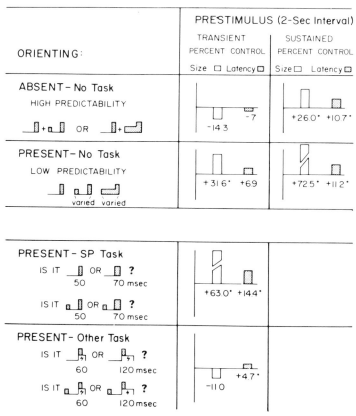

Figure 35.3. Blink size and latency as percent change from control blink when transient and sustained prestimuli preceded a 104 db, 50 msec noise burst by 2 sec. Row 1 shows results of high predictability conditions in which prestimuli did not elicit orienting and row 2 shows results of low-predictability conditions in which orienting was elicited. Orienting was also elicited by the discrimination tasks shown in rows 3 and 4. The task in row 3 directed attention to the startle-eliciting noise burst, but in row 4 attention was directed to a tactual stimulus occurring at the same time as the startle stimulus. Asterisks indicate statistically significant changes ($p < .05$).

affected. At the short 200 msec interval, inhibition was still dominant. However, as the interval lengthened, facilitation became evident following both transient and sustained prestimuli.

To recapitulate (Figure 35.3), with highly predictable conditions in which prestimuli did not elicit orienting (row 1, Figure 35.3), a transient prestimulus had no effect after 2 sec, but a sustained prestimulus produced more facilitation than it did at short intervals. With less predictable conditions in which prestimulus did elicit orienting (row 2, Figure 35.3), a transient prestimulus produced facilitation and a sustained prestimulus produced

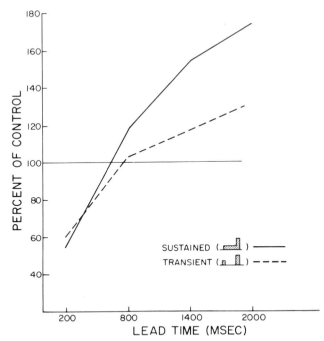

Figure 35.4. Blink size as percent of control blink when transient and sustained prestimuli preceded a 104 db, 50 msec white noise burst by intervals ranging from 200 to 2000 msec.

still greater facilitation. Facilitation by the sustained prestimulus was presumably due to two effects: the effect associated with orienting and the effect due to temporal integration of intensity. The inhibitory effect of transient and the facilitating effect of sustained prestimuli have been seen in other species, cross-modally as well as intramodally (e.g., Ison & Hammond, 1971). However, facilitation by a transient prestimulus that also elicits orienting has been reported only in human subjects.

We were surprised to find that blink was facilitated in the presence of orienting. Although the Laceys (e.g., Lacey & Lacey, 1974) suggest that cardiac deceleration can facilitate sensorimotor behavior, it is frequently assumed (e.g., Obrist et al., 1974) that orienting reduces motor activity, at least irrelevant motor activity. I suggested (Graham, 1975) that blink facilitation might be due to orienting termination rather than to orienting presence; that is, as soon as the startle pulse was detected, no more information was available, orienting terminated, and the release from inhibition produced facilitation. Bohlin and Graham (1977) tested this by preventing the termination of orienting. Subjects were asked to discriminate between the duration of a short, 50 msec startle stimulus and a long,

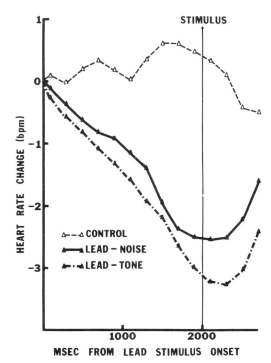

Figure 35.5. Heart rate changes during 2000 msec between the onset of prestimulus and 800 msec following the onset of a startling noise (lead-noise condition) or weak-startle tone (lead-tone condition) or for 2000 msec preceding and 800 msec following the tone (control condition). (Reproduced with permission from Bohlin & Graham, 1977, The Society for Psychophysiological Research.)

70 msec startle stimulus. This is a difficult discrimination and subjects are correct only about 75% of the time.

The termination hypothesis was *not* supported in either of the two experiments. It was possible to show that cardiac deceleration began with prestimulus onset and persisted for 400 msec after startle pulse onset, thus outlasting the blink (Figure 35.5). However (row 3, Figure 35.3), blink was still facilitated. As long as the prestimulus elicited orienting, whether produced by low predictability or by a task requiring attention, blink was facilitated.

These results suggest that orienting does not always inhibit irrelevant motor activity, at least reflex activity. But they do not necessarily imply that orienting produces motor facilitation directly. An alternative hypothesis is that orienting enhances reception of the reflex-eliciting *stimulus*. In the situations described, orienting or attention was directed to the startle stimulus. If reflex facilitation is produced by facilitation of the sensory channel to which attention is directed, then directing attention to a different channel should remove the facilitation.

Louis Silverstein and I tested this hypothesis by having subjects discriminate between durations of tactual stimuli that were presented simultaneously with the startle pulse. The tactual stimuli could not themselves elicit startle and their durations were extended to 60 and 120 msec to make the task approximately the same difficulty as in the Bohlin and Graham experiment. As previously, heart rate decelerated with lead conditions and did not decelerate in the control condition. However (row 4, Figure 35.3), the effects on blink changed. With attention directed to the tactual modality, size of blinking to the combined stimuli was insignificantly smaller under lead conditions than when the combined stimuli were presented alone. Latency was still slightly facilitated. Thus, it appears that orienting can facilitate a reflex, but that it does so by enhancing stimulus reception. Further, it can apparently do so selectively—enhancing an attended channel and having little effect on an unattended channel.

To summarize briefly, it appears that important systems for processing inputs can have strong and specifiable effects on a simple reflex. Facilitation occurs by two different mechanisms: selective attention to a particular sensory channel and nonselective sustained excitation in any channel. Inhibition occurs when there is a transient change in stimulation. The inhibition does not show the dependence on specific sensory quality of sensory-masking effects. Furthermore, it cannot be due to inhibition involving the gamma loop since orbicularis oculi lack muscle spindles (e.g., Lindquist, 1973) and functional tests indicate that the muscle is not subject to the recurrent inhibition that would be expected with feedback from Renshaw cells (e.g., Penders & Delwaide, 1973). Graham and Murray (1977) suggested that it involves a rapidly transmitting system specialized to detect transients. Gersuni (1971) has identified short-time constant, transient-sensitive neurons at all levels of the auditory system. They have also been found in the visual system (e.g., MacLeod, 1978). Why should a stimulus-detecting system produce, across species, such strong suppression of the whole range of flexor activities that make up the startle reaction? An interesting possibility is that inhibition serves to protect what has been called preattentive stimulus processing, that is, it allows a finer stimulus analysis to proceed with minimal interruption during the critical, approximately 250 msec, period needed for stimulus recognition (e.g., Massaro, 1972).

REFERENCES

Berg, K. M. *Elicitation of acoustic startle in the human.* Unpublished doctoral dissertation, Univ. of Wisconsin, 1973.
Bohlin, G., & Graham, F. K. Cardiac deceleration and reflex blink facilitation. *Psychophysiology*, 1977, *14*, 423–430.
Brown, J. W. *Contingent negative variation and cardiac orienting preceding startle modification.* Unpublished doctoral dissertation, Univ. of Wisconsin, 1975.
Davis, M. Sensitization of the rat startle response by noise. *Journal of Comparative and Physiological Psychology*, 1974, *87*, 571–581.

Fleshler, M. Adequate acoustic stimulus for startle reaction in the rat. *Journal of Comparative and Physiological Psychology*, 1965, *60*, 200–207.

Gersuni, G. V. Temporal organization of the auditory function. In G. V. Gersuni (Ed.), *Sensory processes at the neuronal and behavioral levels*. New York: Academic Press, 1971. Pp. 85–114.

Graham, F. K. The more or less startling effects of weak prestimulation. *Psychophysiology*, 1975, *12*, 238–248.

Graham, F. K., & Clifton, R. K. Heart rate change as a component of the orienting response. *Psychological Bulletin*, 1966, *65*, 305–320.

Graham, F. K., & Murray, G. M. Discordant effects of weak prestimulation on magnitude and latency of the reflex blink. *Physiological Psychology*, 1977, *5*, 108–114.

Graham, F. K., Putnam, L. E., & Leavitt, L. A. Lead stimulation effects on human cardiac orienting and blink reflexes. *Journal of Experimental Psychology: Human Perception and Performance*, 1975, *1*, 161–169.

Graham, F. K., & Slaby, D. A. Differential heart rate changes to equally intense white noise and tone. *Psychophysiology*, 1973, *10*, 347–362.

Hoffman, H. W., & Wible, B. L. Role of weak signals in acoustic startle. *Journal of the Acoustical Society of America*, 1970, *47*, 489–497.

Ison, J. R., & Hammond, G. R. Modification of the startle reflex in the rat by changes in the auditory and visual environments. *Journal of Comparative and Physiological Psychology*, 1971, *75*, 435–452.

Lacey, J. I., & Lacey, B. C. On heart rate responses and behavior: A reply to Elliott. *Journal of Personality and Social Psychology*, 1974, *30*, 1–18.

Landis, C., & Hunt, W. A. *The startle pattern*. New York: Farrar & Rinehart, 1939.

Lindquist, C. Reflex organization and contraction properties of facial muscles. *Acta Physiologica Scandinavica Suppl.* 1973, *393*, 3–15.

MacLeod, D. I. A. Visual sensitivity. In M. R. Rosenzweig, & L. W. Porter (Eds.), *Annual Review of Psychology*, 1978, *29*, 613–645.

Massaro, D. W. Preperceptual images, processing time, and perceptual units in auditory perception. *Psychological Review*, 1972, *79*, 124–145.

Obrist, P. A., Howard, J. L., Lawler, J. E., Galosy, R. A., Meyers, K. A., & Gaebelein, C. J. The cardiac-somatic interaction. In P. A. Obrist, A. H. Black, J. Brener, and L. V. diCara (Eds.), *Cardiovascular psychophysiology—Current issues in response mechanisms, biofeedback and methodology*. Chicago: Aldine, 1974. Pp. 136–162.

Penders, C. A., & Delwaide, P. J. Physiologic approach to the human blink reflex. In J. E. Desmedt (Ed.), *New developments in EMG and clinical neurophysiology*. Basel: S. Karger, 1973. Pp. 649–657.

Sechenov, I. M. *Reflexes of the brain* (S. Belsky, trans.). Cambridge, Mass.: MIT Press, 1965.

Sokolov, E. N. *Perception and the conditioned reflex*. New York: Macmillan, 1963.

Stitt, C. L., Hoffman, H. S., Marsh, R., & Boskoff, K. J. Modification of the rat's startle reaction by an antecedent change in the acoustic environment. *Journal of Comparative and Physiological Psychology*, 1974, *86*, 826–836.

Strock, B. D., Graham, F. K., & Zeigler, B. L. Startle and startle inhibition in early development. Presented at the Annual Meeting, Society for Psychophysiological Research, Philadelphia, October, 1977.

VALENTINA ZAVARIN

Modification of Goal-Directed Behavior in Discourse[1]

36

If natural phenomena have an air of "necessity" about them in their subservience to natural law, artificial phenomena have an air of "contingency" in their malleability by environment. . . . Engineering, medicine, business, architecture, and painting are concerned not with the necessary but with the contingent—not with how things are, but with how they might be—in short, with design.

Herbert Simon (1969, ix–xi)

Language is a semiotic system molded to the environment by goals. My perspective on language is mediated by the distinction made by Herbert Simon between natural and artificial phenomena. It entails that the proper perspective for studying an artificial phenomenon such as language is to focus on its design. Traditionally, this was the domain of rhetoric and required consideration not only of units of meaning and structures beyond the sentence, but also of contingency or malleability by the environment; here linguistic behavior and its product—discourse—were studied as structured performances, dependent on learning processes, memory, and goal-directedness. Although presently much effort has been directed toward the identification of units, particularly in psycholinguistics (Aaron, 1975; Clark, 1977; Leont'ev, 1969), in this study I am interested not only in the elements

[1] Research on this paper was partially supported by National Institute of Mental Health Biomedical Research Grant RR-05755, the Langley Porter Neuropsychiatric Institute, University of California, San Francisco, and Research Grant 38500 of the School of Medicine, University of California, San Francisco, and also National Institute of Mental Health Grant MH31360, The Wright Institute, Berkeley.

of design, but more especially, in what can be inferred from discourse about goal-directedness, steps preceding the actual formation of discourse, and the organism–environment perspective.

It was Pavlov's belief that pathology can separate and simplify that which is united and inaccessible in normal processes. It is my contention that the study of linguistic behavior in pathological states such as schizophrenia may be instrumental in segregating certain features that are integral to its design. I thus propose a study of discourse produced by schizophrenic subjects.

Based on the analysis of properties of discourse from a sample of schizophrenic subjects, we propose a working model of this deviant discourse. The conceptualization of aspects of schizophrenic discourse is based on experimental results from 600 Rorschach tests. The analysis of the verbatim protocols has been undertaken in terms of a typology of features (MS typology) formulated by Margaret Singer.[1] In this study we draw upon the semiotic theory of language of Louis Hjelmslev to cluster empirically derived features into a theoretical framework. We then propose an interpretation of our findings in terms of neurolinguistic function units (Luria, 1974, 1976), and relate them to modes of central processing (Pribram, 1976). Further inferences are made about functions of a Content-Addressable Memory that may lead to the production of this deviant discourse (Luria, 1976; Spinelli, 1970).

Hypothesis of Schizophrenic Information Processing in Discourse

Based upon the work of Singer *et al.* (1966) we posit that schizophrenic discourse is characterized by the interruption of discourse (the syntagmatic process) and the intrusion of a paradigm (the paradigm formation) as illustrated by the following excerpt from schizophrenic discourse:

> In the halls of the Justice Department there is an understanding of a *bona fide* agreement between any people scheduled to meet within government circles, government triangles, government rectangles, . . . [Laffal, 1965, pp. 131–132].

[1] This chapter reflects some aspects of the work in progress in conjunction with Margaret Singer. My work has been based on the typology of Margaret Singer (the MS typology). The MS typology is designed to capture formal features of schizophrenic discourse. The units, 41 in number (Singer, 1973), are grouped under five large topids; commitment problems, referent problems, language anomalies, information sequencing, and disruptions. The features of the MS typology have been defined across various levels. Information processing of verbal, visual, and object stimuli has served for a comparative verification of features in Rorschach, TAT, Proverb Test, Object Sorting Test, and Sentence Completion Tests. The networks of features provide a model for (*a*) information processing from various stimuli (verbal, visual, object) as reflected in the linguistic medium; and (*b*) interactive processes of the communication situation. The MS typology has been tested for predictive power in blind matching of parents and offspring in test situations.

Here, the syntagmatic chain is interrupted and the paradigmatic or classificatory system is exposed on the surface of discourse in the paradigm: "government circles, government triangles, government rectangles." Two interpretations have been offered about the phenomenon by which such intrusion is generated. Some researchers believe that such schizophrenic intrusions reflect a passive attitude, "an unreflective yielding to stimuli [Goldstein, 1964]," an "abandonment of *active* direction and taking a *passive* response to stimuli [Chapman, 1973]," a "lapse of attention [McGhie, 1969]," a stimulus flooding, a breakdown of filtering mechanisms, a deterioration of selective attention, and so on. Others have postulated an "active direction" in selecting an inappropriate stimulus to which the schizophrenic responds (Shakow, 1971). We, too, take this view, postulating a "bias" in information processing in schizophrenia; we focus on the possible formal organizations of the intruding "flood," and thus we disagree with the position that schizophrenics "throw in a cluster of more or less related elements [or] collection of fragments (Cameron, 1944, p. 53)."

Hjelmslev's linguistic theory (glossematics) allows us to characterize the "intrusion" in terms of a competing linguistic process (paradigmatic process) and the organization within the intrusion in terms of various paradigms (paradigmatic frames), as proposed in the following sections.

Two Linguistic Processes:
Paradigmatics and Syntagmatics

Two language processes are used in the description of linguistic activity by Hjelmslev (1963, 1970) (see also Lyons, 1969, pp. 70–79; Saussure 1966). Syntagmatic processing is responsible for conjoining units in discourse, as in the example "The dog barks." Paradigmatic processing is responsible for the juxtaposition or classification of units of language, as in the paradigm "sun, moon, star."[2]

---

[2] Paradigmatic and syntagmatic are defined by Lyons as follows: "By virtue of its potentiality of occurrence in a certain context a linguistic unit enters into relations of two different kinds. It enters into *paradigmatic* relations with all units which can also occur in the same context (whether they contrast or are in free variation with the unit in question); and it enters into *syntagmatic* relations with the other units of the same level with which it occurs and which constitute its context. . . . Paradigmatic and syntagmatic relationships are also relevant at the word level, and indeed at every level of linguistic description. For example, by virtue of its potentiality of occurrence in such contexts as *a . . . of milk*, the word *pint* contracts paradigmatic relations with such other words as *bottle, cup, gallon*, etc., and syntagmatic relations with *a, of,* and *milk*. In fact, words (and other grammatical units) enter into paradigmatic and syntagmatic relations of various kinds. 'Potentiality of occurrence' can be interpreted with or without regard to the question whether the resultant phrase or sentence is meaningful; with or without regard to the situations in which actual utterances are produced; with or without regard to the depen-

Luria (1973) also used the two processes in describing the linguistic activity in the becoming of discourse. At first, the message is embodied in private terms dependent on parameters of personal experience (Russian: *smysl*). From a concise state in the form of a "thought syntax [Pick, 1913]," the message goes through various processes of "inner speech [A. Sokolov, 1960]," until it finally acquires meaning (Russian: *znachenie*) as it is translated into the language formed in the social process of communication, and thus becomes discourse. Both syntagmatics and paradigmatics are necessary for the formation of discourse. Syntagmatics is responsible for the organization of the chain of discourse; paradigmatics is responsible for the organization of the lexicon into classes. The two processes are interdependent. Although discourse unfolds under the temporal restraints of verbalization, outside the temporal restraints of discourse, units of language (phonetic, lexical, morphological, structural, semantic, and other semiotic pattern-forming units) are simultaneously processed.

Furthermore, for the realization (the performance) of a linguistic act, syntagmatic processing requires competence in government and agreement between units and in strategies for using the rules and making selections, whereas paradigmatic processing requires competence in lexicon, organizational parameters for the lexical units, and strategies of access.

During past decades, particular emphasis on communication has made the study of the syntagmatic process preeminent. And all communication is syntagmatic. The importance of paradigmatics, largely underplayed until recent years, is now in sharper focus in the studies of lexical semantics (Apresian, 1974; Melchuk, 1974) and cognitive psychology (Palmer, 1977; Rosch, 1979).

Paradigmatic Frames

Investigation of the various internal organizations of paradigmatic intrusions has been carried out on the Singer sample and the results formulated in terms of a lexical theory (Coseriu, 1975), applied to schizophrenic discourse by Zavarin (1977).[3] Among four primitive types of relations (not

---

dencies that hold between different sentences in connected discourse; and so on" [Lyons, 1969, 73–74].

Hjelmslev describes the organization within paradigmatics as "a network of relations between *alternative* terms" and the relations between the terms as disjunctive EITHER–OR relations. In Hjelmslev's theory paradigmatics is opposed to the domain of syntagmatics. Here the network of relations is between *coexisting* parts and the relation is that of conjunction or BOTH–AND relation. Syntagmatics is a sign-process (semiotic process), it has a teleological structure, and its parts have a specific location in the context. Paradigmatics is a sign-system (semiotic system) and its terms are in free variation (Hjelmslev, 1975).

[3] Saussure exemplified paradigmatic relations in frames as associations by content (teaching-instruction-education); he also exemplified associative classes formed between words with common morphological elements (suffixes, prefixes), between basic words and derivatives, between words of the same inflectional pattern, between words with a common sound-image (rhyming words, alliterations) and so on (Saussure, 1966, 122–127).

counting mixed forms) three have been identified in schizophrenic discourse and defined in terms of relational frames (Zavarin, 1977): the privative or antonymic frame, the equipollent or serial frame, and the hyponymic/hyperonymic or subordinate/superordinate frame.[4]

FRAME ONE

The *privative frame* organization subsumes lexical items that are related as synonyms or antonyms. A common response to a stimulus in schizophrenic discourse is by synonyms: "This is a fiddle, violin," or "This is a bat. Die Fledermaus." An example of antonymic intrusion may be observed in an example from Cameron: "I am alive because I was born a human and animal life and normal life [Cameron, 1944, p. 23]." Here, two oppositions overwhelm the syntactic structure of the sentence: "human versus animal" and "animal life versus normal life." [5]

FRAME TWO

The *equipollent* or serial frame organization has been exemplified by the paradigm "circles, . . . triangles, . . . rectangles." Equipollent lexical fields may be exemplified by ordinate (Monday, Tuesday, Wednesday, . . .) or nonordinate lexical groups in terms of their structures (as in the enumeration of geometrical figures).[6]

FRAME THREE

The third type of paradigmatic intrusion in schizophrenic discourse exhibits relations of subordinate (e.g., seagull) to superordinate (e.g., bird)

---

[4] Paradigmatic fields as zones of signification have various internal organizations which may be expressed in terms of frames. We define frame as the form of internal relations of a certain domain. Definition of frame relations within categories is based on the discussions of Greimas in *Sémantique structurale* (1966) and work of Coseriu in "Typologie des champs léxicaux" (1975). Three out of the four types of lexical field organizations are found in schizophrenic discourse. Coseriu discusses four pure types of lexical organization and proposes various mixed organizational structures of lexical fields. The pure organizational frames are *privative, equipollent, hyponymic/hyperonymic*, and the *gradual* or *hierarchical* frames. The gradual or hierarchical frame of lexical organization is not represented in paradigmatic intrusions in schizophrenic discourse. An example of a gradual frame is hot, warm, cool, cold.

[5] Privative frames or antonymic/synonymic frames are both subsumed under polar frames. Privative frames are based on an opposition x/non-x. Privative frames may be exemplified by: high/low; short/long; narrow/wide; to master/to dominate.

[6] Equipollent frames or serial frames may be exemplified by such series as days of the week or months of the year. Colors in English—red, yellow, green—are in equipollent relation. In responses to visual stimuli equipollent frames appear when a series of alternative interpretations (mostly in a nonordinate series) is offered and no oppositional relation is implied (Coseriu, 1975).

and vice versa. On the lexical level it has been discussed as the relation of hyponymy and hyperonymy (Greimas, 1966; Lyons, 1969, p. 476). In schizophrenia, processing of this relational frame has been identified as "overinclusion" and "underinclusion" (Chapman, 1973). This frame has important implications for the generalization process and the "naming" process (Tsvetkova, 1975), and will be discussed in the next section.

Note that it is possible to extend frame categories to other levels, such as the phonetic and morphological levels. Paradigmatic organization according to a similarity of sound may be observed in the examples: *"Imagination* is the worst *nation* in the world" and "He's a *good hood,* in a *broody, moody* way" (Singer, 1965, p. 194). Paradigmatic frames according to the structure of morphological elements (prefixes: sub/terfuge, sub/stitution, or suffixes: conten/tion, applica/tion, posi/tion, contradic/tion, etc.) have been discussed elsewhere (Nöth, 1977; Zavarin, 1977, p. 550).

Breakdown of the Syntagmatic Process: Luria

The linguistic dichotomy of syntagmatics and paradigmatics has been used by Luria in describing the breakdown of linguistic production (Jakobson, 1966; Luria, 1976). In a classification of aphasias, Luria made the following distinction:

> Lesions of the *anterior* parts of the major hemisphere result in a marked deterioration of *syntagmatic* organization of verbal communication while the paradigmatic organization of linguistic codes remains relatively preserved. In contradistinction, lesions of the posterior cortical areas of the major hemisphere result in a breakdown of the paradigmatic organization of linguistic structures in different levels (phonemic level in lesions of the *posterior* parts of the left temporal lobe, articulatory systems in lesions of the lower part of the left postcortical zone, semantic or logicogrammatical level in lesions of the posterior tertiary zone) while the *syntagmatic* organization of the fluent speech remains preserved [Luria, 1974, p. 12].

Interruption of the syntagmatic process (defined in various ways; see Caramazza, 1976) has been discussed by Luria in a taxonomy of aphasia symptoms. His formulations of components in the breakdown of this process are of sufficient generality to be used for language processing disturbances of the type we find in schizophrenia. The following constellation of symptoms that accompanies the breakdown of the syntagmatic process as described by Luria corresponds to the symptomatology in schizophrenic discourse: (*a*) The breakdown of the syntagmatic process is accompanied (*b*) by intrusion of alternative units and (*c*) by the occurrence of disintegration (segmentation) of lexical units. The components in the breakdown of the syntagmatic process are defined by Luria in terms of a neurodynamic system as impairments of motor habits responsible for the flow of "kinetic melodies," of plasticity, of blocking ("capacity to quickly and fluently block the pattern already used [Luria, 1974, p. 8]"), and of the functioning of the "law of strength" (the law that prevents equalization).

We are particularly interested in Luria's analysis of the intrusion of a series of words (in our discussion defined as the paradigmatic intrusion) as an accompanying symptom to syntagmatic breakdown. Luria argues that this peculiar symptom is not a result of forgetting of words and does not represent a search for a word (is not a defect of memory). He specifically proposed that the responsibility for the appearance of the alternative terms is in the fact that "meanings come up with equal probability." [7]

> The patient finds it difficult to select the correct name of the object. This form is based upon a defect in the ability to select the necessary word from among many alternatives. Strange though it seems, this form . . . is characterized not by the poverty of memory but by the *excessive number of associations* which arise. The patient recalls too many designations and is incapable of normal word selectivity. He cannot select the desired meaning because *all of the meanings come up with equal probability* [Luria, 1977, pp. 142–43].

We are impressed also by the fact that a syntagmatic breakdown in Luria's system is accompanied by instances of disintegration of lexical units (lexical units segmented into sublexical units). A rebuslike segmentation of words, as in the following example: content-men-t(tea), is expressed in Luria's system as a breakdown of the "kinetic control."

> (Contentment?)—Well, uh, contentment, well the word contentment, having a book perhaps, perhaps your having a subject, perhaps you have a chapter of reading, but when you come to the word "men" you wonder if you should be content with men in your life and then you get to the letter "t" and wonder if you should be content having tea by yourself or be content with having it with a group or so forth [Lorenz, 1961].

Here, a word is decomposed into sublexical units and processing gives the illusion of a transfer of the meaning units into "schematic space." Whatever interpretation we give, it seems a reasonable supposition that here, too, we have a problem with "movement in time."

We may now formulate a hypothesis of breakdown in schizophrenic discourse in terms of neurolinguistic units and mechanisms that are responsible for them. In schizophrenic discourse there is intermittent breakdown

---

[7] Luria describes this deficit as: "An equalization of excitation evoked by stimuli of different strengths" and defines this phenomenon as follows: "Strong (or significant) stimuli or their traces begin to evoke reactions equal to the reactions of the weak (or insignificant) stimuli or their traces. In the next . . . paradoxical stage weak (or insignificant) stimuli . . . begin to evoke even stronger reactions than strong (or significant) stimuli . . . second or insignificant associations are evoked with the same probability as the principal or significant association" [Luria, 1974, 8].

of syntagmatic structures. In Luria's system of neurodynamics, the mechanisms responsible for this are kinetic control, the "law of strength," and mechanisms plasticity and blocking. Emergence of an excessive number of alternative units is accompanied by a process of segmentation of units.

Failure of the Context-Dependent Processing Mechanism

Paradigmatic or classificatory (context-free) intrusions into the syntagmatic process can be accounted for in terms of a breakdown of the "context-dependent" mode of central processing (Pribram, 1976). The dichotomy of "context-dependent" and "context-free" processing in the system of Pribram parallels the dichotomy of syntagmatics and paradigmatics. In language behavior syntagmatics may be understood as a context-dependent process because it relates preceding linguistic units to following units (and vice versa), and to the extralinguistic dimensions of addressor, addressee, situation, and environment. Paradigmatics may be viewed as context-free processing because it classifies units without reference to preceding or following units or extralinguistic dimensions.

The system of Pribram views the two processes from a different perspective from that of Luria. In comparison with Hjelmslev's immanent linguistic point of view on the two processes, and Luria's emphasis on neurodynamic mechanical properties necessary in the production of linguistic behavior, Pribram focuses on the pragmatic factor of organism-environment relation and the strategy complex during the performance.[8]

Given the two postulated models of central processing, as defined by Pribram, we propose that paradigmatic intrusions (context-free processing) into the syntagmatic process (context-dependent process) may result from deficits in mechanisms necessary for efficient context-dependent processing. It is proposed that in schizophrenia the malfunctioning of the context-dependent process is related to (*a*) a defect in hypothesis formation about the environment (which normally is acquired from experience); (*b*) failures in experiential learning about the probable outcomes of events; (*c*) failure in the strategy and choice-making function; and (*d*) failure in forming an "image of achievement [Pribram, 1971]."

The first and second of these four contentions may be supported by the research results of some 10 studies published under the editorship of Feigenberg (1973). These indicate that schizophrenia exhibits a deficiency in ordering information according to a scale of probable importance or usefulness based on previous experience, and that a probability prognosis

[8] The posterior convexity of the cortex is responsible, according to Pribram, for context-free processing; the fronto-limbic forebrain is responsible for context-dependent processing (Pribram, 1976).

is not carried out efficiently in the behavior of schizophrenics. All cognitive processes are shown to be affected by this defect, including linguistic behavior (Dobrovich, 1973; Polyakov, 1972; Zaitseva, 1973).

The work reported by Feigenberg (1973) describes tests in which schizophrenics perform better than normals. Schizophrenic subjects are free of illusions acquired by normals from experience. In tests that require estimating the weight of large and small cylinders of equal weight, normals judge larger objects to be heavier. Schizophrenics are free of this illusion (the illusion of Charpentier), and more often arrive at a correct answer. In another test verifying prediction gradual focusing is done on projected pictures. Where a picture consists of a light bulb in a basket of pears, normals require more focusing than schizophrenic persons to guess correctly the highly improbable object. Feigenberg defined the deficit in symptomatic processing of information in schizophrenia as probability prognosis defect. It is characterized as failure to make efficient prognosis about probabilities, as a wider resource allocation in comparison with normals; and, finally, as resource allocation that extends to latent and unimportant candidates. Reaching out to latent and highly improbable features may be illustrated by the following example: schizophrenics select a parameter such as "movement" to characterize a "spoon" and see similarity in "spoon" and "automobile" along this parameter. They pick out the features of "noiselessness" in a "beret" while characterizing "trumpet," "umbrella," and "whistle" as "noise-producing objects [Zeigarnik, 1965]." This highly eclectic processing is highly improbable, although not illogical.

Further application of Pribram's context-free processing model allows us to formulate another trait of the paradigmatic process in schizophrenia. In the system of Pribram, the context-free process is involved in recognizing recurrent regularities extracted from the context-dependent process. The result of this process is the formation of invariants. Further, two stages in the context-free process are defined: orienting and habituation. We are particularly interested in the cleavage between the two stages for further specification of paradigmatic processing in schizophrenia (Bernshtein, 1967; Pribram, 1976; Sokolov, 1960, 1977).[9] During the first part of the context-free processing stage (orienting) the following mechanisms are in operation: the match-mismatch process (production of alternatives), the mechanism of criteria application in comparison (various parameters of the paradigmatic process), and strategies to continue the process. All of this is well represented in paradigmatic processing in schizophrenia. The second part of the context-free process is reached when novel things acquire the dimension of familiarity. The effectiveness of this mechanism depends on a "neural model of the environment, a representation, an expectancy, a type

[9] The cleavage is observed in resections of certain forebrain structures: the amygdala and frontal cortex as reported by Pribram (1976).

of memory mechanism against which inputs are constantly matched [Pribram, 1977, p. 105]." If this mechanism is deficient (as may be the case in schizophrenia), the results are (a) input processing does not diminish with repetition and orienting does not disappear; (b) efficiency of output processing is not enhanced in performance; and, most importantly, (c) expectancy is not formed; that is, learned anticipations are not formed that otherwise help predict the outcomes of actions. There is no extrapolation possible since an image of achievement is not constructed to aid extrapolation. There is no model against which new inputs can be matched. Processing occurs as if in a vacuum (Pribram, 1971).

Pribram's definition of the context-free processing complex together with evidence from research of Feigenberg (1973), Wynne, Singer, Bartko, and Toohey (1977), and Zavarin (1977) allow us to formulate the hypothesis that the exposure of paradigmatics in schizophrenia is a manifestation of context-free processing that is not brought to closure. It is proposed that the manifestation of a bias toward paradigmatic processing is related to a modification in behavioral orienting (enhanced orienting) and weakened behavioral habituation. Indeed, it has been shown by Singer (1973, pp. 14–38) that the problem with selectivity, a prolonged search for alternative members, and in general difficulty with commitment are characteristic of the operational mode in schizophrenia. It is further proposed that the mechanism necessary for closure is related to the mechanism responsible for regulating probability prognosis in organism–environment interaction. From another perspective one may relate the deficiency to deviant functioning of mechanisms responsible for invariant formation, for discrimination of foreground/background, and in general deviant functioning of procedures participating in the cognitive generalization activity. In both instances there is a modification in the organizm-environment relation. Specifically, in communication, we note difficulty with semantic plausibility, with real world expectancy, situational and contextual cues, and in general with various real world semantic constraints.

Paradigmatic Processing of Semantic Features; Frame Three

In this discussion, we are looking at the sublexical level of semantic feature organization according to paradigmatic Frame Three (subordinate/superordinate relations). The discussion of feature addition and feature deletion is based on a simple notion that a subordinate ("seagull," for example) encompasses more semantic features than its superordinate ("bird," as it lacks the feature of "seagull-ness"). The theoretical work relating to this discussion has been developed by the Prague School Linguistics under the concept of feature markedness (Clark, 1975; Greenberg, 1966).

Retrieval Hypothesis and Luria's Model of Mnemonic Activity

Pretheoretical orientations in inquiries on human memory focus primarily on retrieval processes. Various versions of retrieval have been constructed.

The analysis of information processing in schizophrenia would be greatly facilitated by an available model of the successive stages through which information flows. A general description of the memory system in terms of some rudimentary congruent operations will be adopted from Luria. Luria describes three stages. From the recording stage we proceed to the recognition stage when new information is matched against past experience, and then to the stage during which information is interpreted and embodied into a system of conceptual connections. Generalization procedures are then effectuated and a schema is stored. This stage is differentiated by the time available, by the nature of the task, the goal, and the materials at hand. The retrieval phenomenon consists of the search for available alternatives and the selection. Both the search and the selection are governed by strategies, by hypotheses about the goal, and by decisions. Inhibition of random or irrelevant items and the search-stopping mechanism complete the schema of mnemonic activity to which only forgetting should be added.

Content-Addressable Memory

(References are made to features numbers of the *MS typology*. See note 1.) The stages in the process may be further elaborated via a Content-Addressable Memory theory (Spinelli, 1970). We do not pretend to make any stronger model hypotheses than those that will be capable of generating predictions as to STEPS between a stimulus, its normative response, and the several varieties of linguistic output typical of schizophrenics. We are here examining the characteristics of the paradigmatic process of the superordinate/subordinate relationship.

Let us look at a hypothetical model of processing and then turn to examples from our deviant discourse. Let us imagine that a Rorschach card is presented as a complex stimulus for which an interpretation is in order. According to the model, the following steps in normal processing are predicted. First, a generalization of the Rorschach design is extracted. Let us formulate it as "symmetrical winglike shapes." The system finds the most appropriate network in which the bat pattern has been learned previously and is now recognized. As a result, the name "bat" is assigned to the Rorschach representation. There are two extreme possibilities: A 100% match in a network that corresponds to the reaction "of course, it is a bat"

Table 36.1
Kent Rosanoff Word Association Test after Infusion of Methylphenidate [a]

Stimulus	Baseline	After 20 min	After 24 hr
Hand	Finger	An implement	Arm
Short	Tall	Measurement	Tall
Fruit	Orange	Food	Orange
Whistle	Sound	Vibration	Sound
Cold	Flu	Temperature measure	Hot

[a] Based on Janowsky et al., 1977, p. 192.

or absence of match when the Rorschach ink blots appear like "nothing you have seen before." These two extremes of memory processing are represented as the endpoints of a continuum of human memory retrieval.

Variability for Matching in Paradigmatics

The next problem raised by Spinelli is the problem of variability admitted for matching or classifying members of a class. This discussion should elucidate some peculiarities of information processing that occur with schizophrenics. Our aim is to define the parameters determined by the range over which things become generalized in schizophrenics. At the same time we are interested in the general questions of semantic possibilities to extend or to narrow the range over which generalization is effectuated in paradigmatic processing. For this purpose let us look at Table 36.1, which exemplifies an extension and a narrowing of the range of a semantic domain.

Table 36.1 summarizes results of an association test. The stimulus words, hand, short, fruit, whistle, cold, are next to their "baseline" words. The next column gives the results after the infusion of methylphenidate. This column exemplifies typical schizophrenic processing, which in the literature has been termed the production of abstract generalities, global terms, and so on (MS typology, feature #119).

A comparison in terms of a paradigmatic process of feature variability can be made between the four columns of Table 36.1. "Hand" has x number of features; its baseline association term reported here has a few more features than "hand" (if we consider "finger" as part of "hand," there is addition of specific features to the ones contained in "hand"). The word "implement" represents an utmost deletion of features in comparison with the stimulus "hand," and "arm" represents deletion of some features from "hand." The variability we observe here between stimulus, baselines, and drug-induced processing, followed by return to a more normal processing after 24 hr, is that of *addition* or *subtraction* of features within a semantic domain.

In view of our immediate purpose, we need only state that a theory of componential semantics is needed for further systematic discussion of semantic parameters and lexical fields. The work in lexical semantics by Apresian and Melchuk in Soviet Russia and in semantic field analysis (Goodglass & Baker, 1976; Zurif, Caramazza, Myerson & Galvin, 1974) may be singled out here as significant effort in this direction.

CASE ONE

Let us imagine that a schizophrenic subject is presented with a bat-shape on a Rorschach card (complex stimulus). He responds to it: "It's anything" (MS typology, feature # 319). The phenomenon here is "feature deletion." We assume that matching does occur, that all networks are involved, and none of the networks is blocked. Two consequences are to be considered here. First, in the specific task, no satisfactory answer is achieved. The solution to the problem is of such generality that it has no semantic value. Second, this type of information processing is lost for the organism; it is not profitable for the organism because it does not contribute to learning the pattern by any of the networks, whereas at the same time no inhibition of networks occurred.

CASE TWO

A schizophrenic subject is presented the same stimulus. He responds: "This is not a cat" (MS typology, feature # 150). In our hypothetical model, matching was solved by selecting (adding) one feature "non-cat-shape." A specific network was called that contained the match for "cats"; no match occurred with the bat-shape. The first network was eliminated from possible searches and the search continued. In case of such processing it would take indefinitely long to hit the right network. What is common in the two cases is a "software" problem: no generalization was performed before matching.

Our recent experimental work suggests that schizophrenics show an unusual semantic feature manipulation. A peculiar processing becomes evident in naming tasks, generalization tasks, or classification tasks, all of which subsume decisions as to what features to choose from the available stimuli and what name to assign to a collection of features. In summary, when confronted with a task such as required by the Rorschach, in which naming is required in response to a stimulus, the typical schizophrenic response is (*a*) omit maximum feature specificity by choosing a large superordinate, a generality, a large abstraction; or (*b*) go to the other extreme

Table 36.2
Addition or Subtraction of Features within a Semantic Domain

Feature deletion	Feature addition	Complex feature addition (lexical level processing)	Feature addition then feature deletion
Feature 319 [a] "materials"	Highly restrictive feature Feature 332 "North pole"-ness (feature at top)	Conjoining of highly restrictive distant semantic features in an expression Feature 312 "Greenland Irishman" (island-shaped; maplike; green surface; human contour)	Feature 196 "ducks/anything"
Feature 140 "anything"	Negative definition Feature 150 "not a cat" "not a dog, not a sheep" "not a crocodile"	Conjoining of features belonging to parts and wholes or parts and parts in one expression Feature 333 "rabbit rug" Local feature: rabbit Global feature: rug "cat rug" Local feature: cat whiskers Local feature: rug	Feature 194 "clouds/I didn't mean it"

[a] Features refer to MS typology (Singer, 1973).

and choose a highly restrictive *single* feature; and (c) give several individual, nonconjoined alternatives (see Table 36.2).

CASE THREE

Complex feature additions and matching without averaging are performed by a schizophrenic subject when he responds: "This is a Greenland Irishman" (MS typology, feature #333). In this expression are conjoined restrictive distant semantic features: island-shaped, maplike, green surface, and human contour (Singer, 1973, p. 155). Here, the features are matched separately and, so to speak, "averaged" or conjoined not on the level of cognitive generalization, but in the lexical expression. Conjoining of features belonging to parts and wholes may be exemplified by the response "rabbit rug" (MS typology, feature #333). Here part of the stimulus matched with "rabbit" and the whole surface of the stimulus image matched "rug." Superposition of features may be exemplified by two nonconjoined alternative responses to a stimulus: "broken mirror/bat" (MS typology, feature # 183). Blends are observed in neologistics, formations by schizophrenics such as "stereosociopathology," which stands for "solid form of study of friendship and diseases," and "intervisitation," which is a blend of "interview" and "visitation" (Forrest, 1969, 1976). Feature addition followed by feature deletion may be illustrated by the alternative responses to a stimulus: "ducks/anything" (MS typology, #196) or "clouds/I didn't mean it" (MS typology #194).

The model, however tentative, can serve as a stepping stone for further investigation of schizophrenic information processing and the verbal output. The surface exposure of the paradigmatic process sheds some light on feature clustering. If features or bits of information are brought together into clusters in the everyday processing of information, clustering occurs either by external control or by internal control. The crucial dimension is the dimension of novelty (old versus new information) and the process of habituation. The process of habituation is of particular importance because it is the organism's way of altering information. We hypothesize that this process is affected in schizophrenia.

Conclusion

I would like to conclude with a thesis that is central to research in language behavior and deviant states. The thesis is that language has a mediating role in the organization of mental processes. Language is a tool, the function of which is to summarize information, to make generalizations, and to form hypotheses. Every linguistic statement is to a large extent prognostic, and reliable prognosis is the art of the organism. Thus, theoretical considerations of goal-directed behavior are vital for the study of discourse.

Naming and, for that matter, predicting are generalization procedures (Goodglass, 1975; Tsvetkova, 1975). They involve cutting out a slice from the environment, taking a stand on an organism-environment relation, establishing relations of this statement to other statements and of this slice of environment-organism to others. On a small scale we have shown this with feature-addition and feature-deletion phenomena in semantic fields.

A statement is in order to place in perspective our efforts in the studies of goal-directed behavior. Prognosis, the mechanisms of computation of probabilities of outcomes, and computations of relevances to the situation or to the outcome, optimization, and "satisficing" conditions, (to use the term coined by Herbert Simon), are parameters crucial in the functioning of all artificial systems in general and language behavior specifically. This perspective, discussed by various researchers (Pribram, Anokhin, & Bernshtein), centered on "what the organism is set to sense" (Pribram) in respect to an environment that is his own. We have proposed that deviant processing in schizophrenia, as demonstrated in discourse, is focused in this relationship.

ACKNOWLEDGMENTS

Research for this chapter was based on findings and data of Dr. Margaret Singer. This paper is part of a larger project in the study of schizophrenic language with Margaret Singer, University of California, San Francisco Department of Psychiatry and the Wright Institute, Berkeley, California.

I wish to thank Drs. Mike Verzeano, George Meyer, Diane McGuinness, Susan Ervin-Tripp, and Jeffrey Samuels for their thoughtful comments, and Dr. Paul Larudee for his critical editorial notes, and I would like to thank Dr. Francis Whitfield for guidance in theoretical problems of glossematics.

REFERENCES

Aaron, D., & Rieber, W. (Ed.). *Developmental psycholinguistics and communication disorders*. New York: New York Academy of Sciences, 1975.
Anokhin, P. K. Cybernetics and the integrative activity of the brain. XVIII International Congress of Psychology. Symposium 2. Moscow, 1966.
Apresian, Jü. D. *Lexical semantics*. Nauka, Moscow, 1974.
Arieti, S. Interpretation of schizophrenia (2nd ed.). New York: Basic Books, 1974.
Bekhtereva, N. P. *The neurophysiological aspects of human mental activity*. New York: Oxford Univ. Press, 1978.
Bernshtein, N. A. *Essays in the physiology of movement and activity*. Moscow, 1967.

Cameron, N. Experimental analysis of schizophrenic thinking. In J. S. Kasanin (Ed.), *Language and thought in schizophrenia*. Berkeley: Univ. of California Press, 1944.
Caramazza, A., Gordon, J. B., Zurif, E. B., & DeLuca. Right-hemispheric damage and verbal problem solving behavior. *Brain and Language*, 1976, 3, 41–46.
Caramazza, A., & Zurif, E. B. Dissociation of algorithmic and heuristic processes in language comprehension: Evidence from aphasia. *Brain and Language*, 1976, 3, 572–582.
Chapman, L. J., & Chapman, J. P. *Disordered thought in schizophrenia*. Englewood Cliffs, N.J.: Prentice-Hall, 1973.
Clark, H. H. Linguistic processes in deductive reasoning. *Psychological Review*, 1969, 76, 387–404.
Clark, H. H. Word associations and linguistic theory. In J. Lyons (Ed.), *New horizons in linguistics*. New York: Pelican, 1975, pp. 271–286.
Clark, Herbert H., & Clark, Eve V. *Psychology and language: An introduction to psycholinguistics*. New York: Harcourt, Brace and Jovanovich, 1977.
Coseriu, E. Vers une typologie des champs lexicaux, *Cahiers de léxicologie*, 1975, 27 (No. 2), 30–51.
Dobrovich, A. B., & Frumkina, R. M. Disturbances of probabilistic organization in verbal behavior. In V. M. Morozov & I. M. Feigenberg (Eds.), *Shizofreniia i veroiatnostnoe prognozirovanie*. Moscow, 1973, pp. 91–101.
Eco, Umberto. *A theory of semiotics*. Bloomington: Indiana Univ. Press, 1976.
Ervin-Tripp, S. *Language acquisition and communicative choice*. Stanford: Stanford Univ. Press, 1973.
Feigenberg, I. M. (Ed.). *Schizophrenia and probability prognosis*. Moscow, 1973.
Feigenberg, I. M. Probability prognosis and superstructure on activity. XVIII International Congress of Psychology. Symposium on Cybernetic Aspects of Integrative Activity of the Brain. Moscow, 1966.
Forrest, D. V. Poiesis and the language of schizophrenia. *Psychiatry*, 1965, 28 (No. 1), 1–18.
Forrest, D. V. New words and neologisms. *Psychiatry*, 1969, 32 (No. 1), 44–73.
Forrest, D. V. Nonsense and sense in schizophrenic language. *Schizophrenia Bulletin*, 1976, 2 (No. 2), 286–301.
Fromkin, V. The non-anomalous nature of anomalous utterances. *Language*, 1971, 47, 27–52.
Fromkin, V. (Ed.). Speech errors as linguistic evidence. The Hague: Mouton, 1973.
Frumkina, R. M. (Ed.). *Probability prognosis in speech*. Moscow, 1971.
Goldstein, K. Methodological approach to the study of schizophrenic thought disorders. In I. S. Kasanin (Ed.), *Language and thought in schizophrenia*. New York: Norton, 1964. (Originally published in 1944.)
Goodglass, H., & Baker, E. Semantic field, naming, and auditory comprehension in aphasia. *Brain and Language*, 1976, 3, 359–374.
Greenberg, J. Language universals. In Th. A. Sebeok (Ed.), *Current trends in linguistics* (Vol. III): Theoretical foundations. The Hague: Mouton, 1966.
Greimas, A. J. Sémantique structurale. Paris: Larousse, 1966.
Halliday, M. A. K., & Hasan, Ruqaia. *Cohesion in English*. London: Langman Group, 1976.
Hjelmslev, L. *Prolegomena to a theory of language*. Madison: Univ. of Wisconsin Press, 1963.
Hjelmslev, L. Language: An Introduction. Madison: Univ. of Wisconsin Press, 1970.
Hjelmslev, L. *Résumé of a theory of language* (Ed. and trans. with an introd. by F. J. Whitfield). Madison: Univ. of Wisconsin, 1975.
Jakobson, R. Selected writings. The Hague: Mouton, 1971.
Jakobson, R. Two aspects of language and two types of aphasic disturbances. In R. Jakob-

son & M. Halle (Eds.), *Fundamentals of language* (Pt. II). The Hague: Mouton, 1956, pp. 53-82.
Janowsky, D. S., Huey, L., Storms, L., & Judd, L. L. Methylphenidate hydrochloride effects on psychological tests. Acute schizophrenic and nonpsychotic patients. *Archives of General Psychiatry*, 1977, *34*, 189-194.
Laffal, F. *Pathological and normal language*. New York: Aberton, 1965.
Lamb, S. M. Epilegomena to a theory of language. *Romance Philology*, 1966, *XIX* (4), 531-573.
Lamendella, J. T. The limbic system in human communication. In H. Whitaker & H. Whitaker (Eds.), Studies in neurolinguistics (Vol. 3). Perspectives in neurolinguistics and psycholinguistics. New York: Academic Press, 1977.
Leont'ev, A. A. *Psycholinguistic units and generation of the speech expression*. Moscow: Nauka, 1969.
Luria, A. R. *The neuropsychology of memory*. Washington, D.C.: Winston, 1976.
Luria, A. R. *Basic problems of neurolinguistics*. Moscow, 1974.
Luria, A. R. *Basic problems of neuropsychology*. Moscow, 1973.
Luria, A. R. Towards the basic problems of neurolinguistics. *Brain and Language*, 1974, *1*, 1-14.
Luria, A. R., & Hutton, T. J. A modern assessment of the basic forms of aphasia. *Brain and Language*, 1977, *4*, 129-151.
Lyons, J. *Introduction to theoretical linguistics*. New York and London: Cambridge Univ. Press, 1969.
Marshall, J. C., & Newcombe, F. Variability and constraint in acquired dyslexia. In H. Whitaker & H. Whitaker (Eds.), *Studies in neurolinguistics* (Vol. 3), 1977, pp. 257-286.
Melchuk, I. A. *Attempt at a theory of linguistic models meaning text*. Nauka, Moscow, 1974.
Melchuk, I. A., & Zholkovsky, A. K. Toward a functioning "Meaning Text" model of language. *Linguistics, an international review*. The Hague: Mouton, 1970, *57*.
McGhie, A. *Pathology of attention*. Baltimore: Penguin Books, 1969.
Newell, A. Artificial intelligence and the concept of mind. In R. C. Schank & K. M. Colby (Eds.), *Computer models of thought and language*. San Francisco: Freeman, 1973.
Nöth, Winfried. Textkohärenz und Schizophrenie, *Zeitschrift fur Literaturwissenschaft und Linguistik* Heft 23/24 Psycholoinguistik, 1977.
Ostwald, Peter. Language and communication problems with schizophrenic patients. A review commentary and synthesis. Paper presented at the symposium on the phenomenology and treatment of schizophrenia, Baylor College of Medicine, Houston, Texas (Dec. 9 and 19, 1976).
Palmer, S. E. Hierarchical structure in perceptual representation. *Cognitive Psychology*, 1977, *9*, 441-474.
Pavlov, I. P. *Pavlovian Clinical Wednesdays*, Moscow: Akedemiia Nauk SSR (Vol. 3), 1957.
Polyakov, Y. E. The use of pathophsychological data in the structure investigations of cognitive processes. Pathological psychology and psychological processes. *Symposium*, 1966, *26*, 87-94. Moscow: Nauka.
Polyakov, Y. F. Pathology of cognitive processes. A. V. Snezhnevskii (Ed.), *Shizofreniia*. Moscow, 1972.
Pribram, K. H. *Languages of the brain: Experimental paradoxes and principles in neuropsychology*. Englewood Cliffs, N.J.: Prentice-Hall, 1971.
Pribram, K. H. Modes of central processing in human learning and remembering. *Brain and learning*, 1976.
Pribram, K. H. Language in a sociobiological frame. *Annals of the New York Academy of Sciences*, 1976, *280*, 798-809.

Pribram, K. H. The linguistic act. *Psychiatry and the humanities*, 1976, *1*.
Pribram, K. H., & Isaacson, R. L. Summary chapter in *The Hippocampus*, 1976, *2*, 429–441.
Rochester, S. R., Martin, J. R., & Thurston, S. Thought-process disorder in schizophrenia: The listener's task. *Brain and Language*. 1977, *4*, 95–114.
Rosch, E. Human categorization. In N. Warren (Ed.), *Advances in cross-cultural psychology* (Vol. 1). New York: Academic Press, 1977.
Saussure, Ferdinand de. *Course in general linguistics*. New York: McGraw-Hill, 1966.
Shakow, K. Some observations on the psychology (and some fewer, on the biology) of schizophrenia. *The Journal of Nervous and Mental Disease*, 1971, *153*, 300–316.
Simon, H. A. The science of the artificial. Cambridge, Mass.: MIT Press, 1969.
Singer, M. T. The Rorschach as a transaction. In M. W. Rickers-Ovsiankina (Ed.), *Rorschach psychology*. New York: Wiley, 1976.
Singer, M. T., & Wynne, L. C. Thought disorder and family relations of schizophrenics. *Archives of General Psychiatry*, 1965, *12*, 187–200.
Singer, M. T., & Wynne, L. C. Principles for scoring communication defects and deviances in parents of schizophrenics: Rorschach and TAT manuals. *Psychiatry*, 1966, *29* (No. 3), 260–288.
Singer, M. T. Scoring manual for communication deviances. Unpublished manuscript, 1973.
Sokolov, A. N. *Inner speech and thought*. New York: Plenum Press, 1972.
Sokolov, E. N. Neuronal models and the orienting reflex. In M. A. B. Brazier (Ed.), *The central nervous system and behavior*. New York: Josiah Macy, Jr., Foundation, 1960, pp. 187–276.
Sokolov, E. N. Orienting reflex as information regulator. *Psychological Studies in the Soviet Union*. Moscow, 1965.
Sokolov, E. N. Psychophysiology of color discrimination. *Voprosy Psikhologii*, 1977, *5*, 98–104.
Spinelli, D. N. Occam: A computer model for a content addressable memory in the central nervous system. In K. H. Pribram & D. E. Broadbent (Eds.), *Biology of memory*. New York: Academic Press, 1970, pp. 293–306.
Thatcher, R. W., & John, E. R. *Functional neuroscience* (Vol. 1): *Foundations of cognitive processes*. New York: Wiley, 1977.
Tsvetkova, L. A. The naming process and its impairment. In E. H. Lenneberg & E. Lenneberg (Eds.), *Foundations of language development: A multi-disciplinary approach* (Vol. 2). New York: Academic Press, 1975, 32–48.
Tulving, E. Episodic and semantic memory. In E. Tulving & W. Donaldson (Eds.), *Organization of memory*. New York: Academic Press, 1972, pp. 382–403.
Uexküll, Jakob von. *Umwelt und Innenwelt der Tiere*. Berlin, 1909.
Uexküll, S. *Bedeutungslehre*. Leipzig, 1940.
Whitaker, H. A case of the isolation of the language function. In H. Whitaker (Ed.), *Studies in neurolinguistics* (Vol. 2), 1976, pp. 1–58.
Wynne, L. C., & Singer, M. T. Thought disorder and family relations of schizophrenics. II. A classification of forms of thinking. *Archives of General Psychiatry*, 1963, *9*, 199–206.
Wynne, L. C., Singer, M. T., Bartko, J., & Toohey, M. L. Schizophrenics and their families: Research of parental communication. In J. M. Tanner (Ed.), *Developments in psychiatric research*. Sevenoaks, Kent: Hodder and Stoughton, 1977.
Zaitseva, Z. M., Liberman, A. E., & Minkin, L. M. Informational structure of the sentence and psychophysiological analysis of normal and pathological speech. In V. M. Morozov & I. M. Feigenberg (Eds.), *Shizofreniia i veroiatnostnoe prognozirovanie*. Moscow, 1973, pp. 102–108.

Zavarin, V. Frame analysis of schizophrenic discourse. Proceedings of the Third Annual Meeting of the Berkeley Linguistics Society (Vol. 3). Berkeley: BLS, 1977, 545–558.

Zavarin, V. *Theory of the formulaic text.* Urbino, Italy: International Center of Semiotics and Linguistics, 1979.

Zavarin, V. *Lexical cohesion in schizophrenic discourse.* (In preparation)

Zeigarnik, B. V. *The pathology of thinking.* New York: Consultants Bureau, 1965.

Zhinkin, N. I. *Mechanisms of speech.* The Hague: Mouton, 1968.

Zurif, E. G., Caramazza, A., Myerson, R., & Galvin, J. Semantic feature representations for normal and aphasic language. *Brain and Language,* 1974, *1,* 167–187.

T. N. USHAKOVA

Neurophysiological Mechanisms of Processing Speech Information in Man

37

Almost all of the participants in our seminar on goal-directed behavior have been examining data obtained from animal experimentation. Given the present state of the art, this is obviously quite natural. But what could be more goal-directed than human behavior? I should like to focus my own remarks on this topic which, along with its difficulties, has its attractive aspects as well.

The speech process is probably the most typical form of human goal-directed behavior. Whether a person is talking or listening, the goal before the person is to express a thought or to understand it. Speech activity is extremely characteristic of the human being and distinguishes the individual from all of the animal world known to us. That activity is highly essential to normal life, as is evident in the case of pathological speech disturbances that are the cause of so much suffering. The importance of studying speech, therefore, would seem to be sufficiently clear

From the physiological viewpoint, the speech process constitutes a highly complex and truly amazing activity of the brain. I am inclined to believe that this activity exceeds by far the complexity of the processes that are occurring in the solar system. I cannot promise that I can resolve these complexities, but as a followup to Thompson's suggestions (Chapter 6, this volume), I should like to present a number of ideas—hopefully without getting lost in factual data.

First of all, I shall define basic concepts. The mechanism of processing speech information is currently understood to comprise three basic links: the perception of verbal stimuli, the central processing of speech signals, and motor realization. When a person is listening, the first and second links are primarily in operation (the person hears and understands speech). The second and third links operate when the individual speaks (proceeding from thought and expressing it through movements). From our viewpoint, the central link in the processing of speech information is the most important one for understanding that the essence of speech activity is closely tied to thought and goal. That link has come to be called internal speech in psychology and psycholinguistics. These latent internal speech processes constitute the subject of our study.

There are presently several approaches to the physiological study of latent speech processes:

1. The recording of individual neuron activity, including that of deep brain structures.
2. Study of EEG recordings made from the cranial surface.
3. Analysis of speech disturbances in focal pathological brain changes.
4. Study of speech as processes of higher nervous activity.

In accordance with the suggestions made here by Professor Lomov, these approaches could be viewed as studies of speech mechanisms at various hierarchical levels—that of the neural cell, the macroprocesses occurring in the large brain regions, and the level of higher nervous activity, including the psychological aspect of phenomena.

The question arises as to the correlation of these levels and how important is it to account for each of them in such studies. A remarkable feature of the studies presented at our seminar is the fact that many investigators have noted the importance of closely correlating electrophysiological and behavioral manifestations. This is an essential concept that, it seems to me, gives a sense and goal to the interpretation of electrophysiological data. That concept is also important for our subject, although it would not seem possible to find behavioral manifestations in a study of internal speech. Therefore, only the general nature of the speech process such as the performance type of speech tests, mental count, and so on, is usually indicated in electrophysiological studies. This kind of correlation is too broad to go beyond establishing statistical correlations.

I propose that we proceed in this regard by analyzing internal speech as processes of higher nervous activity. It seems to me that this would enable us to develop studies of neurophysiological mechanisms of speech behavior.

I. P. Pavlov has suggested that the study of speech phenomena should be approached as studies of higher nervous activity processes. It occurred to him that human speech constituted a special "signal of signals" stimulus. According to I. P. Pavlov, the aggregate of conditional reflex mechanisms, combined with speech, comprises a second human signal system. The laws that govern the first signal system, common to both human and animals, are also applicable to the second signal system. The latter, however, has its own fundamental peculiarities.

Following I. P. Pavlov's original ideas, a number of attempts were made to study the second signal system. Only those works that are directly related to our topic will be reviewed here.

M. M. Kel'tosova, in a study of generalization as a human brain function, demonstrated that special functional structures are formed in a child's brain as the child masters a language and assimilates meaningful words. Each of these functional structures represents a system of time links between individual analyzers that perceive the sound of a word, the appearance of a defined object, its odor, and so on (Figure 37.1). The sketch in Figure 37.1 indicates the type of time links that can be formed during the assimilation of meaningful words. The described functional structures may be considered base units of the second human signal system.

The existing data of the other type indicate that a person in command of a language is affected by words not as isolated signals but as selectively linked stimuli. This linkage of a physiological nature is detected by words close in meaning (synonyms, antonyms, generically related words, monosituational words, and others) as well as similar-sounding words (homophones). This type of data has been obtained in the Soviet Union (in the laboratory of N. I. Krasnogorskii, by Fedorov, and later by A. A. Shvarts, O. S. Vinogradova, and others), as well as in the United States (G. Razian, C. Cofer, J. Foley, J. Lacey, and others). These investigators have usually interpreted their data within the framework of a semantic generalization of semantic fields and verbal networks. From the viewpoint of a higher

Figure 37.1. Diagram of functional structures representing systems of time links between individual analyzers that perceive aspects of words. The structure with receptors indicates the type of links that can be formed during the assimilation of meaningful words.

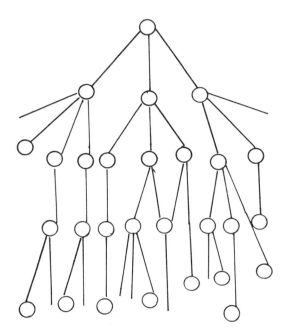

Figure 37.2. *A verbal network of interverbal time links.*

nervous activity process, such data are more naturally interpreted as the result of neural links generated between base functional units. With this interpretation in mind, one might represent multiple "interverbal" time links as a "verbal network" (Figure 37.2). The "verbal network" is the basis for the synthesis of verbal stimuli as a system of special time links. What is the nature of this synthesis and what is its place in speech mechanisms?

Understandably, a constant verbal link is secured with the aid of the "verbal network" mechanism; that is, this is the memory mechanism and apparently one of the language mechanisms upon which speech is based, but it is not the mechanism of speech.

The functioning of the "verbal network" (as is the case of any network) apparently consists of a diffuse dispersion of the nonspeech process. The speech mechanism in which word selection is of necessity made from a person's available vocabulary includes as an important stage a determined concentration of stimulation in the individual elements of the "verbal network."

We shall assume that the most common mechanism of internal speech comprises dynamic restructuring of the time link systems in the verbal network. Such restructuring must occur under the influence of processes that, so to speak, "pile up" on the verbal network.

This assumption has been tested experimentally. The results of two experimental series will be presented here. One experiment recorded changes that took place in the structure of verbal networks when previously nonexistent verbal associations were formulated in the test subject. The other experiment was a study of internal speech characteristics during the formation of a spoken sentence. A description will also be given of a laboratory method for studying latent intraspeech processes.

In the first series, the subject was required to memorize a word link assigned to him and to be able to give the response words when presented with the beginning association word. The number and complexity of associations were selected empirically in order to obtain a time span for their formation. An outline of the experiment's procedure is shown in Figure 37.3.

As can be seen, each association consisted of a cluster. The subject memorized three responses to a single initial word. Six such clusters were used in the experiment. We were interested in the processes occurring after the presentation of the initial word in the response structures of association, as well as in the structures that were linked to them through the "verbal network" mechanism. We therefore investigated the dynamics of functional states in the structures of the response association word (O), its monogeneric word (B), its correlate, generic (P), homophone (☐), and remotely related word (△) in several time periods following the presentation of the initial association word. The experimental procedure will be described later, inasmuch as the method in both series of experiments was the same.

In the second series, a study was made of internal speech during the human performance of more complex activity. In this experiment, the subject composed sentences made from isolated, unrelated words that were presented to him on tape. The sentences had an identical linguistic structure and contained five members with nonprepositional objects. The words from which the sentences were to be composed were presented in random order. The sentences in the experiment were not repeated. A diagram of the sentences formed by the subjects is presented in Figure 37.4.

In this case, we were interested in the dynamic changes that occurred in the functional structures that corresponded to the subject, predicate,

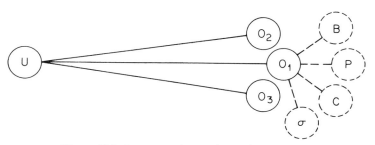

Figure 37.3. Sequence of experimental procedures.

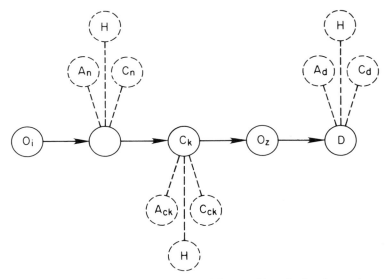

Figure 37.4. Structure of sentences formed by a subject during internal speech.

and object. In addition, in pursuance of the problem at hand, we ascertained the reactions of structures that corresponded to the synonym, antonym, and remotely related word with respect to the subject, predicate, and object.

We used the so-called test stimulus method proposed by E. I. Boiko for studying the internal structure of complex human reactions. The method is essentially a version of the paired associates method, broadly recognized in physiology and neurophysiology. In the experiment, the subject completes two types of activities: (a) the basic activity reproduces the mental activity being examined; and (b) the test indicator activity that consists of a stereotype reaction (as a rule, the manual pushing of a reaction button) upon the presentation of supplemental signals. The basic reaction may be any mental activity selected by the experimenter that can be adequately controlled under experimental conditions. The test stimuli are isolated stimuli that are included in the aggregate of signals at which the subject completes the basic mental activity. The time of the test reactions, measured in thousandths of a second, serves as an objective index of the local functional changes that occur in the mental act being examined.

The test stimulus is applied very closely to the time of the act under examination, and is "addressed" to functionally different elements of the functional structure. Any residual stimulation, inhibition, or general change in the level of the functional state of the neuron groups operating within the basic reactions influences the test reaction that quickly follows behind the functional structure. Therefore, the objective parameters of the test reaction can be used to characterize the state of the local functional structures. The test stimulus plays the role of a "probe" that tests the neural processes that are clearly taking place.

The E. I. Boiko test method was used in its "verbal version" in our experiment, in connection with the development of special methods. Precisely timed verbal stimuli were used as the test stimuli. The selection reactions usually served as the test reactions; that is, the subject reacted with either the right or left hand, depending on the nature of the test stimulus.

The results obtained in the first series of experiments can be interpreted conveniently by referring to the formation stages of the associations being studied. The first stage is when the subject is not yet able to respond to the word link. Nonconsolidated reactions were observed in the second stage. Here, the responses were either correct, wrong, or lacking altogether. The stage of automated reactions was considered to be the third stage.

The data for the first stage of association formation were obtained in two test intervals—a short interval (.5 sec) and a longer one (1 sec)—as shown in Figures 37.5 and 37.6.

The graphs show the testing results of five different functional elements; O—the structure of the response association word; B—the structure of its monogenerically related word; B/sic/—the structure of the related correlate; ⌈—the structure of its homophone; △—the structure of its dis-

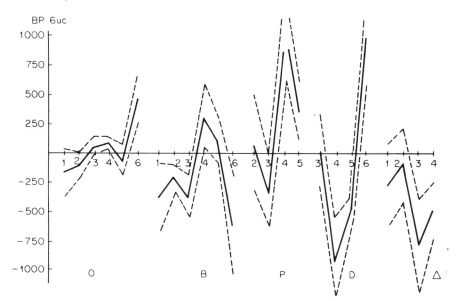

Figure 37.5. Test results for five different functional elements at a .5 sec test interval. Ordinate is the normalized reaction time in milliseconds.

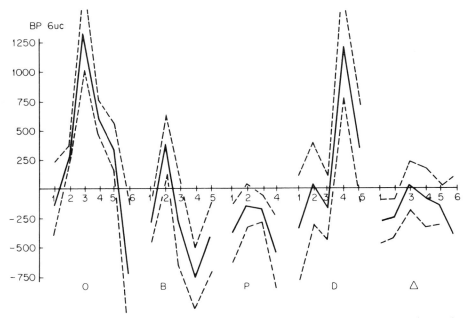

Figure 37.6. *Test results for five different functional elements at a 1.0 sec interval.*

tantly related word. The order in which the association is presented is noted on the axis of the abscissae in each chart. The values of test reactions obtained in the experiments in relation to phonic data are indicated on the ordinate axis.

Practically all of the obtained data were statistically indistinguishable in the short testing interval (.5 sec). However, a statistically reliable "cleavage" between the examined structures was observed in the 1 sec interval. There was a sharp increase in the values of the test indicator in the structures of the response association component (O) and in the structures of its homophone (⌐).

A clear differentiation of the "verbal network" structures was observed in all three intervals in the second stage of association formation (Figure 37.7). Many indicators are reliably distinguishable. The greatest activity was observed in the structures of the response association component and its monogenerically related word; that is, here the word-meaning link begins to play its role. However, the structure of the generically related word and the correlate still exhibit low indices. Warranting attention is the fact that a differentiated activity of the structures can already be seen in the shortest test interval—.5 sec. By the same token, an acceleration in the association process is identified. The dynamics which

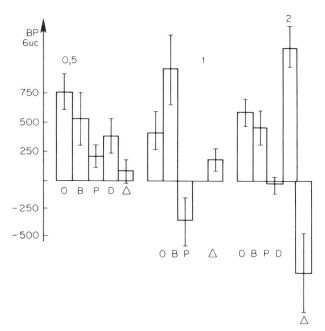

Figure 37.7. Summary reaction time data for three intervals in the second stage of association formation.

appeared in the 1 sec interval in the first stage was observed in the postsecond interval in the second stage.

The data for the third stage of association formation (Figure 37.8) indicate that the differentiated states of the structures are maintained. Although these differences are smaller in absolute values, in most cases they are statistically reliable. There is a rise in activity in the structures of the related word with respect to the response association component. It is important to note that the largest differences in the testing indices are observed in the shortest interval (.5 sec). They are, as it were, extinguished in two other time periods (1 and 2 sec).

In our view, the data presented permit us to conclude that the process of association formation takes place in the human nervous system in a background of previously developed secondary signal links ("verbal networks"). This process exhibits features that are characteristic of the development of any conditional reflex; that is, an initial generalization and relative slowness of the reflex's differentiation and velocity are subsequently increased. At the same time, the process of association formation in the human has its specificity, inasmuch as it occurs on a special functional level—a level of previously developed systems of time links. This

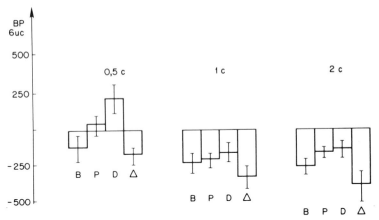

Figure 37.8. The third stage of association formation (see also Figure 37.7).

confirms I. P. Pavlov's ideas about the general and specific aspects of human higher nervous activity.

Our primary interest in the experimental results of the second series was to compare the test indicators obtained by the neutral words and those obtained by words related in meaning to the words comprising a sentence (Figure 37.4—this is a comparison of test indicators of structures A and C with respect to the structure H). This comparison clarifies whether or not the changes that were caused by previously developed time links take place in the functional state of the "verbal network" structures as a sentence is being synthesized.

We were further interested in the nature of the correlation between the experimental indices pertaining to the subject group ( ), the predicate group (CK), and the object group ($\triangle$). A clarification of the indicator characteristics of these three groups would make it possible to characterize the neurodynamic structure that evolves as words are synthesized into a sentence.

For our first comparison, we obtained the data presented in Figure 37.9. The left column indicates the averaged values recorded during the testing of words with related meanings to the words forming the sentence (synonyms and antonyms). The right-hand column reflects the averaged values in all subjects during the testing of words that are "remotely" related. The data for both types are reliably distinguishable at a high significance level ($p < .001$). In our opinion, this indicates that the functional structures of the "verbal network" that were conditioned by the previously developed time links are involved in the process of sentence formation.

Our second comparison was to compare the test data by groups pertaining to the subject, predicate, and object. We found a characteristic

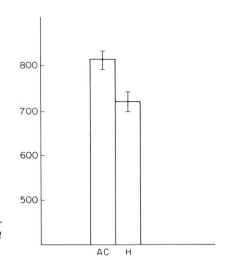

Figure 37.9. Correlations averaged for words with related meanings (left column) and remote meanings (right column).

type of correlation between the recorded indicators. The lowest indicators were observed during the testing of remotely related words, and the highest were observed in the testing of synonyms, where the average indicators were obtained during the testing of antonyms. The values for our data were statistically distinguishable.

We observed that the general case of index correlation in the experiments also had rather correct and constant dynamics (Figure 37.1). Thus, a "correct" index correlation, close to the characteristic type, was observed in the group of indices related to the testing of the predicate structure in the short testing interval (500 msec). The compared indices were distinguishable at a level that is close to statistically reliable. Along with this, the differences between the indices in the next testing interval lost their correctness, and became statistically indistinguishable. This gave us grounds for concluding that the process being observed becomes extinct, as it were.

An examination of the indices related to the testing of the subject structure gives us a different dynamics picture. Here, the "correct" character of index correlation is observed in both testing intervals. However, it is not well defined statistically in the 500 msec interval, but is more closely presented somewhat later in the 1.500 msec interval.

We get a similar picture in testing the structure corresponding to the object. The "correctness" of the index correlation is violated in the short interval and the differences are not well defined statistically. The index correlations become correct and are statistically distinguishable in a later interval of testing.

These data serve as the basis for the hypothesis that the neural processes that take place in the structures corresponding to the subject and

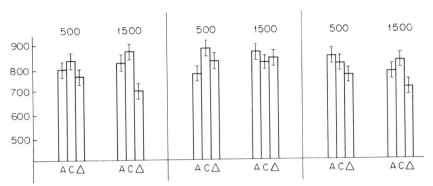

Figure 37.10. Correlations averaged for the words of the subject group (left column) predicate group (middle column) and the object group (right column).

object proceed more slowly and last longer. As a whole, the materials presented in Figure 37.10 disclose variable dynamics in the functional structures that correspond to the different members making up a sentence.

In summing up the results of the materials presented, one might draw the following conclusions:

1. The data obtained disclose the possibility of penetrating the latent intraspeech processes and objectivizing them. The testing method used his produced data about intraspeech processes that are inaccessible to self-observation and that cannot be identified by objective means. It is hoped that the further development of this method will expand its potential in this area.

2. The materials presented at the higher nervous activity level make it possible to characterize the speech processes, both qualitatively and quantitatively. Current activity is characterized by three basic parameters: (*a*) the functional elements that are involved in the process; (*b*) the extent of their involvement; and (*c*) the time occurrence of changes in the functional state of the structures under examination. These types of data might be profitably utilized in comparisons with electrophysiological indices. We are reminded that electrophysiological experiments can produce data on the time characteristics of the process, its quantitative characteristics, and can indicate the area of the brain in which the activity takes place. However, contemporary methods based on the EEG do not allow us to identify the connection between electrical processes and the maintenance of current activity. A consolidation of electrophysiological methods with the ones presented in the chapter would make it possible to fill this gap in the study of brain activity and allow us to approach the problem of dynamic localization of cerebral psychophysiological functions in a more specific fashion.

3. From the theoretical viewpoint, the area of research outlined here would enable us to identify several common conceptual features of higher nervous activity processes that occur in the realization of speech!

First of all, one should postulate that there are functional formations based on systems of time links between the individual analyzers that correspond to the assimilated words of a language. These are the so-called base elements of the second signal system that were investigated by M. M. Kel'tsova. The base elements, in their turn, are joined by multiple-time links that form stable, complexly organized systems ("semantic fields"). The specially formed speech processes apparently take place as dynamic restructuring in the systems of the "semantic field" time links. New syntheses and new differentiations of the base elements of the second signal system are formed in complex interactions. The series presented here merely reflect some forms of these syntheses and the characteristics of those forms that have been outlined are far from being complete. We presume that the basic characteristics discovered in the Pavlovian School in the course of research on animal higher nervous activity will be manifested on the functional level now being investigated. The characteristics of the tonic states in the systems of secondary signal time links seem to be of special interest, inasmuch as this subject is directly applicable to the state of human consciousness. An equally important problem is that of modifications of tonic states, and particularly the phenomenon of switching, as studied in animals by E. A. Asratyan. This is important because the verbal stimuli act as switches that activate their zone in the secondary signal systems of time links.

In accord with the ideas of I. P. Pavlov, higher nervous activity characteristics that are manifested in the aforementioned cases are common to humans and animals. The secondary signal activity, however, doubtless has its specificity. Our data indicated that specificity in the fact that all processes, including those of the verbal stimuli, occur on a special functional level—the level of previously developed systems of time links. We expect that other specific characteristics of the second signal system will be discovered in the course of future research in this area.

ROBERT W. THATCHER

Spatial Synchronization of Brain Electrical Activity Related to Cognitive Information Processing

38

Wilhelm Wundt (1910) was among the first physiological psychologists to formally and systematically study consciousness, subjective experience and mental processes. He strongly advocated that the primary goal of physiological psychology should be to elucidate and analyze the mechanisms underlying mental phenomena in general. It was in this spirit and inspiration that the present studies were undertaken.

Over 50 adults and children have been investigated in a series of averaged evoked potential (AEP) studies requiring relatively high levels of cognitive information processing. These studies include delayed letter matching (Thatcher, 1977a), delayed semantic matching using synonyms and antonyms (Thatcher, 1976, 1977b), delayed matching of letter names (Maisel & Thatcher, 1977), Spanish to English linguistic translations (Thatcher, 1976), delayed form matching and certain mathematical and logical tasks (Thatcher, 1976; Thatcher & Maisel, 1979). In all of these studies AEPs were analyzed from 11 scalp derivations and one channel was used to monitor eye movements. The experimental paradigm was basically the same for all tasks and involved the computer presentation of random dot stimuli (controls), followed by an information display (standard), followed by a second series of random dot displays (intertest-interval stimuli, ITIs), followed by a second information display (test) that matched or mismatched the previous information display. All displays were foveal, 20 msec in duration, and equated for luminance and size. A task varied from 10 to 30 min, contained 24 to 48 trials per session, and the subjects were required to make a delayed differential response during a 4 sec intertrial interval (thus eliminating contamination due to movement).

All of the various tasks shared the general cognitive challenge of delayed matching from sample, involving a general demand on attention, the maintenance of the memory of the standard display, and a subsequent comparison (sometimes at a concrete level and sometimes at more abstract levels). The central aim of these studies was to provide a series of tasks that were short in duration and thus not overly fatiguing and that challenge different aspects of cognitive operations (linguistic, nonlinguistic, and mathematical) and involve different levels of complexity within any given category of cognitive function.

Spatial Synchronization

Neural synchronization refers to the simultaneous or invariant phase relationship between neural elements. Neural spatial synchronization refers to the simultaneous, cooperative, or invariant phase relationship between neural generators in spatially remote sites. Several models of the neural mechanisms of perception, memory, and cognitive function have stressed the concept of spatial synchronization (Barlow & Brazier, 1954; Freeman, 1972, 1975; Gavrilova 1970; Knipst, 1970; Livanov, 1977). A mathematical-experimental model was developed by Freeman (1975, 1977) in which sensory information was represented as invariant phase relations between neural oscillator systems. In the context of this work it can be argued that commonality of an AEP waveform recorded from different regions of the brain reflects a state of maximum information transfer as well as a sharing or commonality of function. A difference in AEP waveform, on the other hand, reflects a difference in function.

In the section to follow the results of varimax factor analyses will be presented. The varimax factor analysis (for details see John, Ruchkin, & Villegas, 1963; Thatcher & John, 1977) is particularly well suited to the extraction of commonality or shared waveshapes recorded from different electrodes or AEPs recorded from the same electrode elicited by different experimental conditions. All of the AEP data that will be presented were amplitude-normalized prior to submission to the varimax factor analysis.[1] In this way, AEPs that load on different factors load exclusively because of differences in *waveshapes* and not because of differences in *amplitude*.

[1] The amplitude normalization process involves first computing the epoch mean voltage $X = 1/N \ \Sigma x_i$, where x_i equals the voltage values in each sample bin. Then a DC level is computed $(X - x_i)$ and normalized $Y^2 = \dfrac{(X - x_i)^2}{(X^2 - x_j{}^2)} \ 1/2$. Unity variance is set for all the AEPs scaling each voltage value so that $Y^2 = 1$.

This analytic distinction between AEP amplitude differences versus AEP waveshape differences is especially important in the functional interpretation of AEPs recorded from homologous hemispheric derivations when they exhibit asymmetries (Thatcher, 1977b; Thatcher & Maisel, 1979). For instance, two AEPs recorded from homologous hemispheric derivations that exhibit the same waveshape but differ only in amplitude do not indicate a qualitative difference in function. That is, since the waveshapes are the same, the AEP neural generators in the two hemispheres exhibit spatial synchrony or the same sequence of excitatory and inhibitory interactions and therefore a commonality of function. In this case, the differences in AEP amplitude are indicative of a quantitative increase in either the number of neural generators or the degree of local synchrony (Thatcher & John, 1977). In contrast, AEP waveform differences from homologous hemispheric locations indicate a qualitative and not simply a quantitative difference in function. That is, if the spatio-temporal sequence of active neural generators (the sequence in which particular subpopulations are turned on and off) differs significantly in one hemisphere compared to the other, then this is indicative of a qualitative difference in function.

The application of the varimax factor analysis involved two types of analyses: (a) an analysis of AEPs elicited from different scalp locations during a particular experimental condition. This analysis is related directly to the issue of spatial synchrony, that is, common waveforms recorded from remote spatial locations (spatial synchrony) are reflected directly in this analysis; and (b) an analysis of AEPs recorded from one scalp location during different experimental conditions. This analysis relates AEP waveform to differences in cognitive function (Thatcher, 1976; Thatcher & Maisel, 1979), but does not directly reflect spatial synchrony.

Shared Waveforms across Remote Scalp Locations

Figure 38.1 shows the results of a varimax factor analysis of 12 AEPs from different scalp locations elicited by random dot control displays in a logic task (Thatcher, 1976; Thatcher & Maisel, 1979). The factor waveshapes are in a column on the left and the AEPs run horizontally at the top of the figure. This figure illustrates an extremely common finding,[2]

[2] The replicability of the factor analyses can be judged by (a) 43 out of 43 subjects analyzed to date show differential loadings on control AEPs versus information AEPs; (b) 38 out of 43 subjects show a unique ITI factor; (c) 7 out of 9 subjects run in the delayed letter-matching task showed higher loadings between match and standard AEPs than between mismatch and standard AEPs; and (d) 39 out of 43 subjects showed an anterior-posterior factor structure with one factor loading heavily on anterior derivations (T_3 through F_z) and another factor loading heavily on posterior derivations (O_1 through $T_{5, 6}$).

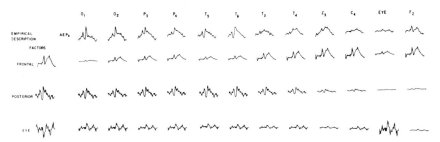

Figure 38.1. Top row are AEPs (N = 24; analysis epoch = 786 msec for this and all other figures) from 12 different derivations elicited by random dot stimuli in the logic task. The three factors (column on left) account for 96% of the variance. Factor loadings are represented by amplitude scaling of the factor waveshapes. The first factor is called a "frontal factor" since it loads most heavily on frontal derivations (T_3, T_4, C_3, C_4, F_z); the second factor is called a "posterior factor" since it loads most heavily on posterior derivations (O_1, O_2, P_3, P_4, T_5, T_6); the third factor is called an "eye factor" since it loads most heavily on the eye lead. (From Thatcher & Maisel, 1978.)

namely, the factor structure of AEPs recorded from the human scalp during noninformation processing is comprised of two basic waveshapes: (a) a posterior waveshape that accounts for most of the variance contributed by bilateral occipital, parietal, and posterior temporal leads; and (b) an anterior or frontal waveshape that accounts for the variance contributed by the bilateral anterior temporal, central, and frontal leads. A third orthogonal factor accounts for the waves generated by the eye lead, thus showing that this anterior-posterior split in waveshape is not an artifact of eye movement. This topographic factor structure shows a considerable amount of spatial synchrony and reveals certain dynamic features when information is presented to a subject. Table 38.1 illustrates such a dynamic change in a subject performing in a delayed letter-matching task. The top half of Table 38.1 is the factor structure of AEPs elicited by the random dot control displays. A clear anterior-posterior split (factors 1 and 2) can be seen. The bottom half is the factor structure elicited by letters of the alphabet in which the anterior-posterior split essentially disappears and is replaced by an occipital-parietal commonality (factor 1), a left hemisphere commonality (P_3, T_5, T_3, F_7; factor 2), and a right hemisphere commonality (T_6, T_4, F_8; factors 3 and 4). In other words, interhemispheric waveshape asymmetries are absent to the noninformation controls, but are present when letters are presented, and in this particular brain state there is considerable spatial synchrony in the left hemisphere that occurs independent of the anterior-posterior plane. As shown elsewhere (Thatcher, 1977b) this factor structure disappears to the succeeding random dot displays (ITIs) and reappears to the matching test stimuli. Because these waves were amplitude-normalized, different loadings on

Table 38.1
Between Derivation Varimax Factor Analysis [a]

Derivations	O_1	O_2	P_3	P_4	T_5	T_6	T_3	T_4	F_7	F_8	Eye	F_z	Factor
Random dot control AEPs													
1	.89	.88	.65	.85	.67	.46	.06	.40	.06	.06	.00	.06	Posterior
2	.06	.04	.15	.07	.11	.05	.63	.44	.91	.83	.34	.67	Anterior
3	.00	.00	.00	.00	.06	.03	.29	.02	.00	.02	.00	.10	
4	.00	.00	.01	.00	.02	.06	.01	.00	.00	.01	.65	.01	Eye
First letter AEPs													
1	.90	.96	.42	.57	.22	.17	.02	.08	.01	.01	.00	.11	Occipital–Parietal
2	.04	.00	.35	.10	.54	.18	.95	.35	.90	.04	.00	.28	Left hemisphere
3	.01	.02	.00	.07	.06	.10	.00	.42	.04	.95	.08	.17	Right hemisphere
4	.00	.00	.15	.16	.07	.52	.00	.06	.00	.00	.14	.32	Right hemisphere
5	.00	.00	.00	.01	.00	.00	.00	.00	.00	.00	.77	.00	Eye

[a] Varimax factor analysis on amplitude-normalized AEPs from 12 derivations (odd numbers correspond to the left hemisphere) for two different experimental conditions in subject D.D. Each row represents the loading of AEPs from different anatomical derivations on a single factor. Each column represents the factor structure for a given derivation. Results show that AEPs from anterior derivations (T_3 through F_z) load on different factors than do AEPs from the posterior derivations (O_1 through T_6). (See Thatcher, 1977a for further details.)

different factors are due to differences in waveshape and therefore reflect differences in function.

Differences in Waveshape during Different Cognitive Tasks

Table 38.2 shows another example of the use of the factor analysis to quantitize AEP waveshape changes related to differences in hemispheric function. The varimax factor analysis was performed on the grand averages from eight subjects participating in a delayed semantic-matching task (Thatcher, 1976, 1977b). The analyses are from AEPs elicited by the random dot controls, first words, ITIs, and the second words of the semantic pairing. In this experiment there were 36 different first words in a session,

Table 38.2
Evoked Potential Factor Loadings for Left and Right Hemisphere Derivations from the Delayed Semantic Matching Task [a]

	P_3: Factors			P_4: Factors		
	1	2	3	1	2	3
Control	.69	.11	.06	.80	.00	.00
Control	.92	.03	.01	.83	.06	.00
First word	.25	.62	.10	.97	.00	.00
ITI-1	.89	.04	.01	.65	.01	.06
ITI-3	.61	.24	.00	.60	.37	.03
Neu.	.40	.31	.25	.91	.03	.00
Ant.	.21	.47	.30	.97	.00	.01
Syn.	.25	.44	.29	.98	.00	.00

	T_5: Factors			T_6: Factors		
	1	2	3	1	2	3
Control	.80	.08	.00	.50	.32	.18
Control	.89	.06	.00	.42	.41	.07
First word	.23	.73	.03	.86	.03	.02
ITI-1	.80	.11	.01	.66	.24	.04
ITI-3	.81	.11	.00	.27	.61	.08
Neu.	.31	.54	.12	.96	.01	.01
Ant.	.21	.69	.10	.95	.01	.03
Syn.	.25	.65	.08	.95	.02	.03

[a] Neu. = neutral second word; Ant. = antonym second word; Syn. = synonym second word. Underlines represent maximum factor loadings for a given condition on a single factor. Factor 1 accounts for the variance due to the control stimuli. Factor 2 accounts for the variance due to the word stimuli in left hemisphere derivations only (P_3 and T_5). No clear factor differentiation between controls and words occurred in the right hemisphere.

38. Brain Electrical Activity Related to Cognitive Information Processing

but only 12 different second words (there was a 3–7 sec delay interval). In other words, the same 12 sec words elicited AEPs in three different semantic contexts (synonym, antonym, and neutral). It can be seen in Table 38.2 that in the left parietal (P_3) and the left posterior temporal (T_5), derivations AEPs elicited by the random dot controls and the AEPs elicited by words loaded on orthogonally different factors (factors 1 and 2), whereas, in P_4 and T_6 or the right side derivations, very little factor differentiation between controls and words is observed. This analysis provides another example of a functional interhemispheric asymmetry in which the "semantic signal-to-noise ratio," represented by differences in the waveshape of evoked responses to randomness versus words, is greater from the left hemisphere than from the right. A similar interhemispheric waveshape asymmetry has been observed in other semantic tasks (such as Spanish to English language translations; see Thatcher, 1976) but has never been observed in nonlinguistic tasks. Thus, these findings are considered as correlates of specialized verbal functions.

A similar within derivation (P_3) analysis is shown in Figure 38.2. This figure shows an example of factor waveshapes derived from one subject performing in the delayed semantic-matching task. The first row of waves are AEPs from the various conditions of the experiment. The factors are the first column of waves on the left. The factor loadings of each of the four factors on the AEPs are represented by scaling the amplitude of the

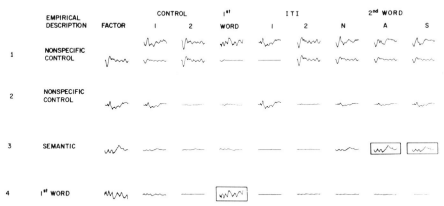

Figure 38.2. *Varimax factor analyses of AEPs from P_3, in a subject performing in the synonym, antonym, and neutral-word experiment. AEPs ($N = 24$), in the top row of waves, are normalized for amplitude. Four orthogonal factors (that account for 87% of the total variance) are in the column of waves on the left. Factor loadings are represented as the factor waveshapes multiplied by the appropriate weighting coefficients. The factors were given empirical descriptions based upon their relative loadings. Note that control and ITI AEPs load on orthogonally different factors than the word AEPs and that synonym (S) and antonym (A) AEPs load on a different factor from the neutral (N) AEP. (From Thatcher, 1977b.)*

factors by their appropriate weighting coefficients (as in Figure 38.1). It can be seen that the control, ITI, and neutral-word AEPs load on factors 1 and 2. Because of their nonspecificity, factors 1 and 2 are called *nonspecific control factors*. The synonym (S) (factor loading = .49) and antonym (A) (factor loading = .66) AEPs load more heavily on factor 3 than the neutral (N) AEP (factor loading = .18). Because of this differential loading, factor 3 is called a *semantic factor*. Factor 4 accounts almost exclusively for the AEP elicited by the first word (factor loading = .89). The results of this varimax factor analysis further demonstrate that overall AEP waveform, independent of amplitude, does differentiate the various conditions of the experiment. Furthermore, this analysis suggests that a more similar functional brain state exists to synonyms and antonyms than to the neutral second words or the first words. Conversely, differences in the overall waveshape of the AEP exist to the different conditions of the experiment even though psychophysical parameters such as stimulus intensity, size, and contour (in the case of the second words) are invariant. Such analysis is at least one prerequisite for concluding that differences in AEP waveshapes reflect differences in mental operations.

Figures 38.3 and 38.4 are three-dimensional representations of the factor structure of the group average of AEPs (AEPs from seven subjects were averaged together) elicited in a logic paradigm. The details of the paradigm are given elsewhere (Thatcher & Maisel, 1979). Briefly, a series of random dot displays were presented followed by a letter (A,B,C,D,), followed by an operator symbol (=, $\neq$, or ⊡ no operation), followed by a letter, thus forming a syllogistic true, false, or no-operation condition. Figure 38.3A shows that in the occipital derivations, factor 1, with a pronounced early component, accounts for the variance due to the control AEPs, and factor 2, with a pronounced late component, accounts for the variance due to the letters and operator symbols. Very little variance is accounted for by factor 3; thus most of the variance due to the various conditions of the experiment is accounted for by only two factors. Since only two factors differentiate the response to random dots versus the response to letters and operators, the occipital regions (O_1 and O_2) are given the functional description of information formatting. In Figure 38.3B a somewhat different picture is seen in the parietal and posterior temporal derivations (P_3, P_4 and T_5 and T_6). In these derivations the factor analysis differentiates between first letters (factor 2) and second letters (factor 3) as well as between the controls (factor 1) and first and second letters. Accordingly, the parietal and posterior temporal derivations are given the functional description of mediating, primarily either secondary or abstract operations since there is a differential loading on orthogonal factors between first letters and second letters and a somewhat different factor structure for operators versus second letters. This type of structure was not observed in the occipital derivations. Note that factor 1, which de-

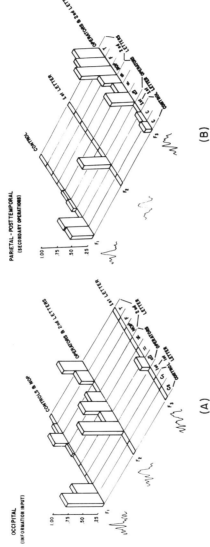

Figure 38.3. Three-dimensional display of factor analysis of grand mean AEPs (7 subjects analysis epoch = 786 msec) from two different topographic regions. The conditions of the experiment are represented on the right, the factor waveshapes are on the left, and the factor loadings are represented by the height of the bars (beginning at .25). Note that factor differentiation between first and second letters occurs only in the parietal-posterior temporal derivations. See text for functional descriptions. (From Thatcher & Maisel, 1978.)

scribes the control space, exhibits an early component, factor 2 only a late component, and factor 3 a combination early–late component structure.

A distinctly different factor structure is seen in anterior derivations (Figure 38.4A). For example, in the central derivations (this same basic factor structure was present in $T_3, T_4, C_3,$ and C_4) there is no differentiation (see factor 1) between controls and first letters and the "equals" and "not equals" operators. On the other hand, factor 2 shows heavy loadings for the no-operation operator and second letters (NOP and false). In this task the no-operation symbol that follows the first letter "closes" or terminates the task in the sense that the subjects know immediately what their response must be. Similarly, the presentation of the second letters "closes" the task and also determines the subject's response. Thus, the no-operator symbol and the second letters have in common task closure or completion. Accordingly, the central derivations (including T_3 and T_4) are given the functional description of mediating task completion or closure.

The frontal derivation (F_z, Figure 38.4B) exhibited a functional structure that was a combination of both the central and the parietal-posterior temporal regions. That is, there was no distinction between first letters and controls. Also, differential loadings occurred to the no-operation symbol and the NOP second letter (and somewhat to the $\neq$). However, some differentiation occurred within the second letters (see factor 2) and between the second letters and the operators.

Summary and Conclusions

Very briefly, the following conclusions can be drawn based on the analyses that were presented:

1. Only the linguistic tasks result in the elicitation of strong and reliable interhemispheric waveshape asymmetries. It is the waveshape of the asymmetry and not the amplitude that reflects a qualitative difference in hemispheric function.
2. The waveshape difference, independent of amplitude, between AEPs elicited by random dot controls versus words or, the "semantic signal-to-noise ratio," is greater from the left hemisphere than the right in right-handed subjects (Table 38.2).
3. Clusters or groupings of AEP waveshapes exhibit a topographic structure that is determined not only by the task, but also by the more subtle aspects of the within task structure (Figures 38.2, 38.3, and 38.4).
4. Shared waveshapes (given the inherent limits of waveshape dif-

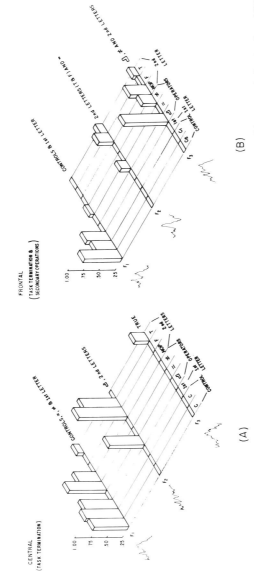

Figure 38.4. Three-dimensional display of factor analysis of grand mean AEPs (7 subjects, analysis epoch = 786 msec) from two different topographic regions. Explanation of the axes is same as in Figure 38.6. Note that, in contrast to the analyses in Figure 38.6 there is no factor differentiation between the random dot controls and first letters in these frontal derivations (i.e., central = T_3, T_4, C_3, C_4 and frontal = F_z). See text for functional descriptions. (From Thatcher & Maisel, 1978.)

ferencing from the scalp) reflect a commonality of function, whereas significant waveshape differences (given amplitude normalization) reflect a difference in function (Tables 38.1 and 38.2; Figures 38.1 to 38.4).

The short latency or early components of the human evoked potential have been experimentally distinguished from the longer latency components (see reviews by Donchin, Kutas, & McCarthy, 1977; John & Schwartz, 1978; Regan, 1972; Thatcher, 1979; Tueting, 1979). For example, stimulus intensity parameters are reflected at approximately 70–110 msec (Armington, 1964; Diamond, 1964; Vaughan, 1966); pattern appearance and disappearance at approximately 100–130 msec (Estevez & Reits, 1977; Harter, 1968; Jeffreys, 1971; Spekreijse); selective attention at approximately 280 msec (Hillyard, Hink, Schwent, & Picton, 1973; Picton & Hillyard, 1974); expectancy, prediction, and information delivery at approximately 280–390 msec (Sutton, Braren, Zubin, & John, 1965; Sutton, Tueting, Zubin, & John, 1967; Tueting, 1978); semantic matching at approximately 400–500 msec (Friedman, Simson, Ritter, & Rapin, 1974; Thatcher, 1976, 1977b); and logical information processing at approximately 600 msec (Thatcher, 1976). Finally, Gomer, Spicuzza, and O'Donnell (1976) and Adams and Collins (1978) showed that the latency of the late positive AEP component varies linearly with serial memory search time in a delayed matching from sample task.

These data, taken as a whole, show that the human AEP reflects different stages of serial information processing, from sensory reception to conscious awareness (Kutas, McCarthy, & Donchin, 1977).

The question remains as to exactly what neurophysiological process corresponds to what mental process, as reflected by changes in AEP components and waveshapes. That the AEP is not purely artifactual or epiphenominal is proven. Since the AEP reflects serial transforms on sensory information, a diagnostic topographic analysis capable of differentiating and grouping people with different cognitive styles and different classes of functional disorders is possible (Thatcher, 1977c). The diagnostic capacities of such functional analyses will likely be of fundamental importance in the development of remediation and therapeutic intervention.

REFERENCES

Adam, N., & Collins, G. I. Late components of the visual evoked potential to search in short-term memory. *Electroencephalography and Clinical Neurophysiology*, 1978, *44*, 147–156.

Armington, J. C. Adaptational changes in the human electroretinogram and occipital response. *Vision Research*, 1964, *4*, 179–192.

Barlow, J. S., & Brazier, M. A. B. A note on a correlator for electroencephalographic work. *Electroencephalography and Clinical Neurophysiology*, 1954, *6*, 321–325.

Diamond, S. P. Input-output relations. *Annual New York Academy of Sciences*, 1964, *112*, 160–171.
Donchin, E., Kutas, M., & McCarthy, G. Electro-cortical indexes of hemispheric utilization. In S. Harnad, R. W. Doty, L. Goldstein, J. Jaynes, & G. Krauthamer (Eds.), *Lateralization in the nervous system*. New York: Academic Press, 1977. Pp. 339–384.
Freeman, W. J. Wave, pulses and a theory of neural masses. *Progress in Theoretical Biology*, 1972, *2*, 86–101.
Freeman, W. J. *Mass action in the nervous system*. New York: Academic Press, 1975.
Friedman, D., Simson, R., Ritter, W., & Rapin, I. The late positive component (P-300) and information processing in sentences. *Electroencephalography and Clinical Neurophysiology*, 1974, *37*, 1–9.
Gavrilova, N. A. Spatial synchronization of cortical potentials in patients with disturbances of association. In V. S. Rusinov (Ed.), *Electrophysiology of the central nervous system*. New York: Plenum Press, 1970.
Gomer, F. E., Spicuzza, R. J., & O'Donnell, R. D. Evoked potential correlates of visual item recognition during memory-scanning tasks. *Physiological Psychology*, 1976, *4*, 61–65.
Harter, M. R. Effects of contour sharpness and check-size on visually evoked cortical potentials. *Vision Research*, 1968, *8*, 701–711.
Hillyard, S. A., Hink, R. F., Schwent, V. L., & Picton, T. W. Electrical signs of selective attention in the human brain. *Science*, 1973, *182*, 171–173.
Jeffreys, D. A. Cortical source locations of pattern-related VEPs (visual evoked potentials) recorded from the human scalp. *Nature*, 1971, *229*, 501–502.
John, E. R., Ruchkin, D. S., & Villegas, J. Signal analysis of evoked potentials recorded from cats during conditioning. *Science*, 1963, *141*, 429–431.
John, E. R., & Schwartz, E. L. The neurophysiology of information processing and cognition. *Annual Review of Psychology*, 1978, *29*, 1–29.
Knipst, I. N. Spatial synchronization of cortical and subcortical potentials in rabbits during formation of conditioned reflexes. In V. S. Rusinov (Ed.), *Electrophysiology of the central nervous system*. New York: Plenum Press, 1970.
Kutas, M., McCarthy, G., & Donchin, E. Augmenting mental chronometry: The P300 as a measure of stimulus evaluation time. *Science*, 1977, *197*, 792–795.
Livanov, M. N. *Spatial organization of cerebral processes*. New York: Wiley, 1977.
Maisel, E. B., & Thatcher, R. W. Evoked potential correlates of delayed matching in adults and children. Abstract #402, *Seventh Annual Neurosciences Convention*, Anaheim, California, 1977.
Picton, T. W., & Hillyard, S. A. Human auditory evoked potentials II: Effects of attention. *Electroencephalography and Clinical Neurophysiology*, 1974, *36*, 191–200.
Regan, D. *Evoked potentials in psychology, sensory physiology and clinical medicine*. London: Chapman Hall, 1972.
Spekreijse, H., Estevez, O., & Reits, D. Visual evoked potentials and the physiological analysis of visual processes in man. In J. E. Desmedt (Ed.), *Visual evoked potentials in man: New developments*. Oxford: Clarendon Press, 1977. Pp. 16–89.
Sutton, S., Tueting, P., Zubin, J., & John, E. R. Information delivery and the sensory evoked potential. *Science*, 1967, *155*, 1436–1439.
Thatcher, R. W. Electrophysiological correlates of animal and human memory. In R. D. Terry & S. Gershon (Eds.), *The neurobiology of aging*. New York: Raven Press, 1976. Pp. 43–102.
Thatcher, R. W. Evoked potential correlates of delayed letter matching. *Behavioral Biology*, 1977a, *19*, 1–23.
Thatcher, R. W. Electrophysiological correlates of hemispheric lateralization during semantic information processing. In S. Harnad, R. W. Doty, L. Goldstein, J. Jaynes, & G. Krauthamer (Eds.), *Lateralization in the nervous system*. New York: Academic Press, 1977b. Pp. 429–448.

Thatcher, R. W. Issues in neurolinguistics: Evoked potential analysis of cognition and language. In D. Otto (Ed.), *Perspectives in event related brain potential research*. Washington, D.C.: GPO, 1979.

Thatcher, R. W., & John, E. R. *Information and mathematical quantification of brain state*. In N. Burch & H. L. Altshuler (Eds.), Behavior and brain electrical activity. New York: Plenum Press, 1974. Pp. 303–324.

Thatcher, R. W., & John, E. R. *Functional neuroscience* (Vol. I). Hillsdale, N.J.: L. Erlbaum Associates, 1977.

Thatcher, R. W., & Maisel, E. B. Functional landscapes of the brain: An electrotopographic perspective. In H. Begleiter (Ed.), *Evoked brain potentials and behavior*. New York: Plenum Press, 1979.

Tueting, P. Event related potentials, cognitive events, and information processing. In D. Otto (Ed.), *Perspectives in event related brain potential research. Washington, D.C.*: GPO, 1979.

Vaughan, H. G. The perceptual and physiologic significance of visual evoked responses from the scalp in man. In *Clinical electroretinography* (Suppl.), *Vision Research*, 1966. Pp. 203–223.

Wundt, W. *Principles of physiological psychology*. New York: Macmillan, 1910.

A. N. LEBEDEV

A Mathematical Model for Human Visual Information, Perception, and Storage

39

A considerable amount of data has been accumulated on the relationship between perception and memory and the characteristics of periodic brain processes (John, 1967; Lindsley, 1960; Livanov, 1972; Pribram, 1975). However, until now, even the approximate quantitative characteristics reflecting that relationship have not been found. We undertook to solve this problem by using Livanov's data (1972) on the spatial organization of periodic brain processes.

We believe that coherent, sustained oscillations of neuron activity with stable phase correlationships are formed under the influence of perceived signals and store information about these signals for a certain time period.

Our hypothesis requires substantiation in view of the well-known phenomena of the apparent instability and random nature of recorded biopotentials of the brain, as well as the discrepancy between the rhythmicity of slow oscillations and the random nature of background neuron impulses of the brain.

Still, one can assume that the stable, sustained oscillations that are capable of storing information are in fact masked by a number of incidental processes. Here, one must seek indirect evidence of instability. We have sought that evidence at the very beginning of our work.

The sustained oscillations of neuron activity are the basis of memory. We saw as the key to this problem the now well-recognized close interconnection between neuron impulses and evoked potentials. We observed that connection together with Kondrat'eva in 1963 in the laboratory of Livanov (1965). To do this, we summarized the reactions of 53 different cortical neurons of a rabbit and compared the resultant poststimulation histogram with an averaged evoked potential on the cranial surface. We saw a "mirror conformity" that indicated that the evoked potential waves are accompanied by a unique type of impulse wave. It was difficult to refrain from the supposition that an analogous picture is also maintained in the case of background activity, although that picture could be even more masked.

What is the essence of this presumed connection? We can see from Figure 39.1 that the neuron firings that coincide with the primary evoked potential are accompanied by a slow negative deflection with a certain lag. The firings are inhibited in the course of negative deflection inasmuch as there is a hyperpolarization of the membrane sectors that generate the impulses. This is known from the works of Creutzfeldt (1967) and others. The slow (at a speed designated by the symbol B in the diagram) increase in negativity ends and is followed by a fall in negativity at a time constant, designated by the symbol C.

Inasmuch as a negative potential deflection on the cranial surface corresponds to the polarization of cortical neurons within the brain, that deflection to a certain degree reflects oscillations in threshold potentials, that is, differences between the current value of the membrane potential and its critical level at which the neurons are fired by impulses. Impulse signals from a mass of independent sources reach the neuron in all cases. This is neuron noise. Random oscillations of the threshold potential that are caused by this neuron noise are subject to a normal distribution. Such are the prerequisites for a mathematical modeling of neuron activity oscillations.

Together with Lutskii, we took stock of those oscillations and composed a differential equation into which we also introduced the free member A, which reflects changes in polarization that are caused by the total effect of impulses on those neurons whose activity, to some extent, depends on the activity of the group of neurons described by the equation. We

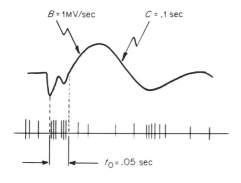

Figure 39.1. Sketch of an averaged evoked potential in response to light flashes in the rabbit cerebral cortex and accompanying unit impulse changes. Polarity: Negative up in the upper trace.

39. A Mathematical Model

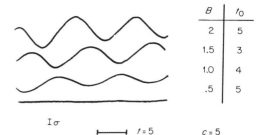

Figure 39.2. Versions of modeled background oscillations of brain biopotential oscillations.

obtained a first-order differential-difference equation with a retarded argument that does not have a precise analytical solution:

$$\frac{du(t)}{dt} = A + B \exp[-u^2(t - t_0)] - \frac{u(t)}{C}$$

where d is the sign of the differential, t is the current time, $u(t)$ is the current value of the average threshold potential, and $u(t - t_0)$ is the value of the threshold potential in the preceding moment. A, B, C are the parameters in the relative units set by the magnitude of the standard deviation of fluctuations caused by "neuron" noise. The physiological explanations of these parameters are previously given. We shall add that the expression after the parameter B reflects the normal distribution of the amplitudes of the oscillations caused by "neuron noise."

We solved the equation on a digital computer by combining the equation's parameters in various ways. The solution models (with parameters in relative units) are given in Figure 39.2 and Table 39.1. We found that the potential oscillations become sustained and stable at a specific ratio of parameters. Stability particularly occurs when parameter B is increased. The value of that parameter in turn depends on the number of synchronously firing neurons. The more intensely they fire, the higher is

Table 39.1
Relationship between the Duration and Amplitude of Background Oscillations of Threshold Neuron Potentials and the Parameters of the Equation $(du(t)/dt) = A + B \exp[-u^2(t - t_0)] - (u(t)/C)$, Where $u(t)$ is the Threshold Potential in Relative Units

Parameters			Period	Duration		Amplitude	
				Rise	Fall	Min.	Max.
3	7	3	12	5	7	.52	1.32
3	7	7	20	10	10	.89	.91
3	7	12	32	16	16	.90	.90
4	20	11	36	16	20	1.10	1.69
9	20	11	37	16	21	1.12	2.62
18	20	11	39	16	23	1.17	3.12

the succeeding delayed polarization of neurons that coincides with the surface negative potential. Thus, the solution to the differential equation shows that stable, sustained oscillations of neuron activity are possible. The consolidation of a certain number of neurons possessing similar parameters of succeeding oscillations in activity is also possible here. Consolidation consists of an exchange of impulses. All the neurons of such a union influence each other to a certain degree by their firings. We modeled such a connection by introducing the free member A of the equation as a function of the activity of individual groups of neurons comprising the union:

$$A = f(t) = \sum_{i=i}^{i=N} \frac{du_i(t)}{dt} K_i$$

where K_i are coefficients, N is the number of neuron groups in the union, and i is the order number of the group. We then used a digital computer to solve a system of such equations by varying the values and signs of the coefficients K_i. We found that the possibility of oscillation stability is preserved in this case too. The configuration of the artificial background oscillations was largely identical to the configuration of the natural waves. The asymmetry of the waves, that is, the difference in the duration of their ascending and descending phases, varied from values that were close to zero to quite large values, depending on the particular combination of parameters. This situation is similar to natural changes in asymmetry. The artificial evoked potentials that followed the modeled sudden depolarization of the current level of the threshold potential also resembled the natural ones. That similarity can be judged by Figure 39.3. Thus, the hypothesis that there is a relationship between a succeeding delayed in-

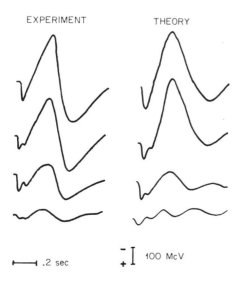

Figure 39.3. Experimental (on the left) and modeled (on the right) evoked potentials in response to light flashes in the rabbit cerebral cortex.

hibition of neuron activity and the preceding firings both in evoked and background oscillations leads us to conclude that stable oscillations of neuron firing are possible. The fact that such stable oscillations do not appear in the autocorrelograms or evoked potentials (which, as is known, are extinguished rather quickly) is explained by the diversity of independent stable oscillations and the many noncontrollable side effects.

The evoked potentials probably reflect a transition process in which new systems that are capable of storing information about perceived signals are formed from preceding systems of stable oscillations. Of course, we mean here only the partial restructuring of the preceding systems. Each system of stable oscillations, generated by a single neuron union (that we shall call a neuron modulus) is distinctive in that the phase of such oscillations that have been changed by a perceived signal does not subsequently change by itself or under the influence of neuron noise; that is, that phase, in conjunction with the phases of other oscillations, stores information about the perceived signal for a certain period of time.

In approximate computations that accounted for the distribution of the widths of background interpulse intervals, Lutskii and I (1972) found that the minimal number of neurons required to maintain stable oscillations was approximately 100–300. These neurons could be located in brain locations quite remote from each other. The initial consolidation of such neurons into a modulus that stores information results from the perceived signal that induces the simultaneous firing of those neurons. The subsequent firing of the modulus neurons, on the average, synchronously sustains the phase's invariability. The random noncorrelated effects of the modulus neurons cannot disrupt it if the number of neurons making up the modulus is greater than the indicated critical value.

Thus, our computations show that a group of neurons consolidated by the mutual coordinated exchange of impulses is capable of stable, sustained, activation oscillations. Moreover, individual neuron groups can fire in random order within certain limits. The periodic consistency of their activity is not disrupted and the oscillation phase is not shifted at random if the number of neurons in that group, or neuron modulus, as we call it, is higher than the critical number—approximately 100–300. We assume that the stable, sustained oscillations of neuron modulus activity is a very simple process in which impressed information is stored. Let us now examine the possible mechanism of information impression and storage.

The wave packets are the storers of information. Within the range of the same stable frequency of oscillations, information may be recorded by a change in their phase in relation to other oscillations of the same frequency. This means that a minimum of two neuron modules are required for information storage. If a phase changes a limited number of times within one period of oscillations, the maximum amount of information that can be stored in the two modules is approximately equal to the logarithm of the maximum number of different phases.

In order to estimate this number, let us consider the following. Livanov (1934) found that, in the first place, evoked potentials are essentially the total reflection of unevenly altered phases of initial background oscillations. In the second place, it has been shown that when the frequency of flashing light stimuli is increased evenly, the changes in the frequency of attached oscillations are uneven. In the experiments of Livanov (1940) the magnitude of difference was approximately 10%. The investigator attributed the wavelike changes in the amplitude of the alpha-rhythm, the so-called spindles, to the mixing of two and three stepwise differing frequencies.

Our work with V. A. Lutskii (Lebedev & Lutskii, 1972, 1973) substantiated M. N. Livanov's idea. According to the estimation we performed with I. A. Komarova, the maximum duration of "pure" spindles is approximately 1 sec (Zabrodin & Lebedev, 1977). At an alpha-rhythm frequency of ten oscillations per second, this corresponds to the duration of the stepwise difference between the periods of the two neighboring frequencies that form alpha-rhythm. The indicated duration is approximately .01 sec, that is, 10% of the average alpha-rhythm period. We assume that the difference between the oscillation phases of the two different modules is a value of the same order. If, for some reason, the difference becomes less than 10%, then both modules are consolidated into one modulus as a result of so-called phase capture.

Frequencies close by are similarly captured or constricted if the difference between their periods is less than the aforementioned critical value. For this reason, the number of differing phases in the course of a single period does not exceed the value determined by the equation $N = 1/\alpha\rho - 1$ where α is the average alpha-rhythm frequency of 10 fluctuations per second, ρ is the average duration of the relative refractory period, or "jog," which is approximately .01 sec in the alpha-rhythm range.

It is probably no accident that the jog's duration approximately corresponds to the duration of the relative refractory period after each neuron impulse, as well as to the duration of the interval during which the postsynaptic potentials can be summed up in the generator zone of a neuron's membrane.

As the perceived signals change the phases of the background oscillations, they give rise, in our hypothesis, to wave packets that accompany the sequences of total neuron firings that are separated by relative refractory intervals. If one observes the background activity of neurons in various animals and in humans, it is not difficult to see that the impulse bundles are most often irregular, and that the intervals between the impulses within the bundle can be shorter than the refractory period, that is, less than .01 sec. However, the distribution mode of the interimpulse intervals, as a rule, still exceeds the indicated values.

We once again emphasize that the graduated difference in phases and frequencies is the average of many neurons that make up the modulus, that is, neurons from different brain regions that have reacted instantly to a perceived signal and then fire in harmony only on the average. Neighboring modules in the wave sequence, which we have called a packet, can even be "exchanged" for neurons subjected to extraneous influences. However, the number of waves (or group firings, in other words) in the packet subsequently remains unchanged until the number of neurons in each module is higher than the previously determined critical number.

The packets differ from each other not only in the number of waves, but also in the mutual disposition of their forward fronts on the time axis. The number of those positions, which may be independent within the previously mentioned limitations, does not exceed the previously calculated degrees of freedom number that is equal to the maximum number of waves in the packet. Consequently, the largest number of possible combinations of packets, each of which differ in position and number of waves, does not exceed N^n, which in logarithmetic units would be $N \log_2 N = 30$ bits of information when $N = 1/\alpha\rho - 1$, as was calculated previously. This is the estimate of the maximum large variety of states in which a system of coherent oscillations can exist in one narrow band of alpha-rhythm frequencies.

Alpha-rhythm consists of a limited number of close frequencies. We assume that stored information is partially duplicated at various frequencies.

At the same time, differing frequencies can store different information. It is generally known that evoked potentials persist only for about 1 sec. Information about perceived signals is stored just as quickly, but not completely disintegrated (Sperling, 1963). It is possible that the portion of information that disintegrates is encoded by oscillations less stable than alpha-rhythm. The stored information that was encoded by alpha-oscillations is more stable. It thus comprises alpha-rhythm in a wavelike form as a result of frequency beat and surface, as it were, in the memory.

The best period, that is, the difference between two adjoining frequencies that differ by approximately 10% (see preceding), can be approximately estimated by the equation $T \cong 1/\alpha^2\rho$ where α is the average alpha-rhythm frequency (10 Hz), ρ is the relative refractory period or "jog" (.01 sec). The beat period corresponds to the wavelike changes in the excitability of neurons that store impressed information. It is possible that the time of information perception is dependent on that. Thus, the packets of bioelectric waves that reflect the stable phase states between the averaged activity of neurons that form groups or modules are capable of storing limited quantities of information. The wavelike slow excitability oscillations of neurons that store impressed information occur as a result of frequency beat.

These are the hypotheses that we have inferred from the stability of succeeding oscillations of averaged neuron central activity. These suppositions should be verified in a psychological experiment that should be preceded by calculating the possible quantitative characteristics of perception and memory.

Computation of Perception Velocity and Volume

The volume of instantaneous perception is the amount of information that can be correctly reproduced by a human 1–3 sec after perception. According to Miller (1956), this amounts to the reproduction of approximately seven information units, for example, random letters or numbers. What constitutes a unit has not been precisely defined. According to our hypothesis, brain rhythms are related directly to information storage. It is known that the alpha-rhythm is retained in autocorrelograms longer than other rhythms without being extinguished (Figure 39.4). Consequently, if information is stored by oscillations in the alpha-rhythm band, then, according to the calculated volume of current information stored by such oscillations, the amount of information retained in each current moment of time preceding the stimulus does not exceed the value determined by the equation

$$K_{max} = \left(\frac{1}{\alpha\rho} - 1\right) \log_m \left(\frac{1}{\alpha\rho} - 1\right)$$

where m is the alphabet, that is, the list of various signals randomly distributed in an equally probably manner, K_{max} is the limiting number of correctly reproduced signals, α is the average frequency of the alpha-rhythm (10 oscillations per second), and ρ is the relative refractory period (.01 sec). One can see from the equation that the greater the diversity of signals, that is, the alphabet, the smaller will be the number of symbols instantaneously stored. This conclusion is confirmed by the data of various investigators as summarized by Cavanaugh (1972).

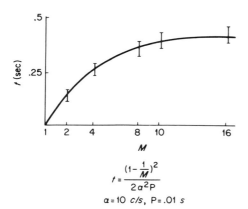

$$t = \frac{(1-\frac{1}{M})^2}{2\alpha^2 P}$$
$\alpha = 10$ c/s, $P = .01$ s

Figure 39.4. Relationship between selection reaction time and the number of alternative signals (our own data).

The perceived signals are recognized, that is, they are not immediately identified at the moment the receptors are activated, but are after a certain lag. This was first studied by Donders (1969). The delay is partially explained by the time required for receptor stimulation and the passage of signals along the nerve paths to the central structures. Donders also considered the time required for the formation and realization of the response reaction.

It is now known that the earliest changes in the cortical biopotentials occur .01–.03 sec after the action of stimulus. About the same amount of time is required for the formation of the response reaction. In addition, a period of time, that is, on the average, equal to one period of neuron activity oscillations, that is, .1 sec, is required for the formation of the wave packets that code the signal. Therefore, the shortest perception lag time is in the range of .1–.2 sec. The irreducible or constant lag time is added, according to Donders (1969), to another alternating perceptive lag whose value we shall calculate from physiological data.

Let K equal the number of equally probable simultaneously perceived signals, M, their alphabet, and T, the frequency beat period, the alpha-rhythm components. The neuron modules coding information about the signals in the memory are stimulated alternately at the indicated time period. The recognition of signal perception occurs at the time of the greatest excitability of neurons in a corresponding modulus at a probability of $P = 1/KM$. The probability of the simultaneous, instantaneous recognition of all K signals, that is, with a time lag of $t_0 = 0$, is equal to $P_0 = P^k$. In the remaining cases, the time lag t_1 occurs at the probability of $P_1 = 1 - P_0$.

Subsequently, the amount of time that elapses from the recognition of one signal to the next is equal, on the average, to

$$t = t_0 P_0 + t_1 P_1 = \frac{1}{1-P} \int_{x=P}^{x=1} T(1-x)^k dx \cdot (1 - P^k) = T \frac{(1-P)^k(1-P^k)}{K+1}$$

The standard deviation of this time is

$$\sigma_t = t \sqrt{\frac{(K+1)^2}{1 - P(2K+1)} + 2P^k - 1}$$

In our calculations, we considered that the presentation times of the signals for recognition do not depend on background oscillations of neuron activity, that is, they are distributed evenly during the beat period in a random fashion. Experience has shown, however, that the distribution of perception time lags differs from the expected even distribution (Lebedev & Lutskii, 1969). This can be explained by the parallelism of comparative operations and by their duplication. One might assume that the signals are recognized after they are successfully collated in most channels. With this proviso, the average time lag does not significantly differ from the calculated value, and the standard deviation decreases in inverse proportion to the square root of the number of parallel operations. That number does not exceed the quotient of dividing all information capable of being placed into operative memory by the amount of information contained in the perceived signals. Therefore, the standard limiting deviation can decrease when the parallel operations of comparison are increased up to the value

$$\sigma_{\min} = t \div \sqrt{\frac{N \log n}{K \log m}}$$

where $N = 1/\alpha\rho - 1$ with the former denotations. Consequently, the real value of the standard deviation lies in the range of

$$\sigma_{\min} \leq \sigma < \sigma t$$

This value depends on the operative memory load of extraneous information bearing no relationship to the perceived signals.

The average value of the selection reaction, in all probability, can also change, depending on the ratio of the critical number of successful parallel operations of comparison in which recognition occurs to the number of all possible parallel operations. This ratio possibly depends on the training of the test subject, his interest in the experimental results, and similar, so-called extrasensory factors.

Such is our hypothesis on the physiological mechanism of information perception and storage, based on the data of Livanov (1972) concerning the spatial-time organization of periodic brain processes. Not everything here has been as yet clarified, and future experiments will indicate the necessary direction of further research.

Experimental Verification of the Hypothesis

In the experiments which we conducted and published earlier, together with Komarova (1975), we determined the volume of instantly perceived random numbers and letter symbols as well as artificial syllables. In all cases, we calculated the volume of information presented and the volume of perceived information. We found that the more information contained in the perceived element (symbol, syllable), the worse that element was reproduced. Subsequently, the average volume of instantaneous perception did not exceed the indicated limit, calculated by electrophysiological characteristics (Table 39.2).

In the experiments conducted together with Lutskii (1969), we determined the relationship between perception lag and the number of perceptive alternatives in a selection-reaction situation. The changing random

Table 39.2
Average (of 10 measurements) Number of Correctly Reproduced Symbols after a Two-Second Exposition of Rows of Random Digital and Letter Symbols Arranged in Random Order

Number of test subjects	Examples of row arrangements		"912 329 353"
	"tuv nizh tal"	"Nym Le"	
	Information	Beat/symbol	
	4	6	3.32
1	7.9	4.7	5.7
2	8.1	4.5	6.7
3	7.2	4.3	7.1
4	7.8	4.3	5.9
5	5.2	2.9	4.7
6	7.0	2.9	6.3
7	4.3	3.3	5.1
8	6.3	4.3	5.8
9	7.6	4.3	6.5
10	8.0	4.4	8.8
Average number of symbols	7.0	4.0	6.3
Standard deviations	1.3	.7	1.1
Perception volume, beats	28.0	24.0	20.9

positions of light dots on the oscillograph screen, delineated into bands of equal intervals, were used as the signals. When a dot coincided with a band, the subject was supposed to press one of two buttons, and when the dot coincided with the interval, the subject was supposed to press the second button.

The reaction time increased monotonically with an increase in alternative positions, and adhered to the equation of perceptive lag that we calculated previously. In another series of experiments, light diodes placed near the buttons were used as the signals. On instruction, the subject in the selection situation was supposed to put out the light on his right as quickly as possible by pressing the right-hand button, and do the same to the left light by pressing the left-hand button, that is, the stimulus and reaction were consistent. In these subjects, we determined the volume of instantly perceived digital symbols and the time required for fully examining a chart with 100 random symbols. The purpose of the examination was to determine how many times a combination of two sequential digits would be found in a chart assigned by the experimenter. The results of these experiments are shown in Table 39.3. They correspond to the calculated data obtained from the equations for perception and memory.

In the experiments of our associate, B. G. Bovin, digital and letter symbols and color cards were used as the stimuli. Since the experiments

Table 39.3
Average (for 10 measurements) Number (K) of Correctly Orally Reproduced Digital Symbols of 10 Equally Possible Symbols with Average Time (t_k) of Examining the Number Chart of 100 Symbols, Measured for 10 Tests, with Average Simple Reaction Time (t_1) and Selection Reaction (t_2) of Two Equally Possible Alternatives (Flashes of Light), Measured for 20 Tests for Each Test Subject

Number of test subjects	Volume and time of perception		Reaction time, msec: Average values		Standard deviation	
	Symbols	Seconds	Simple	Selection	Simple	Selection
1	5.1	10.6	218	324	18	44
2	5.5	11.0	204	312	23	51
3	5.6	14.6	274	368	39	55
4	5.7	13.9	246	356	28	73
5	6.1	15.0	253	341	46	110
6	6.4	12.0	232	330	59	48
7	6.4	10.7	213	327	20	41
8	6.4	13.4	235	303	107	106
9	6.4	10.5	220	315	31	60
10	6.5	12.0	213	325	25	74
11	6.6	12.7	275	386	31	57
12	6.7	10.4	191	305	24	41
13	6.9	13.0	229	316	37	53
14	7.2	11.3	229	309	27	50
15	7.2	11.5	235	383	59	70
16	7.2	13.2	241	380	43	40
17	7.4	14.0	236	375	25	47
18	7.7	14.1	206	391	32	96
19	7.8	14.5	265	391	20	53
20	8.5	12.1	225	302	20	58
21	8.6	13.7	202	311	28	42
22	8.8	10.3	214	304	25	67
23	8.8	10.8	222	302	26	55
Average value	6.9	12.4	229	337	34	60
Standard deviation	1.1	1.5	22	33	19	20
Theoret. average value	6.9	11.4	220	345	—	34

Remarks: $K = 6.9$, $T = 1$ sec for theoretical calculations, and the value (t_2) was calculated to account for the maximum duplication of collating signals with standards. See text for computation equations.

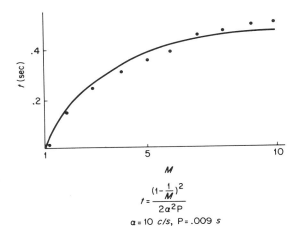

Figure 39.5. Relationship between the selection reaction time and alternative signals. (Data of Merkel, 1885.)

$$t = \frac{(1-\frac{1}{M})^2}{2a^2 P}$$

$a = 10$ c/s, $P = .009$ s

did not have consistent response actions from the signal positions, as was the case in the previous experiments, we accounted for the additional time required for selecting one of the two possible actions by using the perceptive lag equation. The calculations coincided with the experimental data (Figure 39.3). There, in all probability, the neurophysiological mechanisms of selecting motor alternatives, that is, the selection of means of action, were followed by the same characteristics of periodic processes as did the perception processes. The experimental data of Merkel (1885), Hick (1952), and Hyman (1953) (see Figures 39.5 to 39.7) also adhere to the derived equations.

It would seem expedient to continue a further analysis of the proposed neurophysiological mechanisms of perception and memory. Of most interest, in this regard, is the possibility of comparing individual electrophysiological characteristics of the human brain with the individual characteristics of perception speed and volume. In addition, the frequency

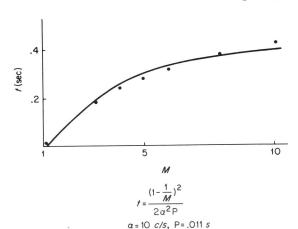

Figure 39.6. Relationship between selection reaction time and the number of alternative signals. (Data of Hick, 1952.)

$$t = \frac{(1-\frac{1}{M})^2}{2a^2 P}$$

$a = 10$ c/s, $P = .011$ s

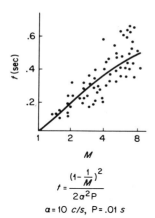

$$t = \frac{(1-\frac{1}{M})^2}{2a^2 P}$$

$a = 10$ c/s, $P = .01$ s

Figure 39.7. Relationship between selection reaction time and the number of alternative signals. (Data of Hyman, 1953.)

processes in the alpha-rhythm band should be analyzed precisely in order to determine the uneven assumed changes in the phases and frequencies of regular oscillations.

The ideas we have been developing would also probably be useful for solving important current problems in psychophysics (Zabrodin & Lebedev, 1977). Of course, there is much that remains unclear in this new area. One should also not exclude completely the possibility that information coding is accomplished not by phase modulation, but by frequency modulation of background oscillations. We have already discussed that possibility (Lebedev & Lutskii, 1969), and our studies there have led to more or less satisfactory results, although they have been of a significantly smaller scope. Thus far, the advantage would seem to lie with the phase hypothesis, although further analysis of all possible directions in the search for the neurophysiological mechanisms of perception and memory is required.

REFERENCES

Cavanagh, J. P. Relation between the immediate span and memory search rate. *Psychology Review,* 1972, *79,* 525.

Creutzfeldt, O. D., & Kuhnt, U. The visual evoked potential: Physiological, developmental, and clinical aspects. *EEG and Clinical Neurophysiology,* 1967 (Suppl. 26), 39–41.

Donders, F. S. On the speed of mental processes. *Acta Physiology, 30; Attention and performance II* (W. G. Coster, Ed.), 1969, pp. 412–431.

Hick, W. E. On the rate of gain of information. *Journal of Experimental Psychology,* 1952, *4,* 11–27.

Hyman, R. Stimulus information as a determinant of reaction time. *Journal of Experimental Psychology,* 1953, *45,* 188.

John, E. R. *Mechanisms of memory.* London, 1967.

Komarova, I. A., Lebedev, A. N., & Litvin, V. E. Individual prediction of the visual efficiency of an operator by his EEG parameters. In *Engineering psychology problems of automatic control systems.* Kiev, 1975, Pp. 68–71. (In Russian)

Lebedev, A. N., & Lutskii, V. A. Neurophysiological principles of simple human reflexes. In *Vital problems in the experimental study of reaction time*. Tartu, 1969. Pp. 138–143. (In Russian)

Lebedev, A. N., & Lutskii, V. A. EEG rhythms—a result of interrelated oscillative neuron processes. *Biophysics*, 1972, *17* (No. 3), 556–558. (In Russian)

Lebedev, A. N., & Lutskii, V. A. Frequency mechanisms of visual perception. In *Biophysics of Vision*. Vilnus, 1973. Pp. 84–105. (In Russian)

Lindsley, D. B. Attention, consciousness, sleep and wakefulness. *Handbook of Physiology*, Washington, 1960, *3* (Sec. I), 1553.

Livanov, M. N. Analysis of bioelectric oscillations in the rabbit cerebral cortex. *Soviet Neuropathology, Psychiatry and Psychohygiene*, 1934, *3* (Nos. 11–12), 98–115. (In Russian)

Livanov, M. N. On the uneven development of certain frequency processes comprising an electrocerebrogram and the berger rhythm, *Fiziol. Zhurn. SSSR*, 1940, *28* (Nos. 2–3), 157–171.

Livanov, M. N. Inhibition in neuron systems of the cerebral cortex. In *Brain reflexes*. Moscow, 1965. P. 70.

Livanov, M. N. *Spatial organization of brain processes*. Moscow, 1972.

Merkel, J. Die Reitenhen Verhaltnisse der Willenstätigkeit. *Philosophisch Studien* (Vol. 2). Leipzig, 1885.

Miller, G. A. The magical number seven: Plus or minus two. Some limits on our capacity for processing information. *Psychology Review*, 1956 (No. 63), 87.

Pribram, K. *Languages of the brain*. Moscow, 1975.

Sperling, G. A. A model for visual memory tasks. *Human Factors*, 1963, *5* (No. 1), p. 19.

Zabrodin, Yu. M., & Lebedev, A. N. *Psychophysiology and Psychophysics*. Moscow, 1977.

Appendix A:
Individual and General Discussions

Following are brief, edited versions of the discussion that followed each paper and a general discussion that was held at the end of the symposium on Neural Mechanisms of Goal-Directed Behavior *held on April 11–18, 1978. Because of the usual problems of tape recorder and translator breakdowns, and because of space limitations, these discussions have been shortened considerably. However, we have attempted to give accurately the substance of what was said and to preserve the unique interpersonal flavor of the meeting, which became increasingly warm and cordial as the symposium progressed.*

Dr. Asratyan

Dr. Gormezano: Professor Asratyan, I'm curious to get some explanation of your bidirectional conditioning hypothesis of goal-directed behaviors—how you use that concept to explain goal-directed behaviors.

Dr. Asratyan: Your question gives me the possibility of completing my paper. In your country, many of your famous researchers say that instrumental conditioning is a very good model to explain motivational behavior. That is quite true. But what is the physiological mechanism of instrumental conditioning? This question is obscure. Pavlov himself gives a hypothetical explanation in contradiction to Konorski's explanation. Pavlov explains that instrumental conditioning causes the elaboration of bilateral connections, two way connections; e.g., flexion is combined with food, two points of the cortex are excited. Thereafter, flexion of the paw evokes an alimentary reflex. Alimentary and flexion reflexes become connected. This can be the basis for goal directed, i.e., motivated, behavior.

In my opinion, the characteristic features of motivation, that is, goal-directed behavior, are based on fundamental Pavlovian ideas. If we assume that instrumental conditioning is a good model of motivated behavior, as I believe your scientists do, then we must explain instrumental conditioning in terms of physiological mechanisms. To do this, we must approach the problem by analyzing the reflexes involved. All the teachings of Pavlov are based on reflexes; however, not as very simple reflexes used by his opponents to caricaturize his ideas, but rather in the sense of goal-directed mechanisms. For Pavlov, and for me, a goal-directed mechanism is the activation of a chain of backward conditioned reflexes. By "backward" I mean as I said above the manner in which alimentary and flexion reflexes become connected. If an animal is motivated, i.e., if the animal shows searching behavior or is searching in several different directions, we say in Pavlovian terms that the excitability of the neural elements of a vitally important unconditioned reflex has been increased. As the animal achieves the correct path or direction, then the *two-way* connection is formed between the conditioned stimulus situation and the biologically important unconditioned reflex. This backward or two-way connection is what I mean by the engram. This issue is very complex. I have brought along some reprints of mine, which I will share with you, which explain this issue in detail.

Dr. I. Gormezano

Dr. Pribram: There is an explanation in terms of memory mechanisms which may fit. Shortly before he died, Clark Hull and I had a very intensive conversation in which he said that he finally decided that his particular approach had really misfired because he would have to have a new hypothesis almost every month and add a new variable to his scheme every month. He was very upset by this toward the end of his life. The particular problem of chaining is what led me to *Plans and the Structure of Behavior*. At Harvard, I went to George Miller and said, "What about chaining of responses? Frontal lobe lesions interfere with the chaining responses." And he said to me, "We can do that any time you want, we do it on the computer all of the time." Computer programs regulate the serial order of behavior, that is, the chaining of responses. Aren't the things that you're trying to explain the same thing that we are trying to explain in cognitive psychology when we talk about primacy and recency effects? The primacy effect is in part dependent on rehearsal. I have evidence that these two domains of explanation are related because in taking out the amygdala and doing a classical conditioning experiment, we found that normal monkeys make anticipatory responses (GSR's), over a widening and widening period of time. After we removed the amygdala, there is a restriction in that period of time in which the anticipatory GSRs are made. Something changes in the conditional stimulus, and it doesn't attach to the response.

Dr. Gormezano: Thank you, Dr. Pribram. With regard to the question of memory concepts and primacy and recency: The paradigm from which I am working has deductive consequences that do not derive from memory concepts. Specifically, if I take all of the known laws of Pavlovian conditioning and apply them to that chained behavior, the jaw movement in the runway, I can generate a great many deductions which do not readily derive from the deductive consequences of memory concepts. When the point is reached where the memory model says something that I can't deduce, that's when I leave the conditioning model. Professor Asratyan characterizes me correctly. I'm the only Pavlovian in the United States in the sense that I know certain kinds of laws and I know how to apply them, and I know how to generate certain kinds of phenomena that to me still have important psychological value. That does not mean that those laws, in the future, would not be subordinated to more general concepts, like memory, images, et cetera, but you're going to have to show me.

Dr. Asratyan: Your experiments on rabbits in this long runway seem to me very interesting. You are able to show that basic Pavlovian notions on conditioning are quite enough to explain these new data. It seems to me you have here a complex conditioned stimulus, not only for protection, but for time, moment, and then space. I do have a question. As the time decreases, the stimulus effect is increasing. How is your complex conditioning stimulus operating?

Dr. Gormezano: Thank you, Professor Asratyan. Professor Asratyan was attempting to make the point that in the simple runway situation, the stimuli that presumably would be governing the conditioned response that is compound to the swallowing reflex would indeed consist of a complex compound of both temporal, spatial, and proprioceptive. No disagreement I assure you, Professor Asratyan. You are absolutely correct. I was attempting, however, to take a component of the situation and analyze it. You call it analytical, I think it's analytical too.

Dr. N. E. Miller

Dr. Shuleikina: Is there a physiological explanation of the main phenomena of the control of visceral functions by the person himself? Also, are you drawing any parallels between your experiments and the data on the training which is done by yoga?

Dr. Miller: To answer the second question first, I do draw parallels with yoga. Now as Professor Anand, who has investigated yogis in India found, some of them control visceral functions in the indirect way that I illustrated. These indirect ways are not unimportant, but they are theoretically somewhat different from direct control. An extreme example of this is the yogis who claim to stop their heart. What they do is perform an exaggerated Valsalva maneuver, which is to take in a deep breath, close off the glottis, breath out very hard without expiring any air and hence build up great pressure in the chest cavity. Because the blood in the veins is returning under low pressure, it is possible for them to collapse the vein. Then, since the heart is getting no blood, the heart sounds disappear because these are made primarily by the action of the blood on the valves on the heart. The pulse, of course, disappears because there is no blood being pumped. Of course, they can only maintain this for a short while, but an electrocardiogram shows that the heart is beating even faster than normal, so this is an indirect control. Anand has found that out of about 300 yogis that he has tested, four seem able to control visceral functions directly. Now this may be a little unfair to the yogis because the Buddhist religion's goal is to get rid of illusions, one of which is the illusion of cause and effect. And so Anand has found that it is difficult to study yogis. They are not interested in having a scientist investigate their powers of control. As to the question of how these phenomena occur physiologically, I do not know exactly. But there are connections from the visceral organs to the highest centers of the brain and from the highest centers of the brain to the visceral organs. So from an anatomical point of view I see no more problems in this than in the physiology of the control of skeletal behavior, which is an area in which many advances have been made, even though we still do not know the physiology in all of its details. You asked me what do yogis do physiologically. I will ask you, what do you do physiologically when you raise your hand?

Dr. Asratyan: I can't agree with you that classical conditioning and instrumental conditioning are two fundamentally different acts. In Pavlov's view, and in mine, there are no principal differences in the act of classical and instrumental conditioning. There is a large family of conditioned reflexes. Each member has its own special features, but they are not fundamentally different.

Dr. Miller: I think that difficulties of language and insufficient discussion of these points has led to a misunderstanding. I am in complete agreement with your point of view. In fact, what I have said is that the ability of visceral responses to be modified by instrumental-training techniques removes what had been in the past thought to be a fundamental difference. In the past in the United States it had been thought that instrumental and classical conditioning were fundamentally different. The two parts of the nervous system related to them were also thought to be fundamentally different, i.e., the somatic nervous system and the autonomic nervous system. The autonomic nervous system was thought to be fundamentally more stupid than the somatic nervous system. The whole thrust of my work would say that there are no fundamental differences of this kind. That is, the two types of nervous system are equally intelligent and modifiable and that the two types of learning are fundamentally the same, although they may have superficial differences. There may be differences between parts of the somatic nervous system and the autonomic nervous system that are analogous to the sensory differences between the lips and the middle of the back. The two-point thresholds are vastly different, but they are different in degree. It is a quantitative difference.

Dr. J. L. McGaugh

Dr. Berger: Dr. McGaugh, I'd like to personally thank you for what I think is both a very troubling paper and a very stimulating one, one that I hope will have an impact on people such as myself who are involved in methodological issues related to learning studies. I wonder whether you have considered as possible explanatory material for some of your work the question of state-dependent learning, even in the posttrial manipulation that you normally use?

Dr. McGaugh: Not in every experiment but in several of the experiments, we have done state-dependent controls sufficient to convince us that it is not a routine requirement for these experiments. For example, even with something as massive in its effects as DDC, reinjecting animals with DDC before the retention test does not restore memory.

Dr. J. Lacey: You use the word feedback several times. Are you postulating a neuronally operative feedback system with some of the formal characteristics of servo amplifier feedback?

Dr. McGaugh: Yes. However, I don't know the nature of that feedback, whether it's neuronally mediated or whether it is some hormone which does pass through the blood-brain barrier downstream from the ones that we're working with.

Dr. J Lacey: I would instantly subscribe to the feedback hypothesis. But then we will have to make some decisions about the formal characteristics of the feedback hypothesis—is it a proportional controlling system, a linear system—how are the set points changed, and so on. I have a fundamental feeling that our analyses of physiological substrates of behavior eventually are going to have to come to systems analysis because each of us can deal with only a small fraction of an extremely complex system. Although we delight in our neurophysiological observations and biochemical manipulations, and our single-cell observations, eventually we are going to have to go to a black box again. (Imagine, neurophysiologists becoming "black box" scientists!) This has become true in such areas as temperature regulation, blood sugar regulation, cardiovascular regulation, and it may become true in such areas as learning and memory.

Dr. McGaugh: There is some implication of this for even black-box neurophysiology. If these support systems do indeed play a role in the normal regulations of learning and memory, then that calls into question *in vitro* systems, e.g., cell cultures and artificial systems. The kinds of learning that these systems may be capable of may not be the kinds of learning that normally occur in the *in vitro* situation, which has not only the neuronal mechanisms but all the rich biological support systems.

Dr. Asratyan: What is the neurophysiological mechanism of memory? Do you think that the formation of memory is something different from elaboration and consolidation of conditioned connections? If so, what is the difference?

Dr. McGaugh: I don't think there is any difference. If there is one mammalian nervous system, we are both talking about it. So at one level it has to be the same. The reason that I'm concerned about using formation of conditioned connections is that I don't know how connections are formed. I know that they are formed with enormous plasticity in behavioral possibilities, and that complicates the picture enormously. The University of California, our parent institution, at Berkeley, was a leading institution in investigations of goal-directed behavior in the '20s, '30s, and '40s. Edward Chase Tolman continued to remind us of the flexibility in the performance of learned responses, and I think was the first one to give a definition of goal-directed behavior, which was plasticity and docility relative to some end—plasticity in that the behavior can change at a moment and docility in the sense that it's teachable. Given that, I share with many people ignorance of the nature of linear connective reflexive systems that can give rise to the flexibility seen in goal-directed behavior. It doesn't mean that goal-directed behavior is not or will not be built out of such reflexes.

Dr. E. R. John

Dr. Thatcher: In the last part of your discussion, you talked about three models of consciousness. All three examples were ones where a system becomes aware of itself. There is a logical problem that people would raise about that. What about the other category of models, for instance, where such as those explicated by Sperry, i.e., a system specialized to be aware of other systems, is not necessarily aware of itself?

Dr. John: I guess that the problem that I have in dealing with the kinds of model that Bob Thatcher just raised is that in order for a system to be aware of information in other places, I can't conceive of it processing that information differently from the way that information is processed in the ensembles that I was describing. So although you may regress the problem to another level and say there is an homonculus, an observer system, a mind, it too must process information with the facilities available in the brain, i.e., neurons, and then immediately you come to exactly the same problem. If you want the different pieces of the system each to be aware of what's happening in the others, you need an integrated process. So you haven't really solved the problem. I don't think there are two approaches to the problem. I think all of them can be reduced to one or another of the three models that I presented today.

Dr. Shvyrkov: In the first place, if I started repeating everything that I agree with in your presentation, it would be a very detailed reproduction. I have several questions, but I'll choose the most important ones. In the introduction to your presentation, you mentioned that goals, complex goal-oriented behavior, emerge on the top strata of the phylogenetic hierarchy. Does it mean that on the inferior levels of this hierarchy we have similar reactions? I ask the question because, in my opinion, the idea of stimulus–response relationships is incompatible with the concept of goal-oriented behavior. I can reduce my question to the following. Do you believe that the concept of stimulus–response is compatible with the concept of goal-oriented behavior?

Dr. John: That's a terribly interesting question. I think that part of my problem is a semantic problem. What I was trying to say was that with phylogenetic development, the relative importance of learned behaviors increases compared to innate behaviors. I was fascinated by Dr. Epstein's presentation. Even *Stentor*, which has no nervous system, shows individualized behavior. I don't think the question is innate versus acquired. I can reconcile both of those as quantitative differences. Some animals have a great deal of innate behavior and very little acquired behavior. Other animals have preponderantly acquired behavior. But neither one of those dependencies excludes the idea of goals. Your question could be reduced to asking is it possible to set a level below which there is no consciousness and no ideas can exist, and at which you can explain behavior simply as a stimulus and response? My last model says that perhaps consciousness is a property of negative entropy, of organized energy and then one would only talk about quantitative distinctions not qualitative distinctions. In other words, while I understand your question I think it's a profound question. I think that it can lead us into a semantic disagreement. I don't believe there is a discontinuity.

Dr. N. Miller: I would like to put in a brief comment. Several times here we've had learned behavior and instinctive behavior as two separate categories. I would like to express the opinion that that's not necessarily so. Let's take the human level—the very reason that learning can be so important with us is that we have such a rich repertoire of innate behavior. If there was an animal that could only do one thing, maybe only retract, you couldn't teach it very much. But if the animal has ten different types of innately organized behavior, then you have much great opportunity to teach it a greater variety of behavior.

Dr. A. N. Epstein

Dr. Shvyrkov: Thank you very much for these interesting thoughts which you shared with us. As far as I could understand your division of behavior into instinctive and motivated, it is in practical terms a division into innate and acquired. If this is not so, if there is a difference, then how would you account for the behavior of the worm and how, motivationally, is his behavior different from that of a rat in the two following cases: The rat follows the maze to food; or the worm climbs to the surface looking for food or grass?

Dr. Epstein: That, of course, is a very good question and is one of the matters about which I am not fully satisfied myself in making this distinction. But I am eager that we use more than one criterion, more than one behavioral characteristic to distinguish these two kinds of behavior. And I therefore made an effort to describe as many of the characteristics of each as I could. The example I gave of that was a situation in which learning is employed, but it is employed in what I suspect may be a fundamentally different way. Rather than being a kind of global capacity that the animal's nervous system brings into virtually any situation, when the animal is displaying instinctive behavior, the learning is constrained to a particular context within sequences of behavior and is limited to a narrow set of possible associations. The example was the learning of a song in a songbird as opposed to the learning of a song in a human being. They're fundamentally quite different. You and I can sing virtually any note, any melody, but the song sparrow literally can only sing the songs of its own species and closely related ones.

Dr. Shvyrkov: I beg your pardon, but perhaps the bird is not a very good example. Many birds learn to copy the singing of other birds. Now as far as motivation goes, if we go back to the worm, we can separate appetitive behavior and consummatory behavior. Appetitive behavior begins when the worm comes to the surface—it may be a whole meter to the ground surface. Then we can separate a second part of its behavior, the search part, when the worm quite goal orientedly looks for the grass. Then we can separate the consummatory part when it eats the grass. If you regard other insects, e.g., bees, although the hive can be viewed as one organism because the bees are genetically identical; nevertheless, each bee has its own life experience. They're individualized since generally learning is possible even with the simplest animals. It's possible that nonmotivated behavior is hardly imaginable. Even instinctive behavior must also be motivated.

Dr. Epstein: I was hoping to convince you that motivated behavior is a product of evolution, a complex product of evolution, and is something different from instinctive behavior. If you consider the several other criteria that I've suggested, you may perhaps appreciate this point. As for learning, I really must differ with you. Perhaps all animals show habituation and sensitization, but learning, like motivation, is a later and complex result of biological evolution. Consider, for instance, the fact that many animals do not have nervous systems. The coelenterates (a very large number of the animals living in the sea are not fish, but coelenterates) have nervous systems, but they have not been shown to exhibit associative learning. Perhaps like learning, motivation is a more recent biological phenomenon.

Dr. P. Teitelbaum

Dr. Hicks: Do your cataleptic animals show the compulsive approaching behavior that Villablanca describes in his acaudate cats?

Dr. Teitelbaum: Yes, if you add locomotion to the orienting, you get the compulsive behavior described by Villablanca. It's simply a question of presence or absence of subsystems. If you place lesions slightly more anteriorly, you can create rat effects like those in caudate kittens. The rat will run after you and chase you if you move, which a normal self-respecting rat would always inhibit. But for a kitten that is perfectly appropriate.

Dr. Asratyan: I want to point out one more coincidence of your data with mine concerning the involvement of higher parts of cortex in the rehabilitation of functions. Some 20 to 25 years ago I surgically damaged different parts of the central nervous system: hemisections of spinal cord, lesions in bulbar areas, some destructions of subcortical structures. Then I studied the dynamics of rehabilitation of function. After I surgically decorticated the animals, rehabilitation functions were destroyed. So your data are quite consistent with mine. But I cannot agree with you that a principal difference exists between conditioned reflexes and operant reflexes. Operant or instrumental conditioning occurs with its own specific features, but generally, they are also reflexes. A conditioned reflex is not a simple uniform reflex. Rather it is a great family of conditioned reflexes—each member of this family has some common features, and each specific features. So operant or instrumental conditioning has its own very important features. But it is a conditioned reflex.

Dr. Teitelbaum: I tend to agree with you. The problem is to find out the specific stimuli that make up the family which constitutes the operant.

Dr. Shuleikina: Did you also destroy the lateral hypothalamus only in the immature animals?

Dr. Teitelbaum: No, but it has been reported that if the lesions are made in the first 2 or 3 days of life there is no effect. But if made at 10 days, the animals become aphagic, though some defects are still not present. However, if the lesions are made at 20 to 21 days, they have the full adult syndrome.

Dr. Shuleikina: We have closely related data, but after the destruction of some peripheral nerves; e.g., chorda tympani and the lingual nerves. Newborn kittens have many dysfunctions of alimentary behavior, especially in the motivational accompaniments of alimentary behavior. The more traumatic disorders of the behavior were in the first days rather than later in the development of the kittens.

Dr. Aleksandrov: You frequently used the term subsystem. What is the difference between subsystem and locomotion? Between subsystem and the reflex or response? What do you infer by system which the subsystem is a part of?

Dr. Teitelbaum: I use the term *subsystem* because I think our level is somewhere between that now used to describe motivated behavior and the level used by Sherrington to describe simple reflexes. But what is astounding to me is that these animals are behaving and maintaining themselves. They run around, they orient, they eat, and drink. It is at a level which I believe is much more complex than has been differentiated before, but is not equivalent to the description of these behaviors in the normal animal.

Dr. Shvyrkov: Dr. Teitelbaum, I completely agree with you when you call locomotion or static support a system, because this is a complex-structured innate organization. I also agree when you call it subsystem, because these subsystems are the elements of the big system of motivated behavior. I would like to criticize your position. Do you believe that the principles of organization of subsystems and the whole system of motivated behavior are different? And that the notion of the goal may be employed only as far as the goal behavior is concerned? I believe also that subsystems have their own goals which correspond to their hierarchical level.

Dr. Teitelbaum: The question highlights the issue of the correspondence, the immediate transformation of these subsystems into motivated behavior. These subsystems have their own set points, which is another way of saying goal, that which they resist displacement from. The difference about them is that they cut across all the normal kinds of motivated behavior that we know. So, support is common to feeding, drinking, mating, attack, and it must have its own laws that function in interaction with these forms of behavior. At this stage of descriptive analysis, I think the two levels are complementary. They proceed in their own organized fashion independently, and we must form the bridging concepts and variables between them.

Dr. C. Cotman

Dr. Shuleyikina: Dr. Cotman, is the growth of synapses accompanied also by growth in the number of dendritic spines?

Dr. Cotman: Yes, there is a remodeling of dendrites which is quite extensive. Not only do fibers grow, but the dendrites go through a transient period where they lose their stereotype configuration, particularly their spines, and then with time these spines grow back again and new synapses are formed on them.

Dr. Satinoff: In your serial lesion paradigm, what does the brain look like after 6 days if you only make the first lesion?

Dr. Cotman: There is a glial reaction and the appearance of a few degenerating terminals. But using the measures of synaptic growth, we find that there is very little.

Dr. R. Miller: In general, what is the correlation between the neural regeneration that you observed and recovery of behavioral function?

Dr. Cotman: We've done some behavioral studies using spontaneous alternation and found that the time course of recovery from hippocampal damage is identical to our synaptic growth response.

Dr. Lebedev: Is it possible there is some interrelationship between the processes of growth after damage and the spontaneous impulses of neurons?

Dr. Cotman: In the hippocampus we don't really have any information on that. In the peripheral nervous system there is a large controversy concerning this issue. If you damage the muscle fiber, what happens to the development of denervation supersensitivity. There are two positions, one is that trophic materials are disrupted and the other is that it's a loss of activity in the muscle. It is clear that activity, per se, is an important component.

Dr. Isaacson: My review of the literature on single versus multiple stage lesions seems to me to indicate that the only difference between the two is the rate at which recovery occurs after each. There are only rare exceptions to the general rule that there is no ultimate difference in the behavior of the animals, say months afterward, whether the lesion is done in single or in multiple stages.

Dr. Cotman: I would like to add one final comment. I think it should be underscored that all of these types of growth responses may not really be for the better for the animal. The recovery process which takes place long after these growth responses are over may very well be just as important. One has to pay really close attention to the time course. I think that perhaps in some sense the brain has to learn to live with this abnormality to come back to some homeostatic condition. As Bob Isaacson said, "It's not just what happens at the site, but it's what happens in the whole system that's really critical."

Dr. L. Hicks

Dr. Marshall: This is more a comment than a question, but maybe you would like to respond to it. One of the things that one finds in studying basal ganglia structures and in the fibers that interconnect them is that the behavioral deficits appear very suddenly after large amounts of damage to the structure, which means you could damage say 60 percent of the structure without being able to detect behavioral impairments, at least with the methods we use. But 70 percent, for example, suddenly in the rat shows you a full syndrome, to use your example. The same thing is true, I think, with the ascending dopamine fibers, and I'd like to defend the notion that 6-hydroxydopamine is a useful tool in studying their behavioral function, precisely because of the fact that one can achieve much more extensive and specific damage to the relevant neural system. But I would like to emphasize here extensive damage to the relevant neural system. One can study the behavioral effects of depletions of the order of 90 to 95 percent of those ascending dopamine fibers which you cannot get simply with the electrolytic lesions.

Dr. Hicks: The controversy over whether intracerebral injections of 6–OH–DA give specific destruction or not has not been resolved, but I would agree that the technique is useful. Your point about the critical amount of tissue destroyed is also well taken.

Dr. David Lindsley: What happens to the spontaneous behavior of these animals with lesions of substantia nigra versus globus pallidus? Do you notice any tremors, or is the general activity level reduced, or what?

Dr. Hicks: In the substantia nigra animals, we often see the head-shaking tremor. The globus pallidus animals resemble the lateral hypothalamic animals described by Teitelbaum.

Dr. Shvyrkov: One question please—behavior is characterizing the organism as a whole. You destroy a local structure which becomes a conceptual model that you use in explaining the changes in the behavior of the entire organism.

Dr. Hicks: The question was if I'm paraphrasing it correctly, how do you evaluate the total behavior of the organism to determine what the effect of the lesion is on the totality of the organism's behavior? I don't know. We were only looking at behavior on these tasks and we are implying that the kinds of behavior which have similar task requirements are also possibly affected. This simply becomes a practical problem of the kinds of behaviors that you can examine in these animals both before and after lesions. I think that what is wrong with most lesion studies is that task analyses have not been made in psychology in such a way that you can specify precisely what the deficit is following the lesion; whether it's sensory, perceptual, motor, motivational, or emotional. Task analyses, I think, have not been done sufficiently well so that we can know this and so then our evaluations remain somewhat gross. But we don't have any mechanism to evaluate the animal's total behavior.

Dr. Gormezano: Dr. Shvyrkov's question was a key question, and it was approached earlier by Dr. Isaacson when he said that the effect of the lesion is not to be understood in the same way that removing two chips from a computer is. Removing a given structure initiated a series of dynamic changes which have to be understood at biochemical, neuroanatomical, and neurophysiological levels. As Dr. Hicks properly points out, this implies an analysis of the task which psychology cannot give us. We really ought to start considering whether our behavioral understanding is built upon the same kind of incorrect fractionation that E. Roy John urges is true of our understanding of brain dynamics. To say that something is sensory or motor or intentional or motivational is to use rubrics that have never satisfied the needs for a physiological analysis of behavior. And I look forward to the systems-theoretical approach to this kind of analysis by our Soviet colleagues to clarify our concepts in approaching physiological problems of behavior.

Dr. Asratyan: It seems to me that all our methods of experimentation have some pluses and minuses. Electrical stimulation is quite unnatural. Even Sherrington said that. The same is true for surgical damage of electrolytic lesions—there are many pluses and many minuses. Interpreting the results of any of our methods depends on the brain of the experimenter, not on the method.

Dr. J. Lacey: I think there is sort of a spiral process that in order to know how to apply the proper method you have to understand what you're dealing with, but in order to gain understanding of what you're dealing with, you have to have the proper methods. So, by just gradually fumbling around we increase our understanding, which then enables us to increase our methods, and I think we are much ahead of where we were geologically an age ago when I was a graduate student. So we have advanced.

Dr. Gormezano: With regard to the two or three positions about the effects of the lesions, I'd like to introduce a fourth, an entirely behavioral position, but not unakin to what Professor Lomov referred to in his introductory comments. Specifically, I'd like to strongly recommend the concept that he put forth regarding the role of behavior in establishing or determining the underlying neural structure and function. Statements, as Dr. Hicks indicated, determining the effects of a lesion as involving sensory, motor, associative, motivational, to me are not meaningless phrases. They represent statements with regard to processes that are anchored in behavioral laws. And accordingly, if I were engaged in the problems of lesions, I would be concerned with whether the effects of the lesions in systematic investigations operate on these four, five, or more processes that are anchored in their determination by behavioral laws. Not unakin, and I think it is an elegant model of this kind of experimental work, is the early research of Teitelbaum and Epstein. Their great value was the detailing of behavioral laws. And not so much factors about the hypothalamus, though they are important, but in the elegant demonstration of how those lesions related to behavioral laws which they sought; the laws were not there, but they sought them in their investigations of lesion effects upon eating and drinking behavior.

Dr. J. Marshall

Dr. Isaacson: Is there any place where you can make a lesion where you don't get recovery; in other words, where you have a permanent sensory neglect? Second, how would the dopamine hypothesis fit with the human instances of sensory neglect, which seem to be due predominantly to parietal cortical damage?

Dr. Marshall: Only one-third of our 6-hydroxydopamine treated animals never recover somatosensory or visual orientation, up to 4 months. There is a population of animals which never recover. The implication is that they don't have some residual dopamine neurons to rely on. But we haven't done the biochemistry to test our prediction. The second question is the human neglect syndrome. Interestingly, the cortical areas from which neglect occurs in humans have a remarkable correspondence to those areas of the cortex which, in the rat, are innervated by dopamine containing neurons—the cingulate cortex, the frontal cortex, and some regions of the parietal areas.

Dr. David Lindsley: In the animals that didn't show any recovery, did you try morphine?

Dr. Marshall: Actually, we didn't test that. The animals that were used in the apomorphine study were among the hardest hit we've ever studied but we only studied them for 2 weeks. I don't know if they would have recovered but they did not recover during those 2 weeks.

Dr. Berger: John, in the data you presented on the recovery of the preference to the right or left food trough in these animals, how does the recovery of that behavior coincide with the rostral–caudal recovery in the tests that you made? Does it coincide with rostral recovery or the caudal recovery?

Dr. Marshall: I can't answer directly. In general, the deficit is most marked within the first week and then disappears. It might be that there is even recovery going on within that first week.

Dr. Asratyan: Have you examined the influence of the cerebral cortex itself on the processes of recovery of function?

Dr. Marshall: I think it's a very good idea. Even if we demonstrated that there's a role of dopamine in recovery, to my mind it certainly does not rule out the possibility that cortical influence may have a tonic effect such that removal of neocortex would have a devastating effect.

Dr. Yu. I. Aleksandrov

Dr. Thompson: Thank you for a very nice presentation of important data. I think your conceptual approach explained for me the very puzzling and interesting studies of Dr. Fetz at the University of Washington. He recorded activity of a neuron in motor cortex that always responded in relation to the monkey making a movement. Then he rewarded the monkey for the activity of the neuron with food and was able to uncouple the neuron so that it would respond completely independently of the movement. That is, it would not respond to the movement itself, but instead would respond so that the animals could get food. As I understand it, this would be consistent with your view.

Dr. Aleksandrov: Yes, it is consistent. The connections of the cortical neurons to the same movement is observed in the case when the movement is incorporated into the same behavioral continuum. Our data stand in seeming contradiction with the results of the research which argue the strict connection of motor cortex neurons to the movement, regardless of the behavioral context. This contradiction only seems to be so, because in our experiments we have a fixed behavioral situation. For example, in the experiments of Evarts, the movement is fixed, in the sense that flexion and extension are fixed with all other elements of the behavioral act. Thus, by correlating the activity of neurons with the flexing movement, he really correlates this activity with the whole behavioral act which leads to the procurement of food.

Dr. Teitelbaum: I want to congratulate Dr. Aleksandrov on his very beautiful analysis and I want to mention that I agree completely with his isolation of subsystems in the preparations that I mentioned yesterday. I did not talk about the isolation of chewing movements of the mouth as a separate subsystem. But there is a great deal of evidence using drugs such as L-dopa that reveals the independent action of the mouth chewing subsystem. Does your fast head movement correspond to head orienting? That is, is it different from the tactile scan along the surface and does the slow postural fixation system correspond to a system of support analogous to the differentiation in locomotion that I pointed out?

Dr. Aleksandrov: I have already mentioned that the result of the quick phase of the movement was the maximum approach of the head toward the carrot. If we analyze the result of this movement from one act to another, we find that in every given act, the head was in a different position with respect to the carrot. The movement terminates near the point in space where the carrot was supposed to be later. Roughly speaking, the fast movement of the head is not an adaptive movement. It is not a precise movement. The slow head movement is a precise, adaptive, adjusting movement and the result of this movement is the difference between the fast and slow phases.

Dr. E. Satinoff

Dr. Asratyan: Do you feel that bar pressing and pecking are not reflexes?
Dr. Satinoff: Yes, I mean that they are not reflexes.
Dr. Asratyan: But what are they then?
Dr. Satinoff: In the sense that I am using the word, I mean that they are operant responses, they are learned responses. It's an interesting question whether what the infants are doing in the thermal gradient is an operant response because it may be a kinesis, a locomotion without any orientation. When the animals are too hot, they move, when they're too cold they move, and when they feel good they settle down. It may not be a higher level of behavior.
Dr. Asratyan: Do you mean that operant reflexes, learned reflexes, are not reflexes?
Dr. Satinoff: If you put it that way, I'm damned if I do and damned if I don't.
Dr. Shvyrkov: I would like to congratulate you on the beautiful data you just presented. Secondly, I would like to pose a question. If as a result of behavioral activity and as a result of vegetative-autonomic activity, the goals are reached, why do you call vegetative activity reflexive? I would like to say that the reflex is a reaction to the outside external stimulus as opposed to goal-oriented activity.
Dr. Satinoff: It's always a response to changes in firing rates of temperature-sensitive neurons. That can be caused by external cold or an internal infection. So there is a stimulus to which these reflexes are responses. I call them reflexes as opposed to operants because they occur in a sleeping animal or an anesthetized animal. Anesthetized animals in the cold shiver and show all the autonomic responses that you see in an awake animal.
Dr. Gormezano: If I may, I think we have an example here of semantic differences which I was hoping to clarify when Professor Asratyan raised the question of Dr. Satinoff in which he attempted to pose or counterpose the operant versus the reflex. Specifically, what Dr. Satinoff was opposing in the language of Western psychologists and physiologists is what you would call in Soviet psychology the unconditioned reflexes which are elicited by a temperature change, innately, without learning experiences, as opposed to those behavioral irregularities which come about through learning, in this case via the conditioning paradigm of the operant or instrumental-conditioning procedures. So with regard to Dr. Shvyrkov's question, Dr. Satinoff is talking about goal-directed behaviors, but referenced here as an operant behavior which occurs through environmental stimuli with expected consequences.
Dr. Miller: If I've understood you right, you have shown that the behavioral regulations start earlier ontogenetically and phylogenetically and involve a different part of the brain. However, I think you also showed they are not totally different in every respect in that once you get the autonomic regulation,

Individual and General Discussions 599

the new part of the brain that does this also is involved in behavioral regulation because cooling the preoptic area can cause bar pressing for a heat lamp. So they are separate but not entirely separate.

Dr. Satinoff: Normally they work together, but they are independent in the sense that you can remove one and see the other function.

Dr. V. B. Shvyrkov

Dr. Thompson: I would like to congratulate Professor Shvyrkov on an extremely important and elegant contribution to the symposium. Perhaps I've misunderstood, but it was my impression that you said that the different categories of neurons that relate to the different aspects of the behavioral act, for example, in visual cortex, are also present in motor cortex, so that different categories of neurons are present in both structures of the cortex. Is that true?

Dr. Shvyrkov: Neurons of different areas of the brain perform completely different physiological functions. However, such functions as movement are functions of the whole mechanism. Therefore, the fact that neurons of different areas have the same temporal structure in the behavior does not exclude, but *requires,* the specific nature of the functions which can be expressed in terms of influencing other neurons.

Dr. Thompson: May I ask one other question? You stated that the activity of certain classes of neurons has a direct causal relation to behavioral acts, not a probabilistic relation. In my view it is very difficult to establish what Professor Lomov has referred to as linear causation. Do you mean that kind of causal relationship?

Dr. Shvyrkov: The goal as the image of the future object determines only the organization of the whole system of relationships of brain neurons. In the stereotype behavior the same goal creates the same system. Then we have neurons which occupy a stabilized place in the system. It is as if they were linearly determined by the goal. In reality the goal determines the organization of activity of neurons and via the organization, activity of every specific neuron.

Dr. Gormezano: Would you expand what you mean in a theoretical sense with regard to the matter in which the goal is represented as an image which determines the functional structure of neural organization?

Dr. Shvyrkov: The notion of the goal can be considered on the one hand as a psychological one and on the other hand as a system category. As a system category, this goal must be presented in the hierarchy of processes. Now we know from the works of Sokolov and other researchers that memory probably is represented by certain organizations of potentiated synapses based on certain molecular mechanisms of interneuron memory. Since the goal is part of the memory, it must be present as a part of hierarchy, which includes the corresponding molecular processes, potentiation of certain synapses of the neuron, the relationships between neurons, and so on; the whole hierarchy.

Dr. Gormezano: One more point of clarification if I may. What do you consider activates the memory? What aspect of the environment activates the memory?

Dr. Shvyrkov: If we go back to the fact that even a receptor is alive and therefore has certain biochemistry, then we can think that its activity is also motivated in the general framework of the organism's motivation and is active, not reactive.

Dr. Epstein: I'd also like to add my appreciation for this exciting and interesting work. It's a pleasure to see an attempt to understand how the brain mediates appetitive behaviors and goal-seeking behavior, what I was speaking about last night as expectancies. I was most curious to know whether you had looked for effects of change in the animal's drive state on these neuronal systems. I assume these animals are hungry when they enter your chamber and wonder whether you have manipulated the animal's state of hunger.

Dr. Shvyrkov: When our animals were hungry, I presume that the informational aspect, the motivation, is a representation or part of life experience. It is acquired in order to survive, according to Darwinism. Then motivation or drive, I guess, is a part of life experience which is a necessity. I believe that the initial emergence of goals in our experiments was connected to the biological necessity that is hunger.

We did not vary hunger, but we could observe that the behavior was cyclical. The rabbit would eat 30 or 40 portions and then would go to sleep. It could be thought that at the beginning or the end of the session the motivation was different. We watched the difference in the activity of the neurons. As long as there is behavior there is certain neural activity. Where there is no behavior, there is no neural activity. The causal connection is no neural activity, no behavior.

Dr. N. Miller: I wonder if you could use Roy John's technique of a pulsating stimulus in order to help you identify neuronal activities. For example, if, when the animal pressed the bar, the light flashed at a certain frequency, then if the animal was having an image of that flashing light and you were very lucky, maybe the visual neurons would be pulsing at that rate. But similarly perhaps, and this is more fanciful, you could have the carrot vibrate at a certain rate or have electric shocks very weak on the carrot at a certain rate and then perhaps, if you were very lucky, if the animal had the image of the carrot, you could identify that by a different frequency of stimulation. It might not work at all, but if it did work, it would be beautiful.

Dr. Shvyrkov: If I understood your remark correctly, it relates to the idea of the rabbit's image of the carrot and the stimuli which we provided. But we did not provide stimuli. The rabbit achieved his goals. We always must consider the whole life experience and we can create the controlled part, but we must take into consideration as many of the components of memory as possible.

Dr. R. F. Thompson

Dr. Asratyan: First, I wish to recall our experiments in 1958 which agree with your result in that we could not elaborate a conditioned response when the interval between the neural and unconditioned stimuli was less than 100 msec. Now, a question concerning the role of the cerebral cortex in conditioning. Can rabbits be conditioned in your situation after removal of the cerebral cortex? In general, your presentation was very nice and I can say that all your theoretical

Individual and General Discussions   *601*

assumptions are in the framework of classical Pavlovian theory and you need not have any other supplementary explanations.

Dr. Thompson: Thank you Dr. Asratyan. I very much appreciate your warm compliment. To answer your question, it has been reported from several laboratories that after extensive decortication of the rabbit, the conditioned nictitating membrane response develops in an apparently normal fashion. However, I would urge that we have caution in interpreting such data. Even though the rabbit can learn this simple response without a cortex, this does not mean that under normal conditions of learning the cortex is not involved.

Dr. Shvyrkov: First, a very small question. Have you seen in the hippocampus the complex spike cells described by Ranck?

Dr. Thompson: We do not have enough data yet to be able to be certain that complex spike cells are or are not pyramidal cells. I must say that many of our antidromically identified pyramidal cells do not show these complex spikes.

Dr. Shvyrkov: I must say that we also did not see complex spike cells in the hippocampus. Now, if I may, I would like to say a few words concerning the presentation. Dr. Thompson, you have a very elegant presentation and I would like to say how it can illustrate the usefulness of this direct contact. The studies of Ranck and others, recording the neural activity in the hippocampus in the behaving rat, showed that the activity of these neurons is obviously connected to certain types of behavior. Ranck supported the idea that the cells are connected with separate consummatory acts, separate appetitive acts. He also described cells that were activated when the rat was playing with a green rubber crocodile. Therefore, they are quite specific to certain kinds of behavior. We confirmed these data on rabbits, but we also showed that activity of hippocampal neurons is also dependent on goal-directed behavior, as I showed in the cortex. We had a question. Why, in the hippocampus, although specific neural activities are connected with separate behavioral results, and yet in the hippocampus certain general functions are performed in common which are necessary for complex behavior, for sequential behavior, and for differential behavior. We could not answer the question. Now, after Dr. Thompson's presentation, I believe that an explanation can be suggested. These fine data indicate that the same behavioral act like nictitation can be accompanied or not accompanied by the neural activity in the hippocampus depending on whether the act is incorporated in a larger system (learning) or in a smaller system (i.e., reflex response). I believe it shows that hippocampal neurons really all become activated in the case when a large, integrated goal-directed system is formed, as when the animal must learn.

Dr. Thompson: Thank you very much. I would only say in response that I'm very grateful for your comment and feel you may have stated the necessary conclusion from our work better than I have.

Dr. Gormezano: I have two comments raised by Dr. Asratyan's remarks. One concerns the possibility of conditioning in the decortical preparation. Dr. Thompson indicated that conditioning had been satisfactorily demonstrated in his preparation. But in addition, there have been investigations in this country involving cats, an animal higher on the phylogenetic scale, which have demonstrated very adequate conditioning. The question then becomes the degree of elaboration of conditioned reflexes that involve the cortex. In this regard, there is now no fundamental difference or disagreement among the American scientists and Soviet scientists. The second point concerns the minimum CS–US interval. This interval has been pushed closer in this country and in fact in my laboratory. Dr. Patterson has been able to obtain very effective conditioning between an indifferent stimulus and a US at an interval of 50 msec when that indifferent stimulus involves stimulation of a brain area.

Dr. Thompson: I would only say that I think my position requires me to argue that if we get conditioning over a shorter interval with an electrical brain shock CS, then we will also get conditioned increases in hippocampal activity with this same shorter interval.

Dr. N. M. Weinberger

Dr. Asratyan: Dr. Weinberger, your presentation was excellent. You say that the salivary glands play a very important role in development of Pavlovian teaching on higher nervous activity. It is true that the salivary glands have some advantages in comparison with motor activity. The salivary glands still continue in this important role, despite the other preparations you discussed. I am also convinced that your preparation, the pupil, is excellent. You can show many obscure aspects of the elaboration of conditioned reflexes. Pavlov was right in saying that the central phenomenon in higher nervous activity is the conditioned reflex. We can use these practical ideas to explain more complicated phenomena, but the conditioned reflex itself still needs detailed investigation. With the help of your methods it seems to me we are now able to continue our research on this more successfully.

Dr. M. M. Patterson

Dr. Berger: I actually have a question that is appropriate to ask not only of your very elegant paper but at least of the two preceding papers and probably has to do with some of the papers yet to come. It has to do with what we have been talking about here as the effective CS–US interval. I refer to recent work, particularly in the United States (and it would be interesting to know if there is parallel work in the Soviet Union), on the phenomenon that is known as conditioned taste aversion or bait shyness. Animals appear to associate the taste or the memory of a novel food with an apparent sickness that is induced after the injection of a chemical compound such as lithium. As a result of this pairing there appears to be a decrease in the future ingestion of the food that had been paired with the stimulus. What is unusual about these many experiments is that the interval between the eating of the novel food and the administration of the subsequent chemical stimulus is not on the order of several seconds, but may be several minutes to several hours.

Dr. Patterson: Thank you very much. It is a very complex question. I think it would be a good one for a general discussion and so I would like to defer it until the general discussion.

Individual and General Discussions 603

Dr. A. P. Karpov

Dr. Thompson: It is very nice work, congratulations. I assume that you are arguing that some of the mechanisms for the change in the neuronal activity as a function of reward are due to descending pathways through the olfactory bulb. Have you tried sectioning these pathways, i.e., isolated the bulb?
Dr. Karpov: No.
Dr. Blass: I too would like to join Dr. Thompson's congratulatory notes on your presentation and would like to ask you if you have studied the characteristics of the other two major olfactory inputs, that is, the trigeminal system and the vomeronasal system. Could you speculate on how they might enter into the functional systems that you're referring to?
Dr. Karpov: As far as [the] trigeminal one is concerned, since 1902 the classification of the different substances which divides them into trigeminal and olfactory at the same time stipulates that both types of stimuli play a role. You can talk about predominantly trigeminal or predominantly olfactory stimuli. We did not attempt to separate out the trigeminal component. Vomeronasal input was not considered.
Dr. Thatcher: I want to congratulate you on your excellent work. I'm sure you're aware of the American scientist, Dr. Walter Freeman, who is also working in the olfactory system. He has presented very interesting results recently at several conferences in which he records using an eight-by-eight array of electrodes simultaneously in the olfactory bulb and also 64 electrodes simultaneously in the cortex. And he finds, [as] you do, that any single neuron is not responsible for the coding or detection of a particular odor, but rather a particular organization or configuration of space is responsible. Furthermore, in commenting on Dr. Thompson's question, Dr. Freeman is convinced that the cortex is not necessary for these patterns since he does not see them in the cortex.
Dr. Karpov: Thank you very much.

Dr. D. Cohen

Dr. Aleksandrov: It is very seldom that one can hear such complete and well-argued analytical research. My first question relates to the following: You have said that the activity of motor neurons occurs in the latent period about 100 msec. Jennifer Buchwald found response to a click stimulus with a latent period of about 18 msec. I showed an illustration on early activation where the latent period was from 16 to 32 msec. These activations were observed in efferents of the spindles of masticulatory muscles. How would you account for such a difference between the latency periods in our research, Buchwald's research, and your research? Do you think it's a consequence of anesthesia or the fact that you were dealing with cardiac muscles or something else? I take liberty on the second question because of the title of our symposium. Would you say that reflex arcs interact or change in order to organize goal-oriented behavior?
Dr. Cohen: With respect to the first question, I think this is merely a matter of specific stimuli that we were using. You have seen that 30 of the 100 milliseconds

is required for retinal processing alone. Also, the latency is partially a function of the particular motor system we are studying where 20 milliseconds is required for just preganglionic to postganglionic processing. Thus, our results are probably not incompatible.

Dr. Lacey: Your electrophysiological descriptions were complete, but your behavioral descriptions were not. There are some theoretical issues in your answers that have been raised several times in this conference. You presented the cardiac response in relative terms. Now these are perturbations from some overall base levels. I don't know how consistent the response is, how much of it depends on the set level of your pigeons. Second, if you translated to absolute terms, what portion of the total dynamic range of the pigeon heart is exhibited in your preparation? For example, . . . 15 beats per minute represents a modest perturbation which would require a different theoretical approach than a larger perturbation. On the other hand, if this is a very large proportion of the dynamic range, that raises another, my third point. I am amazed at the specificity of the efferent path for this response. This is almost incredible. Is that high specificity of the efferent path due to high specificity or to the simplicity of your design? Given another situation with lesser or greater perturbation of heart rate, would the effect change from synergistic to a primarily vagal effect, which would make more sense?

Dr. Cohen: Let's first have some numbers and then speculate about them. If one looks at the response dynamics of the individual animal, at peak rate increases, they will be up on the order of about 35 beats per minute. Now one can manipulate, this number as a function of the training paradigm. We happen to use a standardized paradigm that yields this response magnitude but it could be increased.

Dr. Lacey: Which is what proportion of the dynamic range?

Dr. Cohen: Baseline heart rates are on the order of about 160 to 180 beats per minute. Now, the pigeon is capable of rates up to 450 beats per minute, although these rarely occur in any normal physiological situation. They can occur after vagotomy. The unconditioned response to a reasonably intense foot-shock is a rate increase to about 350 beats per minute. Therefore, the effective dynamic range is approximately 170 beats per minute.

Dr. Asratyan: I am quite with you that heart rate changes are a biological part of every reaction. Changes in heart rate could be taken in a simple behavioral act like a simple conditioned reflex, alimentary or defense. It is almost impossible to analyze functional changes in much of the chain of complicated reaction in the heart beats. Many different agents are taking part. It's not only neuronal but also humoral. Heart beat has special significance because it accompanies all kinds of behavioral acts.

Dr. Cohen: When I first chose this model, I chose it for technical reasons. What I needed was a system where the animal could tell me it was learning without any moving. That's why I chose heart. Having chosen it, I developed an interest in the kinds of questions that you raise, and this has led me to a strong interest in the neural control of the heart more generally.

Dr. Shvyrkov: It seems to me that Dr. Cohen's work has achieved something that no one else in the many years that reflex theory has been in existence has been able to achieve, that is, to track the conditioned reflex arc. It is an outstanding result, which will revalidate the data obtained. I have the following question. Obviously in normal behavior, the changes in the cardiac activity never occurred in an isolated manner. They always occur simultaneously with changes in respiration, flexing of muscles, and many other components. All of them are mutually coordinated. Such coordinated functional co-existence of different systems has insulated the response to a stimulus and excluded the possibility of occurrence of other phenomena because all of them must be coordinated and there must be some special coordinating mechanisms. I believe that the impression of the arc in Dr. Cohen's work is produced by very stable experimental conditions. If we alter the conditions, we would see changes not only in these areas but in others and the connection of this effect in these areas to other areas. You showed that after the flicker of the light, the cardiac frequency changes and you track the pathways which are involved. Why does the cardiac frequency increase and not decrease?

Dr. Cohen: I think the increase is totally appropriate to what the animal would do normally in that kind of situation were we not restraining it experimentally. A bird put in a stressful situation will fly. That's a severe exercise response that requires an increase of cardiac output, the cardioacceleration is merely a component of that. In the rabbit whose normal response might be to freeze, one would expect deceleration. I think the cardiovascular response that one sees can be predicted on the basis of the animal's behavior in that kind of situation.

Dr. M. Gabriel

Dr. Thatcher: I noticed that a major change seemed to occur to the CS− more so than the CS+ both in the cingulate and AV. Initially in the cingulate, the CS+ increased above baseline. Then it tended to stay about like it was in the pretraining condition, whereas the CS− stepped down. Would that indicate to you that the discrimination that was established is largely an inhibitory kind of discrimination, that the neurons responded less to the CS− and that's a more accurate way to characterize change?

Dr. Gabriel: The question of whether the response to the CS+ goes up or the response to the CS− goes down is an interesting one. But our analyses haven't been conducted in such a way that I can answer it very easily. Your visual impression may be accurate, but in the primary time change there is reduction in response to the CS−. I don't know quite what it means to attempt to label that as an inhibitory discrimination though. For my purposes, what's important is that the two signals produce different effects in this area of the brain and what I think we're looking at is some aspect of signal processing that is important for the animal in terms of the resultant behavioral decision to respond.

Dr. Shevchenko: I was quite interested in your report especially because we are working in adjacent areas. I have a very simple question. How many cells did you explore in every area?

Dr. Gabriel: We were doing multiple-unit recording. So we did not look at the activity in individual cells.

Dr. Shevchenko: But if you were exploring ten times as many cells, I think that you would get a result that in the limbic cortex the behavioral response would happen earlier and in the ventral thalamus the neuronal change would occur earlier than the behavioral. Would you clarify this?

Dr. Gabriel: Is that within a single trial or is it in relation to repeated sessions of acquisition? The results that we've found were that all of the placements in AV thalamus responded after the behavioral discriminations was acquired. And I do not think that if we increased our sample of electrodes in AV thalamus that we would find that neuronal changes occurred before the behavior or with the behavior. I'm quite sure that that's the case. It's probably of some importance to say, however, that we've looked at a large number of placements in the cingulate cortex and they are not entirely homogeneous. In fact there really are two populations of cells in the cortex, one population acquires early and induces this discrimination and the second population behaves in a fashion that is identical to what we see in AV—it acquires late. And we have evidence now that the distinction can be correlated nicely with the layer within cingulate cortex. Our placements in the superficial layers of singular cortex are the ones that behave like the placements in AV nucleus, while the placements in the deep layers, 5 and 6, behaved as I showed on the first slide with early acquisition and continued participation. So, there appears to be an important separation that goes on at the cortical level in terms of the time at which a discrimination can be seen.

Dr. Cohen: Did the animals have a significant change in their latency of response to the CS? And if so, did that correlate with the development of change in AV?

Dr. Gabriel: Looking at latency would be an important thing to do to try to find a behavioral correlate with the advent of discriminative activity in AV. And we are in the process of looking for the correlation. I can't answer your question. In general, there is an increasing latency of behavioral response with training in this task. It goes progressively longer. But my impression is that it's a very progressive sort of thing and it's not the kind of thing that would be correlated with the rather subtle advent of discriminative activity in the AV nucleus, but it should be looked into.

Dr. K. H. Pribram

Dr. Shvyrkov: Thank you very much, Dr. Pribram, for your very interesting presentation and most interesting data. I would like to pose two questions with respect to mechanisms of goal-oriented behavior. First, do the arrows on your diagrams mean the sequential excitations in different brain structures? Second, what is the interrelationship of the relationships between the structures of the brain and information about the environment? In other words, are the arrows an indication of biological interrelationships? If so, in what respect does this stand to the "informational" connections?

Dr. Pribram: Thank you for these very penetrating questions. For the first question about whether the arrows represent sequential processing, we've done three studies in monkeys and in children and in adult humans in which we used the evoked potential technique and compared the waveforms of the evoked response. We used the slope of the evoked response as our measure. We found that in monkeys the processing was in fact sequential, in the sense that things would happen in the visual cortex first, then in the temporal cortex, and then in the frontal cortex. In the visual cortex, what is represented seems to be whatever stimulus configuration we present to the monkey. In the temporal cortex, we get differences dependent on the reiforcing contingencies with respect to whether the animal can respond to redness or the diamond shape, or whatever. In the motor cortex, we have differences in the waveform with regard to whether the animal presses the right panel or the left panel; it has to do with motor response. When I say these waveform differences develop sequentially, what I mean is that early in the behavior, the visual cortex already represents the stimulus. Later on, some decisional mechanisms seem to become involved and then the waveforms begin to differentiate in the temporal lobe. And, finally, when the animal is doing very well, then there is a representation in the motor cortex concerning which panel the monkey presses.

Dr. J. Lacey: I'm going to ask you a very complex question, Karl, because it's been a long time since we've been able to talk these things over. You just said different parts of the brain do different things. I want to elicit your thoughts about a concept which hasn't yet really been labeled during this conference, but I'm almost certain will be brought up tomorrow, namely, that the brain is a self-organizing mechanism. In your presentation you placed emphasis on what was happening in the environment. You talked about temporal tags, you talked about spatial tags, and you said, for example, that the temporal cortex and frontal lobe structures have different effects upon the recovery cycles in evoked potential studies. Now, my question is, how does the brain call upon frontal mechanisms or temporal mechanisms to accentuate or diminish the environmental input? How are basal ganglia influenced and by what process so that receptive fields in the visual analyzers can be increased or decreased?

Dr. Pribram: Diane McGuiness and I have suggested that there are at least three discernable mechanisms in the frontal-limbic part of the brain. One deals with phasic responses to any kind of input, whether it be from out in the world or inside. The second system depends very much on what is being done—it is a more tonic type of mechanism. We've called it a set of readiness mechanisms. We've done experiments showing that they relate to the CNV, which may be organized in the basal ganglia. On the basis of a great deal of behavioral work, we have suggested that the readiness systems "keep things going." If I had to specify what the frontal lobes are doing, they are simply sending up flags to interrupt this readiness mechanism or to initiate another readiness mechanism. All of these functions are closely related at lower levels to hippocampal mechanisms. These turn out to be related to an "effect-comfort" dimension.

Dr. Gormezano: Dr. Pribram, I'm interested in getting clarification of your conception of images in the light of several papers presented today and previously. I refer specifically to the Shvyrkov group, who have been referencing a notion of images in their models of goal-directed behaviors. There has been from the Asratyan–Gasanov group a notion of images which is the functional stimulus and the local conditioned reflexes. Although I am anchored by training and disposition to more behavioral analyses, I'm aware of course of the Hebb reverberating circuit analysis and my concern is for perhaps increasing the communication among these concepts as well as enlightening me in the mode of analysis. What is the role of initiating stimuli in the environment to the·evoking of an image of a local reflex? Secondly, what is the role of experience in that image, and thirdly, the question of time parameters with regard to such notions as consolidation?

Dr. Pribram: There is more and more evidence suggesting that individual neurons at the cortical level are encoding events in the frequency, i.e., the wave form, domain: In visual system, it is spatial frequency, in the auditory and somatosensory system, some form of two valued frequency. In the visual system the work has gone along probably faster than anywhere else. Work at the Cambridge laboratories, in Italy, in the Leningrad and in the Bay Area has shown that what has been thought of as feature detectors in the cortex, a "line detector" on the basis of Hubel and Wiesel's original discovery of elongated, orientation selective receptive fields; are in fact spatial frequency selective. Our work on the motor cortex suggests that the motor systems also function in terms of frequency analysis. If we encode events in the waveform domain in the cortex, then when a stimulus comes along either from the outside world or from internal stimulation, we then excite a process which gives rise to any image much as in a hologram. The images aren't localized in the brain any more than they're localized in the hologram. Something excites the wave domain representation which then produces an image and we project that image external to ourselves in the same way that an acoustical image is projected forward and between two speakers in a stereo high fidelity music system.

Dr. U. G. Gasanov

Dr. Verzeano: I've been delighted to see this piece of work which is so elegant in concept, in methods, and in results, and which agrees so well with our own work. Dr. Gasanov's paper is open for discussion.

Dr. Thompson: Thank you, Dr. Gasanov, for a very important paper. I have a small technical question and then an observation that I wish you to comment on. The small question is, in your analyses you computed first-order probabilities, i.e., given that neuron 1 fired, what is the probability of neuron 2 firing, and so on? Did you look at any secondary probabilities?

Dr. Gasanov: No.

Dr. Thompson: Now, the general remark is that it looked to me that the largest change often occurred during extinction. The change in the pattern is very striking. The extinction was not simply a return to the original spontaneous activity, but rather a major change in the pattern of activity.

Dr. Gasanov: Yes, that is correct, it seems to me very important.

Dr. Weinberger: I also congratulate you for a very important piece of work. In the example you gave, before elaboration of the conditioned reflex, the activity circulated in a clockwise fashion. Neuron number 1 excited number 2, number 2 excited number 3, number 3 excited number 1. Whereas after stabilization of a conditioned reflex, neuron 1 excited neuron 3, neuron 3 excited neuron 2, neuron 2 excited neuron 1. Now the question is, whether this reversal of circulating excitation is characteristic of the difference between the early stages of training and the stabilization stage, or whether this is simply an example that has been seen on occasion but is not seen under every experiment.

Dr. Gasanov: I think that such a change in the pattern of circulation is indeed related to the stage of conditioning.

Dr. Goldberg: Dr. Gasanov, it was an elegant demonstration. Do you think you have evidence that these changes in the temporal relationships of discharge during acquisition are due to the relationships of the neurons within this small loop, or rather from their independent relationships to neurons outside of this local circuit?

Dr. Gasanov: I think we see a very small part of a much larger network involved in conditioning.

Dr. Asratyan: In connection with this very last paper and your very excellent paper, Dr. Verzeano, a point of view. Pavlov, in 1911, expressed an idea that when you combine an indifferent stimulus with a relatively important stimulus which we name the reinforcing stimulus, there are two different events. One of them is the formation of a conditioned connection between them. But a second event also occurs. In the framework of neuronal projection of the conditioned stimulus, the neurons compose and complete an image of stimulus. This relates to the possibility of the formation of a cell assembly, a model of stimulus.

Dr. Pribram: I like the distinction between an image forming kind of connection and a conditional one. I think we should get the translation correct here; it's not conditioned but conditional.

Dr. Asratyan: Maybe, maybe.

Dr. Pribram: No, no, a very important difference. Because in one case, it's conditional on the appearance of the reinforcing stimulus. In the other case it would stay afterwards. And I think both your work and the work that we just heard says they're not permanent in that sense. They're conditional on the appearance of the conditional stimulus. And that's a distinction I'd like to maintain that Pavlov made.

Dr. M. Verzeano

Dr. Lebedev: I can't find words for my congratulations on your discovery of the constancy of the spatial and temporal organization of neural activity. Would you comment on the possible contradiction between your results and the notion that the temporal distribution of spontaneous discharges of a neuron are sometimes distributed as random?

Dr. Verzeano: There is no real contradiction here. I presume that the investigators were recording the activity of one single neuron here, and another single neuron there, in unrelated networks. If we put these data on the computer to determine the statistical distribution, we would indeed find a random distribution. A single neuron at certain times will participate in activity which circulates and at other times it will not participate. So if you look at a single neuron at a time, it's a hit or miss proposition.

Dr. J. Lacey: I'd just like to raise a conceptual problem. I haven't been able to get what kind of a network you're conceptualizing. Is this serial processing or parallel processing? What are your criteria for a network? How do you know this is a circulating movement rather than looking at three spots?

Dr. Verzeano: This is a very good observation. What you are actually asking is "how do you know that neuron number one fires neuron number two and that neuron number two fires neuron number three, etc?" The truth is that we don't know any such thing. All we know is that these three neurons keep firing in a regular sequence again and again. It may very well be that impulses coming from another region of the brain trigger these regular sequences. When we say that activity circulates we mean, simply, that it appears that it moves from one point of the network to another, very consistently. We do not assume that it is necessarily directly transmitted from one neuron to the next.

Dr. J. Lacey: So your criteria are geometric?

Dr. Verzeano: They actually involve space, time and velocity

Dr. Shvyrkov: Dr. Verzeano, would you please comment on another possible explanation of your data: that it is not the connections between neurons, not the itinerary of excitation that you see, but rather a reflection of another more general process in which these three or five neurons observed by you are involved, in the same way as other neurons?

Dr. Verzeano: What you say is quite true. These 3 or 4 or 5 neurons are involved in a very general process. If you look simultaneously at the cortex and the thalamus, you find out that the cortical activity is run from the thalamus. Impulses come from those thalamic nuclei which correspond to those particular regions of the cortex. Then, through a short feedback loop, cortical neurons fire back into the thalamus, which then through another circuit fires to another region of the cortex which fires back into the thalamus and so on. The circulation of activity is highly organized.

Dr. Asratyan: When you combine light with electrical stimulation, you are establishing the condition for the formation of a conditioned reflex. The patterns of neuron activity you see might be like Hebb's cell assemblies. This idea belonged to Pavlov. I am very glad to see that your excellent facts concur with Pavlov's ideas.

Dr. D. G. Shevchenko

Dr. Verzeano: You mentioned at one point that you found configurations of neuron activity which were similar in different structures on the brain in relation to behavior. Could you tell us in which regions of the brain you found those similar sequences of activity?

Dr. Shevchenko: This research was conducted both in our lab and in numerous other laboratories around the world. We recorded neuron activity in the visual cortex, reticular formation, the motor cortex, and in the hippocampus.

Dr. Verzeano: The relations that you have shown so beautifully between the activity of neurons and the polarity of the evoked potentials are similar to our own work, but you have gone beyond us in the sense that you combined your recordings with behavior. Very beautiful work.

Dr. Weinberger: You demonstrated that different neurons participate at different phases of the situation. Were these neurons located randomly as a function of depth in the visual cortex or were different types of neurons segregated within the visual cortex?

Dr. Shevchenko: When we recorded these neurons, they were not localized in separate areas of the visual cortex. Similar neurons, as I already said, have been studied in the reticular formations and in other brain areas.

Dr. David Lindsley

Dr. Cohen: I refer to the experiments you did with Dr. Chow on the geniculate. Since that time, the whole story about X and Y cells has developed the whole parallel processing notion. It would be interesting to return to those experiments and determine whether it's the X or Y cells that change or whether both classes do. In particular, since the Y cells seem to be ones that demonstrate plasticity with the developmental perturbation, I've always been very curious as to whether that whole approach will ultimately lead us down the road, indicating that in fact there may be specific classes of neurons that are specialized for plastic change.

Dr. Lindsley: One interpretation that had been made of those data was that they represented corticofugal input to the lateral geniculate. The time relation between the spot and the flash was such that there was plenty of time to go to the cortex and come back. I see subsequently that's been shown by intracellular recording. But still, there remains the interesting question of whether the X or the Y have more response plasticity. We now have the techniques to begin to look at this problem. First of all, if the animal's active, as Dr. Goldberg indicates, you change the response. There is a much more dynamic response in the cells. In many of the experiments that we've been doing, we're also worried about whether the animal's visual attention is constant. We use visual pattern discriminations to ensure the animal's attention. These techniques are evolving in Goldberg's work, that of Mountcastle and others, and we may begin to answer the question about X and Y cells more easily.

Dr. Goldberg: What do you think of the possibility that posteye movement response of those pulvinar neurons is not to movement responses in the colliculus where there's a pre-eye movement response but could be coming either from the frontal eye fields where another posteye movement response has been demonstrated, or from eye movement proprioceptors—the spindles?

Dr. Lindsley: We actually haven't attacked that problem in detail or even through corollary discharge in oculomotor neurons. There really are a number of possibilities. Our interest is in eye movements in relation to visual attention. So we haven't really attempted to track down the answer to your question. But those are good points. The thing that was of some interest to us was that a smaller percentage of cells responded to eye movements in the dark. Now, how do you account for those cells that responded to spontaneous eye movements in the light? Is it that you need a certain level of tonic visual input to increase their excitability so that when oculomotor information enters they are more easily triggered? Or, since there are these interesting reciprocal connections between the pulvinar and extra striate systems, is it possible that it comes through there in a phasic way? I think this is the kind of thing that your work and others will begin to get into more and more.

Dr. M. Goldberg

Dr. Marshall: Dr. Goldberg, these results are very interesting. We have data indicating that, at least in the rat, damage to the superior colliculus does not cause a general inattention to stimuli. There is a difficulty in the head-orientation response to impinging stimuli on the contralateral side. But if the animal is required, for example, to make a movement of the limbs as in locomotion to avoid objects in the path, there is no deficit. This is in contrast to our work in rats with lateral hypothalamic damage and damage to basal ganglia structures which produces what seems to be much more of a clear inattention to stimuli. The latter type of animal I think parallels much more closely what has been described clinically after parietal and frontal lesions in humans.

Dr. Goldberg: I agree with what you said because we looked at collicular lesions in the monkey and the only deficit that we could find was a difficulty with the initiation of an eye movement. The latency of eye movement initiation was 100 or so milliseconds longer in that test. There is an area in the monkey close to the lateral hypothalamus, the substantia innominata, from which we found a very strong projection to the parietal cortex. So I think it's certainly possible.

Dr. Aleksandrov: It was a beautiful presentation. I mentioned the relations of activity of cortical neurons and those of the trigeminal nucleus. The cortical neurons are on the hierarchy of the behavioral level and the neurons of the trigeminal nucleus are on the level of certain subsystems. I think we have here a fine correlation.

Dr. Goldberg: I would agree. I think that there is an interesting distinction. When you were looking at the trigeminal motor nucleus, you were looking at neurons which were very far out on the output side. Our neurons in the superior colliculus are in the input side discharging with a latency of 40 msec over a behavioral act with a duration of 250 msec. But yet the analogy is very pleasing and I agree with you.

Dr. R. L. Isaacson

Dr. Asratyan: I want to comment on the difference between the dorsal and ventral hippocampus. When we use dorsal hippocampus stimulation as a conditioned stimulus for elaborating alimentary or defense reflexes and, in other cases, use ventral hippocampus, there are quite different dynamics for their elaboration. In defensive instrumental reflexes, using stimulation of dorsal hippocampus, there is generalization to the amygdala and septum. They also are responding to our signal stimulus. But during much of the time, amygdala activity is abolished. But activity from the septum is increased. Functional interconnections between dorsal hippocampus and septum are there. It is also not true to say that the hippocampus is a formation which is specialized for Pavlovian inhibition.

Dr. Gasanov: Dr. Isaacson, while I appreciate your theoretical conclusions I'm inclined to think that lesions to separate structures always involve changes of related systems. The structures were involved in certain systematic activities of the brain. Lesions of separate structures do not provide us with sufficient data to find out functional roles of the structures, and in most cases reflect the dysfunction of the area of the brain where the structure plays a subordinate part.

Dr. Isaacson: In terms of understanding the brain or the hippocampus or any other structure, I don't think that any method is going to be useful entirely by itself. The thoughts that I have about nucleus accumbens, for example, derive as much, if not more, from anatomical, electrophysiological, and neurochemical studies of the structures by others rather than from our own data. I don't think that the lesion technique can answer any question in an absolute sense nor do I think that any other method can either. So I think no matter whether you use electrical or chemical stimulation or count receptors or whatever one does, you get a partial answer and that it's when you have a coherence among all of the data that you can come to some reasonable conclusions about how the brain operates. I think there is a great deal of evidence that stimulation, say with dopamine, produces heightened motor activity, heightened locomotion, will actually bring an animal that has been reserpinized out of his lethargic state and into one of mobility and activation without inducing stereotypy. On the other hand, if the same amount of dopamine is placed in the caudate nucleus you get stereotypy without the activation of the organism. In relation to Teitelbaum's findings, our results show that dopamine agonists (those that affect primarily the caudate nucleus) always reduce activity in the adult animal, but when you use drugs which stimulate nucleus accumbens, you never see this reduction in activity. But if you examine the animal's interest in the environment, his curiosity, his exploratory behaviors, all the dopamine agonists that I know will reduce this component of the animal's behavior in a dose-related fashion.

Dr. L. P. Spear

Dr. Pribram: I have only one comment: You cannot understand conditioning very well by looking at presynaptic events.

Dr. Spear: Perhaps from a historical perspective, this is true. But presumably since any event that occurs presynaptically ultimately effects behavior through action on postsynaptic regions, emphasis on presynaptic regions may be an important area for future study.

Dr. Berger: Dr. Spear, I would like to reinforce the approach that you've taken in the utilization of chronic administration of drugs and the analysis of behavior for two reasons. Psychopharmacological agents are usually administered on a chronic basis at the clinical level. Yet, our knowledge of the behavioral profile of these substances is derived almost exclusively from studies of their acute effect. Second, though drugs and chronic administration of drugs are sensitive and selective tools for inducing changes in sensitivity of various functional aspects of the synapse and correlating their effects with behaviors, there is still very little work in this direction. We have found in our laboratories that animals treated with 21 successive injections of chlorpromazine and then during the so-called withdrawal phase are tested for acquisition of a discrimination response show facilitation of that learning.

Dr. Gormezano: Dr. Spear, with regard to your analysis of the role of amphetamine in the retention process, that may be an action on the motor systems or on a general activation system. Is your drug operating on an associative process as opposed to a sensory process or a motor output?

Dr. Spear: We are now running this same procedure using a different retention test or a combination of the active avoidance and past avoidance procedures we've used. Of course the results reported here could be just an effect of locomotor activity. But I don't think so for two reasons: the half-life of amphetamine is such that it should be out of the system 48 hours later. Moreover, in the withdrawal phase, after chronic amphetamine administration which is what we're dealing with here, classical hypoactivity is seen. This withdrawal-induced hypoactivity *per se* would impair, rather than enhance active avoidance responsing.

Dr. Pribram: What do we know about the developmental chronology of these autoreceptors vis-à-vis the receptors on the postsynaptic surface?

Dr. Spear: Very little, and my laboratory is currently involved in this work right now. Presumably these auto-receptors are developing fairly late. We know that postsynaptic receptors develop prior to the time in presynaptic input in certain brain regions. So presumably, they are developing before the auto-receptors, but this has to be clarified.

Dr. Gottschalk: Chronic drug administration may well activate hepatic enzymes which could increase the destruction of some of these pharmacologically active metabolites. In such an event, you may have to find a way to measure what those actual levels are because some changes in sensitivity may be related to lower blood levels.

Dr. Spear: Oh, yes, this in an important consideration.

Dr. Pribram: I think this is terribly important work especially since so many of the children in our schools are chronically being given ritalin, amphetamines, and other drugs and if we really find out that either learning is speeded up or precluded under the drug conditions, it would be a very important finding.

Dr. N. J. Kenney

Dr. Aleksandrov: Could you comment on the situation when a tested animal is in an emergency state and loses a lot of liquid, but the thirst motivation may not emerge. How is this absence of thirst in an emergency explained by your prostaglandin-angiotensin theory?

Dr. Kenney: I don't think that the reduction in water intake following our treatment is due to an increase of emotion. If that were true, all water intake should be blocked, but water intake stimulated by hypertonic saline is not reduced. It also does not reduce food intake, which again would be reduced by hyperactive or hyperresponsive animals.

Dr. Goldberg: What do you know about the synaptic action of either of your agents and do you think that these could be acting by a synaptic mechanism or by something very different from that?

Dr. Kenney: At this point the mechanism of action, especially the synaptic mechanism of action, of both angiotensin and Prostaglandin E is highly controversial. It appears that Angiotensin II, at least in the control of thirst, may be acting as its own neurotransmitter. I could only speculate at this point on whether or not Prostaglandin E acts on synaptic transmission, although some people have argued that angiotensin-induced drinking is mediated by dopamine, and there is some indication that Prostaglandin E inhibits adrenergic systems and dopaminergic systems. That is a possibility.

Dr. K. V. Shuleikina

Dr. Teitelbaum: Is there any evidence that before auditory stimuli become important, kinesthetic and vestibular stimuli are effective? For instance, when the parent bird simply lands on the nest and shakes the nest, does that act as a signal and activator for feeding, perhaps even before the sound and visual stimuli become effective?

Dr. Shuleikina: I agree with you that kinesthetic stimuli play a part. They can play a role with other signals. We did not study this in particular. But I believe that this role is relatively small.

Dr. Verzeano: Were your recordings of neuronal activity all from the cortex or were some from subcortical regions? Following sensory stimulation, the discharge of some of the neurons change. What kind of sensory stimulation was it? How long did it last? And how long did the changes in discharge last?

Dr. Shuleikina: EEG recordings were from several brain areas: from the structures which, according to the literature, relate to serialization of feeding behavior. We recorded activity of the motor cortex, lateral hypothalamus, medial hypothalamic nuclei, medial reticular structures, amygdala, and ventral-posterior and medial nuclei of the thalamus. We stimulated the chorda tympani and the lingual nerve. One sensory stimulus lasted 500 msec to 1 sec, and we used electrical stimulation consisting of 10 pulses. The cortical activity was usually a spike train or just separate cortical discharges.

Dr. Spear: In mammals, it appears that temperature is an important central component in early food-seeking behavior. Animals move toward the warmth of the mother. Might it be temperature sensory cues that are important in this early food-seeking behavior in the birds that you have been looking at?

Dr. Shuleikina: As to temperature, we did not study this specifically. The experience with mammals is that temperature of the developing organism is of great importance and it is also true for birds. But since our experiments were conducted in natural conditions, that is, in the natural interrelationships of the feeding parents to their offspring, under these conditions we might say that the temperature conditions were always prevalent but did not specifically guide our research.

Dr. Satinoff: What was the earliest age that you could record two to four cycles per second activity in subcortical structures in the kittens? Did you try to use other methods than cocaine application to create sensory deprivation? It is known that there are profound effects on appetite with certain concentrations of cocaine.

Dr. Shuleikina: Slow-wave, high voltage activity was observed in the very early stages of the kittens' development. The early stage was technically the first two days of life. The first day the kitten was operated and the second day its behavior was studied. This high voltage activity was present at the first observation. The younger kittens showed this activity generalized and widespread in the brain. Yes, we used another method of deprivation: It was the irreversible cutting of the nerve.

Dr. E. M. Blass

Dr. Shuleikina: I find your research very interesting. Results we have obtained working with human infants and with infant kittens confirm and compliment your results. The odor of the nipple and its taste is a key sensor signal not only for the blind infants but also for those with eyesight. The second fact concerns putting on the nipples substances which have different smells and which destroy the natural smell of the nipples. Kittens do not just stop licking. They usually creep around the cat and fall asleep in regions of the cat they prefer. To a certain extent we could not comprehend this phenomenon and now it all comes together. The last fact concerns our observations on newborn human infants. We examined the sequence of the significance of sensory stimuli in the development of the sucking behavior of a newborn baby. We compared tactile and taste stimuli. It turned out that for the newborn baby, before it has a suckling experience, the most important stimulus for suckling consisted in the taste information. We have a device which made it possible, after removing the nipple from the mouth, to change the taste of the liquid. If, after two or three suckling movements with milk, we suddenly changed to a more salty, bitter tasting liquid, after two or three suckling movements, suckling stopped. If, on the other hand, milk stopped coming into the nipple, the suckling movements continued. But on the sixth to the eighth day, the signal significance of these stimuli was changed. At this time, if we changed the taste of the incoming liquid, then, for a certain period of time, the baby should suck even the bitter liquid. But he very quickly stopped the suckling movements if the influx of the milk stopped.

Dr. Blass: You raised a number of fascinating points on a number of different species. In regards to the trigeminal breakdown, I think the sequence in any mammal is that in order for it to attach to the nipple, the trigeminal nerve must be intact. In the case of the rat, the odor that is being given off from the nipple acts permissively and it allows the tactile qualities of the nipple to gain control over the motoric stereotype response. Second, if you remove the scent from the nipples of the rat and instead of returning them to the nipple, return them to the midline of the animal, every one of the babies will eventually come to the midline, ignore the nipples, and sleep. I think that they are oriented toward a familiar odor. And then finally, I'd like to comment on a number of reports that have come out of California and that are more fully documented in Dutch literature where home births, as opposed to hospital births, are the rule. It is claimed that immediately after birth, the mother will rest for 3 or 4 min. There is a great deal of excitement in the room among those who have witnessed the birth. Even before the umbilicus is severed, the mother will take the baby and offer it the breast, offer it the opportunity to suckle. The babies refuse to do this. Instead what they do is lick the nipple for 5 or 6 min and only then will they suck on it. So the initial bond between the mother and the infant appears to have some chemical mediation, in addition to all of the other ones.

Dr. R. R. Miller

Dr. Shuleikina: Could you tell me, please, precisely what the discriminating element was in your experiment? Were they using both visual and tactile cues?

Dr. Miller: The discriminating element in our apparatus, other than the geometric shape itself, was the floor. The floor on one side of the apparatus, the low-shock density side, was smooth and illumination came up through the bottom. This is necessary in order to count the animals because the tadpoles are close to transparent. The other side of the apparatus was dark and had a very rough texture. It was like sandpaper. The extent to which the animals were responding to the visual as opposed to the tactile cues, we do not know. Tadpoles are filter feeders, they move along the bottom like a vacuum cleaner and they're almost always in contact with the floor. That's why they chose to have a tactile cue as well as a visual one.

Dr. B. D. Berger

Dr. Lomov: Dr. Berger, what is your opinion about the role of the medial hypothalamus; for instance, ventro-medial nuclei?

Dr. Berger: We simply haven't looked yet. I can only say that this is not a non-specific effect of destruction of the brain. We also lesioned animals dorsal to the lateral hypothalamus and in such animals we did not get deficits.

Dr. Lomov: It seems that your results raise an important question about goal-oriented behavior in cooperative animal behavior. I believe that when we're talking about humans it is impossible to comprehend goal-oriented behavior without analyzing social contacts. We conducted certain experiments which indicate that the conditions of social contact affect perception and cognitive functions. Do you have a special method of describing and analyzing this cooperative behavior? How do you isolate the acts which characterize their cooperativeness? Any special manner?

Dr. Berger: In my comments today I have tried not to use the word cooperative behavior. But I've tried to use the word coordinated behavior because I wanted to avoid anthropomorphizing. Yet, our approach is that animal models should contain some of the salient features of various human behaviors. We have studied social components in these behaviors. There is a difference between animals that are raised together and animals that are raised isolated in their ability to learn these kind of tasks. Also, if we train a pair of animals in a situation and then split them up and put a trained animal in with a naive animal, there is great facilitation of the responses. Further learning takes place much more quickly when pairs are identical each day than when they change pairs from day to day.

Dr. Asratyan: If this is a model of relational behavior, then it seems not so difficult to explain the physiological mechanism of it. Three or four decades ago in the Soviet Union, investigators were able to train a dog to flex its leg in order to interrupt a painful stimulus to a second dog. Social interrelations could be studied through the physiological mechanisms of conditioning.

Dr. Berger: Thank you very much for your comments, Dr. Asratyan. One of the reasons that we're interested in this is to try to develop an animal model using the rat that will allow us to do many things, that because of economy and technology, we can't always do in higher animals. In fact, the few studies that have been done in this area have not been done with the rat. One reason the rat has not been used more often may be that the rat is not often regarded as a "social animal." Another reason may be that other behavior tasks that have been attempted were not sensitive enough to reveal the subtleties of social behaviors and thus misleadingly reported "negative results."

Drs. B. Lacey and J. Lacey

Dr. Thatcher: I just wanted to emphasize one thing rather than ask a question because I appreciate very much the work you're doing there. Its relevance is clear in terms of a "systems" perspective. It seems as though you were looking at a dynamic system that is optimizing itself. One subtle aspect of the optimization process is establishment of the best conditions of blood flow, nutrient content, and so on, for optimal performance.

Dr. J. Lacey: Thank you. That is an insightful comment. Indeed, at the back of our work is control systems theory. However, we think it premature to emphasize heavily this aspect. We cannot yet specify the nature of the servo amplification; we do not know whether it's a proportional controller or a linear controller. We cannot specify damping constants, linearity overshoots, et cetera. So it must remain an analogy for the moment.

Dr. Lindsley: John, what is the relation of reaction time to the phase in the cardiac cycle at which the stimulus occurs?

Dr. J. Lacey: I didn't mention that, Don, because it takes us into details of cardiopulmonary physiology. In general, reaction times are slower when stimuli are administered at 350 msec after the R-wave than when they are administered at the R-wave, but this is true only during the inspiratory phase of respiration. It is not true or even reverses in the expiratory phase.

Dr. Lindsley: How about the relationship of time in the cardiac cycle to cortical EEG desynchronization? That brings me to a study that we did on reaction time with warning signals where we got desynchronization at about 300 msec and our reaction times dropped to the lowest level.

Dr. J. Lacey: The duration of alpha blocking in the foreperiods of our experiment seems to depend upon the specific experimental arrangements. There is, however, an article in the literature now that shows that the latency of alpha blocking does depend on where in the cardiac cycle the stimulus is administered, and it seems to coincide with your interpretation.

Dr. Pribram: If you really have a variable set-point in this system, then you're dealing more with feed-forward than with feedback loops.

Individual and General Discussions 619

Dr. J. Lacey: The control theorists are way ahead of us. Most of the work, of course, has been done in negative feedback. There is lots of work being done right now on positive feedback, self-induced oscillators, and so on.

Dr. Pribram: That's not feed-forward. By a feed-forward mechanism I mean the parallel processing mechanism in which you have two inputs, one of which can change the set-point and the other one of which deals with the feedback itself.

Dr. J. Lacey: I don't want to be tempted into a premature answer. Again, we have servo theory in the back of our mind as the explanatory construct. However, there is a long way to go before we can develop quantitative theory that can handle our experimental results.

Dr. Pribram: Drs. Lacey, either one. What is the effect relative to stroke volume? For example, as you decrease heart rate, you increase stroke volume. What would that do to the firing of the baroreceptors?

Dr. J. Lacey: These changes in heart rate are so small that in dog, rat, rabbit, monkey, they would not be accompanied by any changes in cardiac output. We have not yet measured the cardiac output in humans. The effect of increased stroke volume would be to increase baroreceptor firing. But remember, baroreceptor discharge is a function of slopes and rates of change. We believe, but cannot yet prove, that the changes in heart rate are so small that they would not produce changes in cardiac output.

Dr. Pribram: The reason I was asking was that athletes are known to have very slow heart rates. I wonder if that facilitates this sort of sensory-motor integration process?

Dr. J. Lacey: Let's think about the variable set-point. Organisms adapted to a given range of cardiovascular functioning may have a different set-point than organisms adapted to another range in life style. Bea has some data. . . .

Dr. B. Lacey: No, I really did not have any specific data to bring to bear on that, but I was thinking back to some earlier work that John and I did. We have, for example, found low correlations not only between the amount of deceleration and speed of reaction time, but also with momentary heart rate level itself and reaction time, the best relationship, of course, being at the time of the execution of the response. However, we did have some indication at that time that there was a correlation that carried over from the resting heart rate level, so it may be a matter of the capacity of the individual.

Dr. F. Graham

Dr. Thompson: Very beautiful data, Fran. Have you looked at the actual reflex when air is puffed on the cornea to see what the effect of the prestimulus is on the reflex, as opposed to startle.

Dr. Graham: The corneal reflex may not be the same thing as the startle reflex, so when I use air-puff I do not put it directly into the eye.

Dr. Thompson: In the rabbit, we find that a tone presented prior to the elicitation of the corneal reflex causes facilitation of the reflex over these same intervals.

Dr. Graham: Yes, but you had early inhibition in your published data with Young and Cegavske.

Dr. Thompson: Only in the first 20 to 30 msec.

Dr. Ushakova: Two questions, Professor Graham; you said at the beginning of your presentation that blinking is a good test which reflects the orienting reflex. Could you say what other patterns in a newborn baby might reflect the orienting reaction besides blinking? The second question is, could you give me more detail on your data obtained from newborn babies?

Dr. Graham: I don't mean to say that blink reflex is an orienting response. Blink is a part of startle. You can identify orienting by the cardiac changes and by what happens between the two stimuli. But I think blink is the opposite of orienting. Opening of the eyelid might be associated with orienting.

Dr. Asratyan: In our research work we have used the blink reflex for a long time. We combined the blink reflex both with food and with electrical defensive reflexes and by this way we were able to show bidirectional conditioned connections. It seems to me that it is not correct to say that changes in cardiac activity, in respiration and in galvanic reflexes could be considered as elements of the orienting reflex. Such types of changes always occur in each reflex. The orienting reflex is the only reflex in which appropriate orienting behavior occurs.

Dr. V. Zavarin

Dr. Berger: The major tranquilizers change certain significant aspects of what is called abnormal behavior in schizophrenics. I'm curious to find out whether or not their unique patterns of language are modifiable by application of the phenothiazines.

Dr. Zavarin: I'm glad you brought this up. I was thinking during the presentations of Dr. Teitelbaum and Dr. Cotman on dynamics of reconstruction in the brain after lesion. Longitudinal studies of schizophrenic discourse are really needed. It is surprising to realize that there are no such longitudianl studies of schizophrenic language behavior. I believe that the work by the San Diego group demonstrates changes in the semantic processes of schizophrenics with administion of tranquilizers.

Dr. Ushakova: Dr. Zavarin, I would like to thank you for your very interesting data presented today. I would also like to commend you for your courage in making this presentation at a neurophysiological conference on such a complex subject. I think we should be aware of the fact that language is also a function of the brain as well as the behaviors we discussed earlier. There is, however, a large knowledge gap between the study of simple forms of behavior in animals and language behavior. What you presented today is the analysis of the speech *behavior*, what I would call outward speech. This approach is very fruitful, but I would argue that outward speech can reveal much to us about what I term inner speech.

Dr. Zavarin: Thank you. I am very anxious to hear your report about "inner speech" processes.

Dr. J. Lacey: The main purpose of this symposium is to display and exhibit the important physiological mechanisms underlying behavior so that ultimately we can apply our knowledge to the alleviation of man's condition. In dealing with language we're dealing with the very tool we use to analyze ourselves and our data. I would like to put Dr. Zavarin to the test. Since she is able to make exquisite differentiations in different forms of speech behavior, I would like to ask her how often she has detected schizophrenic distortions during this symposium?

Dr. Zavarin: Thank you, Dr. Lacey. I have just one comment: There is not a single schizophrenic distortion that doesn't occur in normal speech. It's the constellation of such distortions that can give us some indication of the schizophrenic condition.

Dr. T. N. Ushakova

Dr. Thompson: I want to compliment Dr. Ushakova for her methods, which are a precise quantitative approach examining the association strengths of the verbal network. I have a theoretical question. As I understand it, your theoretical approach assumes that all the relationships and knowledge of language are learned. This, of course, differs from Chomsky's approach, who argues that deep structure of language is wired into the brain. Would you care to comment on Chomsky's theory?

Dr. Ushakova: Thank you, Dr. Thompson for your interesting comments. I'm familiar with the ideas of Chomsky. It would be flippant if I tried on the basis of my approach to make comparisons. Chomsky, when talking about innate linguistic abilities, does not mean innate in the sense of the words and their interrelations in the verbal network. He means the more in-depth grammatic behavior performed by humans. On the basis of my data, I do attach importance to them. But I don't think on this basis I can make any judgments.

Dr. Berger: I can't understand why one must necessarily link retrieval processes to language ability because there are cases of impairments in retrieval of memory, particularly of short-term memory, that do not impair the ability of speech. And there are also cases of pathologic inability of speech that do not impair the retrieval of nonverbal memory. So the two processes would appear at least at some levels to be independent of one another.

Dr. Ushakova: Thank you, Dr. Berger. In my presentation I argued for the fact that speech and memory are closely interconnected, but what I meant was, of course, long-term memory. I think it is obvious that if we do not know the meanings of the words and they are not interconnected and we cannot define one word from another, then the speech is impossible. It is the long-term memory which serves as the basis for word retrieval. Short-term memory need not play a large part.

Dr. Lebedev: It is known how difficult it is to develop a new method in linguistics. You have developed a new and original method and tested it only in your lab. You learned to discover the sequence of extraction of words and phrases. I think it is a very important result and a very perceptive one. I have three questions in this connection. First, besides the method itself, what exactly is the problem you solved by using this method? Second, in your diagram of the verbal network, I did not see any feedback connections. Third, on one of the slides you had some negative times?

Dr. Ushakova: Your first question is really the loaded one. Nevertheless, I believe that you have yourself to an extent answered it. The problem was to find the correct level of approach to the phenomenon. Here is an example. I stand on this podium and I want to give a scientific description of it. If I do it on the level of the cells of wood, I won't achieve an important result. To describe phenomena, an adequate level should be found. I believe we have done so. For us, the problem is that of determining the functional structures on which the associative connections are established in the memorizing process—how is it organized functionally? Of larger interest for me is the formation of a sense phrase as a clue concept in speech activity. The second question concerns the absence of feedback, I think the feedback is obviously there, although I think it is more important to find the verbal connection of every element and its functional neighbors. Third, there really were negative times, but I mentioned that in our performance testing we referred to spontaneous performance because the words which we used in initial runs give different response times. Therefore, we measured the background and therefore we had negative times.

Dr. Shvyrkov: I have a very small question, absolutely not vicious and not loaded. Obviously the whole process of speech is a product of memory. It is not possible to either speak or understand without memory. Therefore, I believe that the networks shown by you are completely analogous to these networks of life experience which we tried to recreate with rabbits. Since the topic of our seminar is goal-oriented behavior, and the goal is extracted from memory, my innocent question is: What is the role of the goal in the reproduction from memory of separate words and phrases both in comprehension and in speech? Innocent, isn't it?

Dr. Ushakova: Sufficiently loaded. Goal plays a direct role, the same role it plays in the processes you study.

Dr. R. W. Thatcher

Dr. Lebedev: Dr. Thatcher, I believe this research is very important and fundamental in nature. I have only small questions which I must defer because the time is short.

Dr. Ushakova: Dr. Thatcher, I wish to second the congratulations of Dr. Lebedev. I have two brief questions. First, did you have 8 channels of data recording? Second, it was my impression from your data that you found interhemispheric similarities for recognition of semantically symmetrical words in the posterior recording sites. Is that correct?

Dr. Thatcher: Thank you both very much for your very complimentary comments. Our recording system had 12 channels, Dr. Ushakova. If I understand your other question correctly concerning recognition of waveforms for symmetrical words, the answer is yes.

Dr. A. N. Lebedev

Dr. Pribram: Thank you very much for a superb presentation. In *Languages of the Brain*, I warn against using the evoked response as a model because it does assume so many underlying processes. There is a model that was developed here at Irvine by Drs. Luce and Green in which interresponse times were used. In order to make the model effective, they used a Fourier transform of the stimulus in a computer simulation.

Dr. Lebedev: Thank you very much. We discussed this problem during Drs. Luce and Green's visit to Moscow last year.

Dr. Verzeano: I'm very happy to see that Dr. Lebedev is using precise, quantitative mathematical methods to study the phenomenon. We have found that different groups of neurons have different phase relations with the evoked potential, and depend also on the characteristics of the stimulus as well. I wanted to ask you if you found such different phase relations for different groups of neurons?

Dr. Lebedev: Thank you, Dr. Verzeano. To answer your question about groups of neurons, the results of our analyses in general agree with your findings.

Dr. Thompson: It appeared to me that you assume from your data that there is a tight coupling between the evoked potential and the unit histogram. Dr. Shevchenko's data showed that a unit will change its relationship to the evoked potential as a function of the behavioral task or goal. Would you comment?

Dr. Lebedev: There is no contradiction. I have talked about the general picture, but in some details there might be of course different relationships. If in each case we can summarize the activities of all the neurons, we will see strong interrelationships to the evoked potential.

Dr. Pribram: Perhaps what could be done is to use the Luce and Green model, which transforms the interresponse times from unit analysis and see how it correlates with evoked potentials.

Dr. Lindsley: Dr. Lebedev, I've been interested for a long time in what might be called a general diffuse alpha-rhythm that occurs widely over the cerebrum and the localized alpha-process within a particular region of the brain. Have you done any work or given any thought to how one could separate this more diffuse, perhaps reticular, thalamo-cortical regulative general rhythm and the more specific rhythms which seem to develop after an evoked potential, very much like the so-called ringing effect, which seems to have much of the characteristics of an alpha-rhythm but of a local nature or source?

Dr. Lebedev: Thank you for your questions. Of course we have ourselves studied such phenomena. In general, our results are similar to what you describe.

MODERATED BY
JOHN LACEY

Appendix B:
General Discussion

Dr. Thompson: I wish only to raise one simple question. It has to do with the meaning of the term goal-directed behavior. I had more or less assumed that the meaning was similar to what Dr. Epstein meant by motivation, in contrast to such phenomena as instinctive behavior and unconditioned reflexes; however, that may not be the case and I hope to get this clarified in the discussion. If goal-directed behavior means all the behavior of all organisms, then the term has no meaning beyond the more general term, behavior, and it is therefore not useful.

I also have a very brief closing statement for the conference. In comparing our groups of outstanding senior and younger scientists from the Soviet Union and the United States, I think our differences are much less evident than our similarities. I have been extraordinarily impressed. There are, I think, small differences in approach. Perhaps at present American psychobiologists tend to prefer to approach problems in a somewhat more molecular manner. That is to say, we place more of an emphasis on synaptic mechanisms and neurochemical and neurohormonal processes. On the other hand, our Soviet colleagues tend to approach the problems of psychobiology more in terms of systems: the hierarchical neuronal system of Professor Asratyan, the analysis of functional neural systems so elegantly presented to us by Dr. Lomov, Dr. Shvyrkov and their associates, and Dr. Gasanov's elegant work on interneuron systems. However, in my view, both of these general approaches, the molecular approach and the systems approach, are equally important and equally essential. We will never solve the problems of brain behavior until both approaches are put together into a unified science. I think this conference has been an important step in that direction and I thank you all very very much.

Dr. Shvyrkov: I would like to say a few words on the problems of the goal—what is it? To admit the existence of goal-oriented behavior is a very serious matter.

It is not merely the treatment of certain types of data, but rather a deep revolution in biology. This revolution started a long time ago, but we're counting from 1935 when the functional system was created [by the Soviet scientist Peter Anokhin, a student of Pavlov. Editors' note]. When we understand the contents of this revolution, then we will understand the goal. An animal achieves its own goals. To understand its behavior we must observe not what follows a stimulus we present but what *precedes* it and what this activity is precedent to. Everyone admits that behavior is adaptive and ultimately all of those acts are possible which are related to the survival of the animal. As a biologist once said, "Teleology is like a lady of the night; She is a lady without whom no biologist can live but he would be ashamed to go out with her."

When activity is in a certain sense a certain type of organization, it can lead to a certain event. Organization is a set of processes inside the organism. What is the future event? It is an object of the environment toward which the behavior is organized. Today we do not compare energy to energy but organization to organization, a new type of informational interrelationship. Here we need to use psychological terms, since psychology has been dealing for a long time with information and has employed terms like "image," which presumes that the information about any object is also contained in another material substrate in the human brain. Substantively, these are two different things, but informationally they are the same thing. Therefore, we can compare the organization of the elements of the environment to the organization of the elements of the organisms. Between these, we have informational equality. This is the image.

I would like to respond to Professor Pribram that the image is not a plan. A plan would be like a program of action, a program of activity required for achieving this goal of reorganizing the environment. Goal is the image of a forthcoming event and it is an informational reason for activity of the organism. Stimulus-response was the synonym of the causative connection. If we take informational relationships instead, a number of questions emerge. Where does the organization which corresponds to the future goal come from? If it corresponds to the future, then we're looking for the reason because the reason always precedes the consequence. If we presuppose that the reason for this activity is the reproduction of the image from the memory, then we have another problem. What is the reason for reproducing exactly this image? We say another image, and so on.

We study the behavior of a live organism. Life is a goal-oriented organization of chemical processes. What was the direction of organization of chemical processes as life appeared? To the reinforcement of its organization—the very first life creatures supported this organization by conserving their organization from the chemical substances from the environment. The main goal of life is to preserve the entire organization of its processes. A discrete goal, which corresponds to certain objects in the environment is the image we reproduce from memory from the motivational system.

Goals exist in a hierarchy, from the goal to preserve metabolism to the goal of obtaining a certain organization of the environment. When we record the activity of individual nerve cells with a microelectrode from an animal engaged in a goal-directed act, we observe the same relationships of neural unit activity in different structures—the cerebral cortex, the reticular formation, the hippocampus, and even in the mesencephalon. We see the neural mechanism of the goal image. This process is always hierarchical. The biological substrate of the image is the hierarchical neural organization.

In my view, Pavlov's great contribution was the principle of "signality"—the informational relationships between the objects of the environment and the activity of the organism. What does the principle of signality presume? That one object is a signal of another object. In any organism there are many objects (images) and one of the objects is the signaler to the possible emergence of another object. This determines the selection of one of the possible goals, a goal. In this sense, the principle of signality offered by Pavlov works on the informational notion of the connection between the organism and the environment, the organization of the organism. The systems approach is the analysis of this organization. I think that these notions, developed in the functional systems approach, will make it possible to create or recreate the total picture, which is the psychological characterization of processes.

Dr. Gormezano: I have a fundamental concern that I have raised several times during the conference. That is the question of where, during the course of learned behaviors, the neurophysiologists think that modification is occurring within the nervous system. That comment is elicited by the reports here, several of them at least, of purported relatively permanent changes in underlying neural structures, specifically in sensory cortex. I would like to get the physiologists' and neurophysiologists' point of view regarding the loci of change. My early, perhaps outmoded, preconceptions have been that they occurred at the internuncials.

I would also like to respond to the question raised by Dr. Berger earlier regarding toxicosis conditioning. The phenomena of toxicosis, in my judgment, were introduced into the psychological literature in part as a polemic or posture to argue that the laws of learning are dead. Specifically, the fundamental assumption made by Western psychologists with regard to associative process has rested upon the fundamental assumption of contiguity. It is the principle or axiomatic statement that associations are formed when there is almost or strictly simultaneous occurrence of the events to be associated. The assumption comes from British associationistic doctrine, which proposed that sensations become associated when they appeared together simultaneously in the "mind." Aristotle himself advocated that the fundamental foundation for associations was the principle of contiguity. Starting in the 1920s, American experimental psychologists began for the first time to test experimentally the principle of contiguity, employing, would you believe, Pavlovian conditioning procedures. American psychologists quickly learned that conditioning did not appear to occur with strict simultaneity. In fact, it was required that the CS precede the UCS by some temporal interval. American psychologists had two choices. They could have rejected the principle of contiguity. Instead they introduced a theoretical bridge. They proposed, again borrowing from Pavlov, that between the occurrence of the CS and the UCS, there is some class of persistent stimuli. They used Pavlov's notion of a neural stimulus trace and used Guthrie's notions of proprioceptive and kinesthetic stimuli. By the 1960s it was possible to detail the empirical functions required for conditioning. In no case in those 40 years has anybody proposed that the principle of contiguity does not operate. They simply introduced theoretical constructs to bridge the interval. Toxicosis condition studies showed that there was an extensive temporal asynchrony between novel gustatory stimuli and an effective poisoning agent. The users of this paradigm quickly began to argue that contiguity is dead. I understand that not everyone advocates that, but most do. If you will check back through the literature, you will see that this point has been made repeatedly by Garcia and others and it often took the form of a polemic. The fundamental issue was presented as: How can you get associations formed when you have a 1 min, 5 min, 10 min, 20 min, 30 min, 1 hr, 2 hr, 3 hr, or even 5 hr interval between the two events? They argued that this is somehow magical and new. People of my persuasion would argue instead that these results represent simply extensions of already established laws along additional dimensions.

Dr. Patterson: Thank you. Looking up from the spinal cord, this has been a very interesting and enlightening discussion. I agree with Dr. Gormezano that long CS–US intervals do not disprove learning laws. Dr. Gormezano asked the question of the place where learning changes might occur in the brain. I look at it from a systems point of view also. It is my belief that the basic association process is probably a synaptic process occurring over a few synapses. As systems are added in more and more complex brain structures, the synaptic mechanisms probably do exhibit the same properties that we see in simple neural components or subsystems. I think we must be very careful in all our studies, whether it be with modifiable simple nervous systems or with whole brain systems (with which I'm also involved) to delineate the possibilities for higher order systems playing down upon subsystems in a cascading fashion. We should not look for the total behavior at the highest level but rather for an orchestration, for example, the cortex directing and drawing out patterns that have been programmed over much older learning periods into the spinal cord. However, I think that this cascading still follows the same cellular laws that we see in simple systems. It is this aspect that I would like to emphasize and make us all aware of in terms of examining both simple and complex behaviors.

Dr. Gabriel: In several of the interchanges that have occurred between the Soviet and the American scientists, the comment has come up that the Americans have very many facts and not very many theories and that the Soviet scientists have many, many theories but perhaps not as many facts. One of the things that seems to have come across to me from the symposium is that perhaps we Americans should try to adopt more organized conceptual frameworks. In particular, I have two ideas in mind that keep recurring in many of the presentations and seem to cry out for a conceptual framework which we can adopt in common. One of the ideas was just mentioned by Dr. Patterson and I will only repeat the notion and label it. It is the idea of the hierarchical organization of neural control of goal-directed behavior and the consequence that there are superordinate controlling systems and subordinate systems. This kind of thought was one that was introduced at the beginning of this symposium by Dr. Asratyan when he presented his model in terms of hierarchical levels of organization in conditioning. It is a theme that Dr. Teitelbaum and Dr. Marshall touched upon; Dr. Ushakova's presentation in terms of hierarchical levels in the organization of speech; Dr. Hicks' presentation on kinesthetic control and changes in control with overtraining; Dr. Shvyrkov's work on foretelling activation of units correlated with goal as opposed to movment specific units; Dr. Aleksandrov's work on the difference between mesencephalic and cortical units; Dr. Shevchenko's work on the different roles of units at different levels, and many others.

The second idea that has kept recurring in this symposium is the notion that different systems of the brain operate in different stages in the acquisition of goal-directed behavior. Dr. John mentioned the increasing homogeneity of evoked potential waveforms with training and overtraining in the task. Dr. Gasanov mentioned change in the rotational direction of the ringlike organization of neurons with overtraining in a task. Dr. Pribram's theme had to do with the distinction between control and autonomic modes of information processing. Dr. Lindsley's data showed new components of the evoked potential with overtraining in a task. Our own work showed descent of discriminative neuronal activity to thalamic structures as a function of training and overtraining in a task, and so on. I think these are themes that have come up so frequently that we should now try to agree on these general propositions.

Dr. Thatcher: I just thought I'd very quickly add to some of the data that I had presented in terms of Dr. Verzeano's work on the loop action in the brain. In our evoked potential work, the late positive components move from the back of the head to the front of the head in adults in about 50 to 100 msec, and in children, 100 to 200 msec. I don't know exactly how this relates to Dr. Verzeano's work, but the point is that the dynamic picture of the brain you get when you are recording from many areas simultaneously reveals phenomena that cannot be seen very readily with single electrodes. In this conference devoted to goal-directed behavior, many terms were used without any clear definitions—words like goals and images. It seems to me that goals are not necessarily static, that goals themselves change through the course of learning and performance. They're refined and reorganized to some extent as an animal attempts to achieve goals.

Dr. R. Miller: I'd like to speak briefly to two goals for future research. First, I was rather disturbed in some of the discussion about the implicit assumption of causality in the electrophysiological research that has been described over the last week. These issues have not been totally neglected. Dr. Thompson, for example, has in his very exciting work described the role of the hippocampus in learning. On the other hand, he acknowledges that damage to the hippocampus does not seem to affect the overt behavior that follows it. I find this extremely disturbing. The other point I wish to raise is that I am a Darwinian as well as a Pavlovian and I start out with the premise that all behavior is inherently goal-directed but due to abnormalities of an animal's niche, often created by clever researchers; both afunctional and dysfunctional behavior can and does occur, for example, vicious circle behavior. However, I don't think we should forget the adaptive roots of animals' behavior and I think that there is a demand for more functional analysis of behavior.

Dr. Aleksandrov: The basic working hypothesis of the functional systems approach is that the system can be selected only by results, regardless of what elements achieve this result. If we use the approach of stimulus-response, where would you localize each? They can be solely in the visual cortex and the motor cortex because both conduct the impulse in the reflex arc. From our point of view, morphology plays an important role. It turns out that adjacent neurons of the mesencephalic nucleus equally participate in the behavioral act. The different muscles incorporate themselves in a totally different way, dependent on the number of spindles. Besides, I would like to remind you that morphology plays a role in the analysis of hierarchies. Thus, in my research and in the research of Dr. Goldberg it turned out the cortical neurons relate to a more superior level of the hierarchy. The neurons of the mesencephalic nucleus relate to subsystems. I would like to conclude by saying that from the theory of functional systems, morphology is a fixed function. However, function is not determined by morphology.

Dr. Shevchenko: I'd like in my turn to thank the organizers of the conference and all those present for all the good, useful, and interesting things that we heard here. I think the majority of the presentations and the data illustrated testify to the fact that goal-oriented behavior is subject only to goal-oriented research. I would like to cite a simple example from our laboratory. In the course of several decades, study of the evoked potential to neural stimuli was not fully understood. Using the goal-oriented approach to this phenomenon, we showed that the functional significance of the components of the evoked potential is related to the systems processes which secured the brain activity as a *result* of the goal-oriented behavior. Thank you.

Dr. Cotman: In some of the physiological experiments, the question that comes to my mind is distinguishing something that in our science is called corollary discharge from a really causal activity. I'd like to cite the motor cortex. Dr. Shvyrkov indicated that motor neuron discharge in the cortex displayed a variance of activity depending on the goal-oriented behavior. I really wonder what particular layers of the cortex and what types of neurons are being recorded from, what part of the homunculus is really being dealt with, and whether, if one could find a reproducible site, the variation would really disappear or not.

Dr. Lebedev: I see a lot in common between different notions which we discussed here, notions of the image, model, plans and structures of behavior, local reflex, conditional feedback, and so on. Recently, Professor Asratyan and I discussed the possibility of achieving sufficiently convincing new concepts which equal Pavlov's notions. We haven't arrived at any yet. I think that the closest approach to a new theory which will account for the accumulated contradictory data comes from a group of researchers who deal with the analysis of periodical processes in the brain. Among them are Academician Livanov, Drs. John, Pribram, Verzeano, Lindsley, Thatcher, and many others. I would like to urge all others to treat this problem with maximum attention. I don't think that we will be able to have a comprehensive theory of the sort Pavlov dreamt of for at least 10 or 20 years.

Dr. Gasanov: First of all I would like to congratulate the organizers of this conference for the brilliant organization and no less brilliant completion. I believe that from my point of view, the task of our first meeting was not the desire to find mistakes or errors in various theoretical and experimental expressions but to try and understand each other. I believe in this aspect we achieved complete success. With all my desire to be objective, I can't relieve myself of my own tastes and interests in science. Therefore, I want to emphasize the work that was particularly close to me and which I feel very promising—the work of Dr. Thompson, Dr. Verzeano, and Dr. John. I believe that this systems experimental approach to brain activity is productive and useful. I believe that it could be extremely interesting to discuss the strategic goal-directions of neurophysiological research on mechanisms of behavior. I mean the definition, in our research, of the operating mechanisms in neural activity—of neurons which function in any activity of the awake brain and the mechanisms which secure a certain form of behavior. The second question which I deem interesting for discussion was the one of levels of neurophysiological research. Can we find the direct relation to the psychology of human behavior? I think it is a very important fact that there are two important directions in behavioral science, neurophysiology and psychology, and they have tended to develop only in parallel to each other. We must interconnect them.

Dr. Kenney: I came to this conference from a very different perspective than the majority of you. I recognized when the conference began that I would be learning a tremendous amount about the electrophysiological approaches to goal-directed behavior, both from my American colleagues and from my Soviet colleagues. You will recall that Dr. Isaacson and I presented another point of view, that of the neuroendocrinological approach to the study of these types of behaviors. I would like to remind you now that there can be hormonal or chemical mediators of these types of behaviors and I encourage you to consider this aspect as well. One thing I thought I knew at the beginning of this conference, however, was the definition of goal-directed behavior. As Dr. Thompson pointed out at the beginning of this discussion session, we have somehow lost the idea of what goal-directed behavior is. I myself came with the very traditional view that goal-directed behavior was what American psychologists have classically called motivated behavior. I have heard very many other opinions on this in the past week. Since I realize now that we probably will not reach consensus on what goal-directed behavior is, I would like at least some sort of enumeration or delineation of the various definitions that are possible for that term and to get some idea on how people approach the study of goal-directed behavior based on their own definitions.

Dr. Spear: I'd like to reinforce what Dr. Kenney said. I've known for a long time that there are a variety of different approaches to examining nervous system mechanisms of goal-directed behavior and learning. It's my feeling that to understand such complex phenomena we may need to synthesize across research from a variety of different approaches: the neurophysiological approach, the neuroanatomical approach, the neuroendocrinology approach, and the neurochemical approach. The barriers to such syntheses may not be so much spatial or foreign language barriers but barriers between different types of approaches in terms of the languages of the conceptual models that are used. I feel that I have learned a great deal at this conference and I'd like to thank all of those present for sharing their approaches to this area with me.

Dr. B. Lacey: John and I were amused the other day when presented with the title of this symposium and wondering what parts of our work would fit in that might be of interest to you, and we kept saying goal-directed, goal-directed, and then, to paraphrase a character in a play by Molière, who discovered that he had been speaking prose all of his life, we suddenly realized that we had been speaking goal-directed behavior all our lives.

I think it was Dr. Gabriel who was talking about the idea of perhaps selecting some basic theoretical propositions from the American data as suggested by the hierarchical functional systems organization that seems to be representative in the Soviet group. I just wonder how much of the work in science as a whole, not only in neurophysiology or in psychology, has come about from working in a rather loose framework, and from the heuristic value of experimentation. Just a question, an urging, that perhaps when we work in more loosely organized frameworks, we are freer to accommodate things we discover that simply do not fit into a tight theoretical framework.

Dr. D. Lindsley: We have studied behavior in a very broad context ranging from the naturalistic ethological side to the very restricted behavior that many of us have been involved in in the laboratory. We all get enamored about particular techniques and methodologies and I think that we have to recognize that goal-directed behavior has a very basic core in behavioral genetics. We can follow this up through the various reaches of the first signal system and the second signal system up to the higher cognitive levels which we've heard a good deal about in the last session. My own field has been pretty much in the electrophysiological realm and periodically, for my own benefit, I try to see not only where I am but where I think other people are with regard to *behavioral states*. They range from sleep to the highest and most intensive cognitive activity that one could conjure up. Another dimension is the level of sensory input—primarily auditory, visual, and somatosensory. Then there is the EEG; and its degree of the synchrony or desynchrony is an indication of the level of activation electrically in the brain. A particular constellation of these measures or dimensions is necessary for goal-oriented behavior to occur. You obviously have to be aroused, you have to be oriented, alerted, attentional, have a desynchronized EEG, and so forth. At first, when invited to this conference, I thought I haven't been concerned very much with goal-oriented behavior or motivation, per se. However, attention is a process very much concerned with goal-oriented behavior. I couldn't conceive of any goal-oriented behavior without some attention to the goal. Almost by definition, attention is a sine qua non of goal-oriented behavior. I might even suggest that a state of attention or attentiveness could serve as a good operational definition of goal-directed behavior. This slide illustrates the constellation of dimensions and variables that could serve to characterize goal-directed behavior.[1]

Dr. Asratyan: I want to say that this has been an excellent conference. However, I do want to make some remarks. It seems to me that goal as a notion should not be considered axiomatic. It may serve as a basis or determinant for the explanation of different phenomena or for the complexity of various acts. But the notion itself needs strictly scientific experimental and theoretical investigation without any elemental theology. In the light of Pavlovian teaching, goal-

---

[1] Dr. Donald Lindsley showed a most interesting slide in which he summarized a number of behavioral and neurophysiological variables, such as EEG state, CNV, evoked potential components, et cetera, and the intersection or space of values of these variables that could serve to define the properties of goal-directed behavior.

directed behavior could be considered as a complex or special form of most complicated, specialized, and vitally important inborn or unconditioned reflexes and conditioned reflexes. From your point of view it is easy to agree with Dr. Thompson that goal-directed behavior is only a part of behavior in general, namely, the motivational part. Now to answer the question of Dr. Berger. From the point of view of Pavlovian theoretical propositions, the conditioned reflex can be elaborated even by one impulse. Which interval is necessary between the indifferent and reinforcing stimuli depends on the biological value of the reinforcing stimulus. In some cases even 1 sec or 2 sec may be too much. In other cases, a vitally important stimulus can be effective for an interval of several minutes. We have no dogmatic or stereotyped formula. It depends on the biological significance of the stimuli.

Dr. Graham: Well, I just want to say this has been a most pleasant occasion. I am also unsure of what we mean by goal, although I am one of the few living students of Clark L. Hull that remains and he talked a great deal about goal-reinforcement and goal-directed behavior. I was surprised that Dr. Gormezano didn't mention little r.g.

With regard to goal, there are a couple of questions that I would like to raise. The one: Is all behavior goal-directed? I gather that it is not. If it is not all goal-directed, then do we have other sets of laws of behavior? Some laws for goal-directed behavior and some for other behavior? How many goals can be present at one time? One other thing that concerns me is the treatment of the stimulus. I gather that some of you want to be sure that nobody thinks that the stimulus is a prod. Dr. Karpov noted that activity or behavior is determined not by the stimulus, *per se,* but by the goals of the organism, by the desire to accept. So a stimulus is not a prod, it's an energy exchange with the environment. I keep wondering as I listen to this, is it a question where we have a closed loop and we can't decide where we're supposed to start? Because otherwise it is really not clear to me. There's the organism and his goals and his desire to accept and there is energy from the environment. And I don't see how you can say one causes or the other causes.

Dr. Pribram: I also want to echo the problem that Frances Graham just brought up and Richard Thompson started and Thatcher continued. What do we mean by goal-directed behavior? I was the person responsible for keeping the word goal out of *Plans and the Structure of Behavior,* because I don't like the idea of using it in a diffuse way. To give you two examples of what the problem is—in our book we distinguish between image and plan. Now which do you mean by a goal when you're talking about goal—image or plan? You think image. As Dr. Shvyrkov has said, it's when the goal is retrieved from memory that essentially it is an image. That's fine, I would agree with that. But there are many people who use the term goal in terms of a plan, the kinds of behavior that you need to achieve a plan. So one does not communicate clearly when it is unclear as to which sense of the word is being used. I think one of the things this conference has done has opened up the question for precise definition: so from now on we will always be able to say whether, when we say goal, we mean image or plan, a state or a process.

The other issue is whether there is such a thing as behavior that is not goal-directed. I recall a dinner conversation last night where Dave Cohen and I were talking and he said to me that one of the great problems he has with the graduate students who are not doing so well is to point out to them that when they're not making a decision that's also a decision. In *Plans and Structure of Behavior* we pointed out that if an organism is alive he behaves in one way or another, even if he's standing up doing nothing, lying down and sleeping, or whatever. The problem of motivation is essentially a problem of directing which behavior is to be engaged and which is not. What do we mean by behavior when we say it is not motivated behavior? Is there such a thing? [2]

Dr. Berger: I want to say that I think Dr. Gormezano is incorrect in suggesting that there is such a thing as a toxicosis proposition that was introduced to psychology in order to refute the laws of learning. There were data, Dr. Gormezano, that have come to light in the last few years that call to our attention two points that don't refute the laws of learning, but allow us to extend what we already know about learning, as you have suggested. They also allow us to suggest that existing learning theory is not static and not dead, but is dynamic and can include the kinds of things that we see in the empirical laboratory. The first point that toxicosis gave us of special interest was the phenomenon of very long delays: the idea that hours can exist between what appears to be a CS and a US. There is no doubt about the idea of contiguity being called into question. We have to try to understand in some fashion the way in which stimuli can bridge the gap in order to be associated. What are the mechanisms, the neural mechanisms, and the psychological mechanisms that can be called to allow for stimuli that are widely divergent in time to be associated?

The second thing that I think is of particular interest from the taste aversion literature is the calling into question the idea of the arbitrariness of stimuli to be associated. The idea that any stimulus can be associated with any other stimulus. I agree with Dr. Gormezano that this is not necessarily contradictory to basic learning theories. If we look carefully, learning theory has suggested that certain stimuli are more associable with other kinds of stimuli. I introduced this topic into this symposium and was very pleased to see the kind of answer that Dr. Asratyan gave, the kind of understanding of the modern concept of biological relevance and special significance that stimuli can have, one for the other. I also feel that long delays can be accounted for by the notions that we have of learning today.

Dr. Lomov: Distinguished colleagues, I would like to make a couple of general judgments, not because, as Dr. Gabriel said, we prefer theory to facts, but because it seems to me that psychobiology is in a stage of its development when discussions of general questions of a philosophical nature acquire importance. Consequently, misunderstandings will emerge that are due to the fact that we pay a little attention to discussing general questions. Any science which is supported by experiments and critical data has three categories of properties:

---

[2] Earlier in the conference, Dr. Berger raised a question, to Dr. Patterson, about the fact that the CS–US interval can be very long—up to hours—in taste-aversion conditioning, very much longer than in other forms of conditioning. It was agreed that this question be a part of the general discussion. Drs. Gormezano and Asratyan have already addressed this issue in their remarks above.

(*a*) the properties of structures, which are determined by the structure of the objects; (*b*) the functional properties relative to the relevant phenomenon; and (*c*) the system property, which is of a complex nature. I would like to illustrate with a very simple example. We can take any object, for example, this microphone. We can analyze its structure, its composition, its function in the transmission of information, and so on, but this object has also properties which we cannot find, no matter how much we analyze its structure and functions. An example is such a property as its price. This could be comprehended only if we regard it as a part of the economical structure of society. This is just as objective as a physical or informational or any other basis. In thinking about the properties of things, much is determined by their belonging to a certain system. It is very important for the development of science to comprehend the bases of the phenomena which we study. In our area, which is called natural, we started by analyzing the phenomena which determine structure, we started from morphology. As was noted by Goethe, who was not only a poet, but a scientist, when he was talking about research, he first analyzed and then reconstituted a subject, but there was no longer vitality in it. We must pursue understanding from structural qualities to other qualities. The misunderstanding of any property generates further misunderstandings. The study of language made significant progress only when the properties of language were viewed in terms of function and different levels of function.

Something that has been discussed extensively and is of particular importance in psychobiology is the study of the systems basis of different structures. I think this is a most promising feature of our development. Professor Lindsley showed a very good table which illustrated the approach toward understanding how systems properties affect the phenomena. We heard a number of reports which related to the phylogeny and ontogeny of different forms of behavior, from the most elementary to the most complex, and we saw a number of models which were prominent in our discussions. There were attempts to use homeostasis as a functional system, servo mechanisms, and so on. However, in order to understand them, we must first know the basis.

Speaking of our conference, I'm not going to enumerate the names of those who made important contributions because then I would have to enumerate every name. Every presentation has made a good contribution to the problem. Instead, I will try to define some of the directions. A number of presentations related to the problem of structure and function. These presentations definitely enriched our knowledge about neuronal mechanisms and the characteristics of separate neurons and neuronal networks. There were also a number of presentations on biochemical characteristics of brain activity. Finally, there was study of behavioral characteristics which represented a wide spectrum from very elementary behavior to speech behavior and cooperative behavior. The most important thing in my opinion is the fact that everybody attempted to connect the different aspects and implications of their work.

I think we have developed a very good basis for further contacts and for developing our understanding of problems with which we are dealing. I would like to note that we have a lot of common points of contact which must serve as a basis for the further development of our cooperation. Characterizing the conference, I would like to take advantage of the projector and show several slides, as everyone else has done.[3]

In conclusion, I think our conference has been highly successful. There is every reason to look forward not only to further contacts and seminars but also to joint research work. I would like to express my deepest gratitude to the American scientists who participated in this conference, to the organizers, and to my friend, Dick Thompson.

---

[3] Professor Lomov then showed several delightful sketches he had drawn. The first, for example, illustrated the major difference between the American and Soviet delegations —many of the Americans had beards, but the Soviets were all cleanshaven!

Index

A

Accumbens nucleus, 418, 613
Adrenocorticotrophic hormone, fragments, 418–419
Akinesia, and lateral hypothalamic lesions, 129–134
Alimentary behavior, 447
  development in kittens, 451–455
Alpha-methyl-para-tyrosine, somatosensory function, 173
Alpha-rhythm, 575, 623
Amphetamine, 613–614
  and avoidance learning, 430–433
  in learning and memory, 81–88
Amygdala, 586, 612
  in conditioned response, 292–293
Apomorphine, in sensorimotor function, 171–172
Angiotensin II, 614
  in water ingestion, 443–445
Animal learning, social cooperation, 482
Avoidance learning, 154–155, 303, 430
  drug effects, 76
Auditory cortex, 345
  in conditioning, 248–249

B

Basal ganglia, 418, 607
  interaction with limbic system, 420–421
  lesion effects, 594
  in psychomotor performance, 153
Blink reflex, 620
  orienting behavior, 514–516

C

Cardiac cycle, in reaction time, 498–500, 618
Catecholamines, in learning and memory, 77
Caudate nucleus, 418, 592
  in avoidance learning, 157
  in learning tasks, 164
  in maze learning, 161
Cerebral cortex, 610
  in information processing, 327–336
  posterior parietal, 397, 404
  role in conditioning, 600–601
Chlorpromazine, 416
Cingulate cortex, 606
  in conditioning, 304–305
Classical conditioning, 59–60
  spinal reflex, 263–270
Cochlear nucleus, unit responses in conditioning, 248–249
Colchicine, axonal effect, 148
Conditioned reflex, 12
  bidirectional, 585–586
  bilateral conditioned links, 26–35

637

Conditioned response
  mediational, 40–44
  paradoxical, 67–69
Conditioned stimulus, serial compound, 45–46, 48–55
Conditioned stimulus–unconditioned stimulus, effective interval, 602
Conditioning, classical versus instrumental, 588
Consciousness, 590
  neural bases, 110–113
Contiguity, in conditioning, 627–628

D

Dentate gyrus, neuronal structure, 146
Development, longitudinal and cross-sectional approaches, 461–463
Diethyldithiocarbamate, in learning and memory, 76–80
Dopamine, 428, 596–597, 613
  role in sensorimotor deficits, 168–169

E

Engram, and learned behavior, 221–222
Ensemble information, 96, 101–109
Epinephrine, in learning and memory, 78–80
Evoked potential, 375–383, 606, 623, 629
  averaged, 95–98, 555–556, 570
  in cardiac slowing, 500–504

F

Feature detector, single neurons, 94
Fimbria, lesions and synaptic growth, 148
Frontal cortex, evoked potential, 607
Functional system, 8–9

G

Goal, 4–9, 94, 599, 625–627, 632, 633
Globus pallidus
  in avoidance learning, 154
  role in behavior, 163

H

Haloperidol, 428
Heart rate, 604
  changes in reaction time, 496–498
  conditioned changes in 284–286
  nerve pathways in conditioning, 290–292
  slowing and attention, 504–506
Higher nervous activity, 14–15

Hippocampus, 601, 612
  in classical conditioning, 223
  effects of lesions, 409–413
  electroencephalographic activity, 226–227
  multiple unit activity, 227–232
  recovery from damage, 594
  single unit recording, 235–237
  synapse formation, 148
Holography, 320–321
Homeostasis, skeletal responses, learned, 57
Hypothalamus, 418
  in avian conditioned responses, 292–293
  electrostimulation, 18–21
  social behavior, 617
  in thermoregulation, 190–191
6-Hydroxydopamine, 594
  in memory tasks, 83–88

I

Image, 607, 626
  brain process, 200–201
  holographic processing, 320
Instinct, 119–120
Instrumental learning, 59–60
  visceral responses, 61–64

L

Language
  paradigmatic frames, 523–530
  in schizophrenia, 522
Lateral geniculate nucleus, 324, 611
  single cell responses, 387–388
Lateral hypothalamus
  lesions and social behavior, 488–491
  in motivated behavior, 127–129
  in sensorimotor integration, 167
Locus coeruleus, 457–458
  lesions, 135–138

M

Medial geniculate nucleus, neurons in conditioning, 250–253
Memory
  in developing animals, 428
  mechanisms, 586
  neural bases, 75, 599
  speech, 621
  synaptic theories of, 425
Memory deficit, infantile amnesia, 471–472
Motivated behavior, learned characteristics, 122–125

Index

Motivation, brain mechanisms, 119–120
Motoneurons, response in classical conditioning, 225
Motor cortex, neuronal activity, 204–207

N

Neuron
    cortical systems in learning, 341
    learning in, 242–243
    links in conditioned reflex, 346–350
    networks in cognitive function, 354–355
Neurotransmitter, autoreceptors, 426–427
Norepinephrine, in learning and memory, 78–81

O

Olfactory bulb, 603
    neuron activity in food acquisition, 273–277

P

Pentylenetetrazol, effect on neuronal activity, 368
Pons, lesions, 135–138
Potential, slow, 322–323
Prostaglandin E, 614
    and water ingestion, 437–442
Pulvinar nucleus, 611
    single unit responses, 388–394
Pupillary dilation, conditioned reflex, 241, 245

R

Raphe nuclei, lesions, 135–138
Reaction time, 579–582
Representational system, 98–102
Reticular formation, 110

S

Schizophrenia, language, 620–621
Second signal system, 542–543
Septal nuclei, hippocampal projection, 232–234
Septum, 612

Spatial synchronization, 556–557
Startle reflex, 511
Substantia nigra
    in avoidance learning, 155, 163
    lesions, 135–138
Subthalamic nucleus
    in avoidance learning, 157
    in learning, 163
Superior colliculus, 612
    neuron response, 397–402
Sympathetic neuron, in cardiac conditioning, 295–297
Synapse
    axon sprouting, 145–147
    formation of, 148–149
Synaptic growth, 145
Syntagmatics, 523–524

T

Temporal cortex, evoked potential, 607
Thalamus, 610
    anteroventral nucleus, 304, 606
Thermoregulation, behavioral and autonomic response, 189
Theta wave activity, 362
Trigeminal nucleus, 612
    in jaw movements, 179

U

Unconditional reflex, 14–15

V

Verbal network, 544
Visual cortex
    evoked potential, 607
    neuronal activity, 204–207
    simple and complex cells, 324–32
    single cell responses, 387
Visual pathways, in the avian brain, 287–290

Y

Yoga, visceral function control, 587–588

183478